Fundamentals *of* Oil & Gas Accounting

Fundamentals *of* Oil & Gas Accounting

Rebecca A. Gallun

Charlotte J. Wright

Linda M. Nichols

John W. Stevenson

fourth edition

PennWell Corporation
1421 South Sheridan Road/P.O. Box 1260
Tulsa, Oklahoma 74101
800-752-9764
sales@pennwell.com
www.pennwell-store.com
www.pennwell.com

Cover & book design by Morgan Paulus

Library of Congress Cataloging-in-Publication Data

Fundamentals of oil & gas accounting / Rebecca A. Gallun … [et al.].--4th ed.
p.cm
Rev. ed.: Fundamentals of oil & gas accounting / Rebecca A. Gallun, 3rd ed. 1993
Includes bibliographical references and index.
ISBN 0-87814-793-4
1. Petroleum industry and trade--Accounting. 2. Gas industry--Accounting. I. Title: Fundamentals of oil and gas accounting II. Gallun, Rebecca A. III. Gallun, Rebecca A. Fundamentals of oil & gas accounting.

HF5686.P3 G3 2000
657'.862--dc21

00-060655

3 4 5 6 07 06 05

Contents

chapter one

A Profile *of* Petroleum Operations

Today's oil and gas companies may be involved in four different types of functions or segments:

- Exploration and production (E&P)
- Transportation
- Refining and gas processing
- Marketing and distribution

An **integrated oil company** is one involved in E&P activities as well as at least one of the other segments listed above. An **independent oil company** is one involved primarily in only E&P activities.

The oil and gas industry is uniquely characterized by

- a high level of risk
- a long time span before a return on investment is received
- a lack of correlation between the magnitude of expenditures and the value of any resulting reserves
- a high level of regulation

- complex tax rules
- specialized financial accounting rules

These characteristics are most evident in the E&P functions of oil and gas companies. This book describes the accounting for transactions relating to the E&P activities of oil and gas companies.

Exploration and production procedures and processes are essentially the same for any type of oil and gas company. The procedures and steps involved in locating and acquiring mineral interests, drilling and completing oil and gas wells, and producing and selling petroleum products are briefly reviewed in this chapter. A basic knowledge of these procedures is necessary in order to understand their accounting implications which are discussed in the following chapters. A more detailed discussion of these procedures may be found in publications such as *Introduction to Oil and Gas Production,*[1] *A Primer of Oil-Well Drilling,*[2] or *The Petroleum Industry: A Nontechnical Guide.*[3] See the reference section at the end of the book for a list of other readings relating both to the oil and gas industry and to accounting for oil and gas operations.

Brief History of the U.S. Oil and Gas Industry

The modern history of the U.S. oil and gas industry began in the latter half of the 19th century with the first commercial oil drilling venture in Pennsylvania. The product demand at that time was for kerosene, which was used as lamp fuel. The petroleum industry expanded greatly in the 20th century as new uses for oil were developed. The invention of the automobile was responsible for much of the industry's growth.

Early in the 20th century, World War I strained the industry's ability to supply fuel, prompting the decade of the 1920s to be one of great U.S. oil discovery. In that decade, many U.S. companies also began exploring for foreign oil in the Middle East, South America, Africa, and the Far East. The onset of World War II in 1937 once again tested the industry's ability to supply oil. During World War II, drilling began in U.S. waters on offshore structures that resembled the offshore drilling platforms of today. During and following World War II, U.S. companies also increased foreign exploration, especially in the Persian Gulf area.

Following World War II, natural gas was established as a major fuel for industry and home heating. It was also during the post-war period that natural gas transmission pipelines were constructed, linking remote producing areas to large population centers.

The U.S. increased its reliance on foreign oil during the 1950s and 1960s. In 1960, the Organization of Petroleum Exporting Countries (OPEC) was formed. Although OPEC was ineffective during its first decade of existence, its oil embargo of 1973-74 coupled with an increase in the price of crude oil created an energy crisis in the U.S. resulting in shortages of petroleum products. The nation for the first time became acutely aware of the myriad of problems and issues resulting from its dependence on foreign oil.

The energy crisis prompted the passage of the Energy Policy and Conservation Act of 1975, the purpose of which was to encourage energy conservation, reduce reliance on foreign oil, and encourage the development of alternative energy sources. Unfortunately, with the easing of the energy crisis, the nation's attention turned away from energy matters. By 1977, U.S. dependence on foreign sources of petroleum reached an all-time high of 47%. A second round of price increases occurred from 1979 through 1981. The U.S. dependence on foreign sources of petroleum fell to 27% by 1985.

The world oil supply was again disrupted in 1990, with Iraq's invasion of Kuwait. Although the crisis was short-lived, it once more brought to the forefront the nation's dependence on foreign oil, and the need for alternative fuel sources. With the easing of the crisis, U.S. net imports of foreign oil again began to increase. By 1996, the U.S. was importing 46% of its petroleum needs.

Origin of Petroleum

The most widely accepted hypothesis of the origin of oil and gas is the organic theory. This theory holds that petroleum (hydrocarbons) is formed from organic material including marine plants and animals that lived millions of years ago in low-lying areas—normally in the oceans of the world. These plant and animal remains were deposited throughout the years, along with layer after layer of eroded particles of igneous rock. The weight and pressure of the overlying layers caused the eroded rock particles to form **sedimentary rock**. The weight and pressure of overlying layers—and other not fully understood factors, such as chemical and bacterial processes—changed, and still change, the organic material into oil and gas.

After formation, oil and gas move upward through the layers of sedimentary rock due to pressure and the natural tendency of oil and gas to rise through water (salt water is normally contained in the pore space of sedimentary rock). The petroleum migrates upward through porous and permeable rock formations until it becomes trapped by an impervious layer of rock. The impervious rock that prevents further movement of the oil and gas is called a **trap** (Fig. 1–1).

Some of the common types of traps are as follows:

- **Fault trap**—a trap formed when the movement of the earth's crust causes different rock strata to offset or shear off. In a fault trap, a nonporous rock formation that has shifted stops the movement of oil or gas within an offsetting formation that allows petroleum to migrate.
- **Anticline**—a trap formed by the folding of the earth's crust into a dome. This upward folding is caused by pressures developed from the earth's molten core. An impervious or nonporous layer of rock overlying the anticline traps the oil or gas in the anticline structure. Most of the earth's oil and gas reserves are found in anticlines.
- **Salt dome**—a trap created by a bed of nonporous salt pushing upward through overlying formations. Faults, in which oil and gas are trapped, are created along the sides of the upthrusted salt dome.

In order for an oil or gas reservoir to have been formed, four conditions must have been present:

Fig. 1–1 — *Types of Traps*

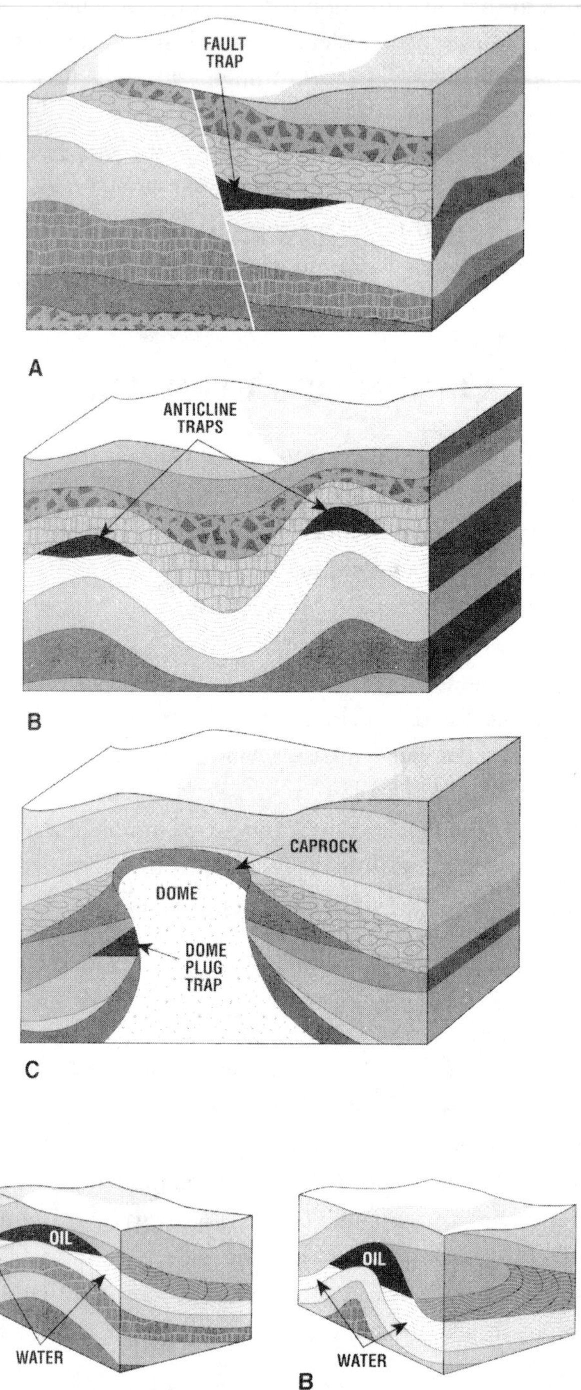

Source: Van Dyke, Kate, *Fundamentals of Petroleum, fourth edition*, Petroleum Extension Service, Austin, Texas, 1997, p.17.

- A source of petroleum, *i.e.,* remains of land and sea life
- Conditions such as heat and pressure resulting in the transformation of the organic material into petroleum
- Porous and permeable rock through which the petroleum was able to migrate after formation
- An impervious rock formation that acts as a trap or cap rock, permitting petroleum to collect

Assuming these four conditions are present, the reservoir rock itself must have porosity and permeability. **Porosity** is the measure of the pore space, *i.e.,* openings in a rock in which petroleum can exist. The greater the porosity, the more fluid (petroleum) the rock can hold. **Permeability** measures the "connectability" of the pores, which determines the ability of the petroleum to flow through the rock from one pore space to another. High porosity is often accompanied by high permeability (Figs. 1–2 and 1–3).

Fig. 1–2 — *Porosity*

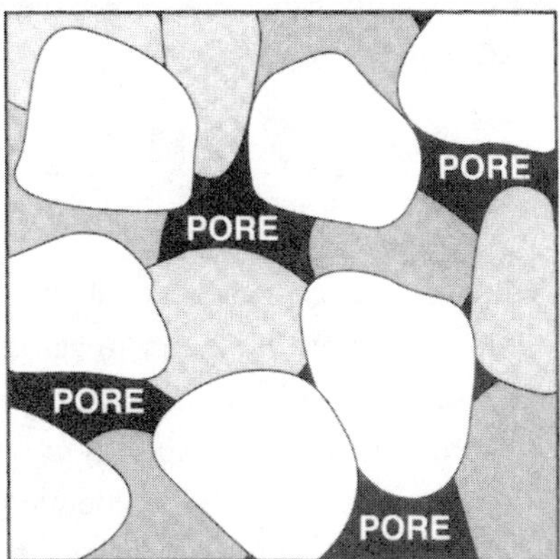

Source: Van Dyke, Kate, *Fundamentals of Petroleum, fourth edition,* Petroleum Extension Service, Austin, Texas, 1997, p.14.

Fig. 1–3 — *Permeability*

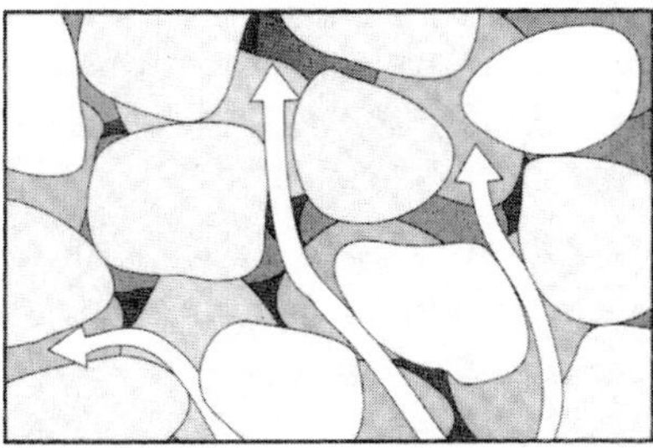

Source: Van Dyke, Kate, *Fundamentals of Petroleum, fourth edition,* Petroleum Extension Service, Austin, Texas, 1997, p.15.

To be commercially productive, a **petroleum reservoir** must have adequate porosity and permeability, and also have a sufficient physical area of rock that contains hydrocarbons (oil and natural gas). In other words, the reservoir must contain enough oil and natural gas to, when produced and sold, cover attendant costs and provide a profit for the producing company.

If an oil and gas reservoir has low permeability, procedures exist to increase the flow of oil or gas through the formation to the well bore. One method used to increase the flow from a "tight" formation is **fracturing**. This method usually involves introducing sand mixed with water or oil into the formation under high pressure to open or clean channels between the pores. Another common method used to increase the permeability of the formation is **acidizing**. Acidizing usually involves introducing hydrochloric acid into the formation to enlarge or reopen the channels between the pores. While these techniques to increase permeability are commonly used, they are effective only for a small area of the formation around well bores.

Exploration Methods and Procedures

Identifying areas that may contain petroleum reserves normally involves geological and geophysical (G&G) exploration studies. G&G exploration involves gathering and evaluating data about geologic structures that may contain petroleum reserves. Geological and geophysical studies are typically of two types:

- **surface surveys**—the study and evaluation of surface structures and features
- **subsurface surveys**—the study and evaluation of underground formations

Surface G&G studies, from which inferences can generally be made regarding subsurface formations, include aerial photography, satellite imaging, imaging radar, and topographical and geological mapping. Subsurface surveys involve accumulation of data to determine properties such as gravitational pull, magnetic field, and response to sound waves of the subsurface rock structures. These properties may indicate the type and depth of rock structures located beneath the earth's surface. The properties are determined using such geophysical devices as gravity meters, magnetometers, and seismographs.

Seismic studies, carried out using seismographs, are the most commonly used and important type of subsurface testing. These studies provide detailed information about subsurface structures by recording the reflection of sound waves on subsurface formations. Recent innovations in seismology, such as 3D and 4D seismic studies, have significantly increased drilling success rates. In one case, drilling success rates increased from 18% to 75% as a result of 3D seismic studies.

By examining all the G&G data collected—including the data collected by seismic studies—a map can be made to indicate formations favorable to the accumulation of petroleum, as well as interest areas that warrant further investigation. A **reconnaissance survey** is a G&G study covering a large or broad area. A **detailed survey** is a G&G study covering a smaller area, called an **area of interest**. Due to the high cost of seismic studies, seismic studies are usually performed after general reconnaissance studies have indicated a formation where high potential for oil and gas accumulation exists. If an area of interest is identified by a detailed survey, the area may be leased, if available.

Even the best G&G techniques cannot guarantee that oil or gas exists in economically producible quantities. The only definite way to determine whether an economically-viable petroleum reservoir exists is to drill wells into the formation. In the past, with traditional G&G technology, approximately 15% of wells drilled in new, unproved areas (areas with no known reservoirs) were successful. With the advent of 3D and 4D seismic technology, success rates have increased dramatically. With current technology, more than 75% of all wells drilled in both unproved areas and proved areas (areas with known reservoirs) are successful.

A successful well is a well that finds reserves in economically producible quantities. However, even though a well is classified as a success, it may nevertheless be unprofitable (chapter 8).

The following flowchart outlines the procedure in exploring for oil and gas:

Fig. 1–4 — *Steps in Finding Oil or Gas Onshore*

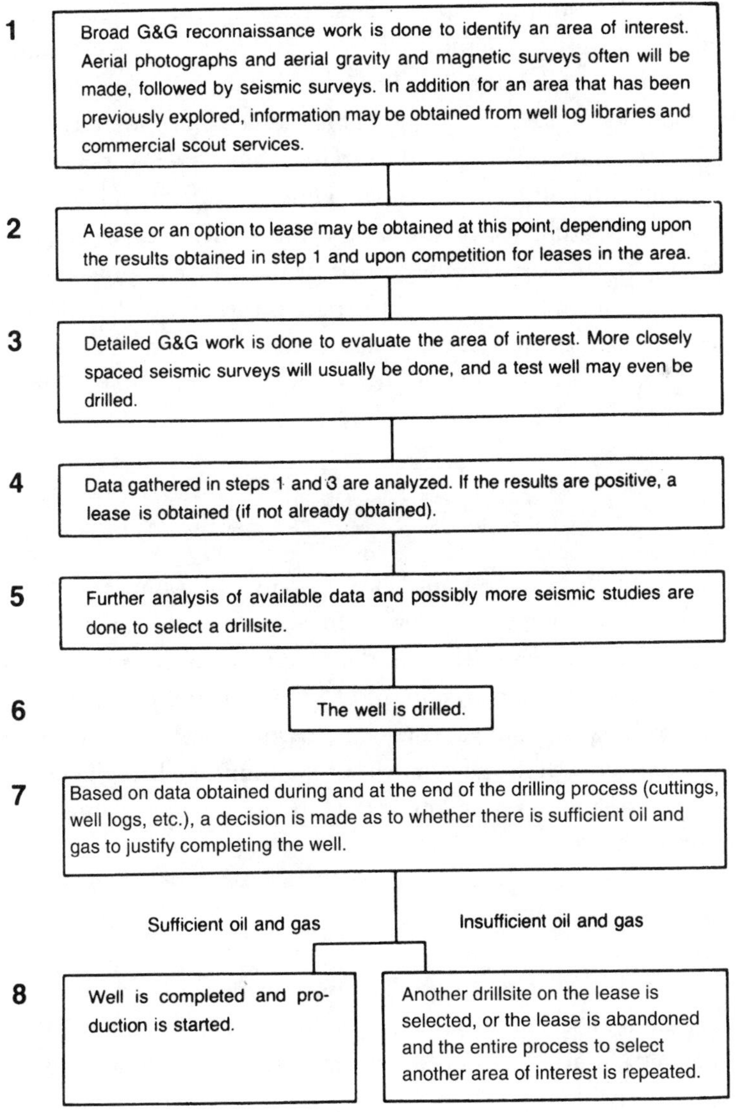

Acquisition of Mineral Interests in Property

After an oil company has identified an area with potential, the company will seek to acquire the right to explore, develop, and produce any minerals that might exist beneath the property, unless it already holds those rights. This right, along with the right to simply share in proceeds from the sale of any minerals produced, is referred to as a **mineral interest** or an **economic interest**. The specific type of mineral interest that is owned determines how costs and revenues are shared. The following section discusses the most common types of mineral interests and the typical methods of acquiring them.

Mineral rights

U.S. law assumes that, for ownership purposes, the surface of a piece of property can be separated from minerals existing underneath the surface. When a piece of land is purchased, one may acquire ownership of the surface rights only, the mineral rights only, or both rights. An ownership of both the surface and mineral rights is called a **fee interest**.

The **surface owner** has the right to use the surface in any legal way that the owner deems appropriate. For example, the surface owner may use the land for farming, ranching, building a residence, building apartments, etc. **Mineral rights (MR)** refer to the ownership, conveyed by deed, of any mineral beneath the surface. If the mineral rights are owned by one party and the surface is owned by another, the surface owner must allow the mineral rights owner, or his lessee, access to the surface area that is required to conduct exploration and production operations. The surface owner is entitled to compensation for any damages that may result from exploration and production operations.

Mr. A owns MR

The U.S. and, to a limited extent, Canada are the only two countries in the world where individual ownership of mineral rights is allowed. In countries outside the U.S. and Canada, ownership of mineral rights resides solely with the government. Therefore, oil and gas companies wishing to obtain a mineral interest in the U.S. must typically do so by executing lease agreements with individuals. Oil and gas companies seeking to obtain mineral interests outside the U.S. must contract with the government of the country where the property is located. The following discussion relates primarily to the U.S., where mineral interests are typically acquired via leasing.

Mineral interests

A **mineral interest (MI)** is an economic interest or ownership of minerals-in-place, giving the owner the right to a share of the minerals produced either in-kind or in proceeds from the sale of the minerals. (**Sharing in-kind** means the company or individual has elected to receive the oil or gas itself rather than the proceeds from the sale of the minerals.) When the owner of the mineral *rights* enters into a lease agreement or contract, two types of mineral *interests* are created—a working interest and a royalty interest. Descriptions of the basic types of mineral interests are given in the following paragraphs. Examples and additional types of mineral interests are provided in chapter 11.

Royalty interest (RI). This type of mineral interest is created by leasing. The RI is retained by the owner of the mineral rights when that owner enters into a lease agreement with another party. The RI receives a specified portion of the minerals produced, free and clear of any costs of exploring, developing, or operating the property. The royalty interest owner is responsible for any severance or production taxes assessed on production from the property.

Most lease agreements indicate that the royalty interest owner is also responsible for a proportionate share of the cost to get the product into marketable condition. Common examples include the cost of on-site treating, transportation costs, and the cost of compression of natural gas. In the past, due to the fact that the amounts were relatively small, allocation of these costs among the various owners was simply not cost effective for the operator. In recent years, with the struggle to keep many domestic operations running at a profit, many operators have begun to charge back a proportionate share of such costs to the royalty interest owners.

Since the royalty interest owner is not responsible for the exploration, development, or production of the property, the interest is referred to as a nonoperating or nonworking interest (non-WI).

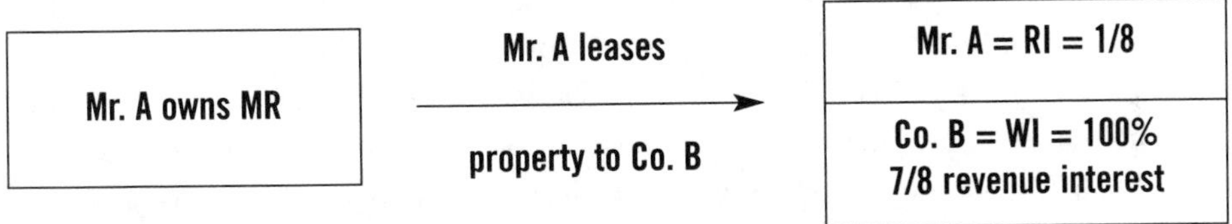

Working interest or operating interest (WI). This interest is created via leasing and is responsible for the exploration, development, and operation of a property. While the working interest owner pays all (100%) of the cost of exploring, drilling, developing, and producing the property, the WI's share of revenue is the amount that remains after deducting the share of the RI and other nonworking interests. Other types of nonworking interests are defined in following paragraphs.

In the simple illustration given, the WI would be a 100% WI with a $7/8$ revenue interest. Specifically, the WI would receive $7/8$ of the revenues and would pay 100% of the costs while the RI would receive $1/8$ of the revenues and pay none of the costs. Note that the working interest is stated in terms of how the costs are to be shared while the royalty interest is stated in terms of its share of gross revenue.

A working interest can be either an undivided interest or a divided interest. An undivided mineral interest exists when multiple owners share and share alike, according to their proportion of ownership in any minerals severed from the ground. In contrast, a divided interest exists when specific parties own specific acreage, minerals, or equipment.

For example, assume that Company A enters into an agreement to lease 640 acres from Farmer Brown who receives a $\frac{1}{8}$ royalty. Company A owns a 100% WI and Farmer Brown owns a $\frac{1}{8}$ RI. Now, Company A sells 50% of the WI to Company B. Company A and Company B each have an **undivided**, 50% working interest in the entire 640 acres. This means that they will each pay 50% of the cost of exploring, developing, and producing the property. Additionally, they will each be entitled to 50% of the revenue net of the royalty (*i.e.*, 50% of $\frac{7}{8}$ of gross sales). Now, instead of selling 50% of the working interest in 640 acres, assume that Company A sells 100% of the WI in 320 acres to Company B. This transaction results in a **divided** working interest. Afterwards, each company owns 100% of the WI in 320 acres or half of the original acreage. Company A and Company B would each be responsible for paying 100% of the cost of exploring, developing, and producing their 320 acres and each would be entitled to 100% of $\frac{7}{8}$ of the gross revenue generated from their own 320 acres.

Joint working interest. A joint working interest is an undivided working interest owned by two or more parties. Sharing the working interest is common in the oil and gas industry since it provides a means for companies to share the costs and risks of operations. Further, since less money is invested in any one property, companies are able to invest in more properties, thus spreading their risk over many properties.

In a joint WI, one of the parties is designated as the **operator** of the property and all the other working interest owners are called **nonoperators**. The operator manages the property and bills the nonoperators for their portion of any costs incurred. Each working interest owner accounts for its proportionate share of revenues and expenses separately (referred to as **proportionate consolidation**). Partnership accounting is not used to account for a joint WI.

Overriding royalty interest (ORI). An ORI is a nonworking interest created from the WI. The ORI's share of revenue is a stated percentage of the share of revenue belonging to the WI from which it was created. As a nonworking interest, it is free and clear of any development and operation costs except for severance taxes. An ORI may be created either by being retained by the WI owner when the working interest is sold or otherwise transferred, or by being carved out. A carved-out ORI is created when the WI owner sells or transfers the ORI and retains the WI.

Retained ORI

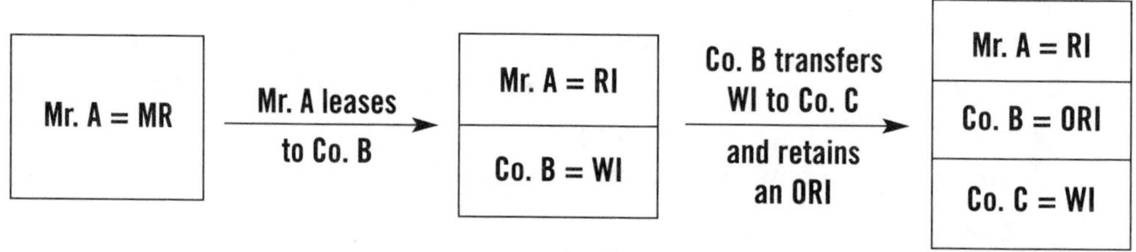

In the example above, if Mr. A leases its property to Co. B retaining a $\frac{1}{8}$ RI and Co. B subsequently conveys its interest to Co. C retaining a $\frac{1}{7}$ ORI, the costs and revenues would be shared as follows:

	Costs	Revenues
Mr. A	0 %	$^1/_8$
Co. B	0 %	$^1/_7 \times ^7/_8$
Co. C	100%	$^6/_7 \times ^7/_8$

Production payment interest (PPI). This interest is a nonworking interest created out of a working interest and similar to an ORI, except a PPI is limited to a specified amount of oil or gas, money, or time, after which it reverts back to the interest from which it was created and ceases to exist. If a PPI is payable with money, the payment is typically stated as a percentage of the working interest's share of revenue. If it is payable in product (*i.e.*, oil, gas, etc.), payment is typically stated as a percentage of the working interest's share of current production. Like ORIs, PPIs are created by carve-out or by retention.

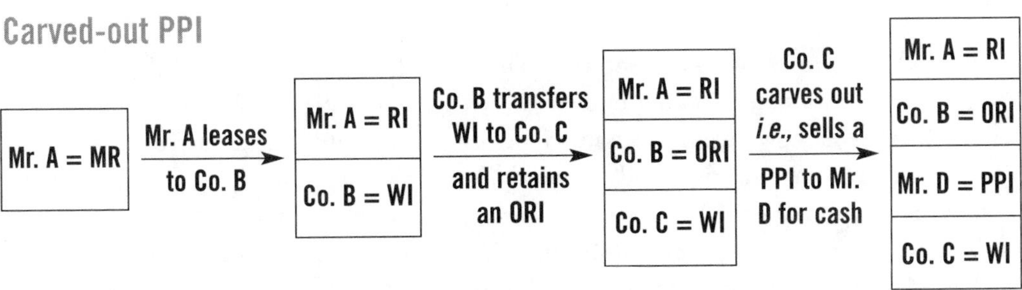

Both an ORI and a PPI are created out of the working interest by the working interest owner. These interests, as well as other interests discussed in chapter 11, are often created by the working interest owner in order to obtain financing or assistance in exploring and developing a property and to spread the risk involved.

Net profits interest (NPI). This nonworking interest is created on onshore property typically from the working interest. Offshore, a NPI is the type of interest that the government, as the mineral rights owner, often retains when leasing an offshore block to a petroleum company. This type of interest is similar to a RI or an ORI except that the amount to be received is a specified percentage of net profit from the property versus a percentage of the gross revenues from the property. The allowed deductions from gross revenues to calculate the net profit are usually specified in the lease agreement. While net profits interest owners are entitled to a percentage of the profits, they are not responsible for any portion of losses incurred in property development and operations. These losses, however, may be recovered by the working interest owner from future profits.

When created from the WI, a NPI may be created either by being retained by the WI owner when the working interest is sold or otherwise transferred, or by carve-out. A carved-out NPI is created when the WI owner sells or transfers the NPI and retains the WI. The first diagram below illustrates a NPI created onshore from the WI by carve-out. The second diagram illustrates a NPI being created offshore when the property is leased by the government to an oil company.

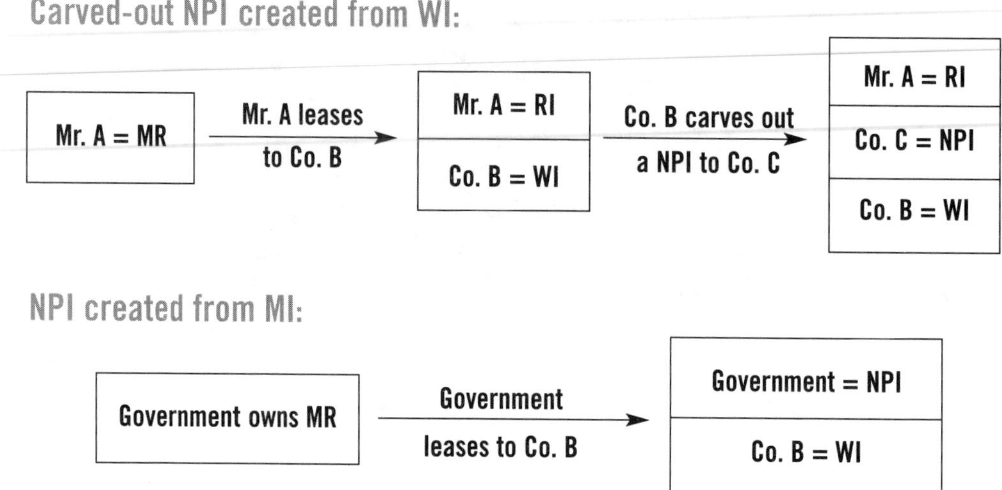

Pooled or unitized working interest. This type of interest is created when working interests as well as the nonworking interests in two or more properties are combined, with each interest owner now owning the same type of interest (but a smaller percentage) in the total combined property as they held previously in the separate property. The properties are operated as one unit, resulting in a more efficient, economical operation.

The terms **pooling** and **unitization** are often used interchangeably and represent the combining of small or irregular tracts of land into a unit large enough to meet state spacing regulations for drilling, or the combining of tracts in a field or reservoir in order to facilitate enhanced recovery projects. (Enhanced recovery projects are more complicated and expensive production techniques used after the first stages of production of a reservoir.) However, the most common usage of the term **pooling** is the combining of undrilled acreage to form a drilling unit, whereas the term **unitization** is commonly used to refer to a larger combination involving an entire producing field or reservoir for purposes of enhanced oil and gas recovery. In most states, pooling or unitization can be forced to facilitate maximum recovery and more efficient production.

Lease Provisions

Petroleum companies must obtain the rights to explore, drill, and produce subsurface minerals before conducting those activities. While these rights may be acquired through an outright purchase of a fee interest or the mineral rights, they are usually acquired through an oil and gas lease agreement (a copy of a typical lease contract is provided in Fig. 1–6).

Generally, oil and gas leases are obtained through the use of a **landman**. Landmen specialize in searching for and obtaining leases. A landman in many cases acts as an agent for an undisclosed principal in trying to obtain a lease at the lowest possible price.

An oil and gas lease embodies the legal rights, privileges, and duties pertaining to the lessor and lessee. The **lessor** is the mineral rights owner who leases the property to another party and retains a royalty interest. The **lessee**, the party leasing the property, receives a working interest. The lessee's working interest provides for investigating, exploring, prospecting, drilling, and mining for oil, gas, and other minerals as well as for conducting G&G surveys, installing production equipment, and producing said products.

Most lease contracts contain the following provisions:

- **Lease bonus**—initial amount paid to the mineral rights owner in return for the rights to explore, drill, and produce. Lease bonus payments usually are a dollar amount per acre. The amount of the lease bonus and other payments made to the mineral rights owner is a result of negotiations or bargaining between the mineral rights owner and the oil company representative.
- **Royalty provision**—specified fraction of the oil and gas produced free and clear of any costs (except severance taxes and certain costs to market the product) to which the royalty interest owner is entitled. Royalty payments are provided in many different formats. The most common method is to provide a share of production. This amount varies greatly, *i.e.,* $^1/_8$, $^1/_4$, etc., of production. On offshore federal leases, complicated formulas may be used for determining royalty payments. In some instances, the mineral rights owner will receive a portion of the net profits as royalty, *i.e.,* net profits leasing. Some offshore leases also provide for a sliding scale royalty whereby the amount of royalty paid is determined by the amount of oil and gas produced.
- **Primary term**—initial term of the lease. The primary term is the maximum time that the lessee has to begin drilling or commence production from the property. Technically, during this term, the lessee must begin drilling within one year from the signing of the lease. However, in the absence of drilling or production, the lessee can keep the lease in effect by making an annual payment called a delay rental payment.
- **Delay rental payment**—yearly payment made during the primary term in the absence of drilling operations in order to retain the lease. In other words, a delay rental payment is an annual payment made to allow the lessee to delay drilling operations for one additional year. Delay rental payments are typically based upon a dollar amount per acre. They must be paid on or before one year from the date the lease was signed and each year following during the primary term if drilling operations have not commenced. Generally, the lease will require the payment to be made to a specified bank. Some short-term leases (two- or three-year primary term) called paid up leases, require the lessee to pay the delay rentals at the inception of the lease. After the primary term, the lease can be held only by drilling or production, *i.e.,* after the primary term, a delay rental payment can no longer keep the lease from terminating. However, once production begins, the revenue from production from the lease keeps the lease in effect whether during or after the primary term. In the case where production commences before the expiration of the primary term, no delay rental payments are necessary. The royalty provides compensation to the RI.

• **Shut-in payments**—if the well is capable of producing oil or gas in paying quantities but is "shut in" (not producing), the lessee may hold the lease by making shut-in payments. Shut-in payments are usually made in natural gas situations where access to a pipeline is not available or an oversupply of gas exists. Shut-in payments are generally not recoverable from future royalty payments.

• **Right to assign interest**—rights of each party may be assigned in whole or in part without the approval of the other party. For example, the working interest owner may carve out a PPI or ORI from the WI without notifying the royalty interest owner.

• **Rights to free use of resources for lease operations**—the operator usually has the right to use, without cost, any oil or gas produced on the lease to carry out operations on that lease.

• **Option payment**—payment made to obtain a preleasing agreement that gives the oil company (the lessee) a specified period of time to obtain a lease from the entity receiving the payment. In addition to specifying the period of time within which the lessee may lease the property, the option contract will also typically specify the lease form, royalty interest, bonus to be paid, etc.

• **Offset clause**—if a producing well is drilled on Lease B close to the property line of Lease A (Fig. 1–5), within a distance specified in the lease contract, the offset clause requires the operator of Lease A to drill an offset well on Lease A, within a specified period of time, in order to prevent the well on Lease B from draining the reservoir. If, however, the leases are within a state with forced pooling or unitization, the leases can be forced to be pooled and operated as one, and the offset clause becomes irrelevant. If the state in question does not have forced pooling or unitization, then the only recourse for the interest owners of Lease A is for the operator to assume the burden of offset drilling.

Fig. 1–5 — *Offset Well Situation*

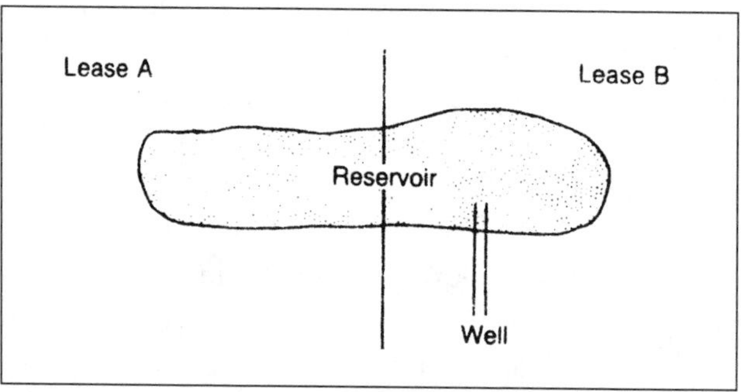

• **Minimum royalty**—a minimum royalty clause provides for the payment of a stipulated amount to the lessor regardless of production. A minimum royalty is similar to a shut-in royalty except that it is commonly recoverable from future royalty payments.

• **Pooling provisions**—modern lease forms provide that if the WI owner forms a pool or unit with other leases, the RI and other non-WI owners may also be forced to combine their interests with the non-WI owners of the other leases forming the unit.

Drilling Operations

Onshore drilling operations include building access roads to the drillsite, preparing the site for the drilling rig, transporting the rig to the site, and drilling the wellbore. Most drilling operations in the U.S. are done by drilling contractors. A **drilling contract** is an agreement between the lessee (WI owners) and a drilling contractor for the drilling of a well. The contract details the rights and obligations of the lessee and the contractor. Drilling contracts generally provide payment on a **day rate** (payment based on the number of days drilled), a **footage rate** (payment based on the number of feet drilled), or a **turnkey basis** (payment of a fixed sum of money based on drilling to a certain depth or stage of completion).

The first step in drilling an oil and gas well is selecting the actual drillsite. Seismic studies, particularly 3D seismic studies, are usually performed and the results of the studies are examined to determine the optimal site for the well. Once the site has been selected and a drilling contract signed, site preparations can begin. The well site is normally surveyed and staked, then access roads are built, and the site is graded and leveled. Reserve and waste pits are also prepared and a water supply is obtained.

After the site is prepared, often the initial 20 to 100 feet of the well will be drilled with a small truck-mounted rig. The drilling rig and related equipment are next moved in and set up, a process called **rigging up**. The well is then ready to be **spudded in**. The **spud date** is the date the rotary drilling bit touches ground.

Routine drilling consists of rotating a drill bit downwards through the formations towards target depth, cutting away pieces of the formation called **cuttings**. During the drilling process, drilling fluid, *i.e.,* **mud,** is constantly circulated down the wellbore. Drilling mud serves several purposes. It raises the cuttings to the surface, lubricates the drilling bit, and keeps formation fluids from entering the wellbore.

Approximately every 30 feet as the hole is deepened, a joint of drill pipe is added, a process called making a **mousehole connection**. Periodically, when the drill bit becomes worn or damaged, all of the drill pipe has to be removed from the hole in a process called **tripping out**. After a new drill bit is attached, the pipe is lowered back into the hole, called **tripping in**. Normally the drill pipe is removed and lowered three joints, *i.e.,* a **stand**, at a time depending upon the height of the derrick or mast. A derrick or mast is a four-legged, load-bearing structure that is part of the drilling rig. The height of the derrick correlates with the depth of the well since the derrick must support the weight of the drill string suspended downhole.

Tripping out is also necessary when casing must be set. **Casing** is steel pipe that is **set**, *i.e.,* cemented into the wellbore. Functions of casing include preventing the caving in of the hole, protecting fresh water sands, excluding water from the producing formations, confining production to the wellbore, and controlling formation pressure.

Although most wellbores are drilled vertically, some wells are drilled at an angle. These wells can be either directional or horizontal wells. **Directional wells** are wells that are normally drilled straight to a predetermined depth and then curved or angled so that the bottom of the wellbore is at the desired location. **Horizontal wells** are also initially drilled straight

Fig. 1–6 — *Oil, Gas and Mineral Lease*

OIL, GAS AND MINERAL LEASE

THIS AGREEMENT made this day of , between

Lessor (whether one or more), whose address is: ,
and , Lessee, WITNESSETH:

1. Lessor in consideration of Dollars
($), in hand paid, of the royalties herein provided, and of the agreements of Lessee herein contained, hereby grants, leases and lets exclusively unto Lessee for the purpose of investigating, exploring, prospecting, drilling and mining for and producing oil, gas and all other minerals, conducting exploration, geologic and geophysical surveys by seismograph, core test, gravity and magnetic methods, injecting gas, water and other fluids, and air into subsurface strata, laying pipe lines, building roads, tanks, power stations, telephone lines and other structures thereon and on, over and across lands owned or claimed by Lessor adjacent and contiguous thereto, to produce, save, take care of, treat, transport and own said products, and housing its employees, the following described land in County, Texas, to-wit:

This lease also covers and includes all land owned or claimed by Lessor adjacent or contiguous to the land particularly described above, whether the same be in said survey or surveys or in adjacent surveys, although not included within the boundaries of the land particularly described above. For the purpose of calculating the rental payments hereinafter provided for, said land is estimated to comprise acres, whether it actually comprises more or less.

2. Subject to the other provisions herein contained, this lease shall be for a term of years from this date (called "primary term") and as long thereafter as oil, gas or other mineral is produced from said land or land with which said land is pooled hereunder.

3. The royalties to be paid by Lessee are: (a) on oil, one-eighth of that produced and saved from said land, the same to be delivered at the wells or to the credit of Lessor into the pipelines to which the wells may be connected; Lessee may from time to time purchase any royalty oil in its possession, paying the market price therefor prevailing for the field where produced on the date of purchase; (b) to pay lessor on gas and casinghead gas produced from said land (1) when sold by lessee, one-eighth of the amount realized by lessee, computed at the mouth of the well, or (2) when used by lessee off said land or in the manufacture of gasoline or other products, one-eighth of the amount realized from the sale of gasoline or other products extracted therefrom and one-eighth of the amount realized from the sale of residue gas after deducting the amount used for plant fuel and/or compression; while there is a gas well on this lease or on acreage pooled therewith but gas is not being sold or used, Lessee may pay as royalty, on or before ninety (90) days after the date on which (1) said well is shut in, or (2) the land covered hereby or any portion thereof is included in a pooled unit on which a well is located, or (3) this lease ceases to be otherwise maintained as provided herein, whichever is the later date, and thereafter at annual intervals on or before the anniversary of the date the first payment is made, a sum equal to the amount of the annual rental payable in lieu of drilling operations during the primary term on the number of acres subject to this lease at the time such payment is made, and if such payment is made or tendered, this lease shall not terminate, and it will be considered that gas is being produced from this lease in paying quantities; and (c) on all other minerals mined and marketed, one-tenth either in kind or value at the well or mine, at Lessee's election, except that on sulphur mined and marketed the royalty shall be fifty cents (50¢) per long ton. Lessee shall have free use of oil, gas, coal, and water from said land, except water from Lessor's wells, for all operations hereunder, and the royalty on oil, gas and coal shall be computed after deducting any so used.

4. Lessee, at its option, is hereby given the right and power to pool or combine the acreage covered by this lease or any portion thereof as to oil and gas, or either of them, with any other land covered by this lease, and/or with any other land, lease or leases in the immediate vicinity thereof to the extent hereinafter stipulated, when in Lessee's judgment it is necessary or advisable to do so in order properly to explore, or to develop and operate said leased premises in compliance with the spacing rules of the Railroad Commission of Texas, or other lawful authority, or when to do so would, in the judgment of Lessee, promote the conservation of oil and gas in and under and that may be produced from said premises. Units pooled for oil hereunder shall not substantially exceed 40 acres each in area, and units pooled for gas hereunder shall not substantially exceed in area 640 acres each plus a tolerance of ten percent (10%) thereof, provided that should governmental authority having jurisdiction prescribe or permit the creation of units larger than those specified, for the drilling or operation of a well at a regular location or for obtaining maximum allowable from any well to be drilled, drilling or already drilled, units thereafter created may conform substantially in size with those prescribed or permitted by governmental regulations. Lessee under the the provisions hereof may pool or combine acreage covered by this lease or any portion thereof as above provided as to oil in any one or more strata and as to gas in any one or more strata. The units formed by pooling as to any stratum or strata need not conform in size or area with the unit or units into which the lease is pooled or combined as to any other stratum or strata, and oil units need not conform as to area with gas units. The pooling in one or more instances shall not exhaust the rights of the Lessee hereunder to pool this lease or portions thereof into other units. Lessee shall file for record in the appropriate records of the county in which the leased premises are situated an instrument describing and designating the pooled acreage as a pooled unit; and upon such recordation the unit shall be effective as to all parties hereto, their heirs, successors, and assigns, irrespective of whether or not the unit is likewise effective as to all other owners of surface, mineral, royalty, or other rights in land included in such unit. Lessee may at its election exercise its pooling option before or after commencing operations for or completing an oil or gas well on the leased premises, and the pooled unit may include, but it is not required to include, land or leases upon which a well capable of producing oil or gas in paying quantities has theretofore been completed or upon which operations for the drilling of a well for oil or gas have theretofore been commenced. In the event of operations for drilling on or production of oil or gas from any part of a pooled unit which includes all or a portion of the land covered by this lease, regardless of whether such operations for drilling were commenced or such production was secured before or after the execution of this instrument or the instrument designating the pooled unit such operations shall be considered as operations for drilling on or production of oil and gas from land covered by this lease whether or not the well or wells be located on the premises covered by this lease and in such event operations for drilling shall be deemed to have been commenced on said land within the meaning of paragraph 5 of this lease; and the entire acreage constituting such unit or units, as to oil and gas, or either of them, as herein provided, shall be treated for all purposes, except the payment of royalties on production from the pooled unit, as if the same were included in this lease. For the purpose of computing the royalties to which owners of royalties and payments out of production and each of them shall be entitled on production of oil and gas, or either of them, from the pooled unit, there shall be allocated to the land covered by this lease and included in said unit (or each separate tract within the unit if this lease covers separate tracts within the unit) a pro rata portion of the oil and gas, or either of them, produced from the pooled unit after deducting that used for operations on the pooled unit. Such allocation shall be on an acreage basis—that is to say, there shall be allocated to the acreage covered by this lease and included in the pooled unit (or to each separate tract within the unit if this lease covers separate tracts within the unit) that pro rata portion of the oil and gas, or either of them, produced from the pooled unit which the number of surface areas covered by this lease (or in each such separate tract) and included in the pooled unit bears to the total number of surface acres included in the pooled unit. Royalties hereunder shall be computed on the portion of such production, whether it be oil and gas, or either of them, so allocated to the land covered by this lease and included in the unit just as though such production were from such land. The production from an oil well will be considered as production from the lease or oil pooled unit which it is producing and not as production from a gas pooled unit; and production from a gas well will be considered as production from the lease or gas pooled unit from which it is producing and not from an oil pooled unit. The formation of any unit hereunder shall not have the effect of changing the ownership of any delay rental or shut-in production royalty which may become payable under this lease. If this lease now or hereafter covers separate tracts, no pooling or unitization of royalty interest as between any such separate tracts is intended or shall be implied or result merely from the inclusion of such separate tracts within this lease but Lessee shall nevertheless have the right to pool as provided above with consequent allocation of production as above provided. As used in this paragraph 4, the words "separate tract" mean any tract with royalty ownership differing, now or hereafter, either as to parties or amounts, from that as to any other part of the leased premises.

5. If operations for drilling are not commenced on said land or on acreage pooled therewith as above provided on or before one year from this date, the lease shall then terminate as to both parties, unless on or before such anniversary date Lessee shall pay or tender (or shall make a bona fide attempt to pay or tender, as hereinafter stated) to Lessor or to the credit of Lessor in Bank at ,
Texas, (which bank and its successors are Lessor's agent and shall continue as the depository for all rentals payable hereunder regardless of changes in ownership of said land or the rentals) the sum of Dollars ($),
(herein called rentals), which shall cover the privilege of deferring commencement of drilling operations for a period of twelve (12) months. In like manner and upon like payments or tenders annually, the commencement of drilling operations may be further deferred for successive periods of twelve (12) months each during the primary term. The payment or tender of rental under this paragraph and of royalty under paragraph 3 on any gas well from which gas is not being sold or used may be made by the check or draft of Lessee mailed or delivered to the parties entitled thereto or to said bank on or before the date of payment. If such bank (or any successor bank) should fail, liquidate or be succeeded by another bank, or for any reason fail or refuse to accept rental, Lessee shall not be held in default for failure to make such payment or tender or rental until thirty (30) days after Lessor shall deliver to Lessee a proper recordable instrument naming another bank as agent to receive such payments or tenders. If Lessee shall, on or before any anniversary date, make a bona fide attempt to pay or deposit rental to a Lessor entitled thereto according to Lessee's records or to a Lessor, who, prior to such attempted payment or deposit, has given Lessee notice, in accordance with subsequent provisions of this lease, of his right to receive rental, and if such payment or deposit shall be ineffective or erroneous in any regard, Lessee shall be unconditionally obligated to pay to such Lessor the rental properly payable for the rental period involved, and this lease shall not terminate but shall be maintained in the same manner as if such erroneous or ineffective rental payment or deposit had been properly made, provided that the erroneous or ineffective rental payment or deposit be corrected within 30 days after receipt by Lessee of written notice from such Lessor of such error accompanied by such instruments as are necessary to enable Lessee to make proper payment. The down cash payment is consideration for this lease according to its terms and shall not be allocated as a mere rental for a period. Lessee may at any time or times execute and deliver to lessor or to the depository above named or place of record a release or releases of this lease as to all or any part of the above-described premises, or of any mineral or horizon under all or any part thereof, and thereby be relieved of all obligations as to the released land or interest. If this lease is released as to all minerals and horizon under a portion of the land covered by this lease, the rentals and other payments computed in accordance therewith shall thereupon be reduced in the proportion that the number of surface acres within such released portion bears to the total number of surface acres which was covered by this lease immediately prior to such release.

6. If prior to discovery and production of oil, gas or other mineral on said land or on acreage pooled therewith, Lessee should drill a dry hole or holes thereon, or if after discovery and production of oil, gas or other mineral, the production thereof should cease from any cause, this lease shall not terminate if Lessee commences operations for drilling or reworking within sixty (60) days thereafter or if it be within the primary term, commences or resumes the payment or tender of rentals or commences operations for drilling or reworking on or before the rental paying date next ensuing after the expiration of sixty (60) days from the date of completion of dry hole or cessation of production. If at any time subsequent to sixty (60) days prior to the beginning of the last year of the primary term and prior to the discovery of oil, gas or other mineral on said land, or on the acreage pooled therewith, Lessee should drill a dry hole thereon, no rental payment or operations are necessary in order to keep the lease in force during the remainder of the primary term. If at the expiration of the primary term, oil, gas or other mineral is not being produced on said land, or on acreage pooled therewith, but Lessee is then engaged in drilling or reworking operations thereon or shall have completed a dry hole thereon within sixty (60) days prior to the end of the primary term, the lease shall remain in force so long as operation on said well or for drilling or reworking of any additional well are prosecuted with no cessation of more than sixty (60) consecutive days, and if they result in the production of oil, gas or other mineral, so long thereafter as oil, gas or other mineral is produced from said land or acreage pooled therewith. Any pooled unit designated by Lessee in accordance with the terms hereof may be dissolved by Lessee by instrument filed for record in the appropriate records of the county in which the leased premises are situated at any time after the completion of a dry hole or the cessation of production on said unit. In the event a well or wells producing oil or gas in paying quantities shall be brought in on adjacent land and within three hundred thirty (330) feet of and draining the leased premises, or acreage pooled therewith, Lessee agrees to drill such offset wells as a reasonably prudent operator would drill under the same or similar circumstances.

7. Lessee shall have the right at any time during or after the expiration of this lease to remove all property and fixtures placed by Lessee on said land, including the right to draw and remove all casing. When required by Lessor, Lessee will bury all pipe lines below ordinary plow depth, and no well shall be drilled within two hundred (200) feet of any residence or barn now on said land without Lessor's consent.

8. The rights of either party hereunder may be assigned in whole or in part, and the provisions hereof shall extend to their heirs, successors and assigns; but no change or division in ownership of the land, rentals or royalties, however accomplished, shall operate to enlarge the obligations or diminish the rights of Lessee; and no change or division in such ownership shall be binding on Lessee until thirty (30) days after Lessee shall have been furnished by registered U.S. mail at Lessee's principal place of business with a certified copy of recorded instrument or instruments evidencing same. In the event of assignment hereof in whole or in part, liability for breach of any obligation hereunder shall rest exclusively upon the owner of this lease or of a portion thereof who commits such breach. In the event of the death of any person entitled to rentals hereunder, Lessee may pay or tender such rentals to the credit of the deceased or the estate of the deceased until such time as Lessee is furnished with proper evidence of the appointment and qualification of an executor or administrator of the estate, or if there be none, then until Lessee is furnished with evidence satisfactory to it as to the heirs or devisees of the deceased and that all debts of the estate have been paid. If at any time two or more persons be entitled to participate in the rental payable hereunder, Lessee may pay or tender said rental jointly to such persons or to their joint credit in the depository named herein; or, at Lessee's election, the proportionate part of said rentals to which each participant is entitled may be paid or tendered to him separately or to his separate credit in said depository; and payment or tender to any participant of his portion of the rentals hereunder shall maintain this lease as to such participant. In event of assignment of this lease as to a segregated portion of said land, the rentals payable hereunder shall be apportionable as between the several leasehold owners ratably according to the surface area of each, and default in rental payment by one shall not affect the rights of other leasehold owners hereunder. If six or more parties become entitled to royalty hereunder, Lessee may withhold payment thereof unless and until furnished with a recordable instrument executed by all such parties designating an agent to receive payment for all.

9. The breach by Lessee of any obligation arising hereunder shall not work a forfeiture or termination of this lease nor cause a termination or reversion of the estate created hereby nor be grounds for cancellation hereof in whole or in part. In the event Lessor considers that operations are not at any time being conducted in compliance with this lease, Lessor shall notify Lessee in writing of the facts relied upon as constituting a breach hereof, and Lessee, if in default, shall have sixty days after receipt of such notice in which to commence the compliance with the obligations imposed by virtue of this instrument. After the discovery of oil, gas or other mineral in paying quantities on said premises, Lessee shall develop the acreage retained hereunder as a reasonably prudent operator, but in discharging this obligation it shall in no event be required to drill more than one well per forty (40) acres of the area retained hereunder and capable of producing oil in paying quantities and one well per 640 acres plus an acreage tolerance not to exceed 10% of 640 acres of the area retained hereunder and capable of producing gas or other mineral in paying quantities.

10. Lessor hereby warrants and agrees to defend the title to said land and agrees that Lessee at its option may discharge any tax, mortgage or other lien upon said land, either in whole or in part, and in event Lessee does so, it shall be subrogated to such lien with right to enforce same and apply rentals and royalties accruing hereunder toward satisfying same. Without impairment of Lessee's rights under the warranty in event of failure of title, it is agreed that if this lease covers a less interest in the oil, gas, sulphur, or other minerals in all or any part of said land than the entire and undivided fee simple estate (whether Lessor's interest is herein specified or not), or no interest therein, then the royalties, delay rental, and other monies accruing from any part as to which this lease covers less than such full interest, shall be paid only in the proportion which the interest therein, if any, covered by this lease, bears to the whole and undivided fee simple estate therein. All royalty interest covered by this lease (whether or not owned by Lessor) shall be paid out of the royalty herein provided. Should any one or more of the parties named above as Lessors fail to execute this lease, it shall nevertheless be binding upon the party or parties executing the same. Failure of Lessee to reduce rental paid hereunder shall not impair the right of Lessee to reduce royalties.

11. Should Lessee be prevented from complying with any express or implied covenant of this lease, from conducting drilling or reworking operations thereon or from producing oil or gas therefrom by reason of scarcity of or inability to obtain or to use equipment or material, or by operation of force majeure, any Federal or state law or any order, rule or regulation of governmental authority, then while so prevented, Lessee's obligation to comply with such covenant shall be suspended, and Lessee shall not be liable in damages for failure to comply therewith; and this lease shall be extended while and so long as Lessee is prevented by any such cause from conducting drilling or reworking operations on or from producing oil or gas from the lease premises; and the time while Lessee is so prevented shall not be counted against Lessee, anything in this lease to the contrary notwithstanding.

IN WITNESS WHEREOF, this instrument is executed on the date first above written.

LESSOR	SS NO. OR TAX ID	LESSOR	SS NO. OR TAX ID

ACKNOWLEDGMENT

STATE OF TEXAS
COUNTY OF

This instrument was acknowledged before me on the day of , ,
by

Notary Public, State of Texas
Notary's name (printed):
Notary's commission expires:

ACKNOWLEDGMENT

STATE OF TEXAS
COUNTY OF

This instrument was acknowledged before me on the day of , ,
by

Notary Public, State of Texas
Notary's name (printed):
Notary's commission expires:

AFTER RECORDING RETURN TO:

Producers 88 (4/76) Revised
With 640 Acres Pooling Provision

Source: COPAS, 1973. Recommended by the Council of Petroleum Accountants Societies of North America. COPAS Bulletin No. 4, "COPAS Forms," January 1974, pp. 33-34. COPAS bulletins may be ordered from Krafbilt, Tulsa, Oklahoma, 1-800-331-7290.

down but then are gradually curved until the hole runs parallel to the earth's surface, with drilling actually achieving a horizontal direction through the formation. Figure 1–7 shows a conventional well, a directional well, and a horizontal well.

Fig. 1–7 — *Types of Wells*

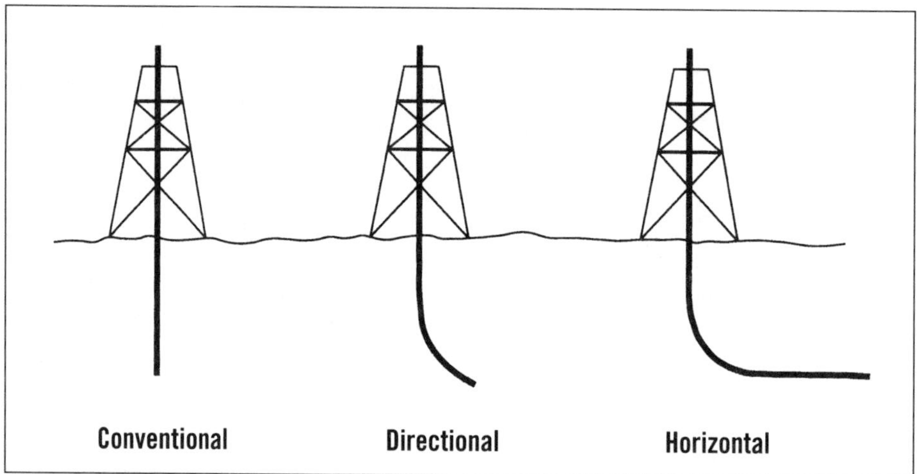

Directional drilling may be used when it is necessary to avoid drilling along a fault line, when the rig cannot be placed over the selected drillsite, when the straight hole misses the reservoir, to sidetrack around a fish (an object lodged in the borehole), or to drill multiple wells from the same offshore drilling platform. Horizontal drilling is advantageous when oil and gas are contained in several vertical, narrow pockets or reservoirs that would be missed by conventional straight drilling. Horizontal wells are also often used as lateral injection wells in enhanced recovery projects.

During drilling operations, the petroleum engineer or geologist examines cuttings and other data to decide whether there is sufficient oil or gas to justify completing the well. As the well is drilled, the cuttings are analyzed to determine fluid content and rock structure. Periodically, a larger piece of formation rock may be obtained by placing a core barrel on the end of the drill stem and cutting a cylindrical core approximately 4 inches in diameter and up to 60 feet in length from the formation. This piece of formation rock, called a **core sample**, is then analyzed to determine formation rock characteristics, sequence of rock layers in the earth, and fluid content of the formation.

After total depth has been reached, the well is **logged** by lowering a device to the bottom of the well and then pulling it back up to the surface. As the device passes up the hole, it measures and records properties of the formations and the fluids residing in them. Based on the results of the analyses and the test methods discussed above as well as other evaluation methods, a decision is made as to whether to complete the well. If the well is judged incapable of providing oil or gas in commercial quantities, it is plugged and abandoned. If the well is judged capable of producing oil or gas in commercial quantities, the well is completed.

Activities incident to completing a well and placing it on production include:

- obtaining and installing production casing (steel pipe lining the wellbore)
- installing tubing (steel pipe suspended in the well through which the oil and gas are produced)
- **perforating** (setting off charges to create holes in the casing and cement so formation fluids can flow from the formation into the wellbore)
- installing the **Christmas tree** (valves and fittings controlling production at the well-head that somewhat resemble a Christmas tree)
- constructing **production facilities** (separators, heater treaters, tanks, etc.) and installing flow lines.

Activities incident to plugging and abandoning a well would include removal of any equipment possible and cementing the wellbore to seal the hole.

Some wells penetrate more than one zone containing oil or gas in commercial quantities. In these cases, the wells may be completed either to produce from only one zone or from multiple zones. In a **multiple completion**, the well is capable of simultaneous production from the multiple zones containing oil or gas.

Determination of whether or not to complete a well entails a comparison of the incremental costs to complete the well with the net cash flows expected to be received from the sale of petroleum products from the well. A well is judged capable of commercial production if the expected net proceeds from production exceed the cost of completing the well. The costs incurred in drilling prior to completion—past costs or sunk costs—are not considered in making the decision to plug and abandon or complete the well.

EXAMPLE

Completion Decision

Lucky Company incurred $300,000 in drilling costs up to target depth. The estimated costs to complete the well (installing casing, Christmas tree, etc.) are $200,000. Lucky's estimated discounted future net cash flow from the sale of the oil and gas from this well is expected to be $350,000. Should the well be completed or abandoned?

Answer:
The well should be completed because the estimated discounted future net cash flow is $350,000 and the incremental costs (after drilling) are $200,000. If the well is not completed, Lucky will have lost $300,000. If the well is completed, Lucky's loss is only $150,000. (See chapter 8 for further discussion concerning the decision to complete a well.)

The activities involved in offshore drilling are somewhat different than in onshore operations. In territorial waters offshore the U.S., operators may acquire mineral leases from state governments or from the federal government. In contrast to most onshore federal leases, offshore federal leases are obtained through a system of closed competitive bidding on available offshore tracts. Normally, the federal government keeps a one-sixth royalty on the tracts. Competitive bidding is also common for state leases.

Drilling operations are much more expensive offshore, consequently, most offshore drilling is done in the form of a joint venture or joint interest operation, *i.e.*, several companies pooling capital resources to explore, develop, and produce an offshore tract. While joint interest operations are also common onshore, they are generally the rule offshore.

In some offshore areas, very little may be known about the types and depths of the subsurface formations. In those areas, a **stratigraphic test well**, a well drilled for information only, may be drilled. Often, such a well will be drilled prior to the bidding process and paid for by multiple companies, who agree to share the information. Exploratory-type stratigraphic test wells are often drilled offshore to determine the existence and quantity of proved reserves and also to determine the location for the permanent development drilling and production platform. (Exploratory wells are wells drilled in area not known to contain oil or gas.)

Exploratory drilling offshore is almost always done from mobile rigs. These temporary rigs can take the form of

- **drilling barges**, which are towed to location
- **jack-up rigs**, which are towed to location, and then four legs are jacked down to rest on sea bottom
- **submersible rigs**, which sink to the ocean floor and are used only in shallow water
- **semisubmersible rigs**, which rest in a partially submerged position
- **drillships**, which look like ordinary ships but contain a derrick

Development wells, which produce a reservoir that has been discovered with exploratory drilling, are often drilled from fixed platforms containing production and well maintenance facilities. These fixed platforms, which are very expensive, are usually made from concrete or steel, depending on location.

The drilling operations of an offshore rig are similar to those of onshore rigs, with the exception of specialized technical adaptations that have been made to deal with the hostile marine environment. In addition, directional drilling is commonly used offshore, since that technique can reach thousands of feet away from the platform, thus allowing the drilling of multiple wells (as many as 40 or more) from the same development platform.

Environmental concerns are of extreme importance in offshore areas, with the prevention of oil spills being the primary concern. Expensive safety equipment is utilized to prevent blowouts and spills.

Recovery Processes

Several types of production processes move the oil from the reservoir to the well. The production processes are commonly divided into three types of recovery methods: primary, secondary, and tertiary.

The initial or **primary recovery** of oil and gas is either by natural reservoir drive or by pumping. Natural drive occurs when sufficient water or gas exists in the reservoir under high pressure. The reservoir pressure provides the natural energy needed to drive the oil to the wellbore. If insufficient natural drive exists, the oil may be pumped to the surface using a beam pumping unit.

When the maximum amount of oil and gas has been recovered by primary recovery methods and the reservoir pressure has been largely depleted, **secondary recovery** may be instituted. Secondary recovery consists of inducing an artificial drive into the formation to replace the natural drive. The most common method is waterflooding, which involves injecting water under pressure into the formation to drive the oil to the wellbore.

Following the second attempt to recover oil, a third attempt, called **tertiary recovery**, may be instituted. (The distinction between secondary and tertiary recovery methods may actually be obscure. Some secondary recovery methods are also considered as tertiary recovery methods and vice versa.) Tertiary recovery involves the use of enhanced recovery methods to produce oil. Tertiary recovery methods include injection of chemicals, gas, or heat into the well to modify the fluid properties and thereby enhance the movement of the oil through the formation. A newer form of tertiary recovery uses microwave technology, which introduces microwaves to reservoirs in northern climates to warm the oil. Tertiary methods may be very expensive, and many methods are still in the developmental stage.

Even with the best recovery methods, a large amount of oil remains locked in the formation. Some experts have estimated that at least 50% of the oil cannot be recovered with current technology. The amount of oil and gas to be recovered in the future from old wells is partially determined by oil and gas prices. Many more secondary and tertiary projects will be attempted when petroleum product pricing makes such recovery economically feasible.

Production and Sales

Fluids produced from a well normally will contain oil, gas, and water. Before the oil and gas are sold, the well fluid must be separated, treated, and measured. Flow lines take well fluid from individual wells to a central gathering point where separators divide the fluid into liquids (oil and water) and gas. Heater treaters then remove the water and other impurities from the oil.

From this point, oil generally goes to stock tanks for storage until it is delivered to a buyer. When the oil is sold, it is measured as it is transferred from the storage tanks to either a truck transport or an oil pipeline. Gas, which is not stored on the lease, is measured as it is gathered, processed, and transferred to a gas pipeline.

The amount of oil transferred from the storage tanks to a transport truck or pipeline is recorded on a document called a **run ticket**. The amount of payment for the oil is based upon information contained in the run ticket (chapter 9). The **gas settlement statement** is used to record similar information for the production and sale of gas.

Common State and Federal Regulations

In the 1920s and the 1930s, overdrilling was commonplace. In that era, overdrilling was so extreme that the legs of one drilling rig might be inside the legs of another rig. Since those days of frantic drilling, many states have established agencies to oversee oil and gas drilling and those agencies have passed regulations to eliminate waste and uneconomical methods of producing oil and gas. For example, the state agency that oversees oil and gas drilling in Texas is the Texas Railroad Commission. In Oklahoma, it is the Oklahoma Corporation Commission. Exploration of federal offshore areas as well as federal onshore lands is regulated by the Department of the Interior.

One common state regulation is related to well spacing. Many states require that wells must be spaced a minimum distance apart to help prevent economic waste and reduce injury to the reservoir. Economic waste occurs when too many wells are drilled into a reservoir, since an increased number of wells drilled does not necessarily increase oil and gas recovery. Further, the reservoir's natural drive may be rapidly depleted if more wells are drilled in one area than in another, causing premature water or gas encroachment. A common spacing requirement is 40 acres for oil wells and 640 acres for gas wells.

Another common regulation relates to drilling permits. Prior to starting the drilling process, whether on private or public land, a drilling permit is generally required from the state, or if on federal lands or water, from the federal government. An application for a permit is made to the appropriate governmental agency outlining the location, depth, etc., of the well. Generally, the drilling permit will not be granted unless the well spacing requirements are met or an exception is granted. In addition to obtaining a permit to drill, various types of reports must be filed, *e.g.*, protection of water-bearing strata, various completion reports, and plug-and-abandon reports. For offshore leases the process of obtaining a drilling permit is much more complicated and time consuming. Permits may be required from both the state and federal government. The application for a federal offshore permit requires more information than is requested for an onshore permit. Information to be provided includes a description of the vessels, platforms, and other required structures, targeted locations for each well, and forms of protection against environmental contamination.

Another important state regulation deals with restriction of production. If demand is adequate, states will typically allow all wells and leases to produce at the maximum efficiency rate (MER). (The MER is the maximum rate at which oil or gas can be produced without damaging the reservoir's natural energy.) In periods of low demand for oil and gas, states frequently restrict production by a proration process. A state agency (regulatory commission) decides the amount to be produced within the state for a given period of time, usually a month, and then prorates this amount among the state's producing fields (field allowable). The field allowable is then prorated to the various leases and wells. The leases and wells will then have an allowable for the month, *i.e.,* the maximum number of barrels of oil or cubic feet of gas that may be produced from the lease and wells during the month.

Pooling or unitization, as discussed earlier, involves combining two or more leases and operating them as one property. The basic purpose is to recover the maximum amount of oil and gas possible in the most efficient and economical manner. The federal and state governments encourage pools and units, with some states imposing mandatory pooling and unitization. Typically, units, especially mandatory units, must be approved by the appropriate state regulatory agencies, or by the federal government if on federal lands.

What Does the Future Hold?

The U.S. oil and gas industry is currently facing a challenge to survive and prosper in the future. The U.S. has less than 3% of the world's reserves of oil in contrast to the Middle East, which is estimated to have 65% of world reserves. Further, the average Middle East well outproduces the average U.S. well by more than 1,500 to 1. These statistics do not paint a bright picture for the U.S. domestic industry. The situation for gas is not as bleak.

Technology is the key if oil production is to have a future in the U.S. Many reserves remain in the ground in wells that have been plugged because remaining reserves cannot be economically produced with current recovery methods. Enhanced recovery methods are currently being researched that may make it economically feasible to produce many of those reserves.

For example, the method of horizontal drilling is now being used to produce from areas that were once abandoned as not economically feasible. The development of horizontal drilling in the oil industry has been compared to the invention of the transistor in the electronics industry. Further, increases in the price of oil in the future, which will happen eventually, will make even currently expensive enhanced recovery methods economically feasible.

Environmental issues will be at the forefront as a concern for the industry for many years to come. The industry has a long tradition of acting responsibly to develop environmental safeguards in order to protect the environment. Hopefully, a balance can be reached between environmentalists and the industry, which will allow for exploration activities to continue and increase in both onshore and offshore U.S. areas.

Finally, as the industry has realized that many of the best opportunities for exploration are in other countries, U.S. oil companies have increased operations in those areas. Companies are working with governments all over the world to explore for and develop new oil and gas reserves. The future of the industry is largely dependent on the successful exploration and production operations in foreign countries. These successes will enable many U.S. oil companies to grow well into the next century.

◆◆◆

This profile of petroleum operations provides the framework for the remaining chapters of this text. Accounting applications of the various activities described above are investigated from the three oil and gas accounting methodologies:

- successful efforts
- full cost
- tax accounting

In addition, the final three chapters present an overview of international oil and gas accounting, the analysis of oil and gas companies' financial statements, and an overview of pipeline accounting. Also included in the book are appendices containing an authorization for expenditure (AFE), authoritative literature by the Financial Accounting Standards Board and the Securities and Exchange Commission, a list of abbreviations used in the book, a list of references, and the financial statements of Kerr-McGee Corporation.

PROBLEMS

1. Describe the organic theory of the origin of oil and gas.

2. Terms
   a. Define the following:
      fault trap
      anticline
      salt dome
      porosity
      permeability

   b. Define the following:
      day-rate contract
      footage-rate contract
      turnkey contract
      horizontal drilling

   c. Explain the following:
      petroleum reservoir
      primary recovery
      secondary recovery
      tertiary recovery

3. List the steps in finding oil and gas.

4. What is the difference between an operating (working) interest and a nonoperating (nonworking) interest?

5. Define the following:
   economic interest in oil and gas
   mineral rights
   mineral interest
   royalty interest
   working interest
   overriding royalty interest
   production payment interest

6. Define and discuss the important provisions of the typical oil and gas lease.

7. What are the drilling operations that give rise to accounting implications?

8. Which of the following would not be a mineral interest?
   a. production payment interest
   b. working interest
   c. overriding royalty interest
   d. surface rights interest
   e. net profits interest
   f. royalty interest
   g. joint working interest

9. Core Oil Company signed a lease contract on January 1, 2006. The primary term specified in the contract was a four-year term.
   a. On what date is the first delay rental payment due?
   b. What is the maximum number of delay rental payments that may be made?
   c. By what date must drilling be commenced in order to keep the lease from terminating?
   d. Assume Core Oil begins drilling a well on January 2, 2007.
      1) Would the first delay rental be necessary to keep the lease from terminating?
      2) If the well is still in process 14 months later, would the second delay rental be necessary?
      3) If, instead, the well was completed and production begun by October 3, 2007, would the second delay rental be necessary?
      4) If production ceased by December 25, 2008, would the third delay rental payment be necessary?

10. Mr. Hopeful owns the mineral rights in a property in Macon County, Georgia. He leases the property to Wishful Oil Company, reserving a $^1/_5$ royalty. Wishful drills a successful well and begins producing oil. Revenue from the first year of operations totaled $20,000 and costs of development and operation totaled $150,000. How much revenue will each party receive? How much of the costs will each party pay?

11. Universal Oil Company owns a working interest in an oil and gas lease. Lacking the funds to develop the lease, Universal assigns the working interest to Droopy Oil Company, reserving $^1/_{32}$ of $^6/_7$ of production. What kind of interest has Droopy acquired? What kind of interest has Universal retained?

12. Lotus Oil Company owns the working interest in a tract of land in Texas. Lacking the funds to develop the property, Lotus assigns Jones Oil 30,000 barrels of oil to be paid out of $^1/_7$ of the WI's share of production in exchange for $600,000 in cash. What type of interest has Jones acquired?

13. Garza Company obtained a lease with a three-year primary term on August 1, 2006.
    a. Drilling operations were commenced on June 1, 2007, and continued until October 15, 2007, when the well was determined to be dry.
        1) Would the first delay rental payment be required?
        2) How many more delay rentals would be necessary to hold the lease without further drilling?
    b. Drilling operations were started on May 1, 2009, and the well was completed on October 12, 2009, as a producer.
        1) Did the lease terminate on August 1, 2009? Explain.
        2) How many years will the lease continue (assuming production in commercial quantities)?

14. Aggie Oil Company incurred $275,000 in drilling costs prior to the time to decide whether to complete the well. Estimated completion costs are $175,000. The expected net cash flows from the sale of the oil and gas from this well are $300,000. Should the well be completed?

15. Horizontal drilling:
    a. Under what conditions would horizontal drilling operations be considered?
    b. Would horizontal drilling operations be more difficult and expensive than the regular vertical drilling process? Explain.
    c. Would horizontal drilling operations be appropriate for most producing formations? Explain.

16. Discuss the following:
    well spacing
    proration
    field and well allowable
    drilling permit

NOTES

1. API. *Introduction to Oil and Gas Production,* Washington, D.C., American Petroleum Institute, 1983.
2. Petex. *A Primer of Oilwell Drilling,* 5th edition revised, Austin: Petroleum Extension Service, 1996.
3. Conaway, Charles. *The Petroleum Industry: A Nontechnical Guide,* Tulsa: PennWell, 1999.

chapter two

Introduction *to* Oil & Gas Accounting

Oil and gas accounting, as discussed in this textbook, relates to accounting for the four basic costs incurred by companies with oil and gas exploration and producing activities. These four basic types of costs are as follows:

- **Acquisition costs**—costs incurred in acquiring property, *i.e.*, costs incurred in acquiring the rights to explore, drill, and produce oil and natural gas. Domestically, these rights are normally acquired by obtaining an oil, gas, and mineral lease as described in chapter 1.
- **Exploration costs**—costs incurred in exploring property. Exploration involves identifying areas that may warrant examination and examining specific areas, including drilling exploratory wells.
- **Development costs**—costs incurred in preparing proved reserves for production, *i.e.*, costs incurred to obtain access to proved reserves and to provide facilities for extracting, treating, gathering, and storing oil and gas.
- **Production costs**—costs incurred in lifting the oil and gas to the surface and in gathering, treating, and storing the oil and gas.

To account for these costs, knowledge of the industry terms and procedures discussed in chapter 1 is imperative. Additional and important terms are defined in the following Glossary as they are used in accounting.

Glossary of Common Terms

Reservoir. A porous and permeable underground formation containing producible oil or gas that is confined by impermeable rock or water barriers and is individual and separate from other reservoirs. Categories of reserves are defined in the following flowchart:

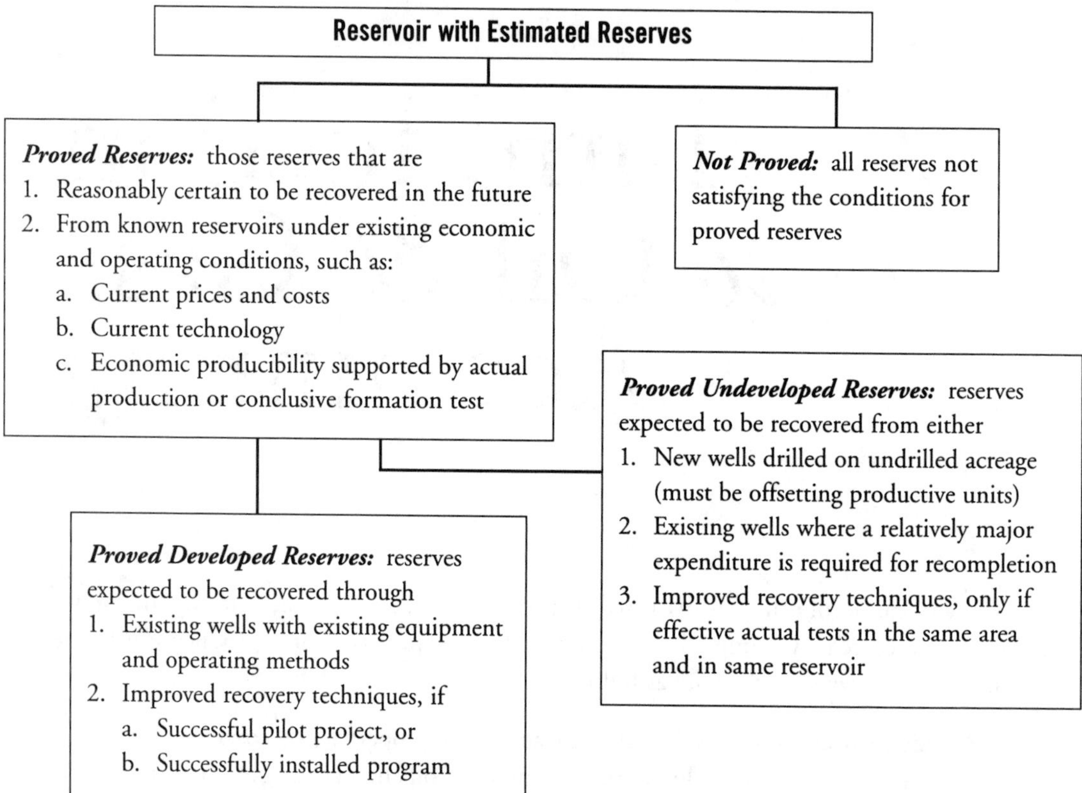

Field. An area consisting of a single reservoir or multiple reservoirs related to the same geological structural feature. There may be two or more reservoirs in a field, separated by intervening impervious strata or geologic barriers. Reservoirs that are in overlapping or adjacent fields may be treated as a single operational field.

Proved area. The portion of a property at a certain depth to which proved reserves have been specifically attributed.

Development well. A well drilled within the proved area of an oil or gas reservoir to the depth of an horizon known to be productive.

Service well. A well drilled for the purpose of supporting production, *e.g.*, a gas injection well, a water injection well, or a salt-water disposal well.

Stratigraphic test well. A well drilled for information only (common offshore). Such wells are drilled to obtain information about subsurface geologic layers and the depths of those layers. Stratigraphic test wells are classified as follows:

1. *Exploratory-type stratigraphic test well:* a stratigraphic test well not drilled in a proved area.
2. *Development-type stratigraphic test well:* a stratigraphic test well drilled in a proved area.

Exploratory well. A well that is not a development well, a service well, or a stratigraphic test well, *i.e.*, a well drilled to find and produce oil or gas in an unproved area, to find a new reservoir in a field with another reservoir which is productive, or to extend a known reservoir.

Oil and gas producing activities. Activities involving the acquisition of mineral interests in properties, exploration, development, and production of crude oil and natural gas.

Proved properties. Properties with known proved reserves.

Unproved properties. Properties with no proved reserves.

Source: *SFAS No. 19, SFAS No. 25,* and *SEC Reg. 4-10.*

The terms *proved reserves, proved developed reserves,* and *proved undeveloped reserves* are especially important in oil and gas accounting. More complete definitions as given in the Securities and Exchange Commission (SEC) Regulation *SX 4-10,* are as follows:

- ***Proved reserves****—Proved oil and gas reserves are the estimated quantities of crude oil, natural gas, and natural gas liquids which geological and engineering data demonstrate with reasonable certainty to be recoverable in future years from known reservoirs under existing economic and operating conditions, i.e., prices and costs as of the date the estimate is made. Prices include consideration of changes in existing prices provided only by contractual arrangements, but not on escalations based upon future conditions.*
- ***Proved developed reserves****—Proved developed oil and gas reserves are reserves that can be expected to be recovered through existing wells with existing equipment and operating methods. Additional oil and gas expected to be obtained through the application of fluid injection or other improved recovery techniques for supplementing the natural forces and mechanisms of primary recovery should be included as "proved developed reserves" only after testing by a pilot project or after the operation of an installed program has confirmed through production response that increased recovery will be achieved.*
- ***Proved undeveloped reserves****—Proved undeveloped oil and gas reserves are reserves that are expected to be recovered from new wells on undrilled acreage, or from existing wells where a relatively major expenditure is required for recompletion. Reserves on undrilled acreage shall be limited to those drilling units offsetting productive units that are reason-*

ably certain of production when drilled. Proved reserves for other undrilled units can be claimed only where it can be demonstrated with certainty that there is continuity of production from the existing productive formation. Under no circumstances should estimates for proved undeveloped reserves be attributable to any acreage for which an application of fluid injection or other improved recovery technique is contemplated, unless such techniques have been proved effective by actual tests in the area and in the same reservoir.

Figure 2–1 illustrates the difference between an exploratory well and a development well. As shown, a well drilled in an unproved area is an **exploratory well.** If an exploratory well finds proved reserves, an area around the well at the depth of the proved reservoir is desig-

Fig. 2–1 — *Examples of Exploratory and Development Wells. All Wells are Assumed to Find Oil at Same Depth or Horizon.*

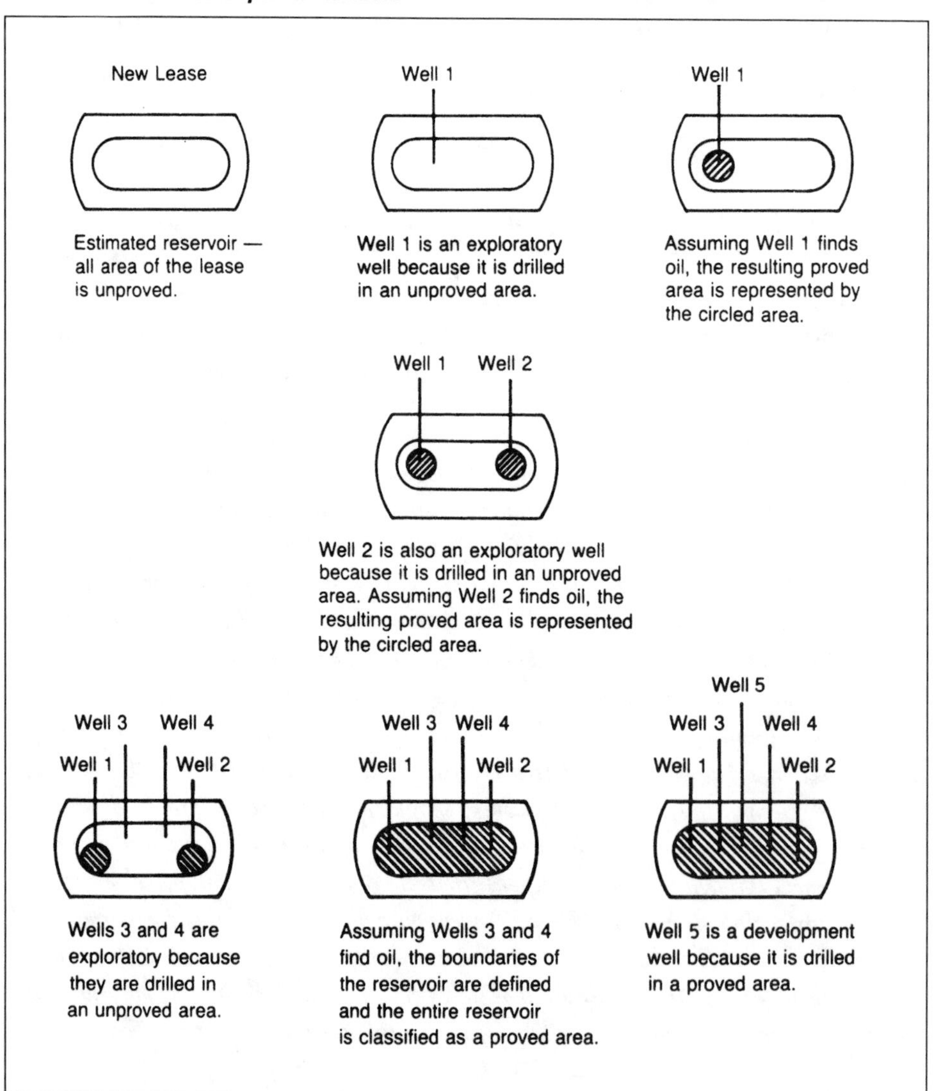

nated a proved area. If additional wells are drilled within the proved area, to the proved depth, the wells are considered **development wells.** Note that a well drilled within a proved area with respect to the surface acreage, but to a shallower depth, is not being drilled in a proved area, and therefore, is considered an exploratory well.

As a result of the first successful well on a property, the property is reclassified from an unproved property to a proved property. However, although the first successful well on a property proves the property, it will prove only a designated area at a specified depth, and not prove the entire area of the property. Therefore, exploratory wells may still be drilled on a property, even if that property has been proved as a result of a previous successful well.

The classification of a well as either an exploratory well or a development well depends upon engineering and geological data and is typically made by engineers and geologists, not accountants. However, in making classifications that will be used for financial reporting, the engineers and geologists follow the definitions of *proved* and *proved developed* reserves given in *SFAS No. 25* and the related SEC Regulation *SX 4-10.* The number of wells that must be drilled in order to classify a reservoir as proved may depend upon the size of the reservoir, whether the area is an explored area, or whether developed reservoirs with similar characteristics exist in the area.[1]

Wells drilled in a new field, wildcat wells, and delineation wells are terms commonly used in the oil and gas industry and refer to wells that are exploratory in nature. A **new field** is an oil and gas field discovered in a location where oil and gas have not been previously found. A **wildcat well** is typically a well drilled in a location where there has been no exploration or production in the general area. A **delineation well** is one drilled to determine the boundaries or the extent of a reservoir. Both wildcat wells and delineation wells are drilled in unproved areas and are always classified as exploratory wells.

An **extension well** is a well that is drilled to test and extend the boundaries of a known, proved reservoir. Extension wells are also included in the definition of exploratory wells; however, these wells are often misclassified since they are frequently drilled on a proved property. The distinguishing characteristic of an extension well is that it is drilled with the intention of *adding* new proved reserves. Wells that are drilled to add proved reserves are exploratory wells while wells that are drilled to produce known reserves are development wells. Thus, even though a well is drilled in a producing region, if the intention is to test for new reserves, then the well is classified as an extension well and accounted for as an exploratory well.

Historical Cost Accounting Methods

The four basic types of costs incurred by oil and gas companies in exploration and production activities must be accounted for using one of two generally accepted historical cost methods: **successful-efforts accounting** (SE) or **full-cost accounting** (FC). In connection with the four basic costs, the fundamental accounting issue is whether to capitalize or expense the incurred costs. If capitalized, the costs may be expensed as expiration takes place either through abandonment (SE only), impairment (SE only), or depletion as reserves are produced. If expensed as incurred, the costs are treated as period expenses and charged against revenue in the current period. The primary difference between SE and FC is in whether a cost is capitalized or expensed when incurred. In other words, the primary difference between the two methods is in the timing of the expense or loss charge against revenue.

The other basic difference between the two accounting methods is the size of the cost center over which costs are accumulated and amortized. For SE, the cost center is a lease, field, or reservoir. In contrast, the cost center under FC is a country. The cost center size has implications in computing depreciation, depletion, and amortization (DD&A) and also in computing ceiling writedowns (chapters 6 and 7).

Successful-efforts accounting is defined in *SFAS No. 19,* "Financial Accounting and Reporting by Oil and Gas Producing Companies" and is generally consistent with financial accounting theory. Paragraph 143 of *SFAS No. 19* states:

> *In the presently accepted financial accounting framework, an asset is an economic resource that is expected to provide future benefits, and nonmonetary assets generally are accounted for at the cost to acquire or construct them. Costs that do not relate directly to specific assets having identifiable future benefits normally are not capitalized—no matter how vital those costs may be to the ongoing operations of the enterprise. If costs do not give rise to an asset with identifiable future benefits, they are charged to expense or recognized as a loss.*

Under SE, a direct relationship is thus required between costs incurred and reserves discovered. Consequently, under SE only successful exploration costs that directly result in the discovery of proved reserves are considered to be part of the cost of finding oil or gas and are capitalized. Unsuccessful exploration costs do not result in an asset with future economic benefit and are therefore expensed.

In contrast, because there is no known way to avoid unsuccessful costs in searching for oil or gas, FC considers both successful and unsuccessful costs incurred in the search for reserves as a necessary part of the cost of finding oil or gas. A direct relationship between costs incurred and reserves discovered is not required under FC. Hence, both successful and unsuccessful costs are capitalized, even though the unsuccessful costs have no future economic benefit.

Specifically, SE treats exploration costs that do not directly find oil or gas as period expenses and successful exploration costs as capital expenditures. Under FC, all exploration costs are capitalized. Under both methods, acquisition and development costs are capitalized and production costs are expensed. Although development costs could include an unsuccessful development well, all development costs are capitalized under SE because the purpose of development activities is considered to be building a producing system of wells and related equipment and facilities, rather than searching for oil and gas. Table 2–1 shows the accounting treatment of these costs under SE compared to FC.

Item	SE	FC
Acquisition costs	C	C
G&G costs	E	C
Exploratory dry hole	E	C
Exploratory well, successful	C	C
Development dry hole	C	C
Development well, successful	C	C
Production costs	E	E
Amortization cost center	Property, field, or reservoir	Country

Table 2–1 — *SE vs. FC*

The divergent accounting treatment of unsuccessful exploratory drilling costs under SE versus FC can have a substantial impact on the income statements of exploration and production companies. A company with a large exploratory drilling program and a normal unsuccessful drilling rate would, under SE, have a significant amount of dry-hole expense. Those dry-hole costs would adversely affect the net income of a successful-efforts company. On the other hand, a full-cost company would capitalize exploratory dry-hole costs, and therefore, these costs would typically have no immediate effect on net income. They would, however, reduce net income through future amortization. The adverse effect on net income of expensing exploratory dry-hole costs under SE may be especially significant for smaller companies.

The following example illustrates the impact of the FC and SE accounting methods on the financial statements of Lucky Company.

EXAMPLE

Financial Statements

Lucky Oil Company began operations on March 3, 1995, with the acquisition of a lease in Texas. During the first year, the following costs were incurred, DD&A (depreciation, depletion, and amortization) taken, and the following revenue was earned.

G&G costs	$ 30,000	
Acquisition costs	50,000	
Exploratory dry holes	1,200,000	
Exploratory wells, successful	400,000	
Development costs	200,000	
Production costs	25,000	
DD&A expense	40,000 (SE)	90,000 (FC)
Revenue	100,000	

Income Statements

	SE		FC	
Revenue		$ 100,000		$100,000
Expenses:				
G&G	$ 30,000		$ 0	
Exploratory dry holes	1,200,000		0	
Production costs	25,000		25,000	
DD&A	40,000		90,000	
Total Expenses		1,295,000		115,000
Net Income		$(1,195,000)		$ (15,000)

Partial Balance Statements

	SE	FC
G&G costs		$ 30,000
Acquisition costs	$ 50,000	50,000
Exploratory dry holes		1,200,000
Exploratory wells, successful	400,000	400,000
Development costs	200,000	200,000
Total assets	650,000	1,880,000
Less accumulated DD&A	(40,000)	(90,000)
Net Assets	$610,000	$1,790,000

As shown in the example, the SE method results in a much greater loss than the FC method: $1,195,000 compared with $15,000. The majority of the $1,180,000 difference is caused by the different treatment of the G&G costs and dry hole costs; under the SE method the costs were expensed, while under the FC method, the costs were capitalized. Another difference is the amount of amortization (DD&A) recognized under each method; under the FC method more costs were capitalized, resulting in a greater amortization expense.

Another major impact can be made on the income statements of successful-efforts companies (but not full-cost companies) by the order in which successful versus unsuccessful wells are drilled.

EXAMPLE

Order of Drilling

Lucky Oil, a successful-efforts company, acquires a lease that has an oil reservoir at 10,000 feet. The reservoir has an unknown fault trap that contains no oil, located at the center of the reservoir. In attempting to locate, define, and develop the reservoir, Lucky drills a total of five wells as shown below at a cost of $300,000 each.

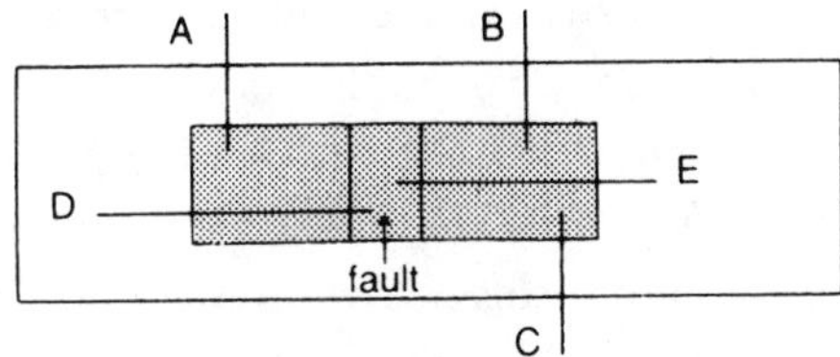

If the successful wells, A, B, and C, are drilled first and are considered to have delineated the reservoir (shaded area) shown in the figure, then the unsuccessful wells, D and E, would be classified as development wells. In this case, Lucky would have no dry-hole expense on its income statement relating to the five wells. If, however, wells D and E are drilled first, they would be classified as exploratory wells and expensed as dry holes. Wells A, B, and C would also be classified as exploratory but would be capitalized because they found proved reserves. Thus, in the situation described, merely changing the order in which the wells were drilled would result in a difference of $600,000 on Lucky's income statement.

If Lucky Company had been a FC company, the order in which the wells were drilled would have had no effect on the income statement.

HISTORICAL DEVELOPMENT OF ACCOUNTING METHODS AND CURRENT STATUS

Accounting for oil and gas producing activities poses many technical and theoretical problems. Characteristics of the industry such as

- high risk
- high cost of investment
- lack of correlation between the size of expenditure and the value of any resulting reserves
- long time span from when costs are first incurred until benefits are received

are four important reasons for the controversy surrounding the accounting procedures used by oil and gas exploration and production companies.

In addition, taxation and regulation have placed demands on companies' accounting and reporting systems. The federal taxation statutes contain many specialized rules applying only to oil and gas operations (*e.g.,* intangible drilling costs and depletion rules). Further, federal and state governments have a long history of regulating the production and pricing of oil and natural gas. Finally, companies typically undertake oil and gas exploration and produce any resulting reserves as joint venture operations. In addition to placing extensive accounting responsibilities on the operator, joint venture accounting requires each of the parties to account for its proportionate share of costs and revenues and to separately report their income to the state and federal taxing authorities. For all the above reasons, accounting for oil and gas operations has complex and specialized accounting rules and procedures.

Prior to the release of *SFAS No. 19* in December 1977, the principal methods of financial accounting for oil and gas producing activities were successful-efforts and full-cost accounting. Both methods, and variations of the methods, were widely used, and the relative merits of each debated for many years.

In 1969, the American Institute of Certified Public Accountants (AICPA) issued *Accounting Research Study No. 11,* "Financial Reporting in the Extractive Industries." This study supported the successful-efforts method of accounting. Despite this study, the Accounting Principles Board chose not to issue an Opinion thus requiring its successor, the Financial Accounting Standards Board (FASB), to deal with the complex and politically sensitive issues surrounding an appropriate method of accounting for oil and gas exploration, development, and production costs.

In 1975, in response to OPEC's embargo and the consequent oil and gas shortage in the United States, the Federal Energy and Conservation Act was passed. This act required that the SEC must either prescribe oil and gas accounting rules or approve oil and gas accounting rules developed by the FASB. In response to a request from the SEC, the FASB began work on the oil and gas financial accounting and reporting problem and in 1977 issued *SFAS No. 19.* This statement, which prescribed the SE method, was to have become effective for fiscal years after December 1978. However, in August 1978, the SEC issued releases stating that until further rules were issued, oil and gas producing companies could use the full-cost method according to *Accounting Series Release (ASR) 258,* or they could use the successful-efforts method according to *ASR 257* or its equivalent, *SFAS No. 19.* (*ASR 257* and *SFAS No. 19* are virtually identical, with the exception of certain reserve definitions. *SFAS No. 25,* "Suspension of Certain Accounting Requirements for Oil and Gas Producing Companies," was later issued that redefined the terms to agree with those in *ASR 257.*) As a result of the SEC's actions, in February 1979, the FASB issued *SFAS No. 25,* which essentially suspended *SFAS No. 19's* requirement that the successful efforts method of accounting be used.

When issuing *ASR 257* and *ASR 258*, the SEC stated that it believed neither full-cost nor successful-efforts provides sufficient information concerning the financial position or operating results of oil and gas companies. The primary deficiency, as perceived by the SEC, was the failure to include in the primary financial statements the most valuable asset an oil and gas company has, namely, oil and gas reserves. The SEC stated that a reserve valuation should be included in the primary financial statements, *i.e.*, the balance sheet and income statement. For this reason, the SEC proposed a new method of accounting under which revenue would be recognized when reserves were discovered versus when they were produced and sold. Assets would be a valuation of the estimated future production of proved oil and gas reserves in place, discounted at a rate of 10%. The method was called *reserve recognition accounting* (RRA) and was intended by the SEC to replace full cost and successful efforts as the basis for the primary financial statements after a trial period in which RRA statements would be presented as supplemental information.

In 1981, the SEC decided that RRA was not the answer and again called upon the FASB to provide a solution to the problem of oil and gas accounting. After much discussion, the FASB issued *SFAS No. 69*, "Disclosures About Oil and Gas Producing Activities," which established required disclosures for oil and gas producing companies.

SFAS No. 69 requires publicly traded companies with significant oil and gas producing activities to disclose supplementary information in their annual financial statements related to the following items:

Historical based:

- Proved reserve quantity information
- Capitalized costs relating to oil and gas producing activities
- Costs incurred for property acquisition, exploration, and development activities
- Results of operations for oil and gas producing activities

Value based:

- A standardized measure of discounted future net cash flows relating to proved oil and gas reserve quantities
- Changes in the standardized measure of discounted cash flows relating to proved oil and gas reserve quantities

Public and nonpublic companies are required to disclose two informational items:

1. Accounting method used in accounting for oil and gas producing activities
2. Manner of disposing of capitalized costs

In 1996, the SEC deleted their rules that applied to the successful-efforts method and added a new rule requiring SE companies to follow *SFAS No. 19* as amended by *SFAS No. 25*.

The current status of oil and gas accounting is that successful-efforts, according to *SFAS No. 19*, and full-cost, according to *ASR 258*, are considered acceptable accounting methods by both the FASB and the SEC. Currently, neither the FASB nor the SEC has any plans to change the acceptability of the two methods. In addition, disclosures prepared according to *SFAS No. 69* must be presented as supplemental information to the financial statements.

While *SFAS No. 19* and *ASR 258* provide the broad guidelines for accounting for exploration and production costs, guidance is needed in accounting for the day-to-day activities. In this regard, accounting for oil and gas transactions has been greatly influenced by the Council of Petroleum Accountants Societies (COPAS). COPAS was formed in 1961, with its activities and projects directed primarily toward issues and problems encountered in joint venture accounting. Since that time COPAS has expanded its horizons to address a wide range of issues dealing with financial reporting, revenue accounting, and tax accounting practices in the oil and gas industry.

COPAS issues "pronouncements" in the forms of procedures, bulletins, and interpretations. The pronouncements issued by COPAS are not authoritative in the same sense as the pronouncements issued by the FASB, the SEC, the Treasury Department, and the IRS. The highest level pronouncement issued by COPAS is an accounting procedure. Generally an accounting procedure is an attachment to a joint operating agreement. These procedures address such issues as which costs are directly shared by the parties to the contract, determination of overhead rates, acquisition and transfer of materials, and conduct of audits and inventories. Since the procedure is a part of the joint operating agreement contract, the parties are bound to follow the terms of the procedure. The procedures do not affect the financial accounting methods or income tax accounting used by the parties.

COPAS often issues bulletins to clarify the specific provisions of various accounting procedures. The purpose of bulletins relating to specific accounting procedures is to provide guidance and interpretation in implementing the procedures. Other bulletins are issued to address and provide advice regarding additional accounting problems that may arise in the industry. If needed, interpretations may be issued that clarify or further address the procedures. In addition, COPAS often sponsors and issues research papers on emerging topics in oil and gas accounting and supports various educational efforts within the industry.

Introduction to Successful-Efforts Accounting

Figure 2–2 presents a flow chart overview of the treatment of the four basic costs under SE. An example follows that gives a brief illustration of typical costs incurred by oil and gas companies and their treatment under SE. As shown in the flowchart and in the example, gross acquisition costs are capitalized as unproved property until either proved reserves are found or until the property is abandoned or impaired. If proved reserves are found, the property is reclassified from unproved property to proved property. Exploration costs are recorded in two different ways, depending upon the type of incurred costs. If the costs are nondrilling, as defined in chapter 3, they are expensed as incurred. If the exploration costs are drilling costs, they are capitalized temporarily as wells in progress until a determination is made whether

Fig. 2–2 — *Successful Efforts Overview*

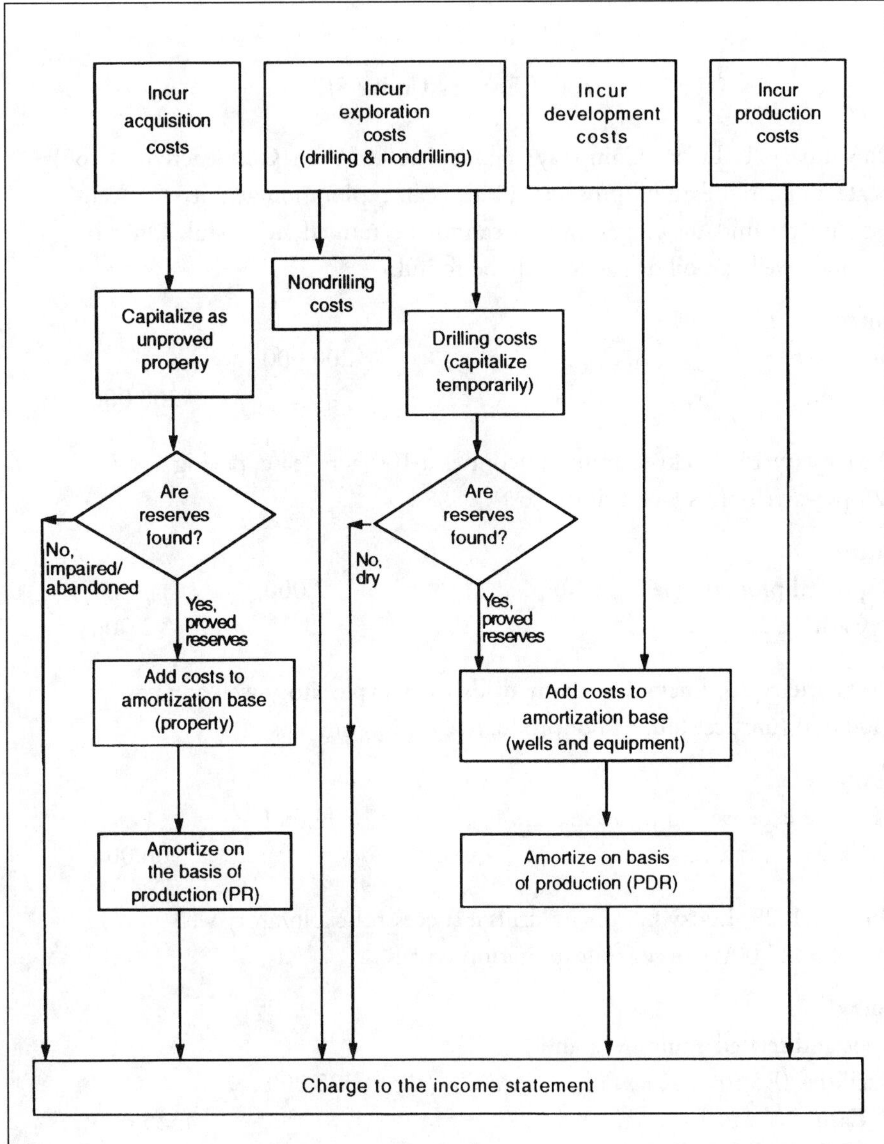

Special thanks to Laura Stevenson and Raphael Bemporad for the flowcharts throughout the book.

proved reserves have been found. If proved reserves are found, the drilling costs are transferred to wells and related equipment and facilities, and are charged to expense, specifically DD&A expense, as production occurs. If proved reserves are not found, *i.e.*, a dry hole, the drilling costs are expensed. Development costs, which include the costs of drilling development wells, are capitalized regardless of whether or not proved reserves are found. All production costs are

EXAMPLE

Overview of Entries Under SE

a. On January 1, Lucky Company spends $200,000 on G&G activities to locate and explore an oil prospect. (This is an exploration activity that cannot directly find oil or gas and so cannot be termed successful. Only by drilling a well can oil or gas actually be found.)

Entry

G&G expense	200,000	
Cash		200,000

b. On January 15, Lucky Company acquires a 100-acre lease, paying a $20-per-acre bonus (acquisition cost).

Entry

Unproved property (100 x $20)	2,000	
Cash		2,000

c. On February 20, Lucky Company drills a dry exploratory well at a cost of $300,000 (unsuccessful or nonproductive exploration cost).

Entry

Dry-hole expense	300,000	
Cash		300,000

d. On March 29, Lucky Company drills a successful exploratory well at a cost of $325,000 (successful exploration cost).

Entry*

Wells and related equipment and facilities (E&F)	325,000	
Cash		325,000

As a result of the successful exploratory well, Lucky must also reclassify the property.

Entry

Proved property	2,000	
Unproved property		2,000

* The same entry would have been made if the well had been a successful development well or a dry development well.

e. On April 10, Lucky Company spends $80,000 on production facilities such as flow lines. (This cost is incurred in preparing proved reserves for production and, therefore, is a development cost.)

Entry

Wells and related E&F.	80,000	
Cash.		80,000

f. On June 3, Lucky Company incurs $40,000 in production costs (production cost).

Entry

Production expense	40,000	
Cash.		40,000

---

expensed as incurred.

A chart of accounts for a company using the successful-efforts method of accounting is presented on the following pages. Each successful-efforts company will have its own chart of accounts. The accounts shown in the following pages are illustrative of the accounts that a typical successful-efforts company might have.

Chart of Accounts for Successful-Efforts Company

00000–00499	**CASH ACCOUNTS**
00001-000	Cash—Houston National Bank
00054-000	Cash—Chase NY
01000–01999	**ACCOUNTS RECEIVABLES**
01010-000	A/R—Lease Revenue
01410-000	A/R—Joint Venture
01411-000	A/R—Joint Venture Cash Calls
01901-000	A/R—Allowance for Doubtful Accounts
02500–02599	**MATERIAL & SUPPLIES**
02501-000	Materials & Supplies (M&S) Inventory
02501-020	M&S—Material Transfers, Pipe
02501-021	M&S—Material Transfers, Equipment

03000–03499	**PREPAIDS**
03001-000	Prepaid Insurance
03101-000	Miscellaneous Prepaids
03500–03999	**ACCRUALS**
03511-000	Accrued Gas Underdeliveries
03521-000	Accrued Pipeline Imbalance
10000–19999	**UNEVALUATED LEASEHOLDS—UNPROVED PROPERTY**
10101-000	Unevaluated Leaseholds
10101-904	Unevaluated Leaseholds, O/H Allocation
10101-905	Unevaluated Leaseholds, Capitalized Interest
10101-907	Unevaluated Leaseholds, Allowance
10105-000	Accumulated Unevaluated Leasehold Improvements
10200-000	Unproved Nonworking Interests
12000–12299	**PRODUCING LEASEHOLDS—PROVED PROPERTY**
12011-000	Producing Leaseholds
12011-904	Producing Leaseholds-O/H Allocation
12019-000	Producing Leaseholds-Capitalized Interest
12031-000	Producing Royalties
12041-000	Producing Property Retirements
12050–12310	**WELLS AND EQUIPMENT**
12051-000	Intangible Drilling Cost
12071-000	Lease & Well Equipment
12310-000–12310-999	**OFFSHORE DEVELOPMENT DRILLING (OFFDEVDRL) (W/P)**
12310-040	Offdevdrl—Title Examination
12310-043	Offdevdrl—Survey, Location, Damages
12310-046	Offdevdrl—Mobile And Demobile
12310-049	Offdevdrl—Contract Drilling
12310-052	Offdevdrl—Turnkey Drilling
12310-055	Offdevdrl—Completion Rig
12310-058	Offdevdrl—Directional Drill, Well Survey
12310-061	Offdevdrl—Water
12310-063	Offdevdrl—Lost/Dam Rental Equipment
12310-064	Offdevdrl—Equipment Rental & Service
12310-067	Offdevdrl—Mud & Chemicals

12310-068	Offdevdrl—Completion W/O Fluids
12310-070	Offdevdrl—Drilling Bits
12310-071	Offdevdrl—Perforating
12310-074	Offdevdrl—Cement & Cement Service
12310-077	Offdevdrl—Well Log/Open Hole
12310-078	Offdevdrl—Well Log/Cased Hole
12310-080	Offdevdrl—Mud Logging
12310-082	Offdevdrl—Fishing Tools & Service
12310-083	Offdevdrl—Wireline Services
12310-086	Offdevdrl—Well Testing
12310-088	Offdevdrl—Tubular & Equipment Repair
12310-089	Offdevdrl—Tubular Inspection Service
12310-091	Offdevdrl—Stimulation
12310-092	Offdevdrl—Sand Control
12310-094	Offdevdrl—Reamers & Stabilizer
12310-095	Offdevdrl—Trucking, Land Transportation
12310-096	Offdevdrl—Boat, Water Transportation
12310-097	Offdevdrl—Helicopter, Air Transportation
12310-098	Offdevdrl—Abandonment Expense
12310-101	Offdevdrl—Company Labor
12310-103	Offdevdrl—Payroll Burden
12310-105	Offdevdrl—Contract Labor
12310-107	Offdevdrl—Consultants
12310-108	Offdevdrl—Meals/Entertainment
12310-109	Offdevdrl—Other Employee Expense
12310-123	Offdevdrl—Fuel & Electric
12310-150	Offdevdrl—Pollution Control
12310-179	Offdevdrl—Insurance
12310-180	Offdevdrl—Legal
12310-189	Offdevdrl—Federal Fuel Use Sales Tax
12310-190	Offdevdrl—State Fuel Use Sales Tax
12310-200	Offdevdrl—Overhead
12310-207	Offdevdrl—Misc., Unclassified
12310-299	Offdevdrl—Intangible JO Share
12310-300	Offdevdrl—Casing
12310-301	Offdevdrl—Tubing
12310-302	Offdevdrl—Wellhead Equipment
12310-303	Offdevdrl—Subsurface Equipment
12310-330	Offdevdrl—Miscellaneous Non-Operating Equipment
12310-399	Offdevdrl—Tangible JO Share
12310-902	Offdevdrl—Account Transfers

12311-000–12311-999	**OFFSHORE FACILITIES (W/P)** (Many of the individual accounts are similar to accounts for offshore development drilling.)
12312-000–12312-999	**OFFSHORE WORKOVER (W/P)** (Individual accounts similar to accounts for offshore development drilling.)
12313-000–12313-999	**OFFSHORE RECOMPLETION (W/P)** (Individual accounts similar to accounts for offshore development drilling.)
12314-000–12314-999	**OFFSHORE DEVELOPMENT WELL PLUG AND ABANDONMENT (OFFDEVPA) (W/P)**
12314-043	Offdevpa—Survey, Road, Location, Damage
12314-046	Offdevpa—Mobile and Demobile
12314-055	Offdevpa—Completion Rig
12314-061	Offdevpa—Water
12314-063	Offdevpa—Lost/Dam Rental Equipment
12314-064	Offdevpa—Equipment Rental & Service
12314-067	Offdevpa—Mud & Chemicals
12314-068	Offdevpa—Completion without Fluids
12314-070	Offdevpa—Drilling Bits
12314-071	Offdevpa—Perforation
12314-074	Offdevpa—Cement & Cement Service
12314-078	Offdevpa—Well Log/Cased Hole
12314-082	Offdevpa—Fishing Tools & Service
12314-083	Offdevpa—Wireline Services
12314-088	Offdevpa—Tubular & Equipment Repair
12314-089	Offdevpa—Tubular Inspection Services
12314-091	Offdevpa—Stimulation
12314-095	Offdevpa—Trucking, Land Transportation
12314-096	Offdevpa—Boat, Water Transportation
12314-097	Offdevpa—Helicopter, Air Transportation
12314-101	Offdevpa—Company Labor
12314-103	Offdevpa—Payroll Burden
12314-105	Offdevpa—Contract Labor
12314-107	Offdevpa—Consultants
12314-108	Offdevpa—Meals/Entertainment
12314-109	Offdevpa—Other Employee Expense
12314-123	Offdevpa—Fuel & Electric

12314-150	Offdevpa—Pollution Control
12314-179	Offdevpa—Insurance
12314-189	Offdevpa—Federal Fuel Use Sales Tax
12314-190	Offdevpa—State Fuel Use Sales Tax
12314-200	Offdevpa—Overhead
12314-207	Offdevpa—Miscellaneous Unclassified
12314-299	Offdevpa—Intangible Joint Operations Share
12314-300	Offdevpa—Casing
12314-301	Offdevpa—Tubing
12314-302	Offdevpa—Wellhead Equipment
12314-303	Offdevpa—Subsurface Equipment
12314-330	Offdevpa—Miscellaneous Non-Operated Equipment
12314-399	Offdevpa—Tangible Joint Operations Share
12317-000–12317-999	**OFFSHORE EXPLORATORY DRILLING (W/P)**
12320-000–12320-999	**ONSHORE DEVELOPMENT DRILLING (W/P)**
12321-000–12321-999	**ONSHORE FACILITIES (W/P)**
12322-000–12322-999	**ONSHORE WORKOVER (W/P)**
12323-000–12323-999	**ONSHORE RECOMPLETIONS (W/P)**
12324-000–12324-999	**ONSHORE DEVELOPMENT WELL PLUG AND ABANDON (W/P)**
12327-000–12327-999	**ONSHORE EXPLORATORY DRILLING (W/P)**
12330-000–12330-999	**OFFSHORE LEASE ACQUISITION—UNPROVED PROPERTY**
12330-400	Offshore Lease Acquisition—General AFE Cost
12330-404	Offshore Lease Acquisition—Lease Bonus
12330-408	Offshore Lease Acquisition—Brokerage
12330-411	Offshore Lease Acquisition—Recording Fees
12330-415	Offshore Lease Acquisition—Unitization
12330-418	Offshore Lease Acquisition—Legal
12330-420	Offshore Lease Acquisition—Division Order Title Exam
14000–14131	**MISCELLANEOUS ASSETS**
13501-000	Plant Equipment & Facilities
14011-000	Transportation-Vehicles
14031-000	Marine Equipment
14051-000	Work Equipment
14071-000	Furniture & Fixtures
14091-000	Computer Equipment-Hardware
14101-000	Computer Equipment-Software
14111-000	Leasehold Improvements

14131-000	Telephone/Communication Equipment
16000–16999	**ACCUMULATED DD&A AND WRITEDOWNS**
18101-000	Long-Term Receivable
18201-000	Note Receivable-Branch
19001-000	Preacquisition Expenses
19021–19091	**DEFERRED FEDERAL INCOME TAX**
20000–20999	**ACCOUNTS PAYABLE**
21000–21999	**SHORT TERM NOTES PAYABLE**
24000–24999	**ACCRUED STATE AND FEDERAL TAXES**
29000–29999	**ACCRUED LIABILITIES**
29001-000	Accrued Miscellaneous Liabilities
29006-000	Accrued Gas Marketing Liabilities
29007-000	Accrued Oil Marketing Liabilities
29010-000	Accrued Rent
29011-000	Accrued Accounting Fees
29111-000	Accrued Gas Overdeliveries
29501-000	Accrued Vacation Pay
30000–31999	**LONG TERM NOTES PAYABLE**
40000–49999	**STOCKHOLDERS EQUITY AND RELATED ACCOUNTS**
40001-000	Common Stock
41001-000	Contributed Capital
42001-000	Retained Earnings
42005-000	Dividends Declared
53000–59999	**REVENUES AND GAINS**
53001-000	Revenue-Oil
53500-000	Revenue-Gas
53600-000	Royalty Revenue
54001-000	Plant Revenue
55001-000	Pipeline Transportation Revenue
59001-000	Miscellaneous Operating Income
59501-000	Gas Balance Settlements/Adjustments
59601-000	Gain on Sale/Retirement of Prop
60000–60999	**LEASE OPERATING EXPENSE (LOE)**
60001-061	LOE—Water
60001-096	LOE—Trucking & Land Transportation
60001-096	LOE—Boats, Water Transportation

60001-097	LOE—Helicopter, Air Transportation
60001-101	LOE—Company Labor
60001-103	LOE—Payroll Burden
60001-104	LOE—Employee Pension
60001-105	LOE—Contract Labor
60001-107	LOE—Consultants & Professional Service
60001-108	LOE—Meals/Entertainment
60001-109	LOE—Other Employee Expense
60001-112	LOE—Chemicals & Treating
60001-114	LOE—Supplies & Tools
60001-116	LOE—Gathering
60001-119	LOE—Transportation
60001-123	LOE—Fuel & Electric
60001-131	LOE—Salt Water Disposal
60001-135	LOE—Groceries & Food
60001-137	LOE—Other Rents
60001-139	LOE—Measurement & Testing
60001-143	LOE—Gas Marketing Service & Repair
60001-147	LOE—Safety & Supplies
60001-150	LOE—Pollution Control
60001-154	LOE—Compressor Parts
60001-158	LOE—Surface Repairs & Maintenance
60001-160	LOE—Communications
60001-162	LOE—Road & Location Maintenance
60001-164	LOE—Controllable Equipment
60001-166	LOE—Subsurface Repairs & Maintenance
60001-170	LOE—Well Workover
60001-172	LOE—Well Services
60001-176	LOE—Abandonment
60001-180	LOE—Legal
60001-182	LOE—Equipment Rental & Service
60001-183	LOE—Compressor Rentals
60001-185	LOE—Ad Valorem Tax
60001-188	LOE—Production/Severance Tax
60001-189	LOE—Federal Fuel Use Sales Tax
60001-190	LOE—State Fuel Use Sales Tax
60001-192	LOE—Prod./Severance Tax (20% prop.)
60001-194	LOE—Field Expense
60001-196	LOE—Facility Operations
60001-198	LOE—Net Profits Interest
60001-200	LOE—Overhead

60001-201	LOE—District Expense
60001-204	LOE—Miscellaneous
60001-205	LOE—Nonoperated Joint Cost
60001-250	LOE—Advances to Operators
60001-995	LOE—Cutback on Insurance
60001-999	LOE—Charges to Joint Owners
61000–61999	**MARKETING AND TRANSPORTATION EXPENSE**
66000–66999	**GENERAL AND ADMINISTRATIVE EXPENSE**
66001-501	G&A—Salaries & Wages
66001-505	G&A—Bonus Compensation
66001-509	G&A—Auto Allowance
66001-514	G&A—Company Vehicle
66001-519	G&A—Parking
66001-525	G&A—Remote Parking
66001-529	G&A—Relocation
66001-533	G&A—Education Reimbursement
66001-538	G&A—Personal Development
66001-542	G&A—Subscriptions/Publications
66001-543	G&A—Professional Dues
66001-545	G&A—Membership Dues
66001-547	G&A—Entertainment
66001-552	G&A—Meals
66001-556	G&A—Lodging
66001-560	G&A—Transportation
66001-564	G&A—Auto/Taxi
66001-568	G&A—Overtime Meal Allowance
66001-572	G&A—Other Employee Expense
66001-576	G&A—Group Insurance
66001-580	G&A—Workers Compensation
66001-584	G&A—Employer FICA
66001-588	G&A—Federal/State Unemployment
66001-592	G&A—Thrift Contributions
66001-594	G&A—Other Employee Benefits
66001-595	G&A—Employee Pension
66001-596	G&A—Office Contract Labor
66001-599	G&A—Contract Labor Benefits
66001-603	G&A—Accounting Fees
66001-608	G&A—Accounting—Tax Fees
66001-612	G&A—Legal Fees, Litigation

66001-616	G&A—Legal Fees, Other
66001-620	G&A—Data Processing Fees
66001-624	G&A—Reserve Engineering Fees
66001-628	G&A—Other Professional Fees
66001-630	G&A—Office Rent
66001-632	G&A—Communications
66001-636	G&A—Reproduction
66001-640	G&A—Postage
66001-644	G&A—Courier/Messenger Services
66001-648	G&A—Office Equip Rentals & Services
66001-652	G&A—Maintenance & Repairs
66001-656	G&A—Moving & Storage
66001-660	G&A—Office Supplies
66001-661	G&A—Data Processing Supplies
66001-664	G&A—Refreshments
66001-668	G&A—Contributions
66001-672	G&A—General Liability Insurance
66001-676	G&A—Property & Casualty Insurance
66001-680	G&A—Miscellaneous Taxes
66001-685	G&A—Ad Valorem Taxes
66001-686	G&A—Outside Service, Other
66001-688	G&A—Other G&A
68000–68999	**DD&A EXPENSE**
70000–70999	**NONPRODUCTIVE EXPENSES**
70000-100	Nonproductive Exploratory Drilling
70000-200	Dry Hole and Bottom Hole Contributions
70000-300	Delay Rentals
70000-400	G&G Services
70000-500	Carrying and Maintenance Costs
70000-600	Abandoned Leases
70000-700	Impairment
79600–85400	**MISCELLANEOUS EXPENSES**
79601-000	Bad Debt Expense
80001-000	Federal Income Tax-Current
81001-000	Deferred Federal Income Tax-Capitalized Interest

INTRODUCTION TO FULL-COST ACCOUNTING

Figure 2–3 presents a flow chart overview of the treatment of the four basic costs under FC. The example used to illustrate SE accounting is also used to illustrate FC accounting. As shown in the flow chart and in the example, acquisition, exploration, and development costs are capitalized under the full cost method, regardless of whether the costs result in a discovery of reserves. As in SE accounting, gross acquisition costs are placed in an unproved property account and are moved to a proved property account if proved reserves are found. If the property is abandoned or impaired, the costs continue to be capitalized but are transferred to an abandoned or impaired

Fig. 2–3 — *Full-Cost Overview*

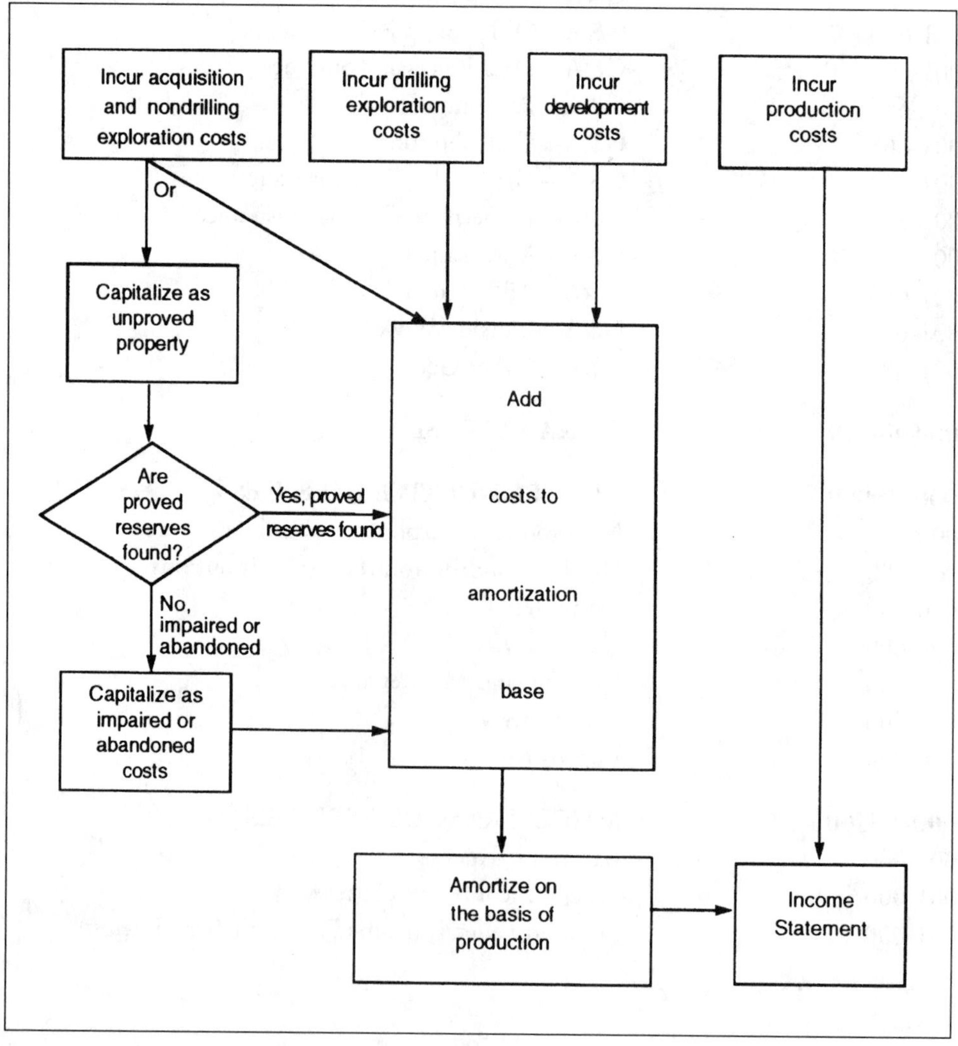

EXAMPLE

Overview of Entries Under FC

a. On January 1, Lucky Company spends $200,000 on G&G activities to locate and explore an oil prospect (exploration cost).

Entry

G&G costs	200,000	
Cash.		200,000

b. On January 15, Lucky Company acquires a 100-acre lease paying a $20-per-acre bonus (acquisition cost).

Entry

Unproved property—acquisition	2,000	
Cash.		2,000

c. On February 20, Lucky Company drills a dry exploratory well at a cost of $300,000 (exploration cost).

Entry

Exploratory dry hole	300,000	
Cash.		300,000

d. On March 29, Lucky Company drills a successful exploratory well at a cost of $325,000 (exploration cost).

Entry

Wells and related E&F	325,000	
Cash.		325,000

As a result of the successful exploratory well, Lucky must also reclassify the property.

Entry

Proved property—acquisition.	2,000	
Unproved property—acquisition		2,000

e. On April 10, Lucky Company spends $80,000 on production facilities such as flow lines (development cost).

Entry

Wells and related E&F.	80,000	
Cash.		80,000

f. On June 3, Lucky Company incurs $40,000 in production costs (production cost).

Entry

Production expense	40,000	
Cash.		40,000

cost account. All acquisition, exploration, and development costs incurred in each country are capitalized into one pool, so that theoretically, individual properties lose their identities. However, companies maintain subsidiary records on individual properties for tax, regulatory, and management purposes. All production costs are expensed as incurred.

A full-cost chart of accounts is presented on the following pages. A full-cost company, in contrast to a successful-efforts company, does not have expense accounts for G&G costs, delay rentals, exploratory dry holes, etc. because these costs are capitalized under the full-cost method of accounting. Impairment and abandoned leasehold costs are also capitalized for FC companies. Other than these differences, the chart of accounts shown on the following pages is essentially the same as the successful-efforts chart of accounts shown earlier.

Chart of Accounts for a Full-Cost Company

00000–00499 **CASH ACCOUNTS**
(For detailed subaccounts, see the SE chart of accounts.)

01000–01999 **ACCOUNTS RECEIVABLES**
(For detailed subaccounts, see the SE chart of accounts.)

02500–02599 **MATERIAL & SUPPLIES**
(For detailed subaccounts, see the SE chart of accounts.)

03000–03499 **PREPAIDS**
(For detailed subaccounts, see the SE chart of accounts.)

03500–03999 **ACCRUALS**
(For detailed subaccounts, see the SE chart of accounts.)

10000–10999 **UNEVALUATED LEASEHOLDS—UNPROVED PROPERTY**

10101-000	Unevaluated Leaseholds
10101-904	Unevaluated Leaseholds, O/H Allocation
10101-905	Unevaluated Leaseholds, Capitalized Interest
10101-907	Unevaluated Leaseholds, Allowance
10105-000	Accumulated Unevaluated Leasehold Improvements
10200-000	Unproved Nonworking Interests
11000–11999	**NONPRODUCTIVE COSTS**
11001-000	Capitalized Nonproductive Drilling
11011-000	Capitalized Dry Hole and Bottom Hole Contributions
11101-000	Capitalized Delay Rentals
11111-000	Capitalized Abandoned Leases
11131-000	Capitalized Impairment
11201-000	Capitalized G&G Services
12000–12299	**PRODUCING LEASEHOLDS—PROVED PROPERTY** (For detailed subaccounts, see the SE chart of accounts.)
12050–12310	**WELLS AND EQUIPMENT**
12051-000	Intangible Drilling Cost
12071-000	Lease & Well Equipment
12310-000–12310-999	**OFFSHORE DEVELOPMENT DRILLING (OFFDEVDRL) (W/P)** (For detailed subaccounts, see the SE chart of accounts.)
12311-000–12311-999	**OFFSHORE FACILITIES (W/P)** (Many of the individual accounts are similar to accounts for offshore development drilling.)
12312-000–12312-999	**OFFSHORE WORKOVER (W/P)** (Individual accounts similar to accounts for offshore development drilling.)
12313-000–12313-999	**OFFSHORE RECOMPLETION (W/P)** (Individual accounts similar to accounts for offshore development drilling.)
12314-000–12314-999	**OFFSHORE DEVELOPMENT WELL PLUG AND ABANDONMENT (W/P)** (For detailed subaccounts, see the SE chart of accounts.)
12317-000–12317-999	**OFFSHORE EXPLORATORY DRILLING (W/P)**

12320-000–12320-999	**ONSHORE DEVELOPMENT DRILLING (W/P)**
12321-000–12321-999	**ONSHORE FACILITIES (W/P)**
12322-000–12322-999	**ONSHORE WORKOVER (W/P)**
12323-000–12323-999	**ONSHORE RECOMPLETIONS (W/P)**
12324-000–12324-999	**ONSHORE DEVELOPMENT WELL PLUG AND ABANDON (W/P)**
12327-000–12327-999	**ONSHORE EXPLORATORY DRILLING (W/P)**
12330-000–12330-999	**OFFSHORE LEASE ACQUISITION—UNPROVED PROPERTY** (For detailed subaccounts, see the SE chart of accounts.)
12340-000–12340-999	**OFFSHORE G&G (W/P)**
12340-400	Offshore G&G-General AFE Cost
12340-420	Offshore G&G-Acquisition
12340-424	Offshore G&G-Processing
12340-429	Offshore G&G-Permits
12345-000–12345-999	**ONSHORE G&G (W/P)**
12345-400	Onshore G&G-General AFE Cost
12345-410	Onshore G&G-General Subcontract Cost
12345-420	Onshore G&G-Acquisition
12345-421	Onshore G&G-Seismic Cost
12345-422	Onshore G&G-Mapping Cost
12345-424	Onshore G&G-Processing
12345-429	Onshore G&G-Permits
12345-430	Onshore G&G-Consultants
12345-432	Onshore G&G-Technical Services
12345-434	Onshore G&G-Bulldozing
12345-436	Onshore G&G-Operations Equipment
12345-438	Onshore G&G-Communication Cost
12345-440	Onshore G&G-Environmental Costs
12345-441	Onshore G&G-Quality Control Costs
12345-442	Onshore G&G-Damages Costs
14000–14131	**MISCELLANEOUS ASSETS** (For detailed subaccounts, see the SE chart of accounts.)
16000–16999	**ACCUMULATED DD&A AND WRITEDOWNS**
18000–18999	**LONG-TERM RECEIVABLES**
18101-000	Long-Term Receivable
18201-000	Note Receivable – Branch

19021–19091	**DEFERRED FEDERAL INCOME TAX**
20000–20999	**ACCOUNTS PAYABLE**
21000–21999	**SHORT TERM NOTES PAYABLE**
24000–24999	**ACCRUED STATE AND FEDERAL TAXES**
29000–29999	**ACCRUED LIABILITIES** (For detailed subaccounts, see the SE chart of accounts.)
30000–31999	**LONG TERM NOTES PAYABLE**
40000–49999	**STOCKHOLDERS EQUITY AND RELATED ACCOUNTS** (For detailed subaccounts, see the SE chart of accounts.)
53000–59999	**REVENUES AND GAINS** (For detailed subaccounts, see the SE chart of accounts.)
60000–60999	**LEASE OPERATING EXPENSE** (For detailed subaccounts, see the SE chart of accounts.)
61000–61999	**MARKETING AND TRANSPORTATION EXPENSE**
66000–66999	**GENERAL AND ADMINISTRATIVE EXPENSE** (For detailed subaccounts, see the SE chart of accounts.)
68000–68999	**DD&A EXPENSE**
79600–85400	**MISCELLANEOUS EXPENSES** (For detailed subaccounts, see the SE chart of accounts.)

The numbers in the above examples and in the other examples and homework problems throughout this book are not intended to be realistic but are lower than actual for easy solution. In addition, the detailed record keeping used in actual practice is not illustrated in the examples and problems. Subsidiary ledgers and cost records would in actual practice be prepared for many of the entries given in the book. For example, if three leases were obtained, subsidiary records would be maintained for each lease, each well, and each owner's interest.

Chapters 3 through 5 detail the accounting for exploration, acquisition, and drilling activities according to SE. Chapter 6 presents the capitalized cost disposition under SE by both DD&A and abandonment of property. Chapter 7 provides a description of FC accounting. Chapters 8 and 9 deal with production and revenue accounting. Chapter 10 presents tax accounting methodology. Joint interest accounting is presented in chapter 11. Accounting for conveyances under SE and FC accounting is described in chapter 12. Chapter 13 explains the disclosures required by *SFAS No. 69* for companies with oil and gas producing activities. Chapter 14 gives an overview of the international dimensions of the industry. Chapter 15 presents financial statement analysis.

PROBLEMS

1. List the costs that are treated the same (*i.e.*, capitalized or expensed) under successful-efforts and full-cost accounting. List the costs that are treated differently.

2. What accounting treatment is given to the following costs under SE: acquisition costs, exploration costs, development costs, and production costs?

3. The successful-efforts method of accounting for oil and gas operations is considered to follow generally accepted accounting principles as pronounced by the FASB. Explain.

4. Define the following:
   - reserves
   - proved reserves
   - proved developed reserves
   - proved undeveloped reserves
   - proved area
   - field

5. Define the following:
   - exploratory well
   - development well
   - delineation well
   - wildcat well
   - extension well
   - service well
   - stratigraphic test well (exploratory and development)

6. Is a delineation well, *i.e.*, a well drilled to establish the perimeters of a reservoir, an exploratory well or a development well?

7. Lease A has a known productive horizon at 15,000 feet. A well is drilled to 10,000 feet. Would the well be classified as an exploratory or development well?

8. Lease B has a producing formation at 8,000 feet. A well is drilled to 11,000 feet and is classified as a development well. Comment.

9. Dry Oil Company, a successful-efforts company, drilled an exploratory well offshore at a cost of $1 million. The well was dry, but Dry Oil felt that the G&G data obtained from

the well was promising and drilled another well close to the first one. Should the first well be expensed or capitalized?

10. Dixie Company incurred the following costs during calendar year 2006:

February 1	Cost of G&G activities to locate an oil prospect, $100,000
March 2	Acquisition costs for a 400-acre lease: lease bonus $50/acre; other costs incurred in acquiring the property, $1,000
May 30	Dry hole costs of an exploratory well, $315,000
June 28	Successful exploratory well costs, $405,000
August 15	Cost of production facilities such as flow lines and separators, $225,000
September 1	Production costs, $50,000

Prepare journal entries for the above transactions using the successful-efforts method of accounting.

11. Given the following costs for Lease A, all incurred during 2008, prepare income statements and unclassified partial balance sheets for a successful-efforts company and a full-cost company.

Acquisition costs	$ 30,000	
G&G costs	80,000	
Exploratory dry holes	1,500,000	
Successful exploratory holes	350,000	
Development wells, dry	200,000	
Development wells, successful	475,000	
Cost of production facilities	250,000	
Production costs	60,000	
DD&A*	55,000 (SE)	125,000 (FC)
Accumulated DD&A	150,000 (SE)	360,000 (FC)
Revenue from sale of oil	225,000	

* Depreciation, depletion and amortization

12. Ebert Company incurred the following costs during the fiscal year ending May 31, 2005.

June 1, 2004	G&G costs, $50,000
August 10, 2004	Lease bonus of $80 an acre on a 1,000 acre lease and other acquisition costs of $2,000
December 15, 2004	Dry-hole costs of an exploratory well, $400,000
January 18, 2005	Successful well costs, $600,000

April 10, 2005	Cost of production facilities, $500,000
April 30, 2005	Production costs, $40,000

Prepare journal entries for the above transaction, assuming that Ebert Company uses the full-cost method of accounting.

13. Revenue and costs for Hopeful Company for the year 2005 are presented below:

Revenue	$ 600,000
G&G Costs	600,000
Acquisition costs	2,000,000
Exploratory dry holes	4,000,000
Successful exploratory wells	3,000,000
Development wells, dry	1,000,000
Development wells, successful	800,000
Production facilities	700,000
Production costs	100,000

	SE	FC
Amortization for 2005	200,000	400,000
Accumulated DD&A	500,000	700,000

Prepare income statements and unclassified balance sheets for a successful-efforts company and a full-cost company and explain the difference in net income.

14. Indicate whether the following costs should be expensed (E) or capitalized (C) depending on whether the company uses successful-efforts or the full-cost method of accounting.

	Successful-Efforts		**Full-Cost**	
	E	**C**	**E**	**C**
Acquisition costs				
G&G costs				
Exploratory dry holes				
Successful exploratory wells				
Development wells, dry				
Development wells, successful				
Production facilities cost				
Production costs				

Notes

1. Burrow, Jackie. "Treatment of Dry Holes Under the Successful-Efforts Method of Accounting—A Defitional Problem," *Journal of Extractive Industries Accounting,* Spring, 1982, pp. 58-59.

chapter three

Nondrilling Exploration Costs Under SE

Exploration involves identifying and examining areas that may contain oil and gas reserves. The principal types of exploration costs are defined in *SFAS No. 19*, par. 17 as follows:

> a. *Costs of topographical, geological, and geophysical studies, rights of access to properties to conduct those studies, and salaries and other expenses of geologists, geophysical crews, and others conducting those studies. Collectively, these are sometimes referred to as geological and geophysical (G&G) costs.*
> b. *Costs of carrying and retaining undeveloped properties such as delay rentals, ad valorem taxes on the properties, legal costs for title defense, and maintenance of land and lease records.*
> c. *Dry hole contributions and bottom hole contributions.*
> d. *Costs of drilling and equipping exploratory wells.*
> e. *Costs of drilling exploratory-type stratigraphic test wells.*

The first three types of costs are often incurred before any drilling activities begin and are nondrilling in nature. In order to distinguish them from types *d* and *e* costs, which are drilling costs, types *a*, *b*, and *c* are referred to in this book as **nondrilling exploration costs.** *SFAS No. 19* specifies that nondrilling exploration costs are to be charged to expense as incurred, as shown in Figure 3–1.

Fig. 3–1 — *Successful Efforts, Exploration Costs*

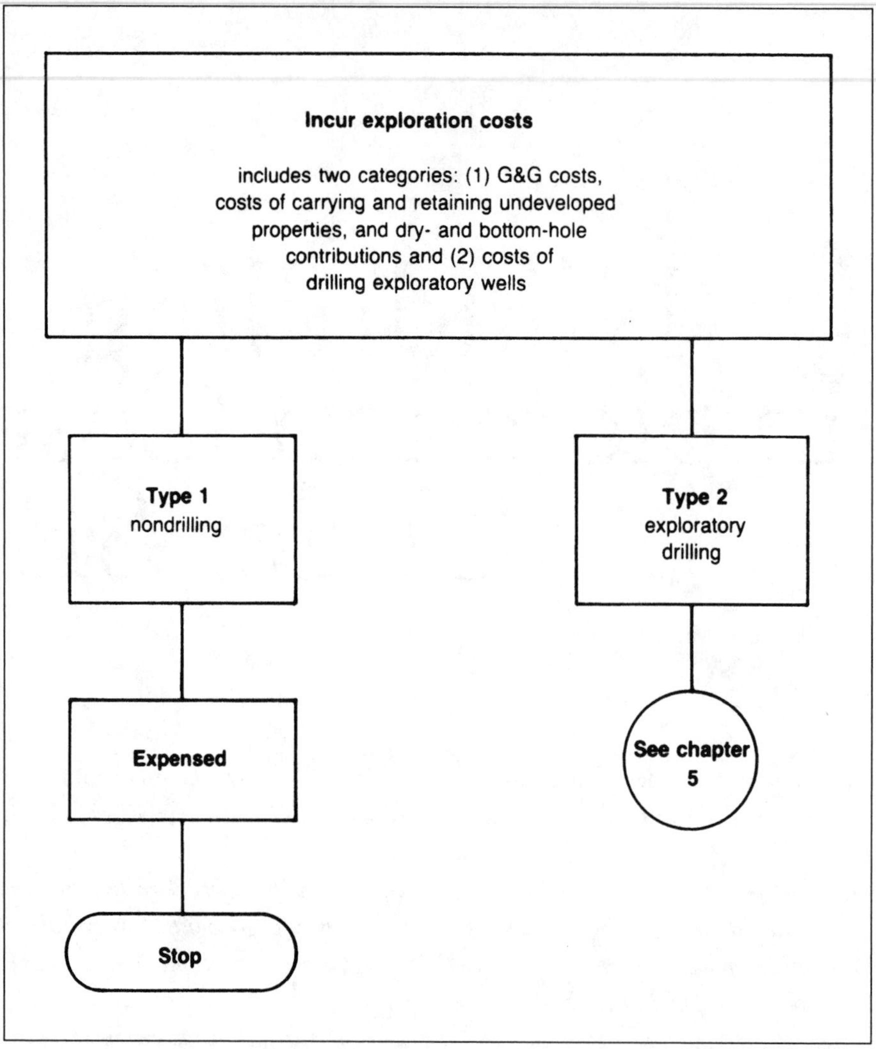

G&G Costs

The purpose of geological and geophysical (G&G) exploration is to locate or identify areas with the potential of producing oil and/or gas in commercial quantities. As discussed in chapter 1, both surface and subsurface G&G techniques are used to locate these areas. Surface techniques are used to evaluate the surface for evidence of subsurface formations with characteristics favorable to the accumulation of oil or gas. Subsurface techniques identify formations capable of containing oil or gas by utilizing the fact that all types of rocks have different characteristics and respond differently to stimuli such as sound or magnetic waves.

A **reconnaissance survey** is a G&G study covering a large or broad area, while a **detailed survey** is a G&G study covering a smaller area, one possibly identified as a result of the reconnaissance survey. By studying the data yielded by the G&G studies, geologists and geophysicists can identify interest areas that may warrant the acquisition of a working interest.

G&G studies may be conducted either before or after a property is acquired. If a company wants to explore an onshore area prior to obtaining a mineral lease on the property, rights to access the property, called **shooting rights,** must first be obtained from the property owner. The exploration company pays a fee for the shooting rights, and typically, the rights are coupled with an option to lease, in order for the company performing the G&G activities to have the right to lease the property if the G&G data are promising.

If shooting rights coupled with an option to lease are obtained and if the surface owner is different from the mineral rights owner, the agreement must be signed with the mineral rights owner. In that case, the surface owner would be compensated for rent and any damages done to the surface. If shooting rights only are obtained and if the surface owner is different from the mineral rights owner, the exploration company may sign a contract with either the mineral rights owner or the surface rights owner.

When offshore tracts are involved, a permit, not usually requiring the payment of a fee, is normally obtained from the U.S. Geological Survey (USGS). If the offshore tract is in state waters, then a permit must be obtained from the state.

G&G costs include all the costs related to conducting G&G studies and the cost of access rights to properties to conduct those studies, including any damages or rent paid to the surface interest owner. These G&G costs must be expensed as incurred regardless of whether they were incurred before or after acquisition of a working interest in the property. G&G costs are similar to research costs because they are incurred to provide information. To a large extent, G&G costs are incurred prior to property acquisition. In many cases, surveyed property is never acquired or, if acquired, is subsequently abandoned. Correlation of G&G costs with specific discoveries (months or years later) is very difficult or impossible, and cannot be made at the time the G&G costs are incurred. Therefore, the FASB concluded that G&G costs should be expensed as incurred.

EXAMPLE

G&G Costs

a. Lucky Company obtained shooting rights—access rights to a property so G&G studies may be conducted—to 10,000 acres, paying $0.30 per acre.

Entry

G&G expense (10,000 x $0.30).	3,000	
Cash.		3,000

b. Lucky Company then hired ABC Company to conduct the G&G work and paid the company $25,000. (This is the normal situation, *i.e.*, a company usually contracts out its G&G work.)

Entry

G&G expense	25,000	
Cash.		25,000

As described above, broad G&G surveys determine interest areas that warrant leasing or further exploration. Detailed surveys may be made to further delineate the more promising areas for possible leasing. Although not required by the SE method, both broad and detailed survey costs are, in many cases, allocated to the leases acquired. However, even if allocated to individual leases by companies using successful-efforts, G&G costs should still be expensed.

Since *SFAS No. 19* was written in 1977, 3D and 4D seismic methods have been developed. These methods are much more expensive and much more accurate than the old 2D seismic studies. The new 3D and 4D methods are used not only to locate new reservoirs, but are also used on existing *producing* reservoirs. When used on existing producing reservoirs, the costs are analogous to development costs. According to a 1999 PricewaterhouseCoopers survey, 63% of the successful efforts companies responding to the survey capitalize G&G costs used on producing reservoirs as a development cost instead of expensing the costs as an exploration expense.[1]

G&G studies may also be conducted on a property owned by another party in exchange for an interest in the property if proved reserves are found—if not found, the G&G costs incurred are reimbursed. In this situation, the G&G costs should be recorded as a receivable when incurred and, if proved reserves are found, should be transferred to become part of the cost of the proved property acquired (*SFAS No. 19*, par. 20). If proved reserves are not found, reimbursement is received.

EXAMPLE

G&G Studies Exchanged for Interest in Property

During 2005, Lucky Company paid for G&G studies to be performed on two leases owned by other parties. The agreements provided that Lucky company would receive $\frac{1}{5}$ of each WI if proved reserves were found and would be reimbursed for G&G costs incurred if proved reserves were not found. The G&G costs incurred by Lucky were as follows:

Lease A	$20,000
Lease B	$30,000

Drilling activity on the leases in 2005 was as follows:

Lease A—Drilling resulted in a dry hole and the lease was abandoned. The owner of Lease A reimbursed Lucky for the G&G costs.

Lease B—Drilling resulted in discovering proved reserves and, as per the agreement, Lucky received $^1/_5$ of the WI.

Entries

Lease A:

Receivable—Lease A	20,000	
Cash.		20,000
Cash.	20,000	
Receivable—Lease A		20,000

Lease B:

Receivable—Lease B	30,000	
Cash.		30,000
Proved property—Lease B	30,000	
Receivable—Lease B		30,000

G&G studies may also be performed before drilling a well to determine the location of the specific drillsite. This type of cost, although involving G&G activity, is considered part of the drilling process and is accounted for as a drilling cost rather than a nondrilling exploration cost. Further discussion of this cost is found in chapter 5.

CARRYING AND RETAINING COSTS

Carrying and retaining costs are incurred primarily to maintain the lessee's property rights, not to acquire those rights. Common carrying and retaining costs include:

- delay rentals
- property taxes
- legal costs for title defense
- clerical and record-keeping costs

Delay rentals are costs paid on or before the anniversary date of the lease in order to delay drilling operations for a year. Ad valorem taxes or property taxes are assessed on the economic interest owned by the working interest and are levied by a governmental agency, *i.e.*, city, county, school district, etc. (These property taxes are incurred to maintain the property, unlike the delinquent property taxes discussed in chapter 4 that may be incurred at the time of acquiring the property.) Legal costs for title defense paid by the lessee include attorney's fees, court costs, etc., incurred when the royalty interest owner is involved in a legal dispute regarding claims to title to the property. (The lessee may be willing to pay these fees to ensure the title problems are resolved satisfactorily.) Lease record maintenance costs are incurred by the land department in maintaining, evaluating, and updating the company's lease records. Employee salaries, materials, and supplies comprise the bulk of these maintenance costs.

Carrying costs do not increase the potential recoverable amount of oil and gas and do not enhance future benefits to be derived from the acquired properties. Instead, carrying and retaining costs are incurred primarily for the purpose of keeping the property interest and keeping a clear title. For these reasons, carrying and retaining costs are expensed as incurred.

The carrying and retaining costs under discussion are related to unproved properties and are usually insignificant in terms of cost. For instance, delay rentals are typically a nominal amount—as little as $1 or $2 per acre. Lease records maintenance cost is generally relatively small in dollar amount. Ad valorem taxes on unproved properties will, if they are assessed, also be a small amount because the existence of minerals in economic quantities is not yet known. Legal costs, in contrast, may be relatively significant.

EXAMPLE

Carrying and Retaining Costs

a. Lucky Company acquired a 500-acre lease in Texas. During the first year, the company did not develop the lease, *i.e.*, no drilling was done. Therefore, to retain the lease Lucky Company paid a delay rental payment of $1,000.

Entry

Delay rental expense	1,000	
Cash.		1,000

b. Lucky Company paid $1,200 in ad valorem taxes, *i.e.*, property taxes assessed on Lucky's economic interest in the lease.

Entry

Ad valorem tax expense	1,200	
Cash.		1,200

c. The land department incurred allocable costs of $5,000 in maintaining land and lease records and allocated $500 of that cost to the property.

Entry

Records maintenance expense.	500	
Cash.		500

The carrying and retaining costs discussed in this chapter are associated with undeveloped properties and are classified as exploration costs. If ad valorem taxes or lease record maintenance costs are associated with proved properties, they are classified as a production cost, not an exploration cost (chapter 8).

Test-Well Contributions

Test-well contributions, diagramed in Figure 3–2 below, result when one company drills a well and another company, which owns the WI in nearby acreage, agrees to pay the drilling company for certain G&G information. For example, per a prearranged agreement, Company B drills a well and provides specified G&G information to Company A (an unrelated company with a WI in acreage nearby). Company A receives the information and, in return, in most cases, pays a test-well contribution to Company B. Since test-well contributions are in essence G&G costs, they are expensed as incurred for the same reasons previously described. The recipient of a test-well contribution (Company B) treats the amount received as a reduction of intangible drilling costs incurred as the well is drilled (chapter 5).

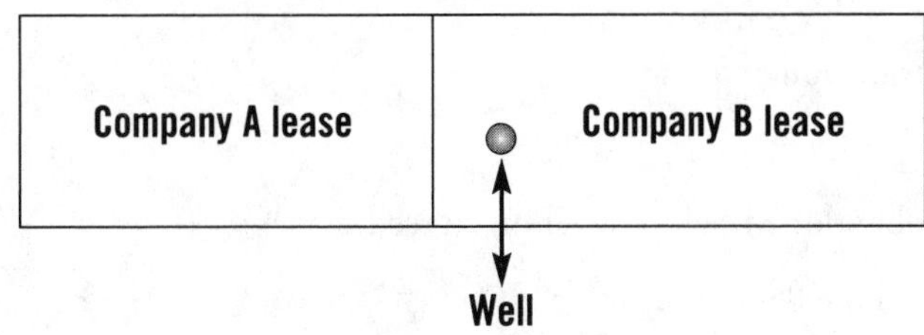

Fig. 3–2 — *Well drilled by Company B. Company A receives G&G information from Company B's well.*

A test-well contribution may be one of two basic types:

- **dry-hole contribution**—payment is made only if the well is dry or not commercially producible
- **bottom-hole contribution**—payment is made when an agreed-upon depth is reached, regardless of the outcome of the well

EXAMPLE

Test-Well Contributions

Several wells were being drilled on leases in close proximity to an undeveloped lease owned by Lucky Company. In order to obtain G&G information from the wells, Lucky Company entered into the following agreements:

Well 1: dry-hole contribution of $15,000
Well 2: dry-hole contribution of $25,000
Well 3: bottom-hole contribution of $10,000
Well 4: bottom-hole contribution of $20,000

The following results were obtained:

Well 1—dry

Entry (dry-hole contribution)

Test-well contribution expense	15,000	
Cash.		15,000

Well 2—completed as a producer

Entry (dry-hole contribution)
None

Well 3—drilled to agreed depth and determined to be dry

Entry (bottom-hole contribution)

Test-well contribution expense	10,000	
Cash.		10,000

Well 4—well abandoned before reaching agreed depth

Entry (bottom-hole contribution)
None

Note: A company receiving a test-well contribution records it as a reduction in intangible drilling costs (chapter 5).

A dry-hole contribution is paid only if the well does not find proved reserves. However, whether a producer is found or not, the driller is obligated to furnish G&G information to the other company involved in the dry-hole contribution agreement. A bottom-hole contribution is paid when the driller reaches the predetermined contract depth, even if proved reserves are not found. The bottom-hole contribution will not be paid if the driller fails to drill to the predetermined depth. Depending on contract terms, the bottom-hole contributor may receive G&G information even if the well is not drilled to contract depth and no payment is made.

The primary reason for making a test-well contribution is that valuable information (well logs, drillstem tests, etc.) can be obtained without the cost of drilling a well. On the other hand, the company receiving the contribution receives money to help offset the cost of drilling the well. A dry-hole contribution is entered into if the company seeks to offset the cost of drilling the well if the well is dry—the assumption being if the well finds reserves, the company drilling the well does not need the reimbursement. In a bottom-hole contribution situation, the company drilling the well wants to minimize its cost regardless of the outcome.

The following example illustrates the accounting for different types of nondrilling exploration costs.

EXAMPLE

Comprehensive Example

a. Lucky Company was interested in a large section of land in West Texas and obtained shooting rights to 6,000 acres for $1.50 per acre. (G&G cost)

Entry

G&G expense (6,000 x $1.50).	9,000	
Cash.		9,000

b. Lucky paid a geological firm $50,000 to conduct a reconnaissance survey on the area. (G&G cost)

Entry

G&G expense	50,000	
Cash.		50,000

c. Based on the results of that study, Lucky acquired one 700-acre lease and immediately commissioned the same geological firm to conduct detailed G&G studies on the lease at a cost of $15,000 (G&G cost).

Entry

G&G expense	15,000	
Cash.		15,000

d. During the first year, Lucky had to pay $2,000 in ad valorem taxes and $10,000 for title defense in connection with the property (carrying and retaining costs).

Entries

Ad valorem tax expense	2,000	
Cash.		2,000
Legal expense—exploration	10,000	
Cash.		10,000

e. By the end of the first year, Lucky had not yet begun any drilling efforts and, wanting to retain the lease, paid the first delay rental of $4,000 (carrying and retaining cost).

Entry

Delay rental expense	4,000	
Cash.		4,000

f. Early in the second year, drilling began on a well on a nearby property. Lucky entered into a bottom-hole contribution agreement to obtain the G&G information from the well. The depth specified in the agreement was reached two months later, and Lucky paid $20,000 as per agreement (test-well contribution).

Entry

Test-well contribution expense	20,000	
Cash.		20,000

Support Equipment and Facilities

Support equipment and facilities may not be directly related to one field or reservoir, but are necessary for efficient exploration, development, and production activities. Included as support equipment and facilities are seismic, drilling, construction, grading, or other equipment, warehouses and division or field offices, repair shops, and vehicles. Support equipment may be used in a single oil and gas producing activity, *i.e.*, exploration, development, or production, or the equipment and facilities may be used to serve two or more of those activities or other activities such as marketing or refining.

The acquisition cost of support equipment and facilities should be capitalized. Any depreciation or operating costs of support equipment and facilities should be classified as an exploration, development, or production cost to the extent the equipment or facility is used for that activity. In the case that equipment and facilities are used in more than one activity, the operating and depreciation costs should be allocated to the appropriate activities based on some measure of usage, such as hours utilized in each activity. Costs of support equipment and facilities used in nondrilling exploration activities should be expensed as incurred. Additional discussion is provided in chapter 6.

EXAMPLE

Support Equipment and Facilities

Depreciation of the seismic equipment used by Lucky Oil Company in West Texas was $10,000 for 2005. Operating costs were $21,000.

Entry to record depreciation

G&G expense—depreciation	10,000	
Accumulated depreciation		10,000

Entry to record operating costs

G&G expense—operating cost	21,000	
Cash		21,000

Offshore and International Operations

In offshore and international locations nondrilling exploration costs, such as shooting rights, ad valorem taxes, and legal costs for title defense are not applicable because a government entity owns the mineral rights; however, the government entity may impose other types of fees and taxes. Offshore G&G surveys are conducted differently than onshore G&G and are significantly more expensive. In international operations G&G surveys may be conducted either onshore or offshore depending on the location of the contract area. In many situations, these costs are borne entirely by the international exploration and production company even when a local, government-owned oil company shares in the working interest. Despite these differences, the costs of G&G and other nondrilling exploration activities incurred offshore and in international locations are accounted for in the same manner as for domestic onshore operations. They are charged to expense as incurred by successful-efforts companies.

PROBLEMS

1. Tiger Oil Company obtained shooting rights on 10,000 acres at $2 per acre and then hired an independent geological firm to conduct the initial G&G work for $60,000. As a result of the G&G work, Tiger Company decided to lease 500 acres (ignore acquisition costs) and hired the same company to perform detailed G&G studies on the 500 acres at a cost of $15,000. Give the entries to record these transactions.

2. Wildcat Oil Company obtained a lease on March 1, 2005. Being short of funds, Wildcat Oil Company did not begin drilling operations during the first year of the primary term and on March 1, 2006, made a delay rental payment of $8,000. On May 12, 2006, the company paid a bottom-hole contribution of $30,000. The information obtained from this well was so encouraging that Wildcat Oil Company decided to begin drilling operations; however, there were some title problems and drilling was delayed. Legal costs incurred for title defense were $50,000. Give the entries.

3. During 2006, the exploration department of Red Ink Oil Company incurred the following costs in exploring South Texas fields. Give the entries.

Shooting rights	$ 12,000
Bottom-hole contribution	40,000
Supplies for exploration (G&G) activities	8,000
Salaries for exploration (G&G) activities	100,000
Mapping costs for exploration (G&G) activities	15,000
Depreciation of exploration (G&G) equipment	20,000
Transportation for seismic crew	5,000
Operating costs for exploration (G&G) equipment	3,000

4. Universal Oil Company obtained the rights to shoot 25,000 acres at a cost of $0.20/acre on May 3, 2007. Universal contracted and paid $80,000 for a reconnaissance survey during 2007. As a result of this broad exploration study, Lease A and Lease B were leased on January 9, 2008. (Ignore acquisition costs.) The two properties totaled 1,500 acres, and each had a delay rental clause requiring a payment of $2 per acre if drilling was not commenced by the end of each full year during the primary term. Detailed surveys costing a total of $30,000 were done during January and February on the leases.

   During July, Universal entered into two test-well contribution agreements: a bottom-hole contribution agreement for $15,000, with a specified depth of 10,000 feet, and a dry-hole contribution of $20,000, also with a specified depth of 10,000 feet. In November both wells were drilled to 10,000 feet. The well with the bottom-hole contribution was successful, but the well with the dry-hole contribution was dry.

   The cost for maintaining land and lease records allocated to these two properties for 2008 was $2,000. Ad valorem taxes were assessed on Universal's economic interest in both properties, amounting to $2,500 for 2008. After preparing their financial statements for 2008, Universal decided to delay drilling on these properties until sometime in 2010.

On April 15, 2010, enough money was left after paying taxes for a well to be drilled on Lease B. Before drilling the well, costs of $7,000 were incurred to successfully defend a title suit concerning Lease B.

Give all entries necessary to record these transactions. Assume any necessary delay rental payments were made.

5. Discuss why each party to a test-well contribution situation would enter into the transaction.

6. Aggie Oil Company purchased seismic equipment on March 1, 2007, costing $100,000. The seismic equipment was used in G&G operations for the remainder of the calendar year, 2007. Compute straight-line depreciation for 2007, assuming a 10-year life and no salvage value, and prepare the entries to record the purchase and depreciation of the equipment.

7. Hard Luck Oil Company had the following transactions in 2007 concerning test-well contributions:
   a. Contracted with Tiger Oil Company, agreeing to pay $50,000 if a well was drilled on Tiger's lease to a depth of 10,000 feet.
   b. Contracted to pay Landa Oil Company $40,000 if a well being drilled on Landa's property was dry.
   c. Agreed to pay Richards Oil Company $100,000 if a well being drilled reached a depth of 7,500 feet.

   Results from the above transactions were the following:
   a. Because of mechanical difficulty, the Tiger well was abandoned at 9,500 feet.
   b. The Landa well was dry.
   c. The Richards well was completed as a producer at 12,000 feet.

   Prepare entries for the above transactions, assuming Hard Luck Oil Company fulfilled its contractual obligations.

8. Basic Oil Company conducted G&G activities on leases owned by Artificial Oil Company and Universal Oil Company. Each agreement provides for Basic Oil Company to receive $\frac{1}{4}$ of each WI if proved reserves are found and to be reimbursed if proved reserves are not found. Basic Oil Company incurred the following G&G costs on Artificial's and Universal's leases:

Artificial	$50,000
Universal	$40,000

   The well drilled on the Artificial Oil lease was successful, and $\frac{1}{4}$ of the WI was assigned. Drilling on the Universal lease resulted in a dry hole, and Basic was reimbursed for the G&G costs incurred. Prepare entries for the above transactions.

9. Core Oil Company obtained a three-year lease on 1,000 acres on May 1, 2004 that contained a $3 per acre delay rental clause. Drilling operations were started on June 15, 2005 and completed on October 16, 2005. The well, determined to be dry, was plugged and abandoned. No further drilling operations were started during the primary term. All required delay rentals were paid. Give all entries relating to the delay rental requirement.

10. Fossil Oil Company entered into two test-well contribution agreements as follows:
    a. On May 17, 2004, a bottom-hole agreement was obtained requiring a payment of $45,000 when the contract depth of 10,000 feet was reached. The contract depth was reached on September 21, 2004, and the required payment was made.
    b. On September 30, 2004, a dry-hole test-well contribution was entered into requiring payment of $50,000 if the well was dry but no payment if the well was successful.
        1) Assume the well is successful.
        2) Assume the well is dry.

    Prepare necessary entries for the above transactions.

11. Dixie Company obtained seismic equipment on January 1, 2004, at a cost of $100,000. The equipment was used in G&G operations for the calendar year, 2004. The equipment has an estimated life of 10 years with a salvage value of $20,000. The company uses the straight-line method in computing depreciation. Record the depreciation for the year 2004.

12. During 2005, Fortunate Oil Company obtained the following leases:

Lease	Acres
A	3,000
B	4,000
C	5,000

In obtaining these leases, Fortunate Oil Company incurred shooting rights on Leases A and C at $0.50 an acre and incurred the following costs:

Salaries for exploration activities	$50,000
Mapping costs for exploration activities	20,000
Minor repairs of G&G exploration equipment	1,000

Give entries for the above transactions.

13. Define the following:
    shooting rights
    G&G costs
    carrying costs
    dry-hole test-well contribution
    bottom-hole test-well contribution

14. Big John Oil Company agreed to conduct G&G studies and other exploration activities on a lease owned by Young Oil Company in exchange for an interest in the property if proved reserves are found. If proved reserves are not found, Big John will be reimbursed for costs incurred.
    a. Big John incurs $200,000 of exploration costs.
    b. Assume proved reserves are found.
    c. Assume instead that proved reserves are not found.

    Give any entries required.

NOTES

1. *1999 PriceWaterhouseCooper's Survey of U.S. Petroleum Accounting Practices*, Denton, TX: Institute of Petroleum Accounting, 1999.

chapter four

Acquisition Costs *of* Unproved Property Under SE

A property may be acquired through the purchase of the mineral rights or the purchase of the fee interest. Land purchased *in fee* means that both the mineral rights and the surface rights are acquired rather than just the mineral rights. However, as discussed in chapter 1, leasing is the typical method of acquiring property.

Acquiring rights in mineral interests through leasing is complicated when, as is often the case, the mineral rights on a property are not owned by the surface owner. In that case, the exploration and production company wishing to lease the property must secure rights from the subsurface owner first, and then in certain states obtain a separate agreement from the surface owner. The surface owner is entitled to receive payment for damages to the surface resulting from the exploration and production company's activities and for conversion of the surface area from its regular activities, but the surface owner cannot deny the exploration company, by rights granted under the lease, access to the land above the leased minerals.

Companies operating in foreign countries must acquire mineral interests by contracting with the local government. While foreign contracts differ dramatically from domestic leases, the financial accounting treatment of the acquisition costs of foreign mineral interests for U.S. GAAP purposes is consistent with the discussion below.

Acquisition costs are costs incurred in acquiring an economic interest in the mineral rights whether through leasing or purchase. The principal types of acquisition costs are, as specified in *SFAS No. 19*, par. 15:

> *...lease bonuses and options to purchase or lease properties, the portion of costs applicable to minerals when land including mineral rights is purchased in fee, broker's fees, recording fees, legal costs, and other costs incurred in acquiring properties.*

A lease bonus, typically the major acquisition amount, is paid by the lessee to the mineral rights owner at the time the working interest is acquired through a lease arrangement. Certain incidental acquisition costs such as brokers' fees, legal costs, and options to lease may also be incurred. Brokers' fees are fees that are paid to real estate agents or to landmen for services in connection with leasing or purchasing property. These services include finding available property, negotiating, and consummating a lease or purchase. Also, brokers generally file the legal papers that result from purchase or lease transactions. Legal fees are paid to attorneys for preparation of legal documents (deeds, leases, etc.) and title examinations, which involve determining if the seller or lessor has good marketable title. A good marketable

Fig. 4–1 — *Successful Efforts, Acquisition Costs*

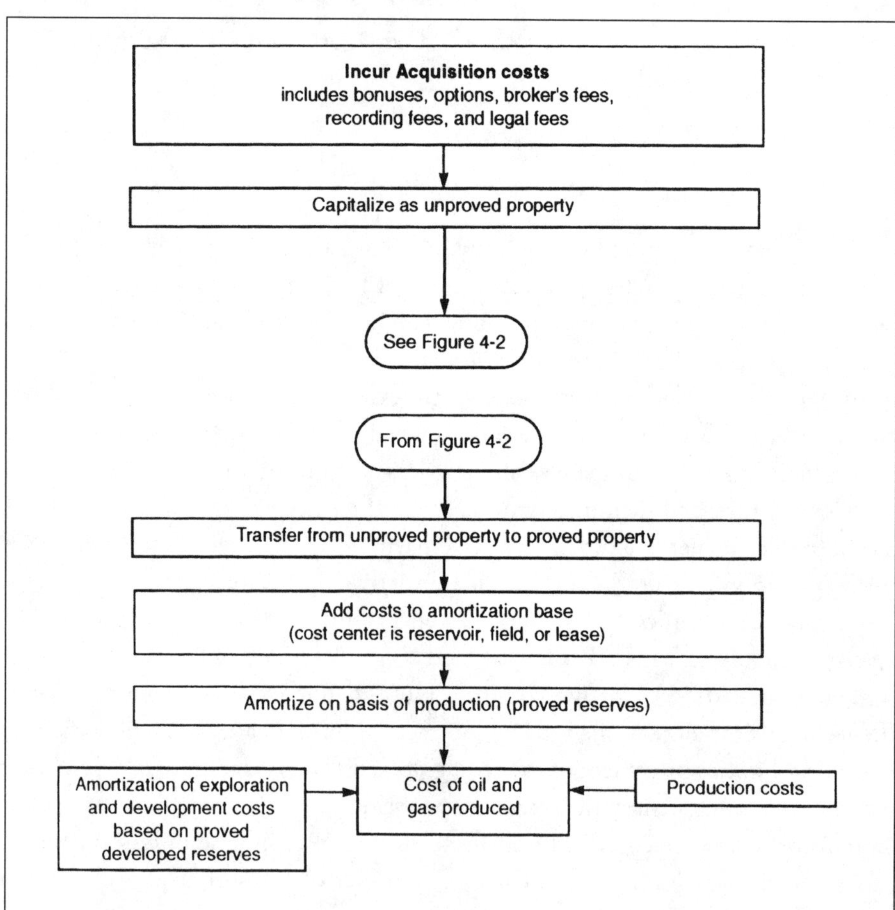

title is determined by tracing the chain of title back to the original land grant. The legal document resulting from the tracing of title is called an *abstract of title*. An option to purchase or lease a property is a contract that gives the prospective lessee a stated period of time in which to decide whether to purchase or lease the property.

Acquisition costs should be capitalized when incurred. See Figure 4–1 and 4–2 for an overview of the accounting for acquisition costs. As discussed earlier, the most significant acquisition cost is normally the lease bonus. Typically, other acquisition costs are relatively insignificant.

Fig. 4–2 — *Successful Efforts, Acquisition Costs*

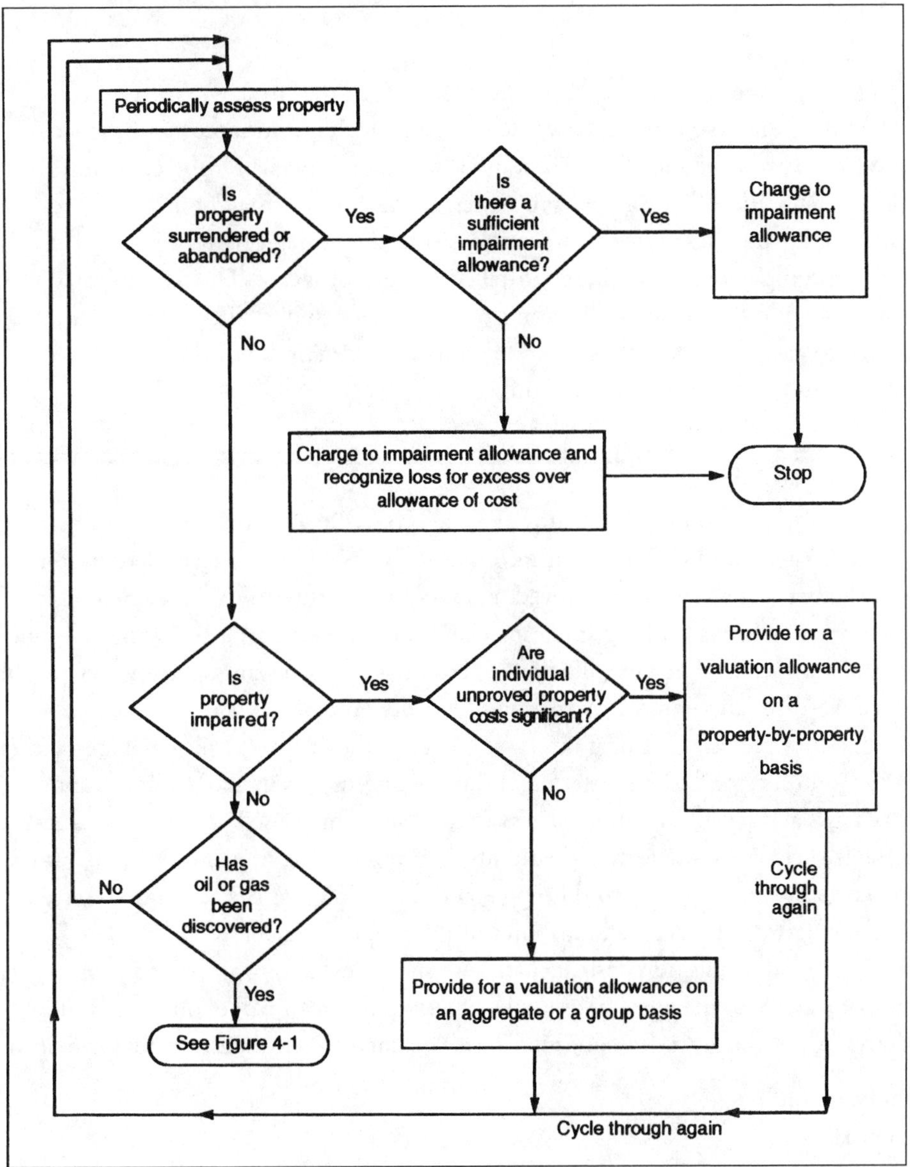

EXAMPLE

Acquisition Costs

Lucky Company acquired a 500-acre unproved property. Acquisition costs included a lease bonus of $40 per acre and recording fees of $1,000.

Entry

Unproved property (500 x $40 + $1,000) . . .	21,000	
Cash.		21,000

Note: A proved property may also be acquired. In that case, proved property instead of unproved property would be debited. Acquisition costs of proved properties would include the costs discussed above and would be capitalized when incurred. When proved properties are acquired it may be necessary to allocate the purchase price between the cost of the mineral rights and the cost of the equipment and facilities installed on the property. This complicated process typically involves allocating to equipment and facilities an amount equal to their fair market value, and debiting the remainder of the acquisition cost to the proved property account.

When a company acquires a lease it is necessary to set the property up in a property database. The database would include such information as location and legal description of the property, whether it is proved or unproved, names of the parties to the agreement, and a listing of all royalty owners, working interest owners, and any other parties with an economic interest in the minerals. The cost allocation percentages and revenue percentages would be determined, verified, and entered by the division order group.

In addition to providing critical information about the property, this database is accessed by both the joint interest billing system and the revenue system in allocating the appropriate amount of costs and revenues to the parties. This database may also support or be accessed by other systems such as the material and equipment inventory maintenance system. While this process is critical to the overall record keeping and functioning of the accounting system, this book does not address the detailed attributes of the property database.

The majority of acquisition costs incurred are simply debited to unproved property when incurred, as illustrated in the previous example. However, the treatment of other acquisition costs is not always so straightforward. Acquisition costs needing further discussion are the following:

- Purchase in fee
- Internal costs

- Options to purchase or lease
- Delinquent taxes and mortgage payments
- Top leasing

PURCHASE IN FEE

When land is purchased in fee, the purchase price should be allocated between the mineral and surface rights acquired based on the relative fair market values (FMVs) of the rights, if known.

EXAMPLE

Purchase in Fee, Both FMVs Known

Lucky Company purchased land in fee for $40,000. A qualified appraiser made the following estimate of the fair market values of the mineral and surface rights:

Mineral rights	$20,000
Surface rights	30,000
	$50,000

The $40,000 acquisition cost would be allocated as follows:

Land (surface rights): $\frac{\$30{,}000}{\$50{,}000} \times \$40{,}000 = \$24{,}000$

Unproved property (mineral rights): $\frac{\$20{,}000}{\$50{,}000} \times \$40{,}000 = \$16{,}000$

$\$40{,}000$

Entry

Land	24,000	
Unproved property	16,000	
Cash		40,000

If the fair market value of only one interest is available, then that interest should be allocated its fair market value and the remainder of the payment allocated to the other interest.

EXAMPLE

Purchase in Fee, One FMV Known

Lucky Company purchased land in fee for $40,000. A qualified appraiser estimated the fair market value of the surface rights to be $30,000. A reasonable estimate of the fair market value of the mineral rights was not possible. The $40,000 acquisition cost would be allocated as follows:

Land (surface rights):		$30,000
Unproved property (mineral rights):	$40,000 – $30,000 =	$10,000

Entry

Land	30,000	
Unproved property	10,000	
Cash		40,000

INTERNAL COSTS

In-house lawyers and landmen (landmen are discussed at the end of this chapter) are often involved in multiple activities including acquisition of mineral rights, examination of records before drilling is commenced, and negotiation and arrangement of joint ventures. Theoretically, the portion of the salaries of in-house lawyers, landmen, or other employees relating to lease acquisition, as well as any other internal or overhead costs relating to lease acquisition should be capitalized. Directly allocating these types of costs to specific leases may be impractical if the costs are insignificant. However, if the costs are material, they should be allocated to individual properties. Two reasonable allocation bases follow:

a. Capitalize the portion of the costs relating to acquisition activities and allocate to the individual leases acquired, based on total acreage acquired.
b. Allocate the portion of the costs relating to acquisition activities on an acreage basis to all prospects investigated, capitalizing the portion of the costs allocated to prospects acquired and expensing the portion of costs allocated to prospects not acquired.

SFAS No. 19 does not address the proper treatment of these types of costs. However, *b* above appears to be theoretically consistent with *SFAS No. 19* because only those successful costs that result in lease acquisition are capitalized. Most companies, however, appear to be expensing these costs because they are generally immaterial in amount.

EXAMPLE

Internal Costs

Lucky Company acquired an 800-acre undeveloped lease at a $10 per acre bonus. Legal costs and recording fees were $200. The salary of an in-house lawyer working solely on lease acquisition was $10,000. The lawyer's salary is allocated to all leases acquired, based on relative acreage acquired. During the accounting period, leases totaling 8,000 acres were acquired.

Entry

Unproved property (800 x $10 + 200 + 800/8,000 x $10,000)	9,200	
Cash		9,200

OPTIONS TO LEASE

Instead of leasing a property before conducting G&G studies, a company will often obtain shooting rights to a property so that G&G studies may be performed before a decision to lease is made. Generally, the shooting rights will be coupled with an option to lease. The option protects the lessee by keeping the right to lease the property open for a period of time and guarantees the company the right to lease the property at a specified price during the specified time period.

The cost of an option to lease is an acquisition cost and should be capitalized temporarily in a suspense account until a decision is made about whether to lease the property. If the property is leased, the option cost should be capitalized as part of unproved property; if the property is not leased, the option cost should be expensed. Shooting rights, in contrast to an option to lease, are a G&G exploration cost and should be expensed as incurred.

Two situations must be addressed when dealing with shooting rights coupled with an option to lease.

a. The payment represents both an acquisition cost and an exploration cost. If the acreage is leased, should the entire payment be capitalized or expensed or, if it can be determined, should the acquisition portion of the payment be capitalized and the exploration portion be expensed?

   Theoretically if the property is leased, the payment should be apportioned and treated as both an acquisition cost and an exploration cost. In practice, however, the entire payment is generally capitalized as an acquisition cost of the property.

b. If none or only a portion of the explored acreage is leased, how should the amount allocated to the option be handled?

If none of the acreage is leased, then the option amount should be expensed. If only a portion of the acreage explored is leased, theoretically only a proportional amount of the option cost based on the relative acres leased should be capitalized as an acquisition cost of the property. In practice, sometimes the entire option amount is capitalized, and sometimes a portion of the option is capitalized and a portion expensed, based on relative acres leased.

A flow diagram of the accounting for options to lease coupled with shooting rights is presented in Figure 4–3.

Fig. 4–3 — *Shooting Rights Coupled with Option to Lease*

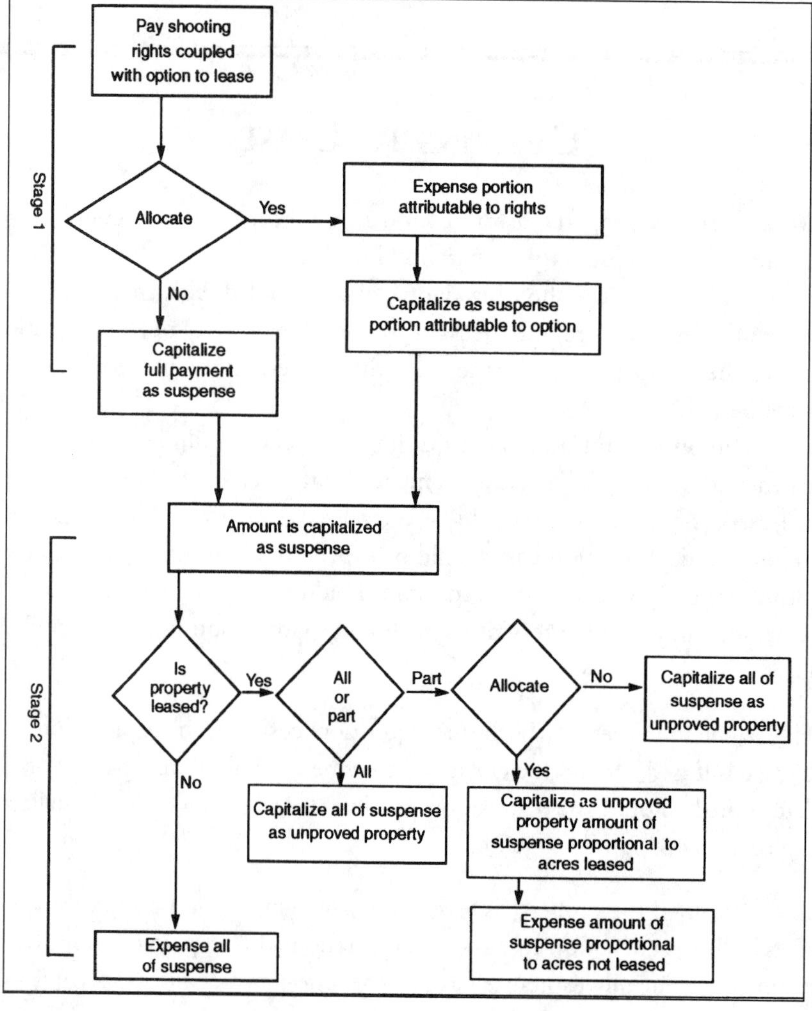

EXAMPLE

Option to Lease

Lucky Company obtained the following rights:

a. Shooting rights only for $2 per acre on 3,000 acres owned by Mr. A.

b. Shooting rights coupled with an option to lease for $15,000 on 3,000 acres owned by Ms. B. These 3,000 acres are located adjacent to the 3,000 acres owned by Mr. A.

c. Shooting rights coupled with an option to lease for $10,000 on 1,200 acres owned by Mr. C in Timbuktu. Lucky Company has no reasonable basis for valuing the shooting rights versus the option to lease.

Entries

MR. A's ACRES

G&G expense ($2 x 3,000)	6,000	
Cash		6,000

MS. B's ACRES—The cost of the shooting rights on Mr. A's acres ($2 per acre) can be used to determine the value of the shooting rights versus the value of the option on Ms. B's acres.

G&G expense ($2 x 3,000)	6,000	
Property purchase suspense (remainder)	9,000	
Cash		15,000

This entry is theoretically correct; however, the entire payment normally is capitalized as shown below.

Property purchase suspense	15,000	
Cash		15,000

MR. C's ACRES—There exists no reasonable basis for apportionment, so the entire payment must be capitalized.

Property purchase suspense	10,000	
Cash		10,000

As a result of the G&G work performed, Lucky Company made the following decisions:

a. Not to lease any of Mr. A's acres.
b. To lease only 1/3 of Ms. B's acres, paid a $20,000 bonus.
c. To lease only 1/4 of Mr. C's acres, paid a $15,000 bonus.

Entries
MR. A'S ACRES
No entry

MS. B'S ACRES—Lucky decides to capitalize only the portion of the suspense that applies to acres leased.

Surrendered lease expense (2/3 x $9,000)	6,000	
Unproved property (1/3 x $9,000 + $20,000)	23,000	
Property purchase suspense (close out) ...		9,000
Cash		20,000

MR. C'S ACRES—Lucky decides to capitalize only the portion of the suspense that applies to acres leased.

Surrendered lease expense (3/4 x $10,000) ...	7,500	
Unproved property (1/4 x $10,000 + $15,000)	17,500	
Property purchase suspense (close out) ...		10,000
Cash		15,000

Note that Lucky Company could have instead capitalized the entire amount in the suspense accounts relating to both Ms. B's and Mr. C's acres. In that case, the entry for Ms. B's acres would have been the following:

Unproved property ($9,000 + $20,000)	29,000	
Property purchase suspense (close out)		9,000
Cash		20,000

DELINQUENT TAXES AND MORTGAGE PAYMENTS

During the initial process of obtaining a lease, the lessee may agree to or be forced to pay such items as the lessor's delinquent taxes or mortgage payments in order to obtain a lease on land with a clear title. Such payments may or may not be recoverable from future delay rental payments or from future royalty payments. Accounting treatment depends upon whether the payments are recoverable.

1. If nonrecoverable, the payments are additional costs of acquiring the property and should be capitalized as an additional leasehold cost, *i.e.*, an acquisition cost.
2. If recoverable, the payments should be held in a receivable or suspense account until recovered. If not recovered, the payments should be charged to surrendered lease expense. (Nonrecovery of the payments would normally occur because the lease had been surrendered.)

In some cases payments such as these may be made during the lease term by the lessee to protect the lessee's interest in the lease. This type of payment would not be an acquisition cost but a cost of maintaining the property, *i.e.*, a nondrilling exploration cost (chapter 3). Consequently, if these payments are nonrecoverable, they should be expensed.

EXAMPLE

Delinquent Taxes and Mortgage Payments

Lucky Company paid delinquent property taxes of $300 on an undeveloped lease as called for in the lease agreement.

a. Assume these taxes were not recoverable.

Entry

Unproved property	300	
Cash		300

Since the taxes were part of the lease agreement and are non-recoverable, the payment would be an acquisition cost.

b. Assume that instead, the taxes were recoverable out of future delay rental payments or royalty payments.

Entry

Receivable from lessor	300	
Cash		300

c. Two months later, a delay rental of $400 was due. Since the $300 delinquent property tax paid by Lucky Company is recoverable from the delay rental, a net payment of $100 is remitted to the lessor.

Entry

Delay rental expense	400	
Receivable from lessor		300
Cash		100

d. Assume that instead of making the delay rental payment Lucky Company had abandoned the lease without recovering the delinquent property taxes.

Entry

Surrendered lease expense	300	
Receivable from lessor		300

Payment of delinquent property taxes by the lessee for the lessor is different from the payment of property taxes discussed in chapter 3. The property taxes discussed in chapter 3 are property taxes assessed by a governmental body on the economic interest owned by the lessee. The delinquent property taxes discussed in this chapter are unpaid taxes paid by the lessee but assessed on the interest owned by the lessor.

Top Leasing

A **top lease**, which occurs infrequently, is a new lease executed before expiration or termination of the existing lease. A top lease is usually sought when the current working interest owner wants to retain a property on which no drilling was done during the primary term. This situation is a two-party top lease. Sometimes, a different working interest owner will negotiate a new lease prior to expiration of the primary term on the current lease. This is called a three-party top lease.

If the existing lease is still in effect when the new lease (same working interest owner) is executed, the book value of the old lease plus the acquisition costs of the new lease will be capitalized as acquisition costs. However, if the existing lease terminates prior to the date on which a new lease is obtained (not a top lease), the book value of the old lease will be expensed, and only the acquisition costs of the new lease will be capitalized.

In the case where a different working interest owner signs a lease agreement with the mineral rights owner before the end of the primary term of an existing lease, the situation is no different from the acquisition of a new lease. Acquisition costs paid by the new lessee are capitalized as described earlier in the chapter.

EXAMPLE

Top Leasing

Lucky Company has two unproved leases with the following costs:

Lease A (500 acres)	$30,000
Lease B (1000 acres)	$40,000

a. A new lease was executed on Lease A by Lucky, the original lessee, prior to the termination of the old lease. The bonus paid for the new lease was $50 per acre.

Entry

Unproved property—Lease A	55,000	
Unproved property—Lease A		30,000
Cash ($50 x 500)		25,000

b. Lease B terminated on January 14, 2007 and a new lease was executed to become effective on January 15. The bonus paid was $60 per acre.

Entries

January 14:

Surrendered lease expense	40,000	
Unproved property—Lease B		40,000

January 15:

Unproved property—Lease B	60,000	
Cash ($60 x 1,000)		60,000

Disposition of Capitalized Costs—Impairment of Unproved Property

Unproved properties should be assessed periodically to determine whether they have been impaired (*SFAS No. 19*, par. 28). The following questions may be asked to help determine whether a property has been impaired:

a. Have there been any dry holes drilled on the lease or on surrounding leases, or has any additional negative G&G information been obtained?
b. How close is the expiration of the primary lease term?
c. Are there any firm plans for drilling?

Dry holes and negative G&G information would normally indicate that the property value in terms of replacement cost has declined, or at least a portion of the original cost has expired and impairment should be recognized. If neither drilling nor production is in progress at the end of the primary term, the lease terminates and the leasehold costs must be expensed. Therefore, a property may also be impaired if the end of the primary term is near and no firm plans for drilling have been established.

Impairment estimation is difficult and subjective. However, the above-mentioned variables should be taken into account in determining whether a property has been impaired and, if impaired, in determining the impairment percentage.

If the results of the assessment indicate that the property has been impaired, a loss should be recognized by providing a valuation allowance. The exact approach used depends upon whether the acquisition costs of the property under consideration are significant. *SFAS No. 19* does not indicate or give any guidance as to what should constitute an individually significant property compared to an individually insignificant property. However, in practice, in determining significance, factors such as cost of the properties, company size, number of unproved leases held, and the company's overall portfolio of unproven properties are taken into consideration.

For example, international operations are characterized by a small number of very expensive contract areas, while domestic operations are characterized by a large number of relatively less expensive leases. Companies may treat each international contract area as individually significant since the costs are large and there are fewer properties. Domestically, the sheer number of properties may make each property in itself insignificant.

One current practice in determining significance is to compare the cost of a property to the total cost of all unproved properties. If the cost of that property exceeds some percentage of the total capitalized costs of all unproved properties, then the property is considered significant. Many companies consider the property significant if the cost of the individual property exceeds 10% to 20% of the total capitalized costs of all the unproved property.[1] However, most SE companies simply use a dollar amount cutoff to determine significance.[2]

The SEC has issued significance guidelines for full-cost companies but not for successful-effort companies. Under the SEC's full-cost guidelines, a property is significant if its costs exceed 10% of the net capitalized costs of the cost center, which for full cost is a country (chapter 7).

If the acquisition costs are significant, the property should be assessed individually. Under individual assessment, each significant lease is examined individually to determine whether it has been impaired. In accounting, determining impairment typically involves comparing cost or carrying value to a market value or a net realizable value. For unproved property assessment in the oil and gas industry, however, few companies use a market value to determine impairment. Instead, as discussed earlier, variables such as dry holes drilled, plans for drilling, and the nearness of the end of the primary term are considered in determining impairment for each significant lease. Domestically, impairment is recognized frequently as either 0% or 100%. In contrast, partial impairment is common in international operations.

EXAMPLE

Impairment of Significant Properties

On January 1, 2007, Lucky Company acquired an undeveloped lease, paying significant acquisition costs of $40,000. During the year, Lucky Company drilled two dry holes on the property. As a result of drilling these dry holes, Lucky Company decided on December 31 that the lease was 25% impaired, *i.e.*, the impaired value was 25% less than cost.

Entry

Lease impairment expense ($40,000 x 0.25) . . .	10,000	
Allowance for impairment		10,000

The allowance for impairment account is a contra asset account. The pertinent part of a balance sheet prepared following this entry would be as follows:

Assets: Unproved property	$40,000
Less: Allowance for impairment	(10,000)
	$30,000

If acquisition costs are not individually significant, assessing impairment on an individual property basis may not be practical. In that case, impairment should be calculated by amortizing individually insignificant properties on a group or aggregate basis. This calculation is similar to that of providing an allowance for bad debts. The percentage used in calculating impairment may be based on the historical experience of the company and factors such as the primary terms of the involved properties, the average holding period of insignificant unproved properties, and the historical percentage of such properties that have been proved.

In the following example, the desired balance of the impairment allowance account is determined by applying the estimated impairment percentage to the total balance of individually insignificant unproved properties being assessed on a group basis. The amount of impairment recognized is the difference between the current balance in the impairment allowance account before impairment and the desired balance. This approach emphasizes the net realizable value of the properties and focuses on the statement of financial position.

EXAMPLE

Impairment of Insignificant Properties

On December 31, 2007, the unproved property account containing leases that are not individually significant had a $100,000 balance and the allowance for impairment account had a $20,000 balance. Past experience indicates that 65% of all insignificant unproved properties are eventually abandoned without ever being proved. Therefore, Lucky Company has a policy of providing at year-end an allowance equal to 65% of the cost of gross unproved properties.

Entry

Lease impairment expense	45,000	
Allowance for impairment, group basis . . .		45,000

Calculations

$100,000 x 0.65 = $65,000

Allowance		
	20,000	← Have
	?	← 45,000
	65,000	← Desired balance

This example provides an allowance equal to the total amount of unproved properties expected to be abandoned, a very conservative approach. An approach yielding an amount closer to the actual amount of unproved properties abandoned during the year might take into account the average holding period for insignificant properties and the average period for which the current properties have been held.[3]

For example, if the average holding period of the insignificant unproved properties in the above problem is five years and the current properties have been held for an average of three years, then the rate would be 13% per year (*i.e.*, 65% x $^1/_5$). Since the properties have been held for an average of three years, the desired balance in the allowance account would be determined as follows:

$100,000 x 13% x 3 years = $39,000

The lease impairment expense would be determined by subtracting the current allowance account balance of $20,000 from the desired balance of $39,000 yielding the required adjustment of $19,000. The $19,000 would be recognized as impairment expense in the current year. Alternatively, a company might amortize insignificant unproved property costs over the average primary term. In the above example, assuming an average primary term of seven years, the impairment recognized would be computed as follows:

\$100,000 x 0.65 = \$65,000	Estimated cost of unproved properties that will be abandoned eventually
\$65,000/7 = \$9,286	Average costs of unproved properties abandoned per year of primary term

Entry

Lease impairment expense	9,286	
Allowance for impairment, group basis . . .		9,286

The above method, in contrast to the first two methods discussed, emphasizes net income measurement. The net realizable value of the unproved properties is secondary. For other examples of these methods and other methods, see the article by King.

All the individually insignificant properties of a company may be amortized together in one group or the properties may be grouped into different groups, with each group being amortized separately. Groupings commonly used in practice include country by country, state by state, by geological prospect, onshore versus offshore, or by year of acquisition. In the following example, Lucky Company operates in two states, Louisiana and Texas. Because its success rate has been different in each state, Lucky decided to group its insignificant unproved properties by state.

EXAMPLE

Impairment of Insignificant Properties Grouped by State

Lucky Company keeps separate unproved property control accounts for the states of Texas and Louisiana containing properties that are not individually significant. As of December 31, 2008, the unproved property account for Texas had a \$250,000 balance and the related allowance for impairment account had a \$70,000 balance. On that date, the unproved property account for Louisiana had a balance of \$150,000 with \$30,000 in the related allowance for impairment account. Past experience indicates that 60% of all insignificant unproved properties in Texas are eventually abandoned, while 65% of Louisiana properties are abandoned. Therefore, Lucky Company has a policy of providing at year-end an allowance equal to 60% and 65% of gross unproved properties in Texas and Louisiana, respectively.

Calculations

Texas

\$250,000 x 0.60 =	\$150,000	
Less: Allowance	(70,000)	
Expense	\$ 80,000	

Allowance—Texas		
	70,000	← Have
	?	
	150,000	← Want

Louisiana		Allowance—Louisiana		
$150,000 x 0.65 =	$97,500		30,000	← Have
Less: Allowance	(30,000)		?	
Expense	$67,500		97,500	← Want

Total expense	
Texas	$ 80,000
Louisiana	67,500
Total expense	$147,500

Entry

Lease impairment expense, Texas	80,000	
Lease impairment expense, Louisiana	67,500	
Allowance for impairment, Texas group . . .		80,000
Allowance for impairment, Louisiana group		67,500

If a property is jointly owned, each working interest owner makes a separate decision as to whether the property is significant and how or whether to impair the property. Each company could legitimately compute a different amount of impairment based on their future plans for the property and their own financial accounting policies and procedures. For example, one company's geologist may, after examining data concerning the property, decide on full impairment while another company's geologist may feel the lease has potential and recommend no impairment.

Unproved property should be assessed individually or in the aggregate at least once a year. Once impairment has been recognized on a property, the property should not subsequently be written back up.

Disposition of Capitalized Costs—Surrender or Abandonment of Property

The actual abandonment of an unproved property may be accomplished in one of several ways. Failure to pay a delay rental payment when due or failure to begin drilling or production operations before the end of the primary term automatically terminates the lease. In some situations, property may be abandoned before the date of the next delay rental payment or before the end of the primary term. One reason for early abandonment would be to establish worthlessness for tax purposes (chapter 10). A formal legal document may be filed in this situation to justify the income tax deduction in the appropriate tax year.

When an individually significant unproved property is surrendered or abandoned, its capitalized acquisition costs should be charged against the related allowance for impairment account. If the allowance account is inadequate, the excess should be charged to surrendered lease expense (*SFAS No. 19*, par. 40). For properties assessed individually, abandonment results in the net carrying value (acquisition cost minus impairment allowance) of the abandoned properties being expensed.

EXAMPLE

Abandonment of Significant Properties

Lucky Company abandoned the following unproved properties, all assessed individually.

Lease	Acquisition Cost	Impairment Allowance
A	$50,000	$50,000 (100% impaired)
B	80,000	20,000 (25% impaired)
C	70,000	0

Entries

LEASE A

Allowance for impairment	50,000	
Unproved property		50,000

LEASE B

Surrendered lease expense.	60,000	
Allowance for impairment	20,000	
Unproved property		80,000

LEASE C

Surrendered lease expense.	70,000	
Unproved property		70,000

When an individually insignificant unproved property assessed on a group basis is abandoned, the cost of the unproved property being abandoned should be charged against the allowance for impairment account, group basis. The allowance account should be adequate; however, in the event it is inadequate, the excess of the abandoned property cost over the allowance balance results in a debit balance in the allowance account. To avoid reporting complications at the end of an interim period, this debit balance should be charged to surrendered lease expense and an appropriate accrual made so that a sufficient allowance balance is established. An insufficient balance in the allowance account may indicate a need for the company to reevaluate its procedure for impairment estimation.

EXAMPLE

Abandonment of Insignificant Properties

During December, Lucky Company abandoned two unproved leases that had been assessed on a group basis. Being near the end of the year, the impairment allowance for individually insignificant properties had only a $30,000 balance. Data for the abandoned leases are as follows:

Lease	Acquisition Cost
D	$10,000
E	25,000

Entries (assume Lease D was abandoned first)

LEASE D

Allowance for impairment, group basis	10,000	
Unproved property		10,000

LEASE E

Allowance for impairment, group basis	25,000	
Unproved property		25,000

The cost of the abandoned lease exceeded the balance in the allowance account by $5,000 resulting in a debit balance in the allowance account. Since the allowance account should never have a debit balance the following entry is required:

Surrendered lease expense	5,000	
Allowance for impairment, group basis . . .		5,000

After the above entry is made the allowance balance is zero. If it is determined that other leases will be surrendered before year end or that the balance in the allowance account is simply insufficient, then the following additional entry is called for. The amount is equal to the balance that should be in the allowance account at the present time.

Surrendered lease expense	XXX	
Allowance for impairment, group basis . . .		XXX

Partial abandonment may occur when a company surrenders part of a property but retains the remaining acreage. This situation occurs frequently in international operations where a contract is likely to cover a large geographical area. The agreement with the government frequently includes provisions requiring, over the exploration period, any acreage that is no longer of interest be relinquished from the contract area. The following example illustrates a partial abandonment.

EXAMPLE

Partial Abandonment

Lucky Company entered into a concession agreement with the government of China covering two 5,000 acre blocks. The contract stipulated that Lucky Company must relinquish 20% of the acreage at the end of three years. The original bonus was $5,000,000. At the end of the third year, Lucky surrendered 2,000 acres and continued to evaluate the remaining 8,000 acres.

Entries

Surrendered lease expense		
($5,000,000 x 20%).	1,000,000	
Unproved property		1,000,000

Surrender or abandonment of proved property is discussed in chapter 6.

POST-BALANCE SHEET EVENTS

Information that becomes available after the end of the period covered by the financial statements but before those financial statements are issued shall be taken into account in evaluating conditions that existed at the balance sheet date, for example, in assessing unproved properties (paragraph 28) and in determining whether an exploratory well or exploratory-type stratigraphic test well had found proved reserves (SFAS No. 19, par. 39*).*

Accordingly, a dry hole completed during the interim period (after balance sheet date but before the financial statements are issued) may indicate that the lease was impaired at year-end. (The condition, *i.e.*, no reserves, existed at balance sheet date, but was unknown at that time.) Any impairment amount should be reflected in the financial statements at fiscal year end by making an adjusting entry recognizing impairment.

EXAMPLE

Post-Balance Sheet Events

Lucky Company has unproved property costs on Lease A of $12,000 at December 31, 2007. On January 18, 2008 prior to the issuance of the audited financial statements, a well on an adjacent lease owned by another party was determined to be dry. No impairment on Lease A had been taken on December 31, 2007. Due to the dry hole on the adjacent property, management now estimates that Lease A was impaired 60% at December 31, 2007. The adjusting entry necessary for the year ending December 31, 2007, would be as follows:

Entry

Impairment expense ($12,000 x 60%)	7,200	
Allowance for impairment		7,200

Disposition of Capitalized Costs—Reclassification of an Unproved Property

When proved reserves are discovered on an unproved or undeveloped property, the property should be reclassified from an unproved property to a proved property (*SFAS No. 19*, par. 29). For a property that has been assessed individually, the net carrying value should be transferred to proved property.

EXAMPLE

Reclassification of Significant Properties

Lucky Company discovered proved reserves on the following unproved properties, both assessed individually.

Lease	Acquisition Cost	Impairment Allowance
A	$150,000	$ 0
B	100,000	25,000

Entries

LEASE A

Proved property	150,000	
Unproved property		150,000

LEASE B

Proved property	75,000	
Allowance for impairment	25,000	
Unproved property		100,000

For a property assessed on a group basis, the gross acquisition cost of the property should be transferred when reclassifying the property. The gross acquisition cost must be used due to the fact that a net carrying value cannot be determined on a separate property basis for properties assessed on a group basis.

EXAMPLE

Reclassification of Insignificant Properties

Lucky Company discovered proved reserves on a lease that had been assessed on a group basis. The acquisition cost of the property was $30,000, and the impairment allowance for individually insignificant properties had a balance of $200,000.

Entry

Proved property..	30,000	
Unproved property		30,000

For a property to be proved, it is not necessary that the entire property be classified as a proved area, only that proved reserves are found on the property. Thus, only a very small portion of a proved property may actually be a proved area. As a result, it is still possible to drill an exploratory well on a proved property.

Occasionally, a property is so large that only the portion of the property to which the proved reserves relate should be reclassified from unproved to proved. *SFAS No. 19* gives as an example a foreign concession covering a large geographical area. Multiple reservoirs may actually exist under the acreage. As a company explores and drills wells in the contract area, some acreage may become proven, other acreage may be retained for future evaluation, and still other acreage may be relinquished. This situation may also occur in other instances, for example in the exploration of certain offshore properties whose evaluation requires multiple wells.[4]

EXAMPLE

Partial Reclassification

Lucky Company has an offshore lease with the federal government that cost $20,000,000. This lease covers four tracts of 6,400 acres each. Proved reserves are found on one tract. Management decides to reclassify only one tract.

Entry

Proved property ($20,000,000 x ¼)	5,000,000	
Unproved property		5,000,000

To reinforce and tie together the material presented in this chapter, a comprehensive example is presented next.

EXAMPLE

Comprehensive Example #1

Lucky Company has the following unproved leases as of December 31, 2007.

Individually significant:

Lease	Acquisition Costs	Impairment Allowance
A	$200,000	$50,000 (25% impaired)
B	150,000	0
C	300,000	0

Individually insignificant:

Lease	Acquisition Costs	Impairment Allowance
D	$ 10,000	(not assessed
E	25,000	individually)
F	15,000	
G	30,000	
Total	$ 80,000	$5,000 (credit balance)

a. At December 31, 2007, it is determined that Lease A should be impaired an additional 30%.

Entry

Lease impairment expense ($200,000 x 0.30)	60,000	
Allowance for impairment.		60,000

b. At December 31, 2007, it is determined that Lease C should be impaired 40%.

Entry

Lease impairment expense ($300,000 x 0.40)	120,000	
Allowance for impairment.		120,000

c. At December, 31, 2007, Lucky Company impairs the insignificant unproved properties. The company's policy is to provide at year end an impairment allowance equal to 40% of the gross acquisition cost of individually insignificant properties.

Entry

Lease impairment expense	27,000	
Allowance for impairment, group basis		27,000

Calculations

$80,000 x 0.40 = $32,000

Allowance		
	5,000	← Have
	?	← 27,000
	32,000	← Balance needed

d. On March 1, 2008, Lucky Company surrendered Lease A.

Entry

Surrendered lease expense	90,000	
Allowance for impairment	110,000	
Unproved property		200,000

e. On March 20, 2008, Lucky Company acquired an unproved lease, Lease H, paying a lease bonus of $21,000 and legal costs of $1,000. The property is individually insignificant.

Entry

Unproved property	22,000	
Cash		22,000

f. On April 13, 2008, Lucky Company discovered proved reserves on Lease C.

Entry

Proved property	180,000	
Allowance for impairment	120,000	
Unproved property		300,000

g. On April 29, 2008, Lucky Company abandoned Lease E.

Entry

Allowance for impaired property, group basis	25,000	
Unproved property		25,000

h. On July 15, 2008, Lucky Company discovered proved reserves on Lease F. (The allowance for impairment, group basis account is not prorated.)

Entry

Proved property	15,000	
Unproved property		15,000

Lucky Company has the following unproved leases as of December 31, 2008, as a result of the above transactions.

Individually significant:

Lease	Acquisition Costs	Impairment Allowance
B	$150,000	$0

Individually insignificant:

Lease	Acquisition Costs	Impairment Allowance
D	$ 10,000	(not assessed
G	30,000	individually)
H	22,000	
Total	$ 62,000	$7,000

i. At December 31, 2008, it is determined that Lease B should be impaired 30%.

Entry

Lease impairment expense ($150,000 x 0.30)	45,000	
Allowance for impairment.		45,000

j. At December 31, 2008, Lucky Company impairs the insignificant unproved properties. Lucky Company retains its policy of providing at year-end a 40% impairment allowance for individually insignificant properties.

Entry

Lease impairment expense	17,800	
Allowance for impairment, group basis		17,800

Calculations

$62,000 x 0.40 = $24,800

Allowance		
	7,000	← Have
	?	← 17,800
	24,800	← Balance needed

Next, a comprehensive example is presented that ties together nondrilling exploration costs and acquisition costs with an emphasis on option costs and delinquent taxes.

EXAMPLE

Comprehensive Example #2

a. Lucky Company obtained the following rights:

Undeveloped Area	Acres	Right Acquired	Cost
A	1,000	Shooting	$ 500
B	2,000	Shooting	1,000
C	1,500	Shooting, option to lease	4,000
D	3,000	Shooting, option to lease	6,000
E	4,000	Shooting, option to lease	9,000

Entries

AREA A

G&G expense	500	
Cash		500

AREA B

G&G expense	1,000	
Cash		1,000

AREA C—Lucky Company decided to treat entire amount as option cost

Property purchase suspense	4,000	
Cash		4,000

AREA D—Lucky Company decided to treat entire amount as option cost

Property purchase suspense	6,000	
Cash		6,000

AREA E—Lucky Company decided to treat entire amount as option cost

Property purchase suspense	9,000	
Cash		9,000

b. Lucky Company hired ABC Company to perform G&G work on areas A-E and paid the company $100,000.

Entry

G&G expense	100,000	
Cash		100,000

c. As a result of the G&G work, Lucky Company decided to lease the following properties:

Lease (Area)	Acres Leased	Bonus per Acre	Total Bonus	Legal Costs, Recording Fees
A	300	$25	$ 7,500	$300
B	500	25	12,500	500
C	300	30	9,000	200
D	1,000	20	20,000	750
E	0	Option expires		
	2,100			

Entries—Lucky Company decided to apportion option cost based on acreage leased

LEASE A

Unproved property	7,800	
Cash		7,800

LEASE B

Unproved property	13,000	
Cash		13,000

LEASE C—leasing 1/5 of total acreage		
Unproved property ($9,200 + 1/5 x $4,000)	10,000	
Surrendered lease expense (4/5 x $4,000) . . .	3,200	
Property purchase suspense		4,000
Cash		9,200
LEASE D—leasing 1/3 of total acreage		
Unproved property ($20,750 + 1/3 x $6,000)	22,750	
Surrendered lease expense (2/3 x $6,000) . . .	4,000	
Property purchase suspense		6,000
Cash		20,750
LEASE E		
Surrendered lease expense	9,000	
Property purchase suspense		9,000

d. Landmen salaries and overhead relating to the above acquisitions were $10,500. The costs were allocated to each lease based on relative acres leased.

$$\frac{\$10{,}500}{2{,}100 \text{ (total acres leased)}} = \underline{\underline{\$5/\text{acre}}}$$

Entries

LEASE A		
Unproved property (300 x $5)	1,500	
Cash		1,500
LEASE B		
Unproved property (500 x $5)	2,500	
Cash		2,500
LEASE C		
Unproved property (300 x $5)	1,500	
Cash		1,500
LEASE D		
Unproved property (1,000 x $5)	5,000	
Cash		5,000

e. During the second year, Lucky Company incurred and paid the following items:

1) Ad valorem taxes of $2,500 on Lease A

Entry

Ad valorem taxes expense (exploration) . . .	2,500	
Cash		2,500

2) Dry-hole contribution, $9,000

Entry

Test-well contribution expense (exploration)	9,000	
Cash		9,000

3) Legal costs for title defense, $1,500 on Lease C

Entry

Legal expense (exploration)	1,500	
Cash		1,500

Note: In contrast, legal fees in connection with a title exam would be an acquisition cost and would be capitalized.

4) Bottom-hole contribution, $3,000

Entry

Test-well contribution expense (exploration)	3,000	
Cash		3,000

5) Nonrecoverable delinquent taxes on Lease A, $1,200 (specified in lease contract)

Entry

Unproved property (acquisition)	1,200	
Cash		1,200

6) Recoverable delinquent taxes on Lease B, $1,800

Entry

Receivable from lessor	1,800	
Cash		1,800

7) Delay rentals on Leases A, C, and D, $6,000

Entry

Delay rental expense (exploration)	6,000	
Cash		6,000

8) Delay rental on Lease B, $2,000

Entry

Delay rental expense (exploration)	2,000	
Receivable from lessor (delinquent taxes)		1,800
Cash		200

f. At the end of the second year, Lucky Company assessed the leases with the following results (assume all of the leases were individually significant):

Lease	Balance in Unproved Property	Percentage Impairment	Amount Impaired	Net CV After Impairment
A	$10,500 ($7,800 + $1,500 + $1,200)	25%	$ 2,625	$ 7,875
B	15,500 ($13,000 + $2,500)	0	0	15,500
C	11,500 ($10,000 + $1,500)	30	3,450	8,050
D	27,750 ($22,750 + $5,000)	20	5,550	22,200

Entries

LEASE A

Lease impairment expense ($10,500 x 25%)	2,625	
Allowance for impairment		2,625

LEASE B

No entry

LEASE C

Lease impairment expense ($11,500 x 30%)	3,450	
Allowance for impairment		3,450

LEASE D

Lease impairment expense ($27,750 x 20%)	5,550	
Allowance for impairment		5,550

g. Early in the third year, Lucky Company abandoned Leases C and D.

Entries

LEASE C

Surrendered lease expense	8,050	
Allowance for impairment	3,450	
Unproved property		11,500

LEASE D

Surrendered lease expense	22,200	
Allowance for impairment	5,550	
Unproved property		27,750

h. Also in the third year, Lucky Company discovered proved reserves on Leases A and B.

Entries

LEASE A

Proved property	7,875	
Allowance for impairment	2,625	
Unproved property		10,500

LEASE B

Proved property	15,500	
Allowance for impairment.	0	
Unproved property		15,500

LAND DEPARTMENT

The land department of an oil company usually is responsible for property acquisition and property administration. The exploration and legal departments are also concerned with these functions. The exploration department is responsible for recommending property acquisition, retention, and development or abandonment. The legal department conducts title examinations and title litigation and approves or prepares any legal documents involved.

The land department acts on information obtained from the exploration department's activities and from land department scouts in acquiring properties. Subscription services and information exchanges provide additional information to the land department. The actual acquisition of properties is negotiated by **landmen**. Landmen also promote trades, joint ventures, unitizations, and various types of sharing arrangements.

The land department is responsible for recording and maintaining basic records on properties, ensuring that all contractual obligations in the lease contract are fulfilled, and preparing various types of reports. An example of an important contractual obligation is the delay rental payment. It is very important that the record system give adequate notice of leases due for rental payment because failure to pay delay rentals when due will result in the lease terminating.

OFFSHORE OPERATIONS

The accounting for offshore acquisition costs are handled in the same manner as for onshore properties. The lease bonus for an offshore tract paid to a federal or state government is usually quite large. Millions of dollars for one offshore block is not uncommon. Because both the bonus paid and the amount of acreage leased is large, the properties are normally considered individually significant for accounting purposes.

PROBLEMS

1. Define the following:
    - top lease
    - bonus
    - option to lease
    - purchase in fee
    - delinquent taxes paid by the lessee
    - impairment
    - internal costs

2. On December 31, 2006, Lotus Oil Company recognized impairment of $100,000 on an individually significant lease. Before the financial statements were issued early the next year, a well was drilled and proved reserves were found. Lotus easily revised their financial statements so that no impairment was recognized on the property. Please comment.

3. Green Company, which uses the successful-efforts method of accounting, owns an individually significant lease, with a cost of $200,000. On December 31, 2007, the lease is not considered impaired. However, prior to completion of the audit, a well on adjacent property is abandoned as a dry hole and the lease is now considered to be 40% impaired. Prepare any necessary adjusting entry.

4. Macon Oil Company paid $500/acre for a lease with a three-year primary term. During the next two years Macon drilled three dry holes on the property. With one more year of the primary term left, Macon still intends to try one more time. Should any impairment be recognized on the property? What if the current going rate for leases near Macon's lease is $600/acre?

5. Fossil Oil Company began operations in 2006 with the acquisition of four undeveloped leases, all individually significant. Give the entries assuming the following transactions. For simplicity, you may combine entries for the different leases for all items marked with an asterisk (*).

Year	Transaction	Lease A	Lease B	Lease C	Lease D
2006	Shooting rights*	$20,000	none	none	$15,000
	G&G costs, broad*	50,000	none	none	90,000
	Lease bonuses*	30,000	$40,000	$60,000	50,000
	G&G costs, detailed*	65,000	30,000	45,000	35,000
	Dry-hole contributions paid*	15,000	none	none	25,000
	Legal costs, title exams*	1,000	5,000	4,000	10,000
	Legal costs, title defense	none	none	50,000	none
	Delinquent taxes (in contract and nonrecoverable)	25,000	none	none	none
Dec 31	Impairment	10%	25%	—	40%
2007	Delay rentals*	2,000	10,000	3,000	8,000
	Property tax*	3,000	none	5,000	none
	Miscellaneous	abandoned lease	drilled & found oil	—	+ 10% impaired

6. On December 31, 2005, Artificial Oil Company's unproved property account for leases that are not individually significant had a balance of $800,000. The impairment allowance account had a balance of $75,000. Give the entries for each of the following transactions occurring in 2005, 2006, and 2007. (All transactions concern insignificant unproved leases.)
   a. Assuming Artificial has a policy of maintaining a 55% allowance, *i.e.*, 55% of gross unproved properties, give the entry to record impairment on December 31, 2005.
   b. During 2006, Artificial surrendered leases that cost $300,000.
   c. During 2006, leases that cost $50,000 were proved.
   d. During 2006, leases costing $310,000 were acquired.
   e. Give the entry to record impairment on December 31, 2006.
   f. During 2007, leases costing $428,000 were surrendered.

7. Given the following data for Gravity Oil Company:

Unproved property—Lease A (significant)	$700,000
Allowance for impairment—Lease A	500,000
Unproved property—Lease B (insignificant)	30,000
Allowance for impairment, group basis	450,000

a. Give the entries to record the abandonment of both Lease A and Lease B.
b. Give the entries assuming instead that both Lease A and Lease B were proved, *i.e.*, oil or gas was discovered on the leases.

8. Sure Thing Oil Company's internal land department incurred costs of $150,000 in acquiring leases. Of the one million acres of prospects, only 450,000 acres were leased.
   a. How much, if any, of the $150,000 incurred by the land department should be capitalized?
   b. If capitalized, what account(s) should be debited?

9. Shut-In Oil Company obtained shooting rights only for $10,000 on 5,000 acres owned by Mr. Q and shooting rights coupled with an option to lease for $12,000 on 4,000 acres owned by Mr. S. The 4,000 acres owned by Mr. S are located adjacent to the 5,000 acres owned by Mr. Q. Ignore any other acquisition costs.
   a. Give the entries to record the rights obtained, assuming there is no apportionment of the cost between the option and the shooting rights.
   b. Give the entry to record the rights obtained from Mr. S, assuming *instead* that the $12,000 was apportioned between the option and the shooting rights.
   c. Give the entry to record the leasing of all 4,000 of Mr. S's acres, assuming that the original cost of $12,000 was *not* apportioned between the option and the shooting rights.
   d. Give the entry to record the leasing of only 1,000 acres from Mr. S, again assuming that the original cost of $12,000 was not apportioned. Also assume Shut-In did not apportion the amount in the suspense account based on the acreage leased.
   e. Give the entry to record the leasing of 1,000 acres from Mr. S, assuming that the original cost of $12,000 was not apportioned. Assume that Shut-In Oil Company apportioned the amount in the suspense account based on relative acreage leased.

10. The following transactions relate to one lease:
   a. On March 10, 2007, Core Oil Company paid delinquent property taxes of $2,000 on an undeveloped lease. Assume that these taxes are recoverable out of future delay rental or royalty payments. Give the entry to record payment.
   b. On February 15, 2008, a delay rental payment of $800 is due. Determine the amount of cash actually paid and give the entry to record payment.
   c. On July 21, 2008, Core Oil Company decided to surrender the lease. Give the entry to record abandonment with respect to the delinquent property taxes. Ignore acquisition costs of the property.
   d. Assume instead that the $2,000 payment of delinquent taxes was not recoverable and was made at the time Core Oil was acquiring the lease. Give the entry to record the payment.
   e. Assume instead that the $2,000 payment was not recoverable and was made by Core Oil six months after acquiring the lease in order to protect Core's investment. Give the entry to record the payment.

11. During 2005, Fortunate Oil Company acquired the following leases:

Lease	Acres	Bonus/Acre
A	2,000	$50
B	3,000	60
C	5,000	40

In acquiring and exploring these leases, Fortunate Oil Company incurred the following additional costs:

Shooting rights,	$ 3,000
Salaries for G&G exploration activities,	40,000
Mapping costs for exploration activities,	10,000
Salary of in-house lawyer working on lease acquisition,	10,000
Minor repairs of G&G exploration equipment,	500
Salaries of landmen working on lease acquisition,	30,000

Fortunate Oil allocates internal costs relating to lease acquisition to specific leases. Assuming Lease A was abandoned at the end of the year, answer the following questions:

a. What was the total nondrilling exploration expense for all three leases for the year?
b. What was the surrendered lease expense?
c. How much was capitalized as unproved property for Lease B?

Hint: Some of these costs must be allocated to the individual leases on some reasonable basis.

12. Porter Oil Company acquired a lease on October 15, 2005, for $200,000 cash. No drilling was done on the lease during the first year. Since Porter wished to retain the lease, a delay rental of $10,000 was paid on October 15, 2006. During November and December of 2006, three dry holes were drilled on surrounding leases. Based on the dry holes, Porter's management decided that the lease was 75% impaired. Porter had still not started drilling operations by the end of the second year and so paid a second delay rental. During November 2007, with less than one year of the primary term left, Porter drilled a dry hole on the lease and decided to abandon the lease. Because the end of Porter's accounting period is December 31 and for income tax purposes, Porter executed a quit claim deed and relinquished all rights to the lease the last day of November 2007. Give the entries.

13. Hard Luck Oil Company owned the following unproved property as of the end of 2005.

Significant Leases		Insignificant Leases	
Lease A	$300,000	Lease C	$ 50,000
Lease B	350,000	Lease D	25,000
Total	$650,000	Lease E	40,000
		Lease F	30,000
		Total	$145,000

Although no activity took place on Lease A during the year, Hard Luck decided that Lease A was not impaired because there were still three years left in that lease's primary term. Two dry holes were drilled on Lease B during the year; but because Hard Luck intended to drill one more well on Lease B in the coming year, it decided that Lease B was only 40% impaired. With respect to the insignificant leases, past experience indicates that 70% of all unproved properties assessed on a group basis will eventually be abandoned. Hard Luck's policy is to provide at year-end an allowance equal to 70% of the gross cost of these properties. The allowance account had a balance of $20,000 at year end. Give the entries to record impairment.

14. Mabel Oil Company decided to explore some acreage in Wyoming before acquiring any leases. Mabel acquired shooting rights only on 15,000 acres owned by Mr. T for $0.10/acre. Mabel obtained shooting rights coupled with an option to lease on 10,000 acres owned by Mr. H for $0.25/acre. After completing G&G surveys at a cost of $85,000, Mabel decided to lease 5,000 acres from Mr. T, paying a bonus of $35/acre. Mabel also decided to lease 5,000 acres from Mr. H, paying the same bonus of $35/acre. Mabel's income statement has seen better days, and although it won't help much, Mabel decides to capitalize every possible cost. Give the entries.

15. Tiger Oil Company has two unproved leases with the following capitalized costs:

Lease C (600 acres)	$20,000
Lease D (1,200 acres)	$50,000

Tiger Oil Company negotiated a new lease on Lease C immediately following the end of the primary term. The lease bonus was $60 per acre on the new lease. Before the end of the primary term, Tiger also obtained a new lease on Lease D at a lease bonus rate of $40 per acre. This lease was to take effect prior to the end of the primary term. Prepare journal entries for the two leases.

16. Aggie Company purchased land in fee for $100,000. The fair market values of the surface and mineral rights were determined by a qualified appraiser as follows:

Surface rights	$ 80,000

Mineral rights	40,000
	$120,000

Prepare a journal entry to record the purchase.

17. Garza Company purchased three leases as follows:

July 1, 2004	Lease A	$100,000
August 15, 2004	Lease B	200,000
October 10, 2004	Lease C	300,000

All the leases are classified as significant.

a. On December 31, 2004, Lease A is determined to be 25% impaired. Leases B and C are not impaired.
b. On December 31, 2005, Lease A is determined to be impaired a total of 75% and Lease C 60%. Lease B is not impaired.
c. On December 31, 2006, Lease A is considered to be 100% impaired and is abandoned. Lease B is 30% impaired and a well on Lease C found proved reserves.

Prepare journal entries for all the above transactions except the initial purchase.

18. Universal Oil Company has an offshore lease that cost $4,000,000. The total acreage is 40,000 acres. Proved reserves are found on a 10,000 acre tract that is included in the 40,000 acres. Management decides to reclassify only the proved area acreage. Prepare the entry to reclassify the acreage.

19. Ebert Company leased 10,000 acres with a lease bonus of $80 per acre. Delay rentals are to be at a rate of $5 per acre. The lease specified that Ebert Company could abandon the lease in 1,000 acre portions. At the end of year 1, Ebert Company abandoned 4,000 acres and paid a delay rental on the remainder. Prepare journal entries for the above activities.

20. Wildcat Company's balance sheet, at 12/31/04, included account balances as follows:

	Cost	**Allowance for impairment**
Lease A	$ 40,000	$ 10,000
Lease B	80,000	24,000
Lease C	100,000	0
Insignificant properties (D-J)	200,000	120,000

During 2005, the following events related to the above unproved properties occurred:

a. Lease A is abandoned.

b. Lease B is surrendered.
c. Leases G & F (insignificant) in the amounts of $2,000 and $3,000, respectively, are abandoned.

Prepare the necessary entries.

21. King Oil Company owned the following unproved property at 12/31/04:

	Significant Leases			**Insignificant Leases**	
	Cost	Impairment		Cost	Total Impairment
Lease A	$300,000	$200,000	Lease C	$20,000	
Lease B	400,000	100,000	Lease D	30,000	
			Lease E	15,000	
			Lease F	10,000	
				$75,000	$32,000

Prepare journal entries for 2005, assuming the following events:

a. Found proved reserves on Leases A & B.
b. Found proved reserves on Lease D.
c. Found proved reserves on Lease C.

22. Willis Company has the following groups of insignificant leases at 12/31/05.

	Group A	**Group B**	**Group C**
Total costs	$200,000	$300,000	$400,000
Total allowance for impairment 12/31/05	40,000	0	80,000
Expected average percentage of impairment	60%	70%	65%

Prepare journal entries to record impairment for each of the groups at 12/31/05.

23. Give the entry to record abandonment in each of the following cases.
    a. Mabel Oil Company abandoned an unproved property that cost $150,000. The property was considered significant and had been impaired $100,000 on an individual basis.
    b. Mabel also abandoned an unproved property that was not considered significant. The property cost $10,000 and the group allowance account had a balance of $400,000 at the time of the abandonment.

24. Glass Company purchased land in fee for $420,000. The land was located in a desolate

portion of West Texas. An appraiser estimated the fair market value of the surface rights to be $200,000. The appraiser was not able to make an estimate of the value of the mineral rights.

Prepare a journal entry to record the purchase.

25. On January 1, 2000, Worldwide Oil Company entered into a concession agreement with the government of Egypt and paid a $3,000,000 signing bonus. The agreement covers 20,000 acres, has a term of five years and requires that exploration begin immediately. At the end of the first three years, Worldwide is required to begin relinquishing acreage at a rate 25% of the contract area per year. However, Worldwide is not required to relinquish proved acreage. On July 16, 2002, Worldwide makes a commercial discovery and determines that a 1,000 acre block is proved. On December 31, 2002, 25% of the initial contract area is relinquished. Worldwide estimates that only another 25% of the original contract acreage will be relinquished.

    Prepare all journal entries that would be required to account for the concession area from January 1, 2000 through December 31, 2002.

NOTES

1. King, Barry G., "Impairment Test for Unproved Properties," *Journal of Extractive Industries Accounting*, Summer, 1982, p. 82.

2. PricewaterhouseCoopers, *Survey of U.S. Petroleum Accounting Practices*, Denton, Texas: Institute of Petroleum Accounting, 1999, p. 66.

3. King, Barry G., "Impairment Test for Unproved Properties," *Journal of Extractive Industries Accounting*, Summer, 1982, pp. 79-89.

4. Moore, Cecil H., and Grier, James, *Accounting Standards and Regulations for Oil and Gas Producers*, Englewood Cliffs, New Jersey: Prentice-Hall, Inc., 1983, p. 707.

chapter five

Drilling *and* Development Costs Under SE

Chapter 3 discussed the *nondrilling* types of exploration costs. This chapter discusses the two types of exploration *drilling* costs and both drilling and nondrilling development costs.

Under SE accounting, a direct relationship is required between costs incurred and specific reserves discovered before costs are ultimately identified as assets. Consequently, with respect to exploration costs, only successful exploration costs incurred in the search for oil and gas are considered to be a part of the cost of finding oil or gas. In contrast to exploration costs, the purpose of development activities is considered to be building a producing system of wells and related equipment and facilities rather than searching for oil and gas. Thus, both successful and unsuccessful development costs are capitalized as part of the cost of the oil or gas.

The board justified the different treatment of exploratory dry holes and development dry holes in paragraphs 205 and 206 of *SFAS No. 19* as follows:

> *In the Board's judgment, however, there is an important difference between exploratory dry holes and development dry holes. The purpose of an exploratory well is to search for oil and gas. The existence of future benefits is not known until the well is drilled. Future benefits depend on whether reserves are found. A development well, on the other hand, is drilled as part of the effort to build a producing system of wells and related equipment and facilities. Its purpose is to extract previously discovered proved oil and gas reserves. By definition (Appendix C, paragraph 274), a development well is a well drilled within the proved area of a reservoir to a depth known to be productive. The existence of future benefits is dis-*

> *cernible from reserves already proved at the time the well is drilled. An exploratory well, because it is drilled outside a proved area, or within a proved area but to a previously untested horizon, is not directly associable with specific proved reserves until completion of drilling. An exploratory well must be assessed on its own, and the direct discovery of oil and gas reserves can be the sole determinant of whether future benefits exist and, therefore, whether an asset should be recognized. Unlike an exploratory well, a development well by definition is associable with known future benefits before drilling begins. The cost of a development well is a part of the cost of a bigger asset—a producing system of wells and related equipment and facilities intended to extract, treat, gather, and store known reserves.*
>
> *Moreover, because they are drilled only in proved areas to proved depths, the great majority of development wells are successful; a much smaller percentage (22 percent in the United States in 1976), as compared to exploratory wells (73 percent in the United States in 1976) are dry holes. Development dry holes occur principally because of a structural fault or other unexpected stratigraphic condition or because of a problem that arose during drilling, such as tools or equipment accidentally dropped down the hole, or simply the inability to know precisely the limits and nature of a proven reservoir. Development dry holes are similar to normal, relatively minor "spoilage" or "waste" in manufacturing or construction. The Board believes that there is a significant difference between the exploration for and the development of proved reserves. Therefore, in the Board's judgment, it is appropriate to account for the costs of development dry holes different from exploratory dry holes.*

Thus, the cost of drilling an exploratory well is expensed if the well is dry and capitalized if the well is successful, while the cost of drilling a development well is always capitalized, regardless of the outcome of the well. Despite this disparity between the final accounting treatment for exploratory drilling and development drilling, preliminary accounting for the costs of drilling a well, *i.e.*, accounting for drilling in progress, is the same regardless of whether the well is an exploratory well or a development well.

Before discussing the financial accounting treatment of these costs in detail, however, the tax treatment must be briefly discussed. The tax treatment of drilling and development costs normally affects the accounts used in financial accounting systems because these costs must be recorded for financial reporting in enough detail to accumulate and provide the information required for tax reporting.

Intangible Drilling and Development Costs versus Equipment Costs

Drilling and development costs are classified as either **intangible drilling and development costs** (IDC) or **equipment costs** (lease and well equipment). The distinction between IDC and equipment costs is very important because for tax purposes all or most of the IDC may be expensed as incurred, whereas equipment costs must be capitalized and depreciated. (The taxpayer makes a one-time election to either expense or capitalize IDC. Regardless of whether the election is to expense or to capitalize IDC, IDC of dry holes may be expensed. This election is per taxpayer and applies to all the taxpayer's properties. Rules for integrated producers are modified somewhat, in that a specified percentage of all productive IDC must be capitalized, regardless of the election made. In contrast, independent producers may deduct all of the IDC in the year incurred. See chapter 10.)

IDC is defined as expenditures for drilling and developing that in themselves do not have a salvage value and are "incident to and necessary for the drilling of wells and the preparation of wells for the production of oil and gas" [U.S. Treas. *Reg. Section 1*, 612-4(a)]. In general, IDC includes the intangible or nonsalvageable costs of drilling up to and including the cost of installing the Christmas tree. The term **Christmas tree** refers to the valves, pipes, and fittings assembly that is used to control the flow of oil and gas from the wellhead. In many cases, the physical arrangement of the valves, pipes, and fittings resembles a Christmas tree. The Christmas tree is located at the top of the well on the surface of the land.

In general, equipment costs include all tangible or salvageable costs of drilling and development up to and including the Christmas tree, plus both intangible and tangible costs past the Christmas tree. In considering whether an item is before or after the Christmas tree, the physical flow of the oil and gas should be considered, not the time of incurrence. Note that neither the word "salvageable" nor the word "tangible" is completely correct in defining which costs are IDC rather than equipment. Some tangible costs such as casing, which are usually not salvageable, are still considered equipment. On the other hand, other tangible costs such as cement or drilling mud, which also are not salvageable, are considered to be IDC. The distinction between IDC and equipment costs appears to be whether a tangible cost in itself has a salvage value.

IDC and equipment costs may be diagramed as follows:

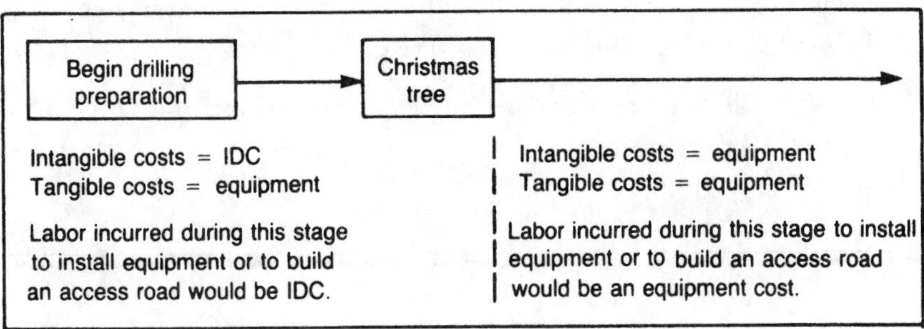

As can be seen from the preceding diagram, the purpose for constructing a lease road determines whether the costs of constructing the road are considered IDC.

EXAMPLE

IDC versus Equipment

a. Lucky Company incurred acquisition (purchase) costs of $40,000 for casing and installation costs of $5,000. Casing is subsurface well equipment and is, therefore, before the Christmas tree.

IDC = $5,000, installation costs
Equipment = $40,000, purchase cost

b. Lucky Company purchased flow lines and storage tanks at a cost of $10,000. Installation costs were $2,000. Flow lines and tanks are nonwell equipment and are past the Christmas tree.

IDC = $0, all costs are considered to be equipment costs since these costs are past the Christmas tree
Equipment = $12,000, purchase cost and installation costs

c. Lucky Company incurred $15,000 in labor costs in building an access road to a drillsite. This cost is related to drilling a well and is, therefore, before the Christmas tree.

IDC = $15,000
Equipment = $0

d. Lucky Company incurred $15,000 in labor costs in building an access road to a producing well. This cost is past the Christmas tree.

IDC = $0, all costs are considered to be equipment costs since this cost is past the Christmas tree.
Equipment = $15,000

Items 1 and 2 below list costs related to drilling an exploratory or development well that are usually considered IDC. These costs are given in the order in which they are usually incurred. Note that the costs listed in Item 2 are *past* the Christmas tree, but because they are intangible and directly related to the process of drilling a well, the costs are considered IDC. Item 3 lists other types of wells for which IDC is incurred.

1. Up to and including installation of the Christmas tree
   a. Prior to drilling
      - G&G to determine specific drillsite
      - Preparation of the site, such as leveling, clearing, and building access roads and disposal pits
      - Rigging up
   b. During drilling
      - Drilling contractor's charges (when the drilling contractor furnishes equipment such as casing, part of the charge is equipment)
      - Drilling mud, chemicals, cement, and supplies
      - Wages and fuel
      - Well testing such as core analysis and analysis of cuttings
   c. At target depth and during completion
      - Well testing such as well logs and drillstem tests
      - Perforating and cementing
      - Swabbing, acidizing, and fracturing
      - Installing subsurface equipment to the wellhead and installation of the Christmas tree
      - If the well is dry, plugging and abandoning costs

2. After the Christmas tree, following completion
   a. Removal of drilling rig
   b. Restoration of land and damages paid to the surface owner

3. Wells other than original exploration or development wells
   a. Intangible costs (those listed above) incurred in deepening a well
   b. Intangible costs incurred in drilling a water or gas injection well
   c. Intangible costs of drilling a water supply or injection well where water is to be used for drilling an exploration or development well or for injection.

In financial accounting, the cost of a machine, for example, includes not only the purchase cost, but also the intangible costs necessary to get the machine up and running. Thus, for financial accounting purposes, the cost of a well includes both the tangible and intangible costs. Consequently, the distinction between IDC and equipment costs has no meaning for financial accounting. However, because of the importance of the classification of costs as either intangible drilling and development costs or equipment costs for tax, the distinction between IDC and equipment costs is usually made in financial accounting as well as tax accounting.

EXPLORATORY DRILLING COSTS

Figure 5–1 diagrams the accounting treatment of all five types of exploration costs. The first three types are nondrilling costs and are discussed in chapter 3. The last two types are (1) the costs of drilling and equipping exploratory wells and (2) the costs of drilling exploratory-type stratigraphic test wells. These drilling exploration costs are discussed in this chapter, and are defined by the SEC in *Reg. SX 210.4-10 (Reg. 4-10)* as follows:

> ***Exploratory well.*** *A well drilled to find and produce oil or gas in an unproved area, to find a new reservoir in a field previously found to be productive of oil or gas in another reservoir, or to extend a known reservoir. Generally, an exploratory well is any well that is not a development well, a service well, or a stratigraphic test well as those items are defined.*
>
> ***Stratigraphic test well.*** *A drilling effort, geologically directed, to obtain information pertaining to a specific geologic condition. Such wells customarily are drilled without the intention of being completed for hydrocarbon production. This classification also includes tests identified as core tests and all types of expendable holes related to hydrocarbon exploration. Stratigraphic test wells are classified as (i)"exploratory-type," if not drilled in a proved area or (ii)"development-type," if drilled in a proved area.*

Exploratory wells and exploratory-type stratigraphic test wells are accounted for as follows:

> *The cost of drilling exploratory wells and the costs of drilling exploratory-type stratigraphic test wells shall be capitalized as part of the enterprise's uncompleted wells, equipment, and facilities, pending determination of whether the well has found proved reserves. If the well has found proved reserves (paragraphs 31-34), the capitalized costs of drilling the well shall become part of the enterprise's wells and related equipment and facilities (even though the well may not be completed as a producing well); if, however, the well has not found proved reserves, the capitalized costs of drilling the well, net of any salvage value, shall be charged to expense (SFAS No. 19, par. 19).*

The following example illustrates typical drilling and completion costs and their accounting treatment under SE. As shown in the example, drilling costs are temporarily classified as (1) wells in progress—IDC or (2) wells in progress—lease and well equipment. (The term *wells in progress* is used in place of the FASB terminology *uncompleted wells, equipment, and facilities.* Note that wells in progress therefore includes not only unfinished wells, but also unfinished equipment and facilities.)

When drilling reaches the targeted depth, a decision must be made as to whether the well has found proved reserves. If proved reserves have been found, both wells in progress account balances will be transferred to wells and related equipment and facilities accounts. In addition, if the well is the first successful exploratory well drilled on the property, the unproved property account will be reclassified or transferred into a proved property account because proved reserves have been found and are now attributed to that property.

If proved reserves have not been found, the well must be plugged and abandoned. Equipment in the hole is salvaged when possible. However, installed casing cannot usually be removed because of either regulatory requirements or physical constraints. If the well is an

Fig. 5–1 — *Successful Efforts, Exploration Costs*

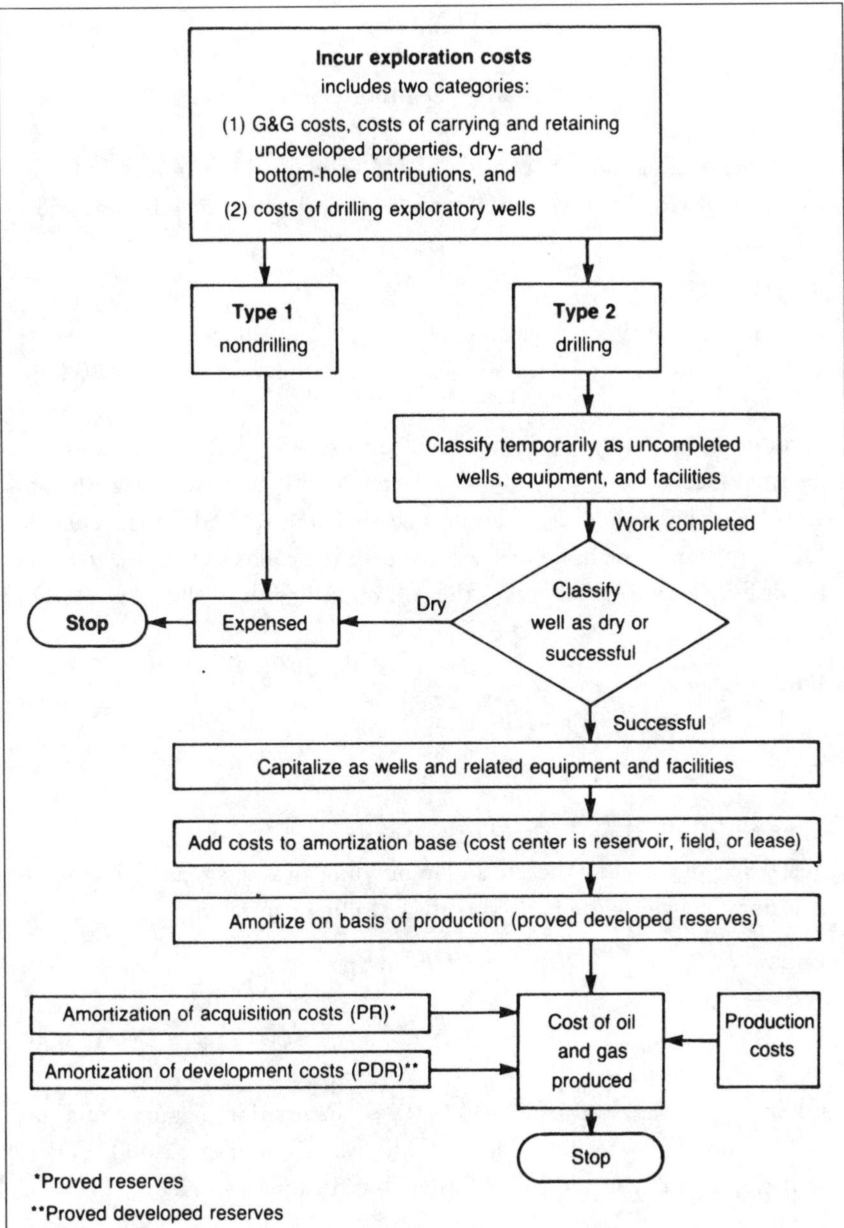

exploratory well as in the following example, the costs of plugging and abandoning, in addition to the capitalized costs in the wells in progress accounts (net of any salvaged equipment), must be written off to dry-hole expense. If the lease is also abandoned, the net amount capitalized as unproved property will be written off as surrendered lease expense or charged to the allowance account, depending upon whether the property is significant or insignificant.

EXAMPLE

Exploratory Drilling Costs

a. On January 2, 2008, as a result of G&G work done in 2007, Lucky Company decided to lease 1,000 acres at $20 per acre. The lease was undeveloped.

Entry

Unproved property	20,000	
Cash		20,000

b. Early in 2008, Lucky Company decided to begin drilling operations and incurred G&G costs of $5,000 to select a specific drillsite. (Even though G&G work has been done to locate a possible reservoir, additional detailed G&G work must be done to select the drillsite. This G&G work to select the drillsite is considered part of the cost of drilling the well and not G&G expense.)

Entry

Wells in progress (W/P)—IDC	5,000	
Cash		5,000

c. In preparing the drilling site, Lucky Company incurred costs of $12,000 in clearing and leveling the site and in building an access road. These activities normally would be performed by a drilling contractor.)

Entry

W/P—IDC	12,000	
Cash		12,000

d. Additional preparation costs of $4,000 were incurred in digging a mud pit and installing a water line. Pipes for the water line cost $2,000. (These activities normally would be performed by a drilling contractor.)

Entry

W/P—IDC	4,000	
W/P—lease and well equipment (L&WE)	2,000	
Cash		6,000

e. Lucky Company purchased pipe and casing for the well at a cost of $6,000.

Entry

W/P—L&WE	6,000	
Cash		6,000

f. Lucky Company had hired a drilling contractor on a footage-rate contract and, as is usual in such an agreement, payment was contingent upon the contractor drilling to a specified depth. The well was spudded (*i.e.,* drilling was begun) early in June and contract depth was reached in late July. Lucky Company paid the contractor $140,000. (In a footage-rate contract, a specified amount is paid per foot drilled. In contrast, in a day-rate contract, a specified amount is paid per day, typically a different amount for a drilling day versus a standby day.)

Entry

W/P—IDC	140,000	
Cash		140,000

g. In evaluating the well, Lucky Company incurred costs of $8,000. A well log was run and a drillstem test was made.

Entry

W/P—IDC	8,000	
Cash		8,000

h. Based on the well log and drillstem test, as well as other tests performed as the well was drilled, Lucky Company decided to complete the well. Casing was set (*i.e.,* installed by cementing between the pipe and well bore) at a cost of $35,000 for casing for the well and $6,000 for cementing services.

Entry

W/P—IDC	6,000	
W/P—L&WE	35,000	
Cash		41,000

i. Lucky incurred acquisition costs of $8,000 and installation costs of $1,000 for a string of production tubing through which the oil and gas will be produced. (Although oil and gas can be produced through casing, tubing is usually used because it is much easier than casing to remove and repair.)

Entry

W/P—IDC	1,000	
W/P—L&WE	8,000	
Cash		9,000

j. Lucky Company incurred acquisition (purchase) costs of $5,000 for a Christmas tree and installation costs of $3,000.

Entry

W/P—IDC	3,000	
W/P—L&WE	5,000	
Cash		8,000

k. Lucky Company incurred $4,000 for perforating and acidizing services. (Perforating involves using a perforating gun to make perforations or holes through the casing and cement so that oil and gas can flow from the formation into the well bore. Acidizing is a method used to increase the permeability of the formation by introducing acid into the formation.)

Entry

W/P—IDC	4,000	
Cash		4,000

l. The work on the well is finished and proved reserves have been found. Two entries are necessary: one to transfer the cost of the well from an unfinished goods account to a finished goods account and one to reclassify the lease as proved.

Entry *l–1*

Wells and related equipment and facilities (E&F)—IDC	183,000	
Wells and related E&F—L&WE	56,000	
W/P—IDC		183,000
W/P—L&WE		56,000

W/P—IDC

	Debit	Credit
b.	5,000	
c.	12,000	
d.	4,000	
f.	140,000	
g.	8,000	
h.	6,000	
i.	1,000	
j.	3,000	
k.	4,000	
	183,000	

W/P—L&WE

	Debit	Credit
d.	2,000	
e.	6,000	
h.	35,000	
i.	8,000	
j.	5,000	
	56,000	

Entry *l–2*

Proved property	20,000	
Unproved property		20,000

m. Lucky Company purchased pipes (flow line to lease tanks), storage tanks and separators (separates the gas from the oil) for a cost of $15,000. Installation costs were $1,000.

Entry

Wells and related E&F—L&WE	16,000	
Cash		16,000

Note that the cost center under SE is a lease or field, not an individual well.

n. If instead, after evaluating the well in part *g*, Lucky Company had decided the well was dry, only costs in parts *a-g* would have been incurred and the entry to record the dry hole would have been:

Entry

Dry hole expense—IDC	169,000	
Dry-hole expense—L&WE	8,000	
W/P—IDC		169,000
W/P—L&WE		8,000

o. Costs of $2,000 were incurred in plugging and abandoning the hole.

Entry

Dry-hole expense—IDC	2,000	
Cash		2,000

It is important to distinguish between abandonment of the well and abandonment of the lease. In this case only the well has been abandoned and therefore no entry would be made relating to the lease, *i.e.,* the unproved or proved property account.

Development Drilling Costs

Development wells are defined by the SEC in *Reg. SX 4-10* as follows:

> ***Development well.*** *A well drilled within the proved area of an oil or gas reservoir to the depth of a stratigraphic horizon known to be productive.*

Although a reservoir is discovered by exploratory wells, drilling additional development wells is typically necessary to produce the reservoir at a satisfactory rate. Development wells may not significantly increase the total amount of oil and gas that is ultimately recoverable from a formation, however, the additional wells do increase the ultimate rate of recovery.

Development costs are accounted for as follows:

> *Development costs shall be capitalized as part of the cost of an enterprise's wells and related equipment and facilities. Thus, all costs incurred to drill and equip development wells, development-type stratigraphic test wells, and service wells are development costs and shall be capitalized, whether the well is successful or unsuccessful. Costs of drilling those wells and costs of constructing equipment and facilities shall be included in the enterprise's uncompleted wells, equipment, and facilities until drilling or construction is completed (SFAS No. 19, par. 22).*

Most of the entries in the previous example (entry *a,* recording property acquisition, through entry *l–1,* classifying the exploratory well as completed and successful) would have been the same if the well had been a development well. Only three entries—*l–2, n,* and *o*—would have been different. If the well had been a development well instead of an exploratory well, it would have, by definition, been drilled in a proved area. Therefore entry *l–2,* reclassifying the acquisition cost of the property from the unproved to proved category, would not have been necessary. Instead, the reclassification to proved property would have been made at an earlier time when proved reserves were first discovered on the property.

For entries *n* and *o* the well was assumed to be a dry hole. Entries *n* and *o* transferred to dry hole expense the cost of drilling, evaluating, and plugging and abandoning the well. If the well had been a development well, these costs would have been capitalized, regardless of the well's outcome, as wells and related equipment and facilities.

Figure 5–2 outlines the accounting for development costs. A brief example that illustrates the accounting for a development well follows.

Fig. 5–2 — *Successful Efforts, Development Costs*

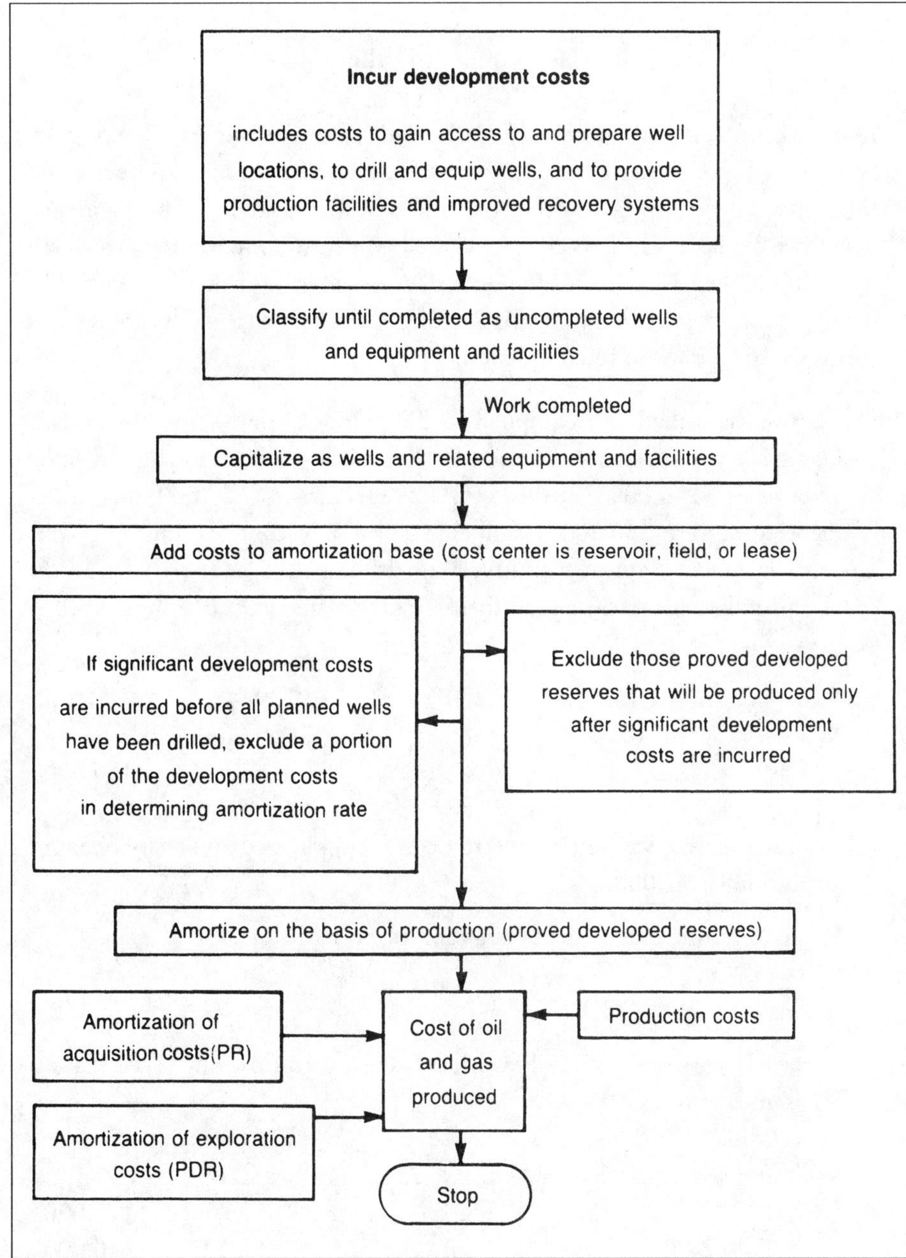

EXAMPLE

Development Drilling

During 2006 and 2007, Lucky Company drilled several successful exploratory wells on Lease A. As a result, Lease A was classified as a proved property and the estimated boundaries of the reservoir were delineated. Lucky Company decided in 2008 to drill an additional well within the proved area—a development well—and hired a drilling contractor under a turnkey contract. The drilling contract specified that the contractor was to perform all services and furnish all materials up to completion.

a. The well is drilled and equipped to the point of completion, and Lucky Company pays the contractor under a turnkey contract the agreed-upon amount of $150,000. Of this $150,000, IDC was $120,000 and equipment costs were $30,000. (Unlike a footage-rate contract and a day-rate contract, under a turnkey contract, the drilling contractor assumes all the responsibility and furnishes all the necessary materials and equipment.)

Entry

W/P—IDC	120,000	
W/P—L&WE	30,000	
Cash		150,000

b. Assume the well was determined to be dry and plugged and abandoned for an additional $2,000.

Entry to record plugging and abandoning costs

W/P—IDC	2,000	
Cash		2,000

Entry to record completion of work on well, *i.e.,* to close out W/P accounts

Wells and related E&F—IDC ($120,000 + $2,000)	122,000	
Wells and related E&F—L&WE	30,000	
W/P—IDC		122,000
W/P—L&WE		30,000

c. Assume instead that the well was successful and that additional IDC of $15,000 and equipment costs of $70,000 were incurred to complete the well.

Entry to record completion costs		
W/P—IDC	15,000	
W/P—L&WE	70,000	
Cash		85,000
Entry to record completion of work on well		
Wells and related E&F—IDC		
($120,000 + $15,000)	135,000	
Wells and related E&F—L&WE		
($30,000 + $70,000)	100,000	
W/P—IDC		135,000
W/P—L&WE		100,000

STRATIGRAPHIC TEST WELLS

The rules for capitalizing or expensing costs of exploratory-type and development-type stratigraphic test wells are the same as the rules for capitalizing or expensing the costs of exploratory and development wells. As a result, dry exploratory-type stratigraphic test wells are expensed; successful exploratory-type stratigraphic test wells are capitalized. All development-type stratigraphic test wells, whether dry or successful, are capitalized. Thus, capitalization is not dependent upon whether the well will be completed as a producer, but upon whether the well has either found proved reserves or is a development well.

AFE'S AND DRILLING CONTRACTS

Costs to be incurred in drilling operations are generally budgeted and detailed in an Authorization for Expenditure (AFE). (See Appendix A.) AFE's provide two useful functions:

- In joint operations, the operating agreement typically requires that the operator get approval from the nonoperators for expenditures relating to the drilling of wells. AFE's are used to fulfill this requirement by informing nonoperators as to drilling plans, providing cost estimates, and obtaining necessary approvals. (See chapter 11 for a discussion of situations where a nonoperator(s) does not agree to approve the AFE.)
- In both joint and sole operations, AFE's are used by the operator for both internal budgeting and cost control purposes.

An AFE should include enough detail for the nonoperators to determine the reasonableness of the estimated costs. Specifically, the AFE should include:

- estimates for intangible drilling costs and equipment costs to be incurred in drilling the well
- completion costs if the well is determined to be successful
- plugging and abandonment costs if the well is determined to be dry

After drilling begins, the operator should monitor actual spending for each cost category. If actual costs exceed the budgeted and approved estimates by a certain amount—typically 10%, or the percentage specified in the joint operating agreement—then the nonoperators must be notified and approval for the cost overruns obtained. In addition to drilling wells, AFE's are used by operators to get approval from the nonoperators for facility construction or other projects where estimated costs exceed the single expenditure limits specified in the joint operating agreement.

After the nonoperators approve an AFE, the operator will typically contact a drilling contractor and negotiate a drilling contract. Drilling contract dollar amounts are usually based on a day rate or a footage rate. Under the day-rate contract, a stated amount is paid per day worked, regardless of the footage drilled. The drilling contractor generally provides a rig and crew and specified contractual services, while the operator provides all materials, supplies, equipment, etc. Under a footage-rate contract, a stated dollar amount is paid for each foot drilled. Normally, payment under a footage-rate contract is contingent upon a specified depth being reached. Under this contract, the drilling contractor generally provides the rig, crew, specified contractual services, and certain materials and supplies. Logging, core tests, drilling mud, and well equipment are typically supplied by the operator.

A drilling contract may also be on a turnkey basis, whereby the contractor agrees to drill to a specified depth for an agreed upon dollar amount. Unlike the other types of contracts, under a turnkey contract the contractor assumes all the responsibility and is in total charge of drilling and completion operations, and provides all labor, equipment, and supplies. Turnkey contracts cost more and are not used as frequently as footage and day-rate based contracts. Turnkey contracts may be used in domestic operations when several investors are buying an interest in the well by paying a designated amount ($5,000, or $10,000, etc.) and need to know the total well cost prior to investing.

In international operations turnkey contracts may be used by an operator who is starting a drilling program in a new country or region and has yet to establish an infrastructure necessary to support drilling (*i.e.*, employees, warehouses, suppliers, contractors). In these cases an operator may pay a drilling contractor to drill the first well or two on a turnkey basis while making the necessary arrangements to establish the necessary infrastructure in the area and then afterward contract utilizing a day-rate or footage-rate basis.

Regardless of the type of drilling contract, the contractor bills the operator for services rendered as drilling progresses and when drilling operations are completed. The billing statements should, at a minimum, detail the drilling costs by categories of intangible drilling costs and equipment costs so that the working interest owners can correctly classify these costs in their accounts.

SPECIAL DRILLING OPERATIONS AND PROBLEMS

Workovers

Workover operations generally involve using a special workover rig to restore or stimulate production from a particular well. A situation in which a workover may be necessary would be an open hole completion where sand from the producing formation has clogged the tubing end, reducing or completely cutting off the fluid flow from the producing horizon. A workover may also be necessary when the casing has been perforated, and rock or sand particles have clogged the openings in the casing. Both of these cases may require a workover to restore production. These types of workover costs are expensed as production expense, specifically lease operating expense (chapter 8), because production has merely been restored. Workover operations may also involve recompletion in the same producing zone in an effort to restore production. This type of workover is also expensed as lease operating expense.

In contrast, an operation whose objective is to add new proved reserves is to be accounted for as a new drilling, *i.e.*, exploratory or development drilling. For instance, a workover operation may involve plugging back and completion at a shallower depth. As an example, a well was producing at 8,000 feet. The reserves played out, so the well was plugged back to 5,000 feet where there was a producing formation and completed in that zone. In another workover operation, a well was drilled to 8,000 feet and casing was set to that depth. The well was then completed at 5,000 rather than 8,000 feet. Later, a workover rig was brought in to dually complete the well at 8,000 feet.

In both of these examples, the costs would be treated as drilling costs and would be subject to SE drilling capitalization rules because the purpose of the workovers was to obtain production from a new formation, not merely to restore production from a formation already producing. In these cases, the costs would be capitalized because the operations were developmental in nature. Similarly, a well might be re-entered and deepened below the casing point in the attempt to obtain production from a deeper horizon. Again, the costs would be treated as drilling costs, with the final accounting treatment dependent upon whether the attempt was successful and whether that portion of the well was classified as development or exploratory.

EXAMPLE

Workovers

Lucky Company had the following expenditures during July 2007:

7/10/2007	Workover costs in connection with well #1036— cleaning and reacidizing producing formation	$ 5,000

7/20/2007	Workover costs on well #1097—testing, perforating, and completion at 8,000 feet. This depth is a new producing formation. Casing was previously set.		
	IDC		$ 20,000
	Equipment		2,000
7/30/2007	Workover completed in deepening well #1102 to a new unproved formation at 9,000 feet. Result was a dry hole at that depth. Lucky continued producing from the formation at 5,000 feet.		
	IDC		50,000
	Equipment		5,000

Entries

July 10, 2007:

Lease operating expense—#1036	5,000	
Cash		5,000

July 20, 2007:

Wells in progress—IDC—#1097	20,000	
Wells in progress—L&WE—#1097	2,000	
Cash		22,000
Wells and related E&F—-IDC—#1097	20,000	
Wells and related E&F—- L&WE—#1097 . . .	2,000	
Wells in progress—IDC		20,000
Wells in progress—L&WE		2,000

July 30, 2007:

Wells in progress—IDC—#1102	50,000	
Wells in progress—L&WE—#1102	5,000	
Cash		55,000
Dry-hole expense—IDC	50,000	
Dry-hole expense—L&WE	5,000	
Workover in progress—IDC—#1102		50,000
Workover in progress—L&WE—#1102 . . .		5,000

Damaged or lost equipment and materials

Equipment or materials may be damaged or lost during the drilling process. Some examples of damaged equipment would be twisted drillpipe or a broken bit. Examples of lost equipment include parts of the drill bit, hand tools (*e.g.*, wrenches), and drillpipe twisted off downhole. Drilling mud may also be lost into a very porous formation with large cracks or fissures.

The costs of the damaged equipment, lost equipment, or lost drilling mud are costs incurred in the drilling process and are handled in the same manner as other drilling costs. Damaged or lost material or equipment, less salvage value for the damaged equipment, is capitalized if the well is a development well. If the well is an exploratory well, the costs are expensed or capitalized, depending upon whether the well is unsuccessful or successful.

Fishing and sidetracking

When equipment is lost in the hole, recovery of the lost equipment may be attempted through **fishing**. If fishing is not successful, drilling through or sidetracking around the lost equipment may be necessary. **Sidetracking** involves plugging the lower portion of a well and drilling around the obstruction. Sidetracking is possible because drillpipe is flexible and allows deviations from the vertical. Sidetracking costs are generally considered part of drilling costs and are capitalized or expensed, depending upon whether the well is a development well or an exploratory well and whether proved reserves are found. If the well is an exploratory well, expensing the costs of the abandoned portion of the well regardless of whether proved reserves are found appears consistent with the theory of successful efforts because that portion of the well was abandoned. However, due primarily to difficulties encountered in attempting to allocate drilling costs between the abandoned portion of the well and the portion which is completed, in practice the cost of the entire well is typically capitalized if the well is successful.

Abandonment of portions of wells

SFAS No. 19 specifies the treatment for "standard" exploratory wells or development wells. However, many drilling situations, such as those described in the previous paragraphs, do not fall under these guidelines. Several "nonstandard" situations are depicted in the drawings that follow. In all the situations, the shaded portion of the well is the portion for which treatment is not explicitly specified in *SFAS No. 19.*

A survey published in the *Journal of Extractive Industries Accounting* [1] reported how companies are handling similar situations in practice. Results of the survey indicate that in most of the situations, there appears to be little agreement as to whether to expense or capitalize the shaded portion of the wells. Some guidance is provided, however, in *COPAS Bulletin No. 10.* For example, in Figure 5–3, situation *A* depicts an initial exploratory well that was abandoned because of mechanical or structural difficulties, with a second well drilled nearby that was suc-

Fig. 5–3 — *Special Drilling Situations*

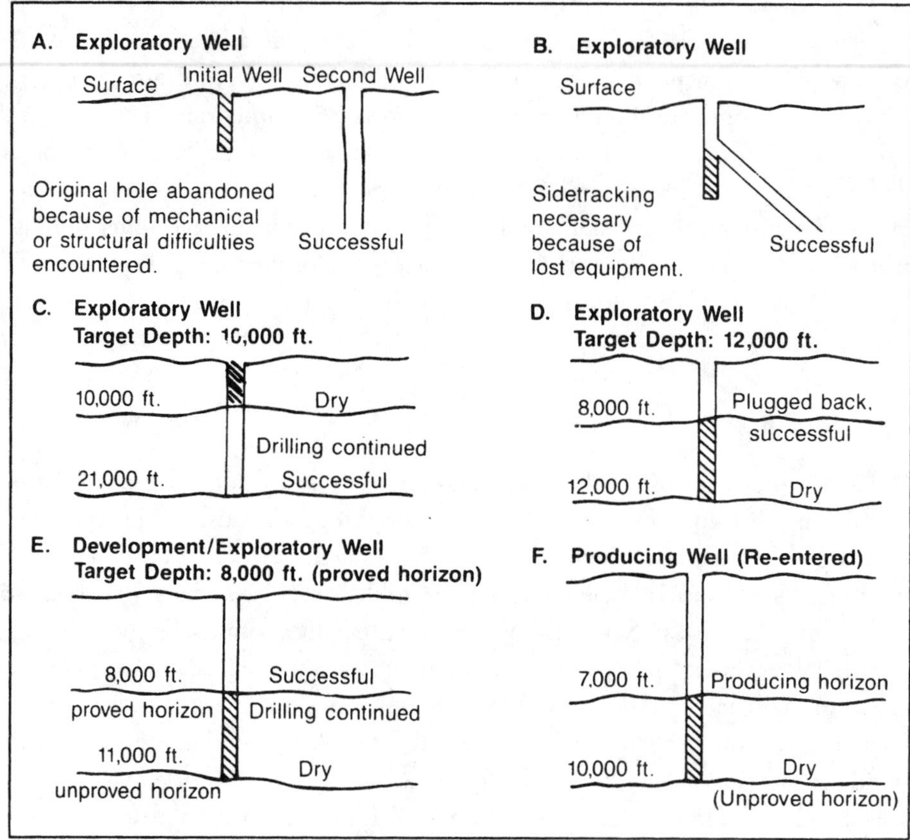

cessful. COPAS *Bulletin No. 10* states that it is preferable to charge the costs of the abandoned hole to exploratory dry hole expense since the abandoned well did not add to the value of the completed well.

Likewise, as illustrated in situation *B*, the Bulletin states that it is preferable to expense the cost of the abandoned portion of an exploratory well that had to be plugged back and sidetracked around because of drilling difficulties. Again, the abandoned portion of the well did not add to the value of the sidetracked hole that was completed.

Situation *D* depicts a situation where the exploratory well was dry at the target depth but was plugged-back and completed at a more shallow depth. The *COPAS* Bulletin suggests that the incremental costs of drilling beyond the depth at which the well was completed should be charged to dry-hole expense. The costs of drilling to the completed depth along with completion costs are capitalized. Similarly, in situation *E*, the incremental costs of drilling beyond the successful target depth to a dry, unproved horizon should be expensed.

If an operator enters a producing well and drills deeper to an unproved horizon but finds no other producing zones as shown in situation *F*, the costs of drilling to an additional depth should also be charged to expense. In all cases in which additional drilling finds another producible zone, the incremental drilling costs along with completion costs should be capitalized.

Situation *C* depicts a well that was dry at the target depth but successful at a lower depth. *COPAS Bulletin No. 10* does not specifically address this situation. However, the entire cost should be capitalized because it is necessary to drill to the target depth in order to reach the lower producible zone.

If a portion of a well is expensed and a portion is capitalized, the question arises as to how the costs should be apportioned. Again, *SFAS No. 19* is silent. Practice seems to be quite varied. Some companies allocate the drilling costs on a per-foot basis, some allocate on a drilling-day basis, some use actual costs incurred, some incremental costs, etc.

Additional Development Costs

In the earlier Development Drilling Example, only development costs directly relating to drilling and completing a well were discussed. *SFAS No. 19* specifies that all development costs should be capitalized as wells and related equipment and facilities. Development costs are defined as follows:

> *...[those costs] incurred to obtain access to proved reserves and to provide facilities for extracting, treating, gathering, and storing the oil and gas. More specifically, development costs, including depreciation and applicable operating costs of support equipment and facilities (paragraph 26) and other costs of development activities, are costs incurred to:*
>
> *a. Gain access to and prepare well locations for drilling, including surveying well locations for the purpose of determining specific development drilling sites, clearing ground, draining, road building, and relocating public roads, gas lines, and power lines, to the extent necessary in developing the proved reserves.*
> *b. Drill and equip development wells, development-type stratigraphic test wells, and service wells, including the costs of platforms and of well equipment such as casing, tubing, pumping equipment, and the wellhead assembly.*
> *c. Acquire, construct,and install production facilities such as lease flow lines, separators, treaters, heaters, manifolds, measuring devices, and production storage tanks, natural gas cycling and processing plants, and utility and waste disposal systems.*
> *d. Provide improved recovery systems.*

Service wells, considered to be development costs per the above definition, are drilled to support production in an existing field. Examples of service wells include gas injection wells, water injection wells, saltwater disposal wells, and water supply wells. Gas or water injection wells may be used for saltwater disposal, pressure maintenance, or secondary or tertiary recovery purposes. Likewise, costs associated with secondary and tertiary recovery methods, *i.e.*, improved recovery systems, are also considered to be development costs.

EXAMPLE

Additional Development Costs

Lucky Oil Company began installing flow lines. The flows lines cost $10,000 and installation charges were $2,000.

Entry

Wells in progress—L&WE	12,000	
Cash		12,000

Installation of the flows lines was completed late in July.

Entry

Wells and related E&F—L&WE	12,000	
Wells in progress—L&WE		12,000

SUPPORT EQUIPMENT AND FACILITIES

As discussed in chapter 3, operating costs or depreciation related to support equipment and facilities used in exploration, development, or production activities should be classified as an exploration, development, or production cost, as appropriate.

EXAMPLE

Support Equipment and Facilities

Depreciation on the automobile used by the drilling foreman totaled $5,000 during the year. Since the automobile is used to support the drilling of several wells, the depreciation must be allocated between all of the wells being drilled during the period. Using a mileage driven basis, $1,000 of the depreciation is allocated to Well No. 1. The entry to record the depreciation would be as follows:

Entry

Wells in progress—IDC	1,000	
Accumulated depreciation		1,000

If Well No. 1 is a development well, the depreciation will ultimately be transferred to wells and related E&F. If the well is an exploratory well, the costs will ultimately be recorded either as wells and related E&F or dry-hole expense, depending upon whether the well finds proved reserves.

G&G TO SELECT DRILLSITE

Geological and geophysical (G&G) costs incurred to determine a specific drillsite are considered to be part of the cost of drilling a well and are initially recorded as wells in progress—IDC. If the well is a development well, the G&G costs will ultimately be recorded as wells and related E&F. If the well is an exploratory well, the G&G costs will ultimately be recorded either as wells and related E&F or dry-hole expense, depending upon whether the well finds proved reserves. The G&G costs incurred to determine a specific drillsite are not considered nondrilling G&G, which, as seen in chapter 3, should be expensed as incurred.

POST-BALANCE SHEET EVENTS

As discussed in chapter 4, post-balance sheet events relate to information about conditions that (a) existed at balance sheet date and that (b) become known after the end of the period but before the financial statements are issued. The drilling of an exploratory well over a fiscal year-end is an example of a potential post-balance sheet event.

As discussed above, the costs associated with drilling an exploratory well are capitalized as wells in progress until it is determined whether the well has found proved reserves. If the well is dry, the capitalized costs are charged to expense. If the well is successful, the capitalized costs are reclassified as wells and related equipment and facilities. Normally, all the costs of an exploratory dry hole must be written off in the year a well is determined to be dry. However, if a well is determined to be dry after year-end but before the financial statements are issued, the costs incurred prior to year-end should be written off in that fiscal year. Any costs incurred in the second year relating to the well should be charged to expense in the second year (Interpretation 36 to *SFAS No. 19*).

EXAMPLE

Post-Balance Sheet Events

Lucky Company incurred the following drilling costs on Well No. 1:

Prior to December 31, 2005:

IDC	$235,000
Equipment	20,000

During January, 2006:

IDC (testing)	30,000

The well was determined to be dry on February 1, 2006, before the financial statements were published. No equipment was salvaged.

Disposition of the Drilling Costs:
IDC costs of $235,000 and equipment costs of $20,000 should be expensed as dry-hole costs for the year ended December 31, 2005. The $30,000 of IDC incurred in 2006 should be expensed in 2006.

Deferred Classification of Costs of an Exploratory Well

The determination of whether a well has found proved reserves is usually made shortly before or after drilling operations are completed. However, sometimes an exploratory well is determined to have found oil or gas reserves but the reserves cannot be classified as proved at the time drilling is completed.

Classification may be delayed and the well carried as an asset if:

1. A major capital expenditure such as a pipeline is required before production can begin, and a sufficient quantity of additional reserves must be found to justify the major expenditure. To find the additional reserves, additional successful exploratory wells need to be drilled. In this situation, the classification may be delayed and the well carried as an asset for an indefinite period of time, even longer than one year, pending determination of whether proved reserves have been found as long as both of the following conditions are met:
   a. The well has found reserves sufficient to justify its completion, if the need for the required major capital expenditure is ignored.

b. Drilling of additional wells is under way or firmly planned for the future.

or

2. All other exploratory wells, i.e., those wells not requiring a major capital expenditure, may be carried as an asset for up to one year following completion of drilling without classification of the found reserves as proved (*SFAS No. 19,* par. 31).

If the exploratory well is of the type described in item 1 above and the specified conditions are not met, or cease to be met without the reserves being classified as proved, then the costs of the well should be expensed. However, as long as the conditions of: (1) sufficient reserves, and (2) firm drilling plans, along with (3) the need for a major capital expenditure are met, the well may be carried in suspense indefinitely.

As stated in item 2 above, wells not needing a major expenditure must be classified as either successful (proved reserves found) or unsuccessful within one year. If classification is still not possible at the end of one year following completion of drilling operations, the well costs must be expensed.

Estimating reserves is done by taking core samples and well logs—electric, magnetic, etc.—in addition to examining mud logging reports and pressure tests. In some cases, the results obtained by reserve engineer examinations may be inconclusive, and further testing may be required. When marginal exploratory wells are involved, proved reserves determination may take a significant period of time and deferred classification may be necessary.

Delays in well classification may also occur in situations similar to the Austin Chalk in Texas. Wells in the Austin Chalk formation have, in many instances, suffered rapid production declines following completion. Great care in estimating reserves is required in such situations to allow for decline in production. Deferred classification is not appropriate, however, based on the chance that some event beyond the control of the company may occur. For example, classification should not be deferred based on the chance that the price of oil or gas will go up in the future, resulting in proving reserves that are currently not commercially producible.

Deferred classification of costs of an exploratory-type stratigraphic test well

If a major capital expenditure is required on an exploratory-type stratigraphic test well, classification may be delayed as long as the following conditions are satisfied:

- a quantity of reserves has been found that would justify completion had the well not been a stratigraphic test well
- drilling of additional exploratory-type stratigraphic test wells has been started or is planned in the near future

Stratigraphic test wells are generally drilled offshore where a production platform would be required to produce any reserves discovered. Thus, a common example of a major capital expenditure related to a stratigraphic test well is construction of a platform that can be justified only if further stratigraphic test wells, which find additional proved reserves, are drilled.

INTEREST CAPITALIZATION

SFAS No. 34 requires capitalizing interest as part of the cost of assets that require a period of time to be prepared for their intended use. Essentially, *SFAS No. 34* requires interest capitalization for all qualifying assets, where qualifying assets are defined as assets constructed by an entity for its own use. Either the interest rate on specific borrowings associated with the qualifying asset may be used or a weighted average interest rate may be used. To obtain the amount of interest to capitalize each period, the interest rate is applied to the average amount of accumulated capital expenditures. Capitalized interest cannot exceed actual interest costs. The interest capitalization period begins when the three following conditions are met:

1. Expenditures for the asset have been made
2. Activities necessary to get the asset ready for its intended use are in progress
3. Interest cost is being incurred

The term *activities*, used in condition 2 above, is to be construed broadly encompassing technical and administrative activities such as obtaining permits. The interest capitalization period should end when the asset is substantially complete and ready for productive use.

Applying this statement to an industry as unique as the oil and gas industry creates interpretation problems. In fact, according to a survey of successful-efforts companies, application of this statement has been quite varied.[2] The starting point used has ranged from the time a prospect is acquired to the spud-in date. The stopping point, which is less varied, has ranged from the time proved reserves are found to the time when production begins. Activity cost has included leasehold costs and tangible and intangible drilling costs or IDC and tangible equipment only.

Specifically, capitalized interest is computed as follows:

Average accumulated expenditures during construction x Interest rate x Construction Period

Average accumulated capitalized expenditures is computed by adding the beginning balance and ending balance of capitalized expenditures and dividing by two. A simple example of one interpretation of interest capitalization for an oil and gas company follows:

EXAMPLE

Interest Capitalization

Lucky Company has unproved property costs of $60,000 for Lease A at January 1, 2007. During the year 2007, drilling costs are incurred on Lease A in the amount of $300,000. A 10%, $400,000 note is outstanding during the entire year and was specifically obtained for the acquisition and drilling program related to Lease A. Compute the interest capitalization amount and prepare the entry to record the interest.

Interest to be capitalized during 2007:

Average accumulated expenditures: $\frac{\$60{,}000 + \$360{,}000}{2} \times {}^{12}/_{12}{}^{*} = \underline{\underline{\$210{,}000}}$

Interest costs to be capitalized: $\$210{,}000 \times 0.10 = \underline{\underline{\$21{,}000}}$

Entry

W/P—IDC	21,000	
Interest expense		21,000

*Because the property was acquired on January 1, 2007, the capitalization period is a full year or $^{12}/_{12}$.

In the above entry, W/P—IDC is debited although it is unlikely that the same amount would be considered IDC for tax. In practice, when taxes are prepared, companies generally make relevant adjustments to the accounts to accommodate tax and financial accounting differences such as this one.

OFFSHORE AND INTERNATIONAL OPERATIONS

The principles of accounting for offshore and international drilling and development costs are the same as for domestic properties. However, the significant cost of operating offshore and/or international locations creates special accounting concerns. Several COPAS Bulletins deal solely with offshore operations and allocation of the associated costs. Drilling is often substantially more expensive because of location and technical adaptations that may be necessary, especially for offshore locations. Drilling crews must be housed, fed, and transported to and from their base location to the drillsite. Costs related to support equipment and facilities such as warehouses, repair facilities, docks, boats, and helicopters for transportation are much higher than for domestic onshore operations. In addition, most if not all, international operations are taxed according to local laws and importation of equipment and supplies is subject to payment of customs fees and duties.

Problems

1. Define the following terms:
   dry hole
   wells and related E&F—lease and well equipment
   wells and related E&F—IDC
   day-rate contract
   footage-rate contract
   turnkey contract
   AFE

2. Which of the following would be IDC?
   a. labor costs to build a road to the drillsite
   b. labor costs to build a road to a producing well
   c. cost of a drillstem test
   d. cost of surface casing
   e. installation costs of surface casing
   f. cost of drilling mud
   g. damages paid to landowner
   h. cost of flow lines, tanks, and separators
   i. installation costs of flow lines, tanks, separators
   j. drilling contractor's charges under a footage-rate contract
   k. drilling contractor's charges under a turnkey contract
   l. cost of cement

3. List and briefly discuss the three basic types of drilling contracts.

4. An exploratory-type stratigraphic test well that was drilled offshore discovered proved reserves. However, it was decided that the permanent platform should be placed in a different location. How should the costs of the well be handled?

5. Royalty Oil Company drilled an exploratory well on a lease located in a remote area. The well found reserves, but not enough to justify building a necessary pipeline. The company does not plan to drill any additional exploratory wells at this time. How should the costs of the well be handled?

6. Universal Oil Company drilled an exploratory well that found reserves, however, the reserves could not be classified as proved at that time. No major capital expenditure was required. How should the costs of the well be handled?

7. Near the end of 2007, King Oil Company drilled an exploratory well that found oil, but not in commercially producible quantities unless the price of oil went up from $25 per barrel to $30 per barrel. King decided to defer classification of the well for up to one year because its financial advisors felt that there was a high probability that oil prices would go up during the next year. Please comment.

8. a. Aggie Oil Company drills an exploratory well during 2006 that finds oil, but not in commercially producible quantities at current oil prices. Since proved reserves are not found, Aggie expenses the cost of the well in 2006. Early in the next year, but after Aggie's financial statements have been published, the price of oil goes up so that the reserves found by the exploratory well during 2006 become commercially producible. Should the costs of the well be reinstated?
   b. Assume the same situation except that the price of oil goes up early that next year before Aggie's financial statements are published. Should the costs of the well be reinstated in this case?

9. Tiger Oil Company had an exploratory well in progress at the end of 2008. Total costs incurred by 12/31/08 were $300,000. During January of 2009, drilling was continued and costs of $200,000 were incurred. Total depth was reached and the well was determined to be dry by the end of January. Assuming Tiger's financial statements are not published until early February, what costs, if any, should be expensed for 2008 and for 2009?

10. Hard Luck Oil Company incurred the following costs during 2008:
    a. began drilling an exploratory-type stratigraphic test well, incurred $50,000 of IDC
    b. began drilling an exploratory-type stratigraphic test well, incurred $80,000 of IDC and $10,000 in equipment costs
    c. began drilling a development-type stratigraphic test well, incurred $100,000 of IDC
    d. began drilling a development-type stratigraphic test well, incurred $200,000 of IDC and $30,000 in equipment costs

    The following results were obtained late in 2008:
    a. well was determined to be dry
    b. well found proved reserves
    c. well found proved reserves
    d. well was determined to be dry

    Prepare the necessary entries.

11. Formation Oil Company began in 2005 with the acquisition of four individually significant unproved leases. Give the entries, assuming the following transactions. You may combine entries for items marked with an asterisk (*).

Year	Transaction	Lease A	Lease B	Lease C	Lease D
2005	Lease bonuses*	$50,000	$40,000	$70,000	$55,000
	G&G costs*	60,000	50,000	75,000	90,000
	Drilling costs		Well 1 (exploratory)	Well 1 (exploratory)	Well 1 (exploratory)
	IDC	None	300,000	250,000	150,000
	Equipment	None	125,000	50,000	40,000
	Drilling results	None	Well dry; property impaired 25%	Drilling completed; results undetermined*	Drilling not completed
2006	Delay rentals*	4,000	none	3,000	none
	Drilling costs	Well 1 (exploratory)	Well 2 (exploratory)		Well 1
	IDC	275,000	225,000	none	60,000
	Equipment	50,000	50,000	none	40,000
	Drilling results	Drilling completed; proved reserves	Dry; abandoned lease	Still undetermined; greater than 1 year	Drilling completed; proved reserves
2007	Delay rentals*	None	—	3,000	none
	Drilling costs	Well 2 (development)			Well 2 (development)
	IDC	300,000	—	none	250,000
	Equipment	80,000		none	100,000
	Drilling results	Dry	—	none	Drilling completed; proved reserves

* No major capital expenditure required.

12. Record the following transactions:
    a. Brock Oil Company incurred costs of $30,000 in preparing a drillsite.
    b. The contractor was paid $400,000 on a day-rate contract (all intangible).
    c. Equipment (casing) costs of $75,000 were incurred.
    d. Costs of $70,000 were incurred in evaluating the well.
    e. Brock Oil decided to complete the well and incurred costs of $45,000 (perforating and fracturing), $60,000 (cementing), and $100,000 (equipment) in completing the well.

13. Hard Luck Oil Company incurred the following costs during the years 2006 and 2007.

2006

a. Contracted and paid $50,000 for G&G surveys during the year.
b. Leased acreage in four areas as follows:
   1) Miller lease—500 acres @ $50 per acre bonus; other acquisition costs, $2,000
   2) Ebert lease—800 acres @ $100 per acre bonus; other acquisition costs, $3,000
   3) Ewing lease—200 acres @ $60 per acre bonus; other acquisition costs, $500
   4) Johnson lease—600 acres @ $30 per acre bonus; other acquisition costs, $800

   Each lease had a delay rental clause requiring payment of $2 per acre if drilling was not commenced by the end of one year. Also, each of the above leases was considered individually significant.

c. The company also leased 10 individual tracts for a total consideration of $60,000. The tracts are considered to be individually insignificant and are the first insignificant unproved properties acquired by Hard Luck.
d. The company incurred $1,000 in costs to maintain lease and land records in 2006. Also, costs of $8,000 were incurred to successfully defend a title suit concerning the Miller lease.
e. During 2006, the company incurred the following costs in connection with the Miller lease when drilling an exploratory well:

Roads, location, damages, etc	$ 18,000
G&G costs to locate the specific drillsite	2,000
Drill pipe	18,000
Conductor casing	10,500
Wellhead equipment	28,000
Contractor's charges and drilling fee (no equipment)	737,000
Equipment rentals	30,000
Water, fuel, power, lubricants	70,000
Drill bits	19,000
Electric logging	20,000
Cement	20,000
Cementing services	15,000
Casing crews	40,000
Surface casing	32,600

   Completion costs in connection with the above well were as follows:

Production casing	$151,500
Christmas tree	22,000
Tubing	37,500
Labor for installing casing	20,000

Labor for installing Christmas tree	$ 1,000
Perforating	40,000
Flow lines and tanks	45,000
Labor for installing flow lines	8,000

f. An exploratory well was drilled on the Ebert lease in 2006 on a turnkey basis to 9,000 feet. The contractor's charge was $300,000, which included $40,000 for "casing. At the end of 2006, a decision had not been made to complete or abandon the well. A major capital expenditure was not required.
g. At the end of 2006, the Johnson lease was impaired by 40% and the Ewing lease by 20%. The company has a policy of maintaining an allowance for impairment equal to 60% of individually insignificant leases.

2007

a. Delay rentals were paid on the Ewing and Johnson leases.
b. Late in 2007, the company abandoned the Ewing lease and two of the individually insignificant leases, which cost a total of $8,000 when acquired. The Johnson lease is now considered to be a very valuable lease because a large producer was found on adjacent property.
c. At year-end, the company could still not decide whether to complete or abandon the well on the Ebert lease.

Prepare journal entries for the above transactions.

14. During the calendar year, 2005, Garza Company had the following transactions on an unproved property:
    a. Drilled Hays #1 with IDC costs of $310,000 and equipment cost of $42,000.
    b. The well was determined to be dry and was plugged and abandoned at a cost of $10,000. Salvaged equipment placed in inventory was valued at $8,000.
    c. Prepare journal entries for the transactions.

15. During 2005, Willis Oil Company incurred the following costs in connection with the Grove lease.
    a. Acquired the 800 acres Grove lease at a lease bonus of $70 per acre and other acquisition costs of $10,000.
    b. Incurred the following costs in connection with Grove #1:

G&G costs to locate wellsite	$ 3,000
Surface damages	15,000
Surface casing	7,000
Contractor's fee/day workrate	175,000
Equipment rentals	200,000
Drilling fluids	35,000
Fuel	9,000

Drillbits	$ 20,000
Cementing services	5,000
Casingcrews	6,000
Roustabout labor	8,000
Hauling and transportation	7,500
Production casing	36,000
Tank battery	11,000
Lines and connections	5,500
Pumping unit motor and accessories	50,000
Casinghead and connections	6,500
Tubing	8,500
Separating and treating equipment	7,100
Measuring equipment	300
Downhole pump and rods	5,600

Record the above transactions. **Hint:** The type of equipment installed when the well reached target depth indicates whether the well was successful or dry.

16. The Allen Oil Company incurred the following costs and had the other transactions shown below for the years 2005 and 2006. The company uses the successful efforts method of accounting.

2005

a. Paid $100,000 for G&G costs during the year.
b. Leased acreage in three individually significant areas as follows:
   1) Adams lease—1,000 acres @ $60 per acre bonus, other acquisition costs, $3,000.
   2) Grove lease—800 acres @ a lease bonus of $70 per acre, and other acquisition costs of $10,000
   3) Borden lease—600 acres @ $60 per acre bonus and other acquisition costs of $8,000.
c. The company also leased 20 individual tracts for a total cost of $80,000. These leases are considered to be individually insignificant and are the first insignificant unproved properties acquired by Allen.
d. Paid $5,000 in costs to maintain lease and land records in 2005. Also, paid $30,000 to successfully defend a title suit concerning the Grove lease.
e. Paid the following costs in connection with Grove #1:

G&G costs to locate a wellsite	$ 3,000
Location and road preparation prior to spudding-in the well	8,000
Surface damages	15,000
Surface casing	7,000
Contractor's fee/daywork rate	175,000

Equipment rentals	$200,000
Drilling fluids	35,000
Fuel	9,000
Drill bits	20,000
Cementing services	5,000
Roustabout labor	8,000
Hauling and transportation	7,500
Production casing	36,000
Tank battery	11,000
Flow lines and connections	5,500
Pumping unit motor and accessories	50,000
Casing head and connections	6,500
Tubing	8,500
Separating and treating equipment	7,100
Measuring equipment	300
Downhole pump and rods	5,600
Testing and acidizing	11,000

f. An exploratory well was drilled on the Borden lease in 2005 on a turnkey basis to 8,000 feet. The contractor's charge was paid for $400,000, which included $60,000 for casing. At the end of 2005, a decision had not been made to complete or abandon the well. No major capital expenditure was required. At the end of 2005, the Adams lease was impaired 60% and the Borden lease by 30%. The company's policy is to maintain an allowance for impairment at 70% of the cost of insignificant leases.

2006

a. Delay rentals of $2,000 were paid on the Adams lease, $1,200 on the Borden lease, and $3,000 on insignificant leases.
b. During 2006, the Adams lease was abandoned and three of the individually insignificant leases (cost $8,000) were also abandoned.The Borden lease is now considered to be a very valuable lease because a large producer was discovered on adjoining land.
c. At year-end (2006), the company had not made a decision to complete or abandon the Borden well.

REQUIRED: Prepare journal entries for the above transactions.

17. Support equipment used to drill a development well cost $13,000 and has a 10-year life with a salvage value of $1,000. The equipment was used for three months in drilling Hope #1. Record depreciation. Ignore the wells-in-progress account and use the appropriate final account.

18. An exploratory well that was later determined to be dry had the following costs that are appropriately assigned to the well:

Depreciation on support equipment . . . . $ 3,000
Operating costs of the support equipment
(including fuel, maintenance, labor, etc.) 10,000

Record the above amounts.

19. Cruser Company paid a seismic crew $2,000 to complete a G&G survey to select a drill-site where Hope #1 would be spudded-in. Record the above transaction.

20. Knight Company had the following transactions in 2005. Record the transactions.
    a. Acquired an undeveloped lease, $40,000
    b. Paid a drilling contractor as follows:
       Footage rate for drilling $250,000
       Equipment costs (casing) 75,000
       Equipment costs (tanks, flow lines and labor to install equipment) 80,000
    c. Paid costs in evaluating the well, $20,000
    d. Completion costs for fracturing and perforating, $25,000

21. Aggie Oil Company incurred and paid the following costs during 2006:

	Lease A Unproved	**Lease B Unproved**	**Lease C Proved**	**Lease D Proved**
Acquisition costs	$ 30,000	$ 35,000	Purchased in 2005	Purchased in 2005
Well:	Exploratory	Exploratory	Development	Development
Drilling contractor's charges—day-rate	180,000	200,000	$160,000	$190,000
Casing	30,000	35,000	28,000	22,000
Production equipment		75,000		80,000
Well logs	10,000	15,000	12,000	14,000
Drilling results	Well dry; abandon lease	Found proved reserves	Well dry	Found proved reserves

Record the above transactions.

22. The Willis Company has unproved property costs of $40,000 at January 1, 2006. During the year 2006, Willis incurred $400,000 drilling costs on Lease A. An 8%, $500,000 note is outstanding during the entire year and was obtained to finance the drilling program.

    Compute the interest capitalization amount and record the interest.

23. Indicate whether the following types of expenditures are capitalized (C) or expensed (E) under the acceptable GAAP methods for each well drilled.

	Successful-Efforts		Full-Cost	
	IDC	L&WE	IDC	L&WE
Exploratory well: Successful				
Dry Hole				
Development well: Successful				
Dry Hole				

24. Ramsey Company hires a drilling contractor to drill a well to the depth of 8,000 feet at a cost of $300,000. The $300,000 cost includes $5,000 for surface casing. Any drilling to be completed after reaching the 8,000 foot depth is to be paid at a day rate of $4,000 per day. The 8,000 foot depth is reached on September 14, 2007, and the additional drilling to 10,000 feet is completed on October 27, 2007.

Ramsey incurred additional costs as shown on the attached schedule.

REQUIRED: Complete the schedule showing whether the cost is IDC or L&WE.

Description	Amount	IDC	L&WE
1. Rig drilling—turnkey contract	$ *		
2. Rig drilling—day rate	*		
3. Location and road preparation	8,000		
4. Drilling water	7,500		
5. G&G—Select wellsite	22,400		
6. Electric logging	1,200		
7. Bits	13,000		
8. Surface damages	6,300		
9. Mud (after 8,000')	40,000		
10. Production casing	35,000		
11. Cementing services	11,000		
12. Casing crews	5,500		

13. Tank battery	11,500
14. Flow lines and connections	5,000
15. Separators	7,000
16. Cement	48,000
17. Installing separators	4,800
18. Tubing	30,000
19. Christmas tree	20,000
20. Measuring equipment	500
21. Installing Christmas tree	3,000
22. Equipment rental	10,000

* To be determined from the information given in the problem statement.

NOTES

1. Institute of Petroleum Accounting, "Survey of Successful-Efforts Accounting Techniques," *Journal of Extractive Industries Accounting*, Summer, 1982, pp. 91-119.

2. Gallun, Rebecca A., Pearson, Della, and Seiler, Robert, "Capitalization of Interest by Oil and Gas Companies," *Journal of Extractive Industries Accounting*, Spring 1983, pp. 63-70.

chapter six

Proved Property Cost Disposition Under SE

The previous chapters outlined which costs should be capitalized and which costs should be expensed under successful-efforts accounting (SE). Chapter 4 discussed the acquisition and disposition of unproved property costs. In chapter 5, accounting for the incurrence of drilling and development costs was presented. This chapter deals with the disposition of capitalized costs of proved properties and wells and related equipment and facilities (E&F).

The T-accounts below show the types of costs capitalized as proved properties and wells and related equipment and facilities. Costs are moved to the proved property classification after proved reserves are discovered on the property. Costs are moved to the wells and related E&F classification only after a well has been successfully completed or a dry development well has been plugged and abandoned. Therefore, the cost of wells in progress is not included in wells and related E&F. Note that the wells and related E&F account also include the cost of installed equipment and facilities.

Proved Property	
Acquisition costs of property classified as proved, net of impairment	

Wells and Related E&F	
Completed development drilling costs Completed successful exploratory drilling costs Lease equipment and facilities	

Figure 6–1 is a flowchart summary of the four basic types of costs incurred in the exploration and production segments of oil and gas companies and their accounting treatment. Trace through this flowchart for further clarification or as a refresher of how costs become part of proved properties and wells and related equipment and facilities.

Fig. 6–1 — *Successful Efforts Summary*

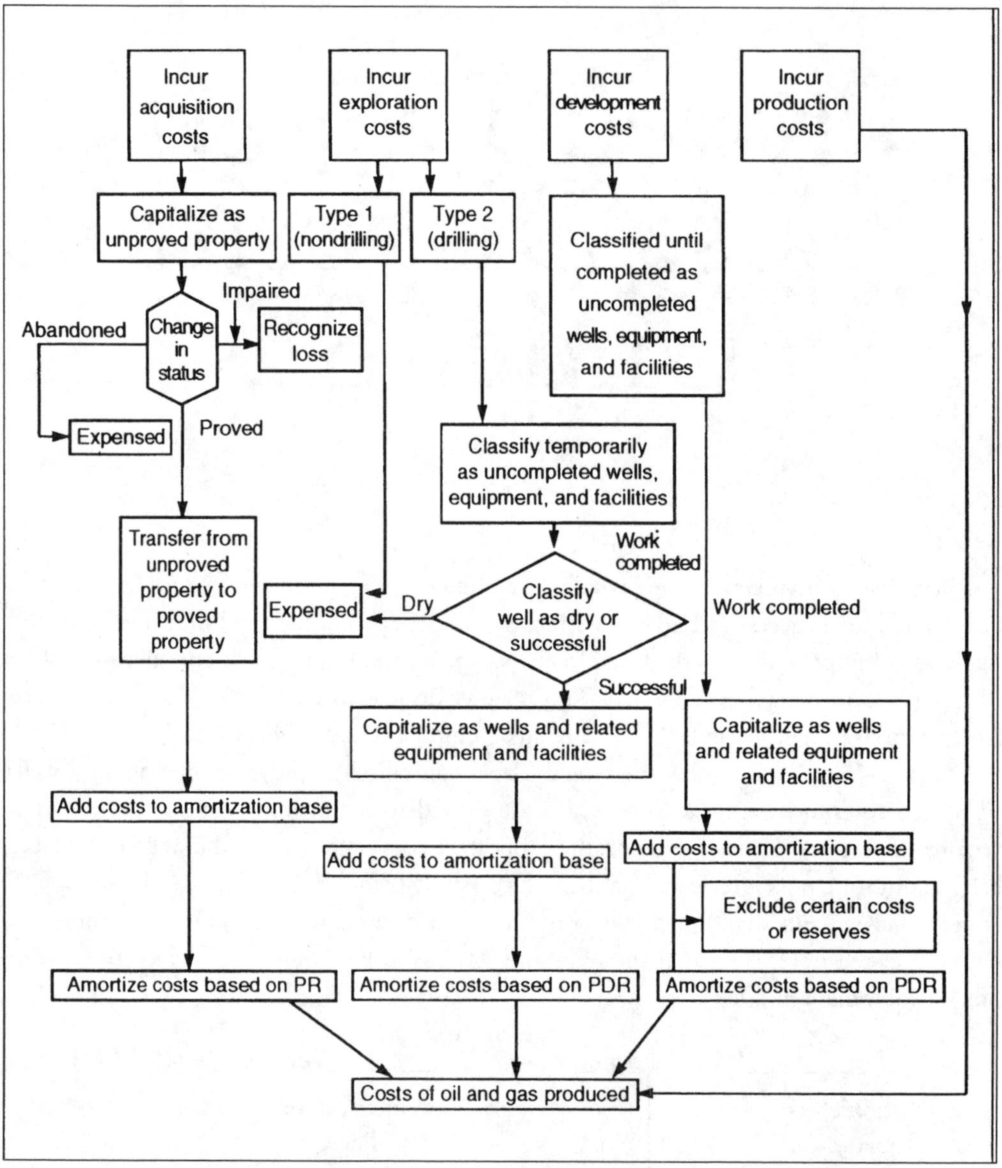

COST DISPOSITION THROUGH AMORTIZATION

In financial accounting, tangible property and equipment is depreciated, intangible property is amortized, and natural resources are depleted. For oil and gas financial accounting, industry professionals commonly refer to depreciation, depletion, and amortization of proved property and wells and related equipment and facilities as DD&A. Both the SEC and the FASB use the term amortization in their written rules.

In particular, for SE companies, acquisition costs of proved properties and the costs of wells and related equipment and facilities are amortized to become part of the cost of oil and gas produced (*SFAS No. 19,* par. 27). Acquisition costs are amortized over **proved reserves.** (See definitions of reserve categories in chapter 2.) Wells and related equipment and facilities are amortized over **proved developed reserves.**

Acquisition costs represent expenditures made on behalf of the entire cost center and thus apply to all reserves that will be produced from that cost center. Proved reserves are the reserves reasonably certain of being produced from a property and include both reserves that will be produced from wells already completed and from wells to be drilled in the future. Therefore, proved reserves should be used to amortize acquisition costs.

Proved developed reserves are reserves that will be produced from existing wells and equipment. Wells and related equipment and facilities should be amortized using proved developed reserves because, by definition, those are the reserves that will be produced as a result of the costs already incurred for completed wells and equipment. The remaining proved reserves, *i.e.*, the proved undeveloped reserves, are excluded in amortizing wells and related equipment and facilities because those reserves will be produced only as a result of incurring additional future costs. When a property is fully developed, proved reserves and proved developed reserves are the same.

The cost center for accumulating costs to be amortized is a property or some reasonable aggregation based on a common geological feature, such as a field or reservoir (*SFAS No. 19,* par. 30, 35).

Both acquisition costs and the costs of wells and related equipment and facilities are amortized using the unit-of-production method, which is as follows:

Unit-of-production formula

$$\frac{\text{Book value at year end}}{\text{Estimated reserves at beginning of year}} \times \text{Production for year}$$

Equivalent formula

$$\frac{\text{Production for year}}{\text{Estimated reserves at beginning of year}} \times \text{Book value at year end}$$

In the unit-of-production formula, the book value at year-end is used. Book value at year-end is the total cost accumulated to year-end, minus accumulated DD&A at the beginning of the year. Consistency would suggest a book value at year-end would be used with a reserve estimate also made at year-end, so that all added reserves found by incurring the costs during the year would be included in the DD&A computation. Instead, the denominator calls for estimated reserves as of the beginning of the year. However, rather than using a reserve estimate determined as of the beginning of the year, the most current estimate—preferably an estimate determined as of the end of the year, which includes current year discoveries—should be used. The production during the year is then added to the year-end reserve estimate to convert the year-end reserve estimate of reserves in place at year-end into an estimate of reserves in place at the beginning of the year. The resulting estimate of beginning-of-the-year reserves would, therefore, include the reserves discovered by costs incurred during the year. In this way, the most up-to-date estimate is used, and one that reflects the additional reserves found by costs incurred during the year.

In determining the reserves and production to use in calculating DD&A for working interest (WI) owners, remember that WI owners (or indeed any economic interest owner) are entitled to only a portion of total reserves and total production. For example, if there is a $^1/_8$ royalty interest then the WI owners would be entitled to $^7/_8$ of total reserves and $^7/_8$ of production. In this basic situation, either the WI's share of gross production and gross reserves or total gross production and gross reserves could technically be used in the DD&A calculation because the $^7/_8$'s, $^5/_6$'s, etc., cancel. However, in other situations, determination of the WI's portion of gross production and gross reserves is more complicated and the $^7/_8$'s, $^5/_6$'s, etc. do not cancel. Consequently, in computing DD&A throughout this book, only the economic interest's share of reserves and production is used. In the above formulas and in the discussion of amortization throughout this chapter, proved reserves, proved developed reserves, and the current year's production refer to the WI owners' proportionate share of reserves and production unless stated otherwise.

Reserve and production data are typically measured in either barrels (bbl) or thousand cubic feet (Mcf). Oil reserves and oil production are estimated and measured in terms of barrels. A barrel (bbl) is 42 gallons of oil measured at a standard temperature of 60° Fahrenheit. Gas reserves and gas production are commonly estimated and measured in terms of Mcf. An Mcf is 1,000 cubic feet of gas measured at a temperature of 60° Fahrenheit under pressure of 14.73 psia (pounds per square inch absolute).

EXAMPLE 1

DD&A

Lucky Oil Company drilled the first successful well on Lease A early in 2008. The company plans to develop this lease fully over the next several years. Data for the lease as of December 31, 2008, are as follows:

Leasehold cost (acquisition costs—proved property)	$ 50,000
IDC (wells and related E&F)	90,000
Lease and well equipment (wells and related E&F)	30,000
Production during 2008	5,000 bbl
Total estimated proved reserves, December 31, 2008	895,000 bbl
Total estimated proved reserves recoverable from the well, December 31, 2008 (*i.e.*, proved developed reserves)	95,000 bbl

DD&A Calculations

a. First, reserves as of the beginning of the year are determined.

Proved Reserves (barrels):

Estimated proved reserves, 12/31/08—end of year	895,000
Add: Current year's production	5,000
Estimated proved reserves, 1/1/08—beginning of year	900,000

Proved Developed Reserves (barrels):

Estimated proved developed reserves, 12/31/08—end of year	95,000
Add: Current year's production	5,000
Estimated proved developed reserves, 1/1/08—beginning of year	100,000

b. Next, DD&A is calculated using the first unit-of-production formula given previously.

For Leasehold (proved property):

$$\frac{\text{Book value at year end}}{\text{Estimated reserves at beginning of year}} \times \text{Production for year}$$

$$= \frac{\$50{,}000}{900{,}000 \text{ bbl}} \times 5{,}000 \text{ bbl} = \underline{\underline{\$278}}$$

Alternatively, using the equivalent formula the calculation would be:

$$\frac{\text{Production for year}}{\text{Estimated reserves at beginning of year}} \times \text{Book value at year end}$$

$$= \frac{5{,}000 \text{ bbl}}{900{,}000 \text{ bbl}} \times \$50{,}000 = \underline{\underline{\$278}}$$

For IDC and Lease and Well Equipment (wells and related E&F):

$$\frac{\text{Book value at year end}}{\text{Estimated reserves at beginning of year}} \times \text{Production for year}$$

$$= \frac{\$120{,}000}{100{,}000 \text{ bbl}} \times 5{,}000 \text{ bbl} = \underline{\underline{\$6{,}000}}$$

Alternatively, using the equivalent formula the calculation would be:

$$\frac{\text{Production for year}}{\text{Estimated reserves at beginning of year}} \times \text{Book value at year end}$$

$$= \frac{5{,}000 \text{ bbl}}{100{,}000 \text{ bbl}} \times \$120{,}000 = \underline{\underline{\$6{,}000}}$$

Entry to record DD&A:

DD&A expense—proved property	278	
DD&A expense—wells	6,000	
Accumulated DD&A—proved property . . .		278
Accumulated DD&A—wells		6,000

EXAMPLE 2

DD&A

Data for Lucky Oil Company's partially developed lease as of December 31, 2009, are as follows:

Cost Data:

Lease bonus	$ 500,000
Other capitalized acquisition costs	40,000
Total leasehold costs at year–end	$ 540,000

Accumulated DD&A on leasehold costs at beginning of year	$ 40,000
IDC at year–end	650,000
Accumulated DD&A on IDC at beginning of year	120,000
Lease and well equipment at year–end	275,000
Accumulated DD&A on equipment at beginning of year	50,000

Reserve and Production Data:

Estimated proved developed reserves, 12/31/09	1,750,000 bbl
Estimated proved undeveloped reserves, 12/31/09	2,200,000 bbl
Production during year	50,000 bbl

DD&A Calculations

a. To calculate DD&A, first reserves as of the beginning of the year are determined. Proved reserves equal proved developed reserves plus proved undeveloped reserves.

Proved Reserves (barrels):

Estimated proved developed reserves, 12/31/09	1,750,000
Add: Estimated proved undeveloped reserves, 12/31/09	2,200,000
Estimated proved reserves, 12/31/09	3,950,000
Add: Current year's production	50,000
Estimated proved reserves, 1/1/09	4,000,000

Proved Developed Reserves (barrels):

Estimated proved developed reserves, 12/31/09	1,750,000
Add: Current year's production	50,000
Estimated proved developed reserves, 1/1/09	1,800,000

b. Second, year-end costs are determined.

Proved Property Costs:

Leasehold costs at year-end	$ 540,000
Less: Accumulated DD&A on leasehold costs	(40,000)
Net leasehold costs	$ 500,000

Wells and Related E&F:

IDC at year-end	$ 650,000
Less: Accumulated DD&A on IDC	(120,000)
Net IDC	$ 530,000

Lease and well equipment at year-end	$ 275,000
Less: Accumulated DD&A on L&WE	(50,000)
Net lease and well equipment	$ 225,000

c. Third, DD&A is calculated.*

For Proved Property Costs:

$$\frac{\text{Current year's production}}{\text{Estimated proved reserves, 1/1/09}} \times \text{Book value at year-end}$$

$$= \frac{50{,}000}{4{,}000{,}000} \times \$500{,}000 = \$6{,}250$$

Wells and Related E&F:

$$\frac{\text{Current year's production}}{\text{Estimated proved developed reserves, 1/1/09}} \times \text{Book value at year-end}$$

$$\text{IDC: } \frac{50{,}000}{1{,}800{,}000} \times \$530{,}000 = \$14{,}722$$

$$\text{L\&WE: } \frac{50{,}000}{1{,}800{,}000} \times \$225{,}000 = \$6{,}250$$

Total DD&A = $6,250 + $14,722 + 6,250 = $27,222

Entry to record DD&A

DD&A expense	27,222	
Accumulated DD&A—proved property . . .		6,250
Accumulated DD&A—IDC		14,722
Accumulated DD&A—L&WE		6,250

* Although this example shows DD&A expense computed separately for IDC versus lease and well equipment, separate computation is not necessary.

DD&A when oil and gas reserves are produced jointly

Most reservoirs contain both oil and gas. In those cases, it is necessary to:

Convert oil and gas reserves and oil and gas produced to a common unit of measure based on relative energy content. Energy content is measured by the British thermal unit (Btu) and varies somewhat from reservoir to reservoir. Most companies use a generally accepted industry average, converting the Btu content of a barrel of oil to the Btu content of an Mcf of gas at an approximate rate of six to one. In other words, one barrel of oil is approximately equal to 6 Mcf of gas in terms of energy content. The conversion can be made by either dividing the Mcf of gas by six to get barrels of oil equivalent (BOE), or multiplying the barrels of oil by six to get equivalent Mcf's.

However, if the relative proportion of oil to gas extracted in the current period from the reservoir is expected to remain the same, then amortization may be computed based on only one of the two minerals—either oil or gas.

Or if either oil or gas clearly dominates both the reserves and the current production based on relative energy content, then amortization may be computed based on the dominant mineral only (*SFAS No. 19,* par. 38).

Therefore, if oil and gas reserves are produced jointly, one of three different amortization methods may be used, assuming the conditions given above are satisfied:

1. common unit of measure—converting to common energy unit
2. same relative proportion—using either oil or gas
3. dominant mineral—using the dominant mineral

Additionally, measurement and pricing units other than Mcf's may be used for gas. For example, gas may be measured and priced in terms of MMBtu (million British thermal units) rather than Mcf. If a measurement unit other than Mcf is used for gas, the conversion ratio must be calculated based on the units of measurement being used. However, the underlying conversion process is the same process as described above and as illustrated in the following example.

EXAMPLE

Joint Production DD&A

Lucky Oil Company has a *fully developed* producing lease that has both oil and gas reserves. Data for the lease are as follows. (In a fully developed lease the proved reserves and proved developed reserves are the same amount.)

Net capitalized costs, December 31	$2,200,000
Estimated proved developed reserves, December 31:	
Oil	400,000 bbl
Gas	1,800,000 Mcf
Production during the year:	
Oil	50,000 bbl
Gas	240,000 Mcf

1. Assume DD&A is determined based on a common unit of measure, either **equivalent barrels** or **Mcf's.** If BOEs are to be used, then gas would be converted to BOEs; if equivalent Mcf's are to be used, then oil would be converted to equivalent Mcf's. Both total annual production and total proved and proved developed reserves would be converted. These alternative calculations are illustrated below:

BOE:

Total production during year:

Oil		50,000 bbl
Gas	240,000/6 =	40,000 BOE
Total		90,000 BOE

Total proved developed reserves, December 31:

Oil		400,000 bbl
Gas	1,800,000/6 =	300,000 BOE
Total		700,000 BOE

Equivalent Mcf:

Total production during year:

Oil	50,000 x 6 =	300,000 equivalent Mcf
Gas		240,000 Mcf
Total		540,000 equivalent Mcf

Total proved developed reserves, December 31:

Oil	400,000 x 6 =	2,400,000 equivalent Mcf
Gas		1,800,000 Mcf
Total		4,200,000 equivalent Mcf

In computing amortization, the next step is to determine proved developed reserves at the beginning of the year:

Proved developed reserves, beginning of year (in BOE):

Proved developed reserves, 12/31	700,000 BOE
Add: Production	90,000 BOE
Proved developed reserves, 1/1	790,000 BOE

Alternatively, proved developed reserves, beginning of year (in equivalent Mcf):

Proved developed reserves, 12/31	4,200,000 equivalent Mcf
Add: Production	540,000 equivalent Mcf
Proved developed reserves, 1/1	4,740,000 equivalent Mcf

Calculate DD&A:

Using BOE:

$$\frac{\$2,200,000}{790,000 \text{ BOE}} \times 90,000 \text{ BOE} = \$250,633$$

Using equivalent Mcf:

$$\frac{\$2,200,000}{4,740,000 \text{ eq. Mcf}} \times 540,000 \text{ eq. Mcf} = \$250,633$$

Entry

DD&A expense	250,633	
Accumulated DD&A		250,633

2. *Instead* assume the relative proportion of oil to gas extracted in the current period is expected to remain the same and that Lucky Company decides to calculate DD&A using only gas.

Calculations

To calculate DD&A, first reserve estimates as of the beginning of the year are determined.

Proved developed reserves:

Proved developed reserves, 12/31	1,800,000 Mcf
Add: Current year's production	240,000 Mcf
Proved developed reserves, 1/1	2,040,000 Mcf

DD&A computed using only Mcf:

$$\frac{\$2{,}200{,}000}{2{,}040{,}000 \text{ Mcf}} \times 240{,}000 \text{ Mcf} = \$258{,}824$$

Entry

DD&A expense	258,824	
Accumulated DD&A		258,824

3. *Instead* assume that oil is clearly the dominant mineral and that Lucky Company decides to calculate DD&A based on the dominant mineral. Therefore, only oil reserves should be used in computing DD&A.

Calculations

To calculate DD&A, first reserve estimates as of the beginning of the year are determined.

Proved developed reserves:

Proved developed reserves, 12/31	400,000 bbl
Add: Current year's production	50,000 bbl
Proved developed reserves, 1/1	450,000 bbl

DD&A computed using only oil barrels:

$$\frac{\$2{,}200{,}000}{450{,}000 \text{ bbl}} \times 50{,}000 \text{ bbl} = \$244{,}444 \text{ (rounded)}$$

Entry

DD&A expense	244,444	
Accumulated DD&A		244,444

DD&A on a field-wide basis

As noted previously, the cost center for successful-efforts accounting is a property or some reasonable aggregation of properties based on a common geological feature, such as a field or reservoir (*SFAS No. 19,* par. 30, 35). Properties must be related geologically to be combined for amortization purposes, such as where several properties cover a single reservoir or a single field. When a field or reservoir is used as a cost center, multiple leases or properties will be grouped together.

If all the properties in a field or reservoir amortization base are proved properties, then all of the leasehold costs will be aggregated and amortized over the total proved reserves of the field or reservoir. If the field or reservoir grouping contains some properties that are proved and some properties that are unproved, only the proved property leasehold costs would be amortized. The total capitalized costs for wells and related equipment and facilities would also be aggregated and amortized over the total proved developed reserves of the field or reservoir.

(Note that the term amortization used in this chapter to refer to amortization of proved properties using the unit-of-production formula is also used in chapter 4 to refer to impairment of unproved properties assessed on a group basis.)

EXAMPLE

Field DD&A

Lucky Company has interests in three leases, all obtained in January 2006, which cover the same field. Two of the leases have been explored and proved. The well on the third lease, Lease C, is incomplete and the lease remains unproved. (The costs of the well on Lease C would thus be in wells in progress accounts versus wells and related E&F accounts.) Lucky decides to use the field as the amortization base for DD&A. Information for each lease as of December 31, 2006, is as follows:

	Lease A	Lease B	Lease C
Leasehold costs	$ 50,000	$ 40,000	$ 43,000
IDC costs	102,000	81,000	80,000
Lease and well equipment costs	40,000	22,000	21,000
Production during 2006	8,000 bbl	6,000 bbl	0
Estimated proved reserves, 12/31/06	950,000 bbl	650,000 bbl	0
Estimated proved developed reserves, 12/31/06	100,000 bbl	50,000 bbl	0

Field-wide production and reserves:

Production		Proved Reserves, 12/31/06	
Lease A	8,000 bbl	Lease A	950,000 bbl
Lease B	6,000 bbl	Lease B	650,000 bbl
Total	14,000 bbl	Total	1,600,000 bbl

Proved Developed Reserves, 12/31/06	
Lease A	100,000 bbl
Lease B	50,000 bbl
Total	150,000 bbl

Proved property DD&A: Leasehold costs to be amortized equal $90,000, the costs of leases A and B. Lease C is unproved as of 12/31/06 and so is not included in amortization.

$$\frac{\$90,000}{1,600,000 + 14,000^*} \times 14,000 = \underline{\underline{\$781}}$$

Wells and related E&F DD&A: IDC and well and equipment costs to be amortized equal $245,000, the costs of IDC and equipment on leases A and B. The well on Lease C is uncompleted, and as such is not included in amortization.

$$\frac{\$245,000}{150,000 + 14,000^*} \times 14,000 = \underline{\underline{\$20,915}}$$

* The "14,000 barrels" was added in the denominator of both of the above calculations in order to convert end of the year reserve estimates to beginning of the year reserve estimates.

Estimated future dismantlement, site restoration, abandonment costs, and salvage values

According to *SFAS No. 19,* par. 37, estimated future dismantlement, site restoration, abandonment costs, and estimated salvage values must be taken into account in determining DD&A rates. This paragraph prescribes only that these costs must be taken into account in determining DD&A rates, not how the costs are to be estimated or whether they should be recorded. These costs can be quite significant, and as such their accounting treatment is important.

The following example illustrates two different methods of accounting for estimated future dismantlement, site restoration, and abandonment costs. In the first method, the costs are estimated and recorded as part of the wells and related E&F and as a liability at the time the wells are drilled and equipment installed. In the second method, these costs are accrued as a liability as production and amortization take place. In the second method, the costs are never added to the cost of the underlying asset.

EXAMPLE

Future Dismantlement, Restoration and Abandonment Costs

During 2006, Lucky Oil Company completed development of an offshore lease. Total capitalized drilling and development costs were $70,000,000, and it was estimated that future dismantlement costs would be $15,000,000 and future salvage value would be $2,000,000. (Ignore leasehold costs.) Following are two methods used in practice for recording these costs:

a. The estimated future costs are added to the cost of the asset and a liability is recorded when the equipment or facilities are put in place. Accordingly, the drilling and development, future dismantlement, restoration, and abandonment costs are recorded as follows:

Entry

Wells and related E&F	85,000,000	
Cash		70,000,000
Liability		15,000,000

Assuming proved developed reserves of 3,000,000 barrels *at the beginning* of the year and production of 30,000 barrels, amortization of the costs would be:

$$\frac{\$85{,}000{,}000 - \$2{,}000{,}000}{3{,}000{,}000 \text{ bbl}} \times 30{,}000 \text{ bbl} = \underline{\underline{\$830{,}000}}$$

Entry

DD&A expense	830,000	
Accumulated DD&A		830,000

b. Instead assume only actual drilling and development costs are recorded. However, even though not recorded, future dismantlement, restoration, and abandonment costs must still be included in the DD&A calculation as follows:

Entry

Wells and related E&F	70,000,000	
Cash		70,000,000

DD&A Calculation

$$\frac{\$70{,}000{,}000 - \$2{,}000{,}000}{3{,}000{,}000 \text{ bbl}} \times 30{,}000 \text{ bbl} = \$680{,}000$$

$$\frac{\$15{,}000{,}000}{3{,}000{,}000 \text{ bbl}} \times 30{,}000 \text{ bbl} = \underline{150{,}000}$$

$$\text{Total DD\&A } (\$680{,}000 + \$150{,}000) = \underline{\underline{\$830{,}000}}$$

Entry

DD&A expense	830,000	
Accumulated DD&A		680,000
Liability		150,000

Note that the total DD&A amounts are the same regardless of the method.

If the lease had not been fully developed, proved reserve quantities would have been different from proved developed reserve quantities. In this situation, companies typically treat future dismantlement, restoration, and abandonment costs as being related to the wells and related equipment and facilities; thus, these costs are amortized using proved developed reserves as opposed to proved reserves.

The treatment of future dismantlement, restoration, and abandonment costs has received a considerable amount of attention from accounting authorities during the past few years.[1] The FASB is planning to issue a standard in 2000. In February 1996, the FASB issued an Exposure Draft entitled *Accounting for Certain Liabilities Related to Closure or Removal of Long-Lived Assets.* The Board subsequently decided to re-evaluate the issues and eventually re-named the project *Accounting for Obligations Associated with the Retirement of Long-Lived Assets.* The FASB has indicated that they plan to require estimated future dismantlement, restoration, and abandonment costs be added to the cost of the underlying asset and a liability be recognized at the time the wells are drilled or the equipment is installed. The costs are to be estimated at fair market value or using a probability weighted present value approach.

Exclusion of costs or reserves

Generally, the capitalized cost of all wells and related equipment and facilities in a cost center (*i.e.*, property, field, or reservoir) are to be amortized over the proved developed reserves recoverable from that cost center. This requirement poses a problem in situations where significant costs of a development project have been incurred but a portion of the proved reserves are still undeveloped.

For example, assume that an offshore platform capable of producing 5,000,000 barrels of oil is constructed at a cost of $5,000,000. However, at the time the platform goes into service only 2 of the anticipated 10 wells (each capable of producing $^1/_{10}$ of the proved reserves) have been drilled. The $5,000,000 cost would be capitalized; however only $^2/_{10}$ of the reserves are proved developed ($^8/_{10}$ of the proved reserves are proved undeveloped). Since the cost of wells and related equipment and facilities are to be amortized over proved developed reserves, unless an adjustment is made, the $5,000,000 cost would be amortized over only a portion ($^2/_{10}$) of the related reserves. Consequently, unless a portion of the cost of the platform is excluded from the DD&A calculation, costs and reserves will be mismatched in the DD&A calculation.

Therefore, in situations where a significant development expenditure (*i.e.,* the platform) has been made, but a portion of the related proved reserves are not yet developed, *SFAS No. 19* requires a portion of those development costs be excluded in determining the DD&A rate. The exclusion of a portion of the significant development costs is to continue until additional wells are drilled and all of the related reserves are proved developed (*SFAS No. 19,* par. 35).

Similarly, if proved developed reserves will be produced only after significant additional development costs are incurred, *e.g.,* an improved recovery system, then those proved developed reserves must be excluded in determining the DD&A rate (*SFAS No. 19,* par. 35). In this case, the mismatching occurs because the reserves are known but the costs have not yet been incurred. Note, however, that proved developed reserves are defined to be reserves that are expected to be recovered from existing wells with existing equipment and operating methods. Thus, if the proved reserves are developed, then the related development costs have normally already been incurred. However, when an improved recovery system is involved, reserves may be classified as proved developed reserves after testing by a pilot project or after the operation of an installed program has confirmed that increased recovery will result. Thus, by definition, the situation of proved developed reserves being produced only after significant future development costs are incurred should not arise unless an improved recovery system is involved.

EXAMPLE

Exclusion of Costs

Lucky Oil Company has constructed an offshore drilling platform costing $18,000,000. At the end of 2007, only 2 wells have been drilled at a cost of $3,000,000, with 16 more wells to be drilled in the future. Proved developed reserves at year-end were 3,000,000 barrels, and 300,000 barrels were produced during the year.

DD&A Calculation: A portion of the drilling platform must be excluded in computing DD&A. Of the 18 total wells to be drilled, only 2 have been completed; therefore, 16 will be completed in the future. In this case, $^{16}/_{18}$ of the \$18,000,000 platform cost would be excluded.

$$\text{DD\&A expense} = \frac{(\$18{,}000{,}000 \times {}^{2}/_{18}) + \$3{,}000{,}000}{3{,}000{,}000 + 300{,}000} \times 300{,}000 = \underline{\underline{\$454{,}545}}$$

In the preceding example, the amount of development costs to be amortized was determined based on the ratio of wells already drilled over total wells, both drilled and yet to be drilled. Other reasonable methods that may be used include drilling costs incurred over total expected drilling costs or proved developed reserves over total proved reserves.

The following example illustrates the situation in which proved developed reserves will be produced only after significant additional development costs are incurred.

EXAMPLE

Exclusion of Reserves

Lucky Oil Company has an offshore lease that has proved developed reserves of 50,000,000 barrels at the beginning of the year. Of those 50,000,000 barrels, 10,000,000 are associated with significant development costs to be incurred in the future. Total capitalized drilling and equipment costs (*i.e.,* wells and related equipment and facilities) at the end of the year are \$3,000,000. Production during the year was 250,000 barrels.

DD&A Calculation

Proved developed reserves associated with the future development costs must be excluded in calculating DD&A.

$$\text{DD\&A expense} = \frac{\$3{,}000{,}000}{50{,}000{,}000 - 10{,}000{,}000} \times 250{,}000 = \underline{\underline{\$18{,}750}}$$

Note that because the reserve estimate is already as of the beginning of the year, production is not added in the denominator of the DD&A calculation.

Under no circumstances should future development costs be included in computing DD&A expense under successful-efforts accounting.

Depreciation of support equipment and facilities

When support equipment and facilities are acquired, they must be analyzed to resolve two critical issues:

1. Will the equipment or facility serve only one cost center?
2. What activities (property acquisition, exploration, drilling, production, etc.) will the equipment or facility support?

Support equipment and facilities that service a particular field or other area constituting a cost center should be capitalized to that cost center and depreciated using the unit-of-production method over the proved developed reserves of the cost center. This situation poses little difficulty either from a theoretical or practical standpoint. When support equipment and facilities cannot be identified with a single field or cost center, the unit-of-production method may not be appropriate. For example, a warehouse may service numerous fields. Since the warehouse is not related to a particular field or cost center, the unit-of-production is not used. Instead the straight-line, unit-of-output method (based on miles driven or usage), sum-of-the-years-digits, or some other acceptable method should be used.

Similar issues arise when support equipment and facilities are used for more than one activity (*i.e.,* property acquisition, exploration, development, and production) on multiple cost centers. For example, a warehouse might store equipment used in both drilling and production activities for all of the properties in a large geographical area. Similarly, a truck might be used by the foreman to travel between drilling locations and production locations covering a large geographical area. A field office building could be used to support all types of activities for numerous properties.

In each of these cases, the equipment or buildings should be depreciated using a method other than unit-of-production. The depreciation would then be allocated to the activities served based on usage and either capitalized or expensed, depending on the activity. For example, if a truck is serving production activities, drilling activities, and development activities in multiple fields or cost centers, it would be depreciated using straight-line, unit-of-output, or some other method. The depreciation would then be allocated to the activities being served (*i.e.,* production, drilling, and development) and to the different properties. The portion allocated to production activities would be written off as operating expense. The portion allocated to nondrilling development costs would be capitalized to lease and well equipment for each of the cost centers the truck is serving. The cost would then be depreciated along with the other capitalized costs for the cost center using the unit-of-production method. The portion allocated to drilling would be capitalized to wells in progress for the wells being drilled in the various cost centers being served. The wells in progress accounts would subsequently be cleared to dry hole expense (if the wells to which the cost was allocated were dry exploratory wells) or lease and well equipment if the wells being drilled were development wells, service wells, or successful exploratory wells.

Support equipment and facilities serving multiple activities within a single cost center are typically capitalized directly to the cost center and amortized with wells and related E&F using unit-of-production over the life of the cost center. Although depreciation of support equipment and facilities should theoretically be allocated to exploration, development, or production as appropriate, in practice, the depreciation on the support equipment and facilities is not separated and allocated between multiple activities. Instead, it is expensed as DD&A expense. This treatment may result in minor differences in the timing of expense recognition, however, these differences are not likely to be material.

Buildings and equipment that cannot be related to specific activities (*i.e.,* the home office building) should be depreciated using straight-line, unit-of-output, or some other method. The depreciation would then become part of general overhead.

EXAMPLE

Depreciation of Support Equipment and Facilities

Assume Lucky owns the following assets that are used to support operations in various cost centers during 2005:

Warehouse: Purchased on 1/1/03 for $100,000 with an expected life of 20 years. Approximately 80% of the equipment stored at the warehouse is ultimately used in production operations and 20% in drilling operations. During the current year, 10 production operations were served and 5 wells were drilled. Ignore salvage value.

$$\text{Depreciation for the year is: } \frac{\$100{,}000}{20} = \underline{\underline{\$5{,}000/\text{yr}}}$$

The $5,000 depreciation would be allocated to production operations and drilling operations as follows:

Production:	$5,000 x 0.80 =	$4,000
Drilling:	$5,000 x 0.20 =	$1,000

The $4,000 related to production operations would then the allocated to the specific properties served and recorded as operating expense. The $1,000 related to drilling operations would be allocated to the specific wells being drilled. The $1,000 would be capitalized to wells in progress and cleared according to whether the wells were exploratory-type or development-type wells and whether they were successful or dry.

Automobile: Purchased on 5/3/03 for $30,000. It is estimated that the automobile will be driven a total of 100,000 miles during its useful life. Accumulated depreciation is $5,000. This year the automobile was driven 5,000 miles related to property acquisition (50% of the properties under consideration were leased), 10,000 miles related to production activities, and 5,000 miles related to the drilling of development wells.

$$\text{Depreciation for the year is: } \frac{\$30{,}000}{100{,}000} = \underline{\underline{\$.30/\text{mile}}}$$

Property acquisition: 5,000 miles x $.30 = $1,500. The $1,500 would be allocated to the properties that were considered and then would be capitalized if the properties were leased or charged to expense if the properties were not leased.

Production: 10,000 miles x $.30 = $3,000. The $3,000 would be allocated to the producing properties being served and then charged to operating expense.

Drilling: 5,000 miles x $.30 = $1,500. The $1,500 would be allocated to the wells that were being drilled and then capitalized (since the wells were development wells).

Cost disposition—nonworking interests

Nonworking interests should theoretically be amortized over proved reserves using the same unit-of-production formula utilized for working interests in proved properties. However, individual nonworking interests may be insignificant in value, and the reserve quantity information necessary to compute unit-of-production amortization is often not obtainable from the related working interest owner. Paragraph 30 of *SFAS No. 19* states with regard to royalty interests:

> *When an enterprise has a relatively large number of royalty interests whose acquisition costs are not individually significant, they may be aggregated, for the purpose of computing amortization, without regard to commonality of geological structural features or stratigraphic conditions; if information is not available to estimate reserve quantities applicable to royalty interests owned (paragraph 50), a method other than the unit-of-production method may be used to amortize their acquisition costs.*

Thus, in contrast to working interests, nonworking interests may be amortized using a method other than unit-of-production, such as straight-line; further, if the nonworking interests are not significant, the interests may be aggregated without regard to commonality. In practice, nonworking interests are commonly aggregated and amortized on a straight-line basis over eight or ten years (*COPAS Bulletin No. 10,* p.90).

The dollar amount of nonworking interest acquisition costs to be amortized depends upon the method of acquisition. Nonworking interests may be acquired in several different ways. When mineral rights are purchased separately from the surface rights, the purchase price is the amount to be amortized. If purchased in fee, the cost of the property acquired in fee should be allocated between the surface and the mineral rights on a fair market value (FMV) basis. The cost allocated to the mineral rights is the amount to be amortized.

An overriding royalty interest (ORI) is created out of a working interest by being either retained or carved out. An ORI is carved out when the WI owner sells an ORI to another party or when the WI owner conveys an ORI to another party for some other reason, such as compensation for services rendered. If awarded for services rendered, the party receiving the ORI should assign the ORI the fair market value of the services rendered. An ORI is created by being retained when the WI owner sells the WI and retains an ORI, or when the WI owner transfers the WI to another party willing to develop the lease and retains an ORI. The amount assigned to a retained ORI depends upon how it was created. For more details, see chapter 12. An example of an ORI awarded as compensation follows:

EXAMPLE

DD&A on ORI

Paul Jones, a landman, received a $^1/_{10}$ ORI for his services in obtaining a lease for Lucky Company. The FMV of Jones' services as a landman is $4,000. Total proved reserves related to the ORI at the end of the first year of production (2007) were 18,000 barrels. The ORI's share of production during the year was 2,000 barrels. Prepare journal entries for Paul Jones and compute DD&A for 2007.

Entry to record ORI at acquisition

Investment in ORI	4,000	
Revenue		4,000

Entry to record DD&A

DD&A expense—ORI	400	
Accumulated DD&A—ORI		400

DD&A Computation

$$\frac{2{,}000}{18{,}000 + 2{,}000} \times \$4{,}000 = \underline{\underline{\$400}}$$

EXAMPLE

Multiple Nonworking Intersts

Lucky Oil Company has multiple small nonworking interests located in Texas. The reserve information necessary to compute unit-of-production DD&A is not available. Further, the interests are individually insignificant and the benefit received from computing DD&A for the interests on an individual basis does not justify the cost. The interests have been assigned a total cost of $60,000. Lucky amortizes such interests straight-line over 10 years.

DD&A Computation

$$\frac{\$60{,}000}{10} = \underline{\underline{\$6{,}000}}$$

Another type of nonworking interest is a production payment interest (PPI). A PPI, which generally has a shorter life than the total life of the reservoir, is not always an economic interest for which amortization may be recognized. A PPI that is an economic interest is one that is acquired by purchasing a portion of proved reserves. The cost paid for the proved reserves should be amortized as production takes place, using total purchased reserves as the base. An illustration of a PPI follows. (PPIs are discussed in greater detail in chapter 12.)

EXAMPLE

DD&A on PPI

Paul Jones purchased a 10,000 barrel PPI for $150,000 from Lucky Company. The PPI is to be paid at the rate of $\frac{1}{4}$ of the WI's share of production from Lease B. Lease B is burdened with a $\frac{1}{6}$ RI. Total gross proved reserves at 12/31/07 were 40,000 barrels. Total gross production from Lease

B during 2007 was 6,000 barrels. DD&A and entries for Jones for 2007 are as follows:

DD&A Computations

Production for 2007	6,000 bbl
Less: $\frac{1}{6}$ RI ($\frac{1}{6}$ x 6,000)	1,000 bbl
Production for WI and PPI	5,000 bbl
Production to PPI ($\frac{1}{4}$ x 5,000)	1,250 bbl

DD&A for PPI:

$$\frac{1,250}{10,000} \times \$150,000 = \underline{\underline{\$18,750}}$$

In the DD&A computation, 10,000 barrels are used rather than the 40,000 barrels of proved reserves because 10,000 is the number of barrels purchased by Paul Jones. Thus, DD&A is based on barrels received during the year relative to the total number to be received.

Entries

Investment in PPI	150,000	
Cash		150,000
DD&A expense—PPI	18,750	
Accumulated DD&A—PPI		18,750

Revision of DD&A rates

Reserves estimates that effect DD&A rates are required to be reviewed at least annually. Any resulting revisions to DD&A rates should be accounted for prospectively as a change in estimate (*SFAS No. 19,* par. 35) *i.e.,* no changes should be made to adjust accounts to what they would have been had the new estimate been used throughout the period. If a company reports on a yearly basis, the revised reserve estimate is used in the year-end DD&A calculation. If a company reports on a quarterly basis, then the effect of a change in an accounting estimate should be accounted for in the period in which the change is made. According to *APB No 28,* no restatement of previously reported interim information should be made for changes in estimates; however, interim periods are considered an integral part of the annual period rather than a discrete time period.

The following example illustrates one method used in revising DD&A rates when a new reserve estimate is obtained. In the method shown, the fourth quarter DD&A amount—the quarter in which the revised estimate was received—is determined by using the new reserve estimate converted to an estimate as of the beginning of the year. Note that, as would be the case with a change in estimate, DD&A amounts for the first three quarters remain unchanged. DD&A for the year equals the sum of the four quarterly amounts.

EXAMPLE 1

Revision of DD&A Rates

Lucky Oil Company reports on a quarterly basis. On December 2, 2006, the company received a new reserve report dated November 30, 2006, concerning a fully developed lease in Texas. The reserve report showed proved developed reserves of 450,000 barrels. The last report, dated December 31, 2005, showed reserves of 400,000 barrels. Net capitalized costs as of December 31, 2005, were $1,000,000. Production and amortization through the third quarter of 2006 were as follows:

Quarter	Production	Amortization	Calculations
1	20,000	$50,000	$1,000,000/400,000 x 20,000
2	16,000	40,000	$1,000,000/400,000 x 16,000
3	22,000	55,000	$1,000,000/400,000 x 22,000

October and November production	10,000
December production	13,000
Total fourth quarter production	23,000

DD&A Calculations

For fourth quarter:
Reserve estimate as of the beginning of the year using the new estimate:

Reserve estimate, November 30, 2006	450,000 bbl
Add: Production 1st quarter	20,000
Production 2nd quarter	16,000
Production 3rd quarter	22,000
Production during October and November	10,000
Reserve estimate, January 1, 2006	518,000 bbl*

* The new reserve estimate is an estimate of reserves in place as of November 30, 2006. Thus, to obtain estimated reserves as of the beginning of the year (518,000 barrels), production from the first of the year through November must be added to the new reserve estimate.

DD&A, fourth quarter: $\dfrac{\$1,000,000}{518,000} \times 23,000 \text{ bbl} = \underline{\underline{\$44,402}}$

The fourth quarter computation of DD&A expense treats the fourth quarter as a discrete time period.

DD&A for full year:

Amortization for the first three quarters	$ 50,000
	40,000
	55,000
Amortization for the fourth quarter	44,402
Amortization for the full year	$189,402

Amortization for the year includes the first three quarters' computations (unchanged), plus the fourth quarter amount computed using the new reserve estimate.

Another widely accepted method for revising DD&A rates involves computing the entire DD&A for the year by using the new estimate and then subtracting the previously recognized quarterly amounts to arrive at the current quarterly amount to be recognized. The following example illustrates this method using exactly the same data as in the previous example:

EXAMPLE 2

Revision of DD&A Rates

Lucky Oil Company reports on a quarterly basis. December 2, 2006, the company received a new reserve report dated November 30, 2006, concerning a fully developed lease in Texas. The reserve report showed proved developed reserves of 450,000 barrels. The last report, dated December 31, 2005, showed reserves of 400,000 barrels. Net capitalized costs as of December 31, 2005, were $1,000,000. Production and amortization through the third quarter of 2006 were as follows:

Quarter	Production	Amortization	Calculations
1	20,000	$ 50,000	$1,000,000/400,000 x 20,000
2	16,000	40,000	$1,000,000/400,000 x 16,000
3	22,000	55,000	$1,000,000/400,000 x 22,000
	58,000	$145,000	

October and November production	10,000
December production	13,000
Total fourth quarter production	23,000

Production for year:

Quarter	Production
1	20,000
2	16,000
3	22,000
4	23,000
Total production for year	81,000

DD&A Calculations

Reserve estimate as of the beginning of the year using the new estimate:

Reserve estimate, November 30, 2006	450,000 bbl
Add: Production 1st quarter	20,000
Production 2nd quarter	16,000
Production 3rd quarter	22,000
Production during October and November	10,000
Reserve estimate, January 1, 2006	518,000 bbl

The beginning of the year reserve estimate, 518,000 barrels, is computed exactly the same as in the previous example.

DD&A for full year: $\frac{\$1,000,000}{518,000} \times 81,000 \text{ bbl} = \$156,371$

For fourth quarter:

Amortization for the first three quarters	$ 50,000
	40,000
	55,000
	$145,000

DD&A for year	$156,371
Less: Total DD&A for first three quarters	(145,000)
DD&A for fourth quarter	$ 11,371

DD&A for the first three quarters is unchanged. The fourth quarter DD&A amount is determined by subtracting total DD&A for the first three quarters from DD&A for the full year.

Cost Disposition through Abandonment or Retirement of Proved Property

When proved property or wells and related equipment and facilities are abandoned, no gain or loss is normally recognized until the last well ceases to produce and the entire amortization base is abandoned. Until that time, any well, item of equipment, or lease that is abandoned and that is part of an amortization base should be treated as fully amortized and charged to accumulated DD&A. However, if the abandonment or retirement results from a catastrophic event, a loss should be recognized (*SFAS No. 19,* par. 41).

> *Normally, no gain or loss shall be recognized if only an individual well or individual item of equipment is abandoned or retired or if only a single lease or other part of a group of properties constituting the amortization base is abandoned or retired as long as the remainder of the property or group of properties continues to produce oil or gas. Instead, the asset being abandoned or retired shall be deemed to be fully amortized, and its cost shall be charged to accumulated depreciation, depletion, or amortization. When the last well on an individual property (if that is the amortization base) or group of properties (if amortization is determined on the basis of an aggregation of properties with a common geological structure) ceases to produce and the entire property or property group is abandoned, gain or loss shall be recognized. Occasionally, the partial abandonment or retirement of a proved property or group of proved properties or abandonment or retirement of wells or related equipment or facilities may result from a catastrophic event or other major abnormality. In those cases, a loss shall be recognized at the time of abandonment or retirement. (SFAS No. 19, par. 41)*

EXAMPLE

Well Abandonment

Lucky Oil Company abandons a well with total capitalized costs of $500,000. The well is located on a lease with 10 other producing wells. Total accumulated DD&A for wells and related equipment and facilities on this lease was $1,500,000. No equipment was salvaged.

Entry to record abandonment

Accumulated DD&A	500,000	
Wells and related E&F		500,000

Since the well is amortized as part of a larger group of equipment, it is assumed that the accumulated depreciation related to that individual well is not determinable; thus, the entire cost of the well, net of any salvage value, is charged against the accumulated DD&A account. However, the entry would have been different if the well were the last one in the amortization base and the entire amortization base were being abandoned. See below:

EXAMPLE

Lease Abandonment

Lucky Oil Company abandons a lease with capitalized acquisition costs of $70,000 and capitalized drilling and equipment costs of $200,000. Equipment worth $30,000 was salvaged.

a. The lease was part of a field that had total accumulated DD&A for proved property of $300,000 and total accumulated DD&A for wells and related equipment of $500,000. This lease was not the last producing lease in the field.

Entry

Materials and supplies (salvage)	30,000	
Accumulated DD&A—proved property . . .	70,000	
Accumulated DD&A—wells	170,000	
Proved property		70,000
Wells and related E&F		200,000

Again, since this lease is amortized as part of a larger group of properties, it is assumed that the accumulated depreciation related to that lease is not determinable; thus, the entire cost of the lease, net of any salvage value, is charged against accumulated DD&A.

b. The lease was the last producing lease in the field. All other properties have already been abandoned. The balance in accumulated DD&A for proved property was $65,000 and accumulated DD&A for wells and related equipment totaled $155,000.

Entry

Materials and supplies (salvage)	30,000	
Accumulated DD&A—proved property . .	65,000	
Accumulated DD&A—wells	155,000	
Surrendered lease expense	20,000	
Proved property		70,000
Wells and related E&F		200,000

When the last well on the last lease in the field or cost center is plugged and the lease is abandoned then all of the remaining asset accounts and the accumulated DD&A accounts are written off and a gain or loss recognized.

If a flood, fire, blowout, or some other catastrophic event occurs, gain or loss recognition is required. In that case the accumulated DD&A balances would be apportioned between the equipment and IDC that were lost and the remaining equipment and IDC. The net difference between the accumulated DD&A allocated to (a) the lost equipment and IDC and (b) the original cost of the lost equipment and IDC, would be recognized as a loss.

When a portion of a proved property or group of properties being accounted for as a single cost center is sold by a successful efforts company, the sale may be treated as a normal retirement with no gain or loss recognized so long as non-recognition of gain or loss does not have a material effect on the unit-of-production amortization rate for the cost center. If there would be a material change or distortion in the amortization rate then a gain or loss should be recognized (*SFAS No. 19,* par. 47j).

EXAMPLE

Sale

Lucky Company sold a pump costing $2,000 for $200 salvage. The entry to record the sale of the pump would be:

Entry

Cash	200	
Accumulated DD&A—Equipment	1,800	
Equipment		2,000

If a single piece of equipment is sold, gain or loss recognition is not required, assuming the amortization rate is not significantly affected. The piece of equipment would be treated as if it were fully depreciated and the cost charged against the accumulated DD&A account.

To illustrate the concepts relating to proved property cost disposition under SE as well as to tie together the material learned in the previous chapters, the following comprehensive example is presented:

COMPREHENSIVE EXAMPLE

a. On February 2, 2005, Lucky Company acquired a lease burdened with a $^1/_6$ royalty for $100,000. The property was undeveloped and the acquisition costs were considered to be individually significant.

Entry

Unproved property	100,000	
Cash		100,000

b. On February 2, 2006, a delay rental of $3,000 was paid.

Entry

Delay rental expense	3,000	
Cash		3,000

c. Several dry holes were drilled on surrounding leases during 2006. As a result, on December 31, 2006, Lucky Company decided the lease was 40 percent impaired.

Entry

Impairment expense ($100,000 x 40%) ..	40,000	
Allowance for impairment		40,000

d. During 2007, Lucky Company drilled a dry hole at a cost of $350,000 for IDC and $35,000 for equipment. The equipment has no salvage value.

Entry to accumulate costs

W/P—IDC	350,000	
W/P—L&WE	35,000	
Cash		385,000

Entry to record dry hole

Dry-hole expense—IDC	350,000	
Dry-hole expense—L&WE	35,000	
W/P—IDC		350,000
W/P—L&WE		35,000

e. Undiscouraged, Lucky Company drilled another exploratory well in February at a cost of $500,000 for IDC and $175,000 for equipment. The well found proved reserves.

Entry to accumulate costs

W/P—IDC	500,000	
W/P—L&WE	175,000	
Cash		675,000

Entry to record completion of well

Wells and related E&F—IDC	500,000	
Wells and related E&F—L&WE	175,000	
W/P—IDC		500,000
W/P—L&WE		175,000

Entry to reclassify property as proved

Proved property	60,000	
Allowance for impairment	40,000	
Unproved property		100,000

f. A total of 12,000 barrels of oil were produced from the successful well during 2007. Related lifting costs were $4 per barrel. (Lifting costs, which are production costs, should be expensed as lease operating expense. Note that production costs are based on total production from the lease; whereas, revenue and the calculation of DD&A expense for Lucky would be based only on the working interest's share of production.)

Entry

Lease operating expense ($4 x 12,000)	48,000	
Cash		48,000

g. During December Lucky Company began drilling a third exploratory well. Accumulated costs by December 31 were IDC of $100,000 and equipment costs of $15,000.

Entry

W/P—IDC	100,000	
W/P—L&WE	15,000	
Cash		115,000

h. The reserve report as of December 31, 2007 and production during 2007 for Lucky Company were as follows:

Proved reserves	900,000 bbl
Proved developed reserves	300,000 bbl
Production	10,000 bbl

DD&A calculation

(assume the lease constitutes a separate amortization base)

Acquisition costs:

$$\frac{\$60,000}{900,000 + 10,000} \times 10,000 = \underline{\underline{\$659}}$$

Wells and related E&F:

$$\frac{\$675,000}{300,000 + 10,000} \times 10,000 = \underline{\underline{\$21,774}}$$

Entry

DD&A expense—proved property	659	
DD&A expense—wells	21,774	
Accumulated DD&A—proved property		659
Accumulated DD&A—wells		21,774

i. During 2008, Lucky Company completed the third well at an additional cost of $300,000 for IDC and $175,000 for equipment. The well was successful.

Entry to record additional costs

W/P—IDC	300,000	
W/P—L&WE	175,000	
Cash		475,000

Entry to record successful well

Wells and related E&F—IDC	400,000	
Wells and related E&F—L&WE	190,000	
W/P—IDC		400,000
W/P—L&WE		190,000

j. During 2008, a total of 36,000 barrels of oil were produced. Related lifting costs were $5 per barrel (expense lifting costs).

Entry

Lease operating expense (36,000 x $5) . . .	180,000	
Cash		180,000

k. The reserve report as of December 31, 2008 and production during the year for Lucky Company were as follows:

Proved reserves	1,470,000 bbl
Proved developed reserves	970,000 bbl
Production	30,000 bbl

DD&A Calculation

Acquisition costs:

$$\frac{\$60{,}000 - \$659}{1{,}470{,}000 + 30{,}000} \times 30{,}000 = \underline{\underline{\$1{,}187}}$$

Wells and related E&F:

$$\frac{\$675{,}000 + \$590{,}000 - \$21{,}774}{970{,}000 + 30{,}000} \times 30{,}000 = \underline{\underline{\$37{,}297}}$$

Entry

DD&A expense—proved property	1,187	
DD&A expense—wells	37,297	
Accumulated DD&A—proved property		1,187
Accumulated DD&A—wells		37,297

l. A disaster struck and Lucky Company abandoned the lease. No equipment was salvaged.

Entry

Accumulated DD&A—proved property . . .	1,846	
Accumulated DD&A—wells	59,071	
Surrendered lease expense	1,264,083	
Proved property		60,000
Wells and related E&F—IDC		900,000
Wells and related E&F—L&WE		365,000

The following comprehensive example illustrates accounting for the disposition of costs on a field-wide basis rather than a lease basis:

EXAMPLE

Comprehensive Field DD&A

Lucky Oil Company computes DD&A on a field-wide basis. Balance sheet data as of 12/31/07 for Lucky's Texas field are as follows:

Unproved properties, net of impairment		$ 200,000
Proved properties	$ 500,000	
Less: Accumulated DD&A	200,000	
Net proved properties		300,000
Wells and related E&F—IDC	2,100,000	
Wells and related E&F—L&WE	800,000	
Less: Accumulated DD&A—wells	(850,000)	
Net wells and related E&F		$2,050,000

Lucky's activities during 2008 were as follows:

Unproved properties acquired	$ 50,000
Delay rentals paid	6,000
Test well contributions paid	30,000
Lease record maintenance, unproved properties	10,000
Title defenses paid	20,000
Unproved properties proved during 2008, net of impairment	60,000
Impairment of unproved properties	40,000
Exploratory dry hole drilled	300,000
Successful exploratory well drilled	500,000
Development dry hole drilled	350,000
Service well drilled	275,000
Tanks, separators, etc., installed	100,000
Development well, in progress 12/31/08	160,000

	Oil (bbl)	Gas (Mcf)
Production	50,000	300,000
Proved reserves, 12/31/08	900,000	3,000,000
Proved developed reserves, 12/31/08	500,000	1,800,000

Additional data: A truck serving this field in a production capacity was driven 4,000 miles during 2008. Total estimated miles for the truck are 50,000. The truck cost $12,000 and salvage value is estimated to be $0. The truck is support equipment and facilities.

Additional data: Lucky also owns a building that houses the corporate headquarters. The operations conducted in the building are general in nature and are not directly attributable to any specific exploration, development, or production activity. Since the building is not related to exploration, development, or production, the building is depreciated using straight-line depreciation. The depreciation is charged to general and administrative overhead and not to the field. The cost of the building was $400,000. The building has an estimated life of 40 years.

DD&A Calculations

	Production		*Proved Reserves*	
Oil:	50,000	50,000 bbl	900,000	900,000 bbl
Gas:	300,000/6 =	50,000 BOE	3,000,000/6 =	500,000 BOE
		100,000 BOE		1,400,000 BOE

	Proved Developed Reserves	
Oil:	500,000	500,000 bbl
Gas:	1,800,000/6 =	300,000 BOE
		800,000 BOE

Leasehold

Costs to amortize:

Proved properties, net at 12/31/07	$300,000
Properties proved during 2008	60,000
	$360,000

$$\frac{\$360{,}000}{1{,}400{,}000 + 100{,}000} \times 100{,}000 = \$24{,}000$$

Wells

Costs to amortize:

Wells and related E&F, net at 12/31/07	$2,050,000
New successful exploratory well	500,000
New development well	350,000
New service well	275,000
New tanks, etc.	100,000
Truck	12,000
	$3,287,000

$$\frac{\$3,287,000}{800,000 + 100,000} \times 100,000 = \underline{\underline{\$365,222}}$$

Truck

Since the truck serves only one cost center, in practice it would typically be capitalized directly to the cost center and depreciated using unit-of-production over the life of the cost center as shown above.

Building

$$\frac{\$400,000}{40} = \underline{\$10,000}$$

Entry

DD&A expense—proved property	24,000	
DD&A expense—lease and well equipment . . .	365,222	
G&A Overhead—(building depreciation)	10,000	
Accumulated DD&A—proved property		24,000
Accumulated DD&A—lease and well equipment		365,222
Accumulated depreciation—building		10,000

SE Impairment

The potential exists in all industries for the recorded net value of assets to exceed their underlying value. The practice of recording impairment (*i.e.,* writing assets down when their net book value exceeds their underlying value) has been required by the SEC for companies using the successful-efforts method for a number of years, even though no specific standard was ever issued.[2] In 1995, the FASB changed this situation by issuing *SFAS No. 121, Accounting for the Impairment of Long-Lived Assets and for Long-Lived Assets to Be Disposed Of.* This standard requires companies in all industries to apply an impairment test. (Companies using the full-cost method of accounting are not currently required to apply this standard. See discussion in chapter 7.) *SFAS No. 121* provides broad guidelines as to when assets should be tested for impairment, how to measure the impairment, and how to account for any impairment.

Companies are required to test assets for impairment only when events or changes in circumstances indicate an asset may be impaired. Examples of such events in the oil and gas industry include significantly lower revisions to reserve estimates or significantly lower prices or higher costs. When such events or changes in circumstances exist, companies are required to screen for impairment by comparing the carrying value of the asset with the estimated *undiscounted* future net cash flows expected to result from the use and ultimate disposition of the asset in question. If the carrying value of the asset exceeds its estimated future net cash flows, impairment has been triggered and an impairment loss must be recognized.

The impairment loss, however, is measured differently from the method used to screen for impairment. The impairment loss is determined by subtracting the carrying value of the asset from the fair value of the asset. The fair value of the asset should determined using quoted market prices in active markets. However, if quoted market prices are not available, alternative valuation techniques may be used. In the oil and gas industry, almost all SE companies use the *discounted* future net cash flows associated with the asset to determine the asset's fair value.

Note that to screen for impairment, *undiscounted* future net cash flows are used, but to measure the amount of the impairment, *discounted* future net cash flows are usually used. To discount the future net cash flows, companies are required to use a discount rate "commensurate with the risks involved." In the oil and gas industry, many SE companies use either the 10% rate specified in *SFAS No. 69* disclosures (chapter 13) or their cost of capital. In applying the impairment standard, assets are required to "be grouped at the lowest level for which there are identifiable cash flows that are largely independent of the cash flows of other groups of assets." Although there is some variability in the oil and gas industry as to how assets are grouped, the majority of oil and gas companies using SE are grouping their assets by the field or reservoir.[3]

Any resulting impairment must be recognized on the income statement and may not be recovered in future periods. Further, the reduced value of the impaired asset becomes the asset's new cost.

EXAMPLE

Impairment

Lucky Company owns three oil and gas fields located in Texas and Oklahoma. Late in 2010, oil and gas prices throughout the world decreased by 30%. Lucky Company's managers do not expect prices to recover significantly any time in the near future. Data for Lucky's fields as 12/31/2010 are as follows:

	Field A	**Field B**	**Field C**
Net capitalized cost	$500,000	$200,000	$800,000
Expected undiscounted future net cash flows	600,000	130,000	700,000
Fair value (discounted future net cash flows)	440,000	95,000	520,000

Impairment screen:

Lucky must screen for impairment for all three fields because an event occurred that indicates assets on all the fields may be impaired. To determine if impairment has been triggered, the carrying value of the assets must be compared to associated undiscounted future net cash flows.

	Field A	**Field B**	**Field C**
Expected undiscounted future net cash flows	$ 600,000	$130,000	$ 700,000
Net capitalized cost	500,000	200,000	800,000
Difference	$ 100,000	$(70,000)	$(100,000)
Has impairment been triggered?	No	Yes	Yes

Measurement of impairment:

	Field B	**Field C**
FV/Discounted future net cash flows	$ 95,000	$ 520,000
Net capitalized cost	200,000	800,000
Difference	$(105,000)	$(280,000)

Entry

Loss on producing properties	385,000	
Accumulated capitalized cost reduction . . .		385,000

Disclosures

Under SE there are no additional disclosures other than those required by *SFAS No. 69.* For further discussion of required disclosures, see chapter 13.

Problems

1. Define the following:
   future dismantlement
   restoration and abandonment costs
   common unit of measure based on energy

2. Duster Oil Company drilled its first successful well on Lease A in 2006. Data for the Lease A as of 12/31/06 are as follows:

Leasehold costs	$ 50,000
IDC of well	200,000
L&W equipment	75,000
Production during 2006	8,000 bbl
Total estimated proved reserves, 12/31/06	792,000 bbl
Total estimated reserves recoverable from well, 12/31/06	102,000 bbl

REQUIRED: Compute amortization for 2006.

3. Should amortization always be computed using a common unit of measure based on relative energy when oil and gas reserves are produced jointly? If no, under what circumstances would amortization not be based on units of energy? What basis would be used?

4. Both oil and gas are produced from BlowOut Oil Company's lease in Texas. Additional information, 1/1/07:

Unrecovered IDC (unamortized IDC)	$900,000
Proved property costs, net	100,000
L&W equipment, gross	300,000
Beginning of year accumulated DD&A equipment	50,000

Estimated proved reserves, 12/31/07	
Oil	200,000 bbl
Gas	1,000,000 Mcf
Production during 2007	
Oil	10,000 bbl
Gas	300,000 Mcf

Assuming the lease is fully developed, compute amortization:

a. assuming oil is the dominant mineral
b. using a common unit of measure based on BOE

5. The following are costs incurred on Murphy lease:

Acquisition costs	$ 80,000
Well 1 costs	225,000
Well 2 costs	200,000
Well 3 costs	275,000

Tanks, separators, flow lines, etc	100,000
Total wells and related E&F	800,000
Total lease, wells and related E&F costs	880,000

Treat each of the following independently:

a. A fourth well, an exploratory well, was drilled at a cost of $175,000 and was determined to be dry. Give the entry to record the dry hole.

b. Give the entry to record abandonment of Well 2. Equipment costing $20,000 was salvaged. Accumulated DD&A on wells and related E&F was $300,000. Wells 1 and 3 are still producing.

c. Give the entry to record abandonment of the entire Murphy lease. Assume the lease constituted a separate amortization base with accumulated DD&A on leasehold costs of $30,000 and accumulated DD&A on wells and related E&F of $300,000.

d. Give the entry to record abandonment of the entire Murphy lease, assuming instead that amortization had been computed on a field-wide basis with accumulated DD&A on leasehold costs of $400,000 and accumulated DD&A on wells and related E&F of $2,500,000.

6. What purpose does a cost center serve?

7. During 2005, Young Oil completed the last well from their drilling and production platform off the coast of Texas. Unrecovered costs on December 31, 2005, were $25 million, including $5 million in acquisition costs and $20 million in drilling and development costs. Total proved developed reserves were estimated to be 600,000 barrels as of January 1, 2005. Production during 2005 was 30,000 barrels. At the end of the life of the reservoir, dismantlement costs are estimated to be $14 million and salvage value is estimated to be $1 million. Compute DD&A for 2005.

8. Describe the two methods currently used by oil and gas companies to account for future dismantlement, restoration and abandonment costs. Discuss the financial statement consequences of the use of each method.

9. Under what circumstances should development costs be excluded in determining DD&A? Under what circumstances should a portion of proved developed reserves be excluded in determining DD&A? Is the exclusion of the development costs or proved developed reserves dependent upon the choice of the company, or is it required by *SFAS No. 19?*

10. Droopy Oil Company just completed (December 28, 2007) the successful testing of a tertiary recovery pilot project and as a result has determined that 900,000 barrels of oil should be classified as proved developed reserves. However, 200,000 of the 900,000 bar-

rels will be produced only after significant future development costs are incurred. Calculate DD&A for Droopy's wells and related E&F, assuming net capitalized drilling and equipment costs of $1,850,000 and production of 40,000 barrels.

11. During 2007, Wildcat Oil Company constructed an offshore production platform at a cost of $25,000,000. A total of 16 wells are planned. As of 12/31/07, only 2 out of the 16 wells had been drilled. Calculate DD&A given the following information.

Leasehold costs	$ 300,000
Drilling costs	2,200,000
Platform costs	25,000,000
Proved reserves, 12/31/07	1,800,000 bbl
Proved developed reserves, 12/31/07	900,000 bbl
Production	100,000 bbl

12. The following information as of 12/31/06 relates to the first year of operations for Basic Oil Company. From the data, (1) prepare entries and (2) prepare an income statement for Basic Oil Company for 2006, assuming revenue to the company from oil sales is $200,000. Expense lifting costs as lease operating expense.

Transactions, 2006	**Lease A**	**Lease B**	**Lease C**
*a. Acquisition costs of undeveloped leases ($^1/_8$ RI)	$ 60,000	$ 30,000	$ 20,000
*b. G&G costs	60,000	70,000	90,000
*c. Drilling costs	200,000	200,000	250,000
d. Drilling results: Proved reserves Proved developed reserves	Drilling completed 700,000 bbl 300,000 bbl (as of 12/31)	Drilling completed dry	Drilling not completed
e. Production	10,000 bbl		
f. Lifting costs	$ 50,000		
g. December 31	Recorded DD&A	Impaired lease 40%	
Transactions, 2007			
h. Assume on January 2 of the second year (2007) that disaster struck both Lease A and Lease B. Give the entries to record abandonment of Lease A and B. Assume equipment costing $15,000 was salvaged from Lease A. Assume this is not a post-balance sheet event that would give rise to changes in the balance sheets or income statements of previous years.			

* May combine entries for different leases.

13. Balance sheet data for Gusher Oil Company as of 12/31/05 is as follows for Lease A:

Leasehold costs	$ 100,000
Less: Accumulated DD&A	(20,000)
Net leasehold costs	$ 80,000
Wells and related E&F—IDC	$ 1,100,000
Less: Accumulated DD&A—IDC	(300,000)
Net wells and related E&F—IDC	$ 800,000
Wells and related E&F—L&WE	$ 700,000
Less: Accumulated DD&A—L&WE	(50,000)
Net wells and related E&F—L&WE	$ 650,000

Gusher's activities during 2006 related to Lease A were as follows:

Exploratory dry hole drilled	$ 275,000
Development dry hole drilled	300,000
Tanks, separators, etc., installed	125,000
Production	100,000 bbl
Proved reserves, 12/31/06	1,020,000 bbl
Proved developed reserves, 12/31/06	900,000 bbl

Calculate DD&A for 2006 assuming Lease A constitutes a separate amortization base.

14. Gusher Oil Company computes DD&A on a field-wide basis. Balance sheet data as of 12/31/05 for Gusher's Macon field are as follows:

Unproved properties, net of impairment		$ 200,000
Proved properties	$ 500,000	
Less: Accumulated DD&A	(100,000)	
Net proved properties		400,000
Wells and related E&F—IDC	3,000,000	
Wells and related E&F—L&WE	1,400,000	
Less: Accumulated DD&A—Wells	(1,300,000)	
Net wells and related E&F		3,100,000

Gusher's activities during 2006 were as follows:

Unproved properties acquired	$ 35,000
Delay rentals paid on unproved properties	10,000
Test well contributions paid on unproved properties	20,000
Title exams paid on unproved property	8,000
Title defenses paid	15,000

Unproved properties proved during 2006, net of Impairment		50,000
Exploratory dry hole drilled		275,000
Successful exploratory well drilled		400,000
Development dry hole drilled		300,000
Service well		325,000
Tanks, separators, etc., installed		125,000
Development well, in progress 12/31/06		140,000

	Oil (bbl)	Gas (Mcf)
Production	100,000	500,000
Proved reserves, 12/31/06	1,020,000	5,000,000
Proved developed reserves, 12/31/06	900,000	4,700,000

REQUIRED: Using BOE,

a. Calculate DD&A for 2006.
b. Calculate DD&A for 2006, assuming that part of the field, a proved property with gross acquisition costs of $30,000 and gross equipment cost of $70,000 and IDC of $215,000, was abandoned during 2006.

15. King Oil Company has the working interest in a fully developed lease located in Texas. As of 12/31/07, the lease had proved developed reserves of 1,200,000 barrels and unrecovered costs of $12,000,000. During the third quarter of 2008, a new reserve study was received that estimated proved developed reserves of 1,500,000 as of August 1, 2008. Calculate DD&A for each quarter, assuming the following production and using the first method illustrated in the chapter.

Quarter	Production
1	30,000
2	35,000
July	10,000
August	15,000
September	12,000
4	40,000

16. Roberts Oil Company had the following account balances for the years shown relating to a proved property:

	12/31/05	12/31/06
Proved property cost	$ 30,000	$ 30,000
Accumulated DD&A—proved property	6,000	
Wells and Related E&F—IDC	350,000	450,000

	12/31/05	12/31/06
Accumulated DD&A—Wells and RE&F*—IDC . . .	60,000	
Wells and related E&F—L&WE	250,000	325,000
Accumulated DD&A—Wells and RE&F—L&WE . . .	55,000	

* Related Equipment & Facilities

	2005	2006
Proved reserves, 12/31	800,000 Mcf	900,000 Mcf
Proved developed reserves, 12/31	500,000 Mcf	700,000 Mcf
Production during 2005 and 2006	40,000 Mcf	60,000 Mcf

Compute DD&A for the year ended 12/31/06.

17. Dixie Oil Company had the following information and account balances for the years shown relating to Lease No. 1.

	12/31/05	12/31/06
Proved property—cost	$ 40,000	$ 40,000
Accumulated DD&A—proved property	4,000	
Wells and related E&F—IDC	400,000	600,000
Accumulated DD&A—wells and RE&F—IDC . . .	60,000	
Wells and related E&F—L&WE	300,000	420,000
Accumulated DD&A—wells and RE&F—L&WE . . .	45,000	

	2005	2006
Proved reserves, 12/31—Oil	30,000 bbl	50,000 bbl
Gas	450,000 Mcf	600,000 Mcf
Proved undeveloped reserves, 12/31—Oil	10,000 bbl	12,000 bbl
Gas	200,000 Mcf	120,000 Mcf
Production during 2005 and 2006—Oil	5,000 bbl	7,000 bbl
Gas	50,000 Mcf	70,000 Mcf

REQUIRED: Compute DD&A for the year ended 12/31/06 using:

a. a common unit of measure based on equivalent Mcf
b. gas as the dominant mineral
c. same relative proportion

18. Sharon Montez purchases a $^1/_8$ ORI for $10,000. Proved reserves at year-end 2006 are 20,000 barrels and production for 2006 was 3,000 barrels. Prepare journal entries for Sharon and compute DD&A for 2006.

19. Duncan Oil Company has capitalized costs on Lease R, Lease S, and Lease T as of 12/31/2003 as follows:

	Lease R	Lease S	Lease T
Net capitalized cost	$ 300,000	$ 600,000	$ 400,000
Expected undiscounted future net cash flows	200,000	630,000	380,000
Fair value (discounted future net cash flows)	160,000	540,000	320,000

Duncan has no other capitalized costs. The leases are located in different counties in Texas.

REQUIRED: If necessary, test the assets for impairment and make any necessary journal entries assuming production costs tripled late in the year on all three leases and that production costs are not expected to decrease.

20. Sure Thing Company has a working interest in Lease A. As of 12/31/05, the lease had reserves as follows:

Proved developed reserves:	Oil	120,000 bbl
Proved reserves:	Oil	180,000 bbl

Account balances at 12/31/05 were as follows:

Proved property	$ 120,000
Accumulated DD&A—proved property	40,000
Net proved property	$ 80,000
Wells and related E&F	$1,600,000
Accumulated DD&A—wells and RE&F	300,000
Net wells and related E&F	$1,300,000

A new reserve estimate on October 31, 2006 was as follows:

Proved developed reserves:	Oil	100,000 bbl
Proved reserves:	Oil	150,000 bbl

Calculate DD&A for each quarter of 2006, assuming the following production, using both methods described in the book.

Quarter	Production
1	10,000 bbl
2	12,000 bbl
3	14,000 bbl
October	2,000 bbl
November	4,000 bbl
December	6,000 bbl

21. The Bryant Company had the following costs at 12/31/05 relating to Lease A:

Proved property	\$ 50,000
Accumulated DD&A—proved property . . .	30,000
Well No. 1: Wells and related E&F	400,000
Well No. 2: Wells and related E&F	350,000
Accumulated DD&A—wells	500,000

In January, 2006, Well No. 1 quit producing and was abandoned. In March, 2006, Well No. 2 quit producing, and the well and the lease were abandoned.

REQUIRED: Prepare journal entries for the abandonments.

22. On January 1, 2008, Mary Mabel purchased a 40,000 barrel PPI from Bad Luck Company for \$700,000. The PPI will be paid out of $^1/_5$ of the WI's share of production from Lease number 1003. The lease is burdened with a $^1/_7$ RI. Gross production from Lease number 1003 during 2008 was 28,000 barrels. Gross proved reserves at 12/31/08 were 130,000 barrels. Give the entries by Mary Mabel to record the purchase of the PPI and DD&A expense for 2008.

23. Hopeful Company entered the oil and gas business in 2004 with the acquisition of one field. Hopeful proved the field during 2004. At the end of 2004 prices were high and costs low. During 2005, Hopeful continued exploration and development activities in the field, but towards the end of 2005, oil and gas prices plummeted. During 2006, due to continued low prices, Hopeful suspended all exploration and development activities. Prices at the end of 2006 declined significantly once again. Data for Hopeful's one field are as follows:

	2004	**2005**	**2006**
Capitalized costs before any write-downs	\$ 300,000	\$ 900,000	\$ 900,000
Accumulated DD&A	20,000	100,000	180,000
Expected undiscounted future net cash flows	290,000	820,000	700,000
Fair value (discounted future net cash flows)	200,000	580,000	520,000

REQUIRED: Make any necessary journal entries for the above years relating to impairment.

24. Bonnel Company uses the successful-efforts method. Recently, the company acquired a truck costing \$60,000 with an estimated life of five years (ignore salvage value). The foreman drives the truck to oversee operations on seven leases all in the same general geographical area. The foreman keeps a log of his mileage in order to determine how the truck is utilized. Analysis of the log indicated that $^1/_3$ of the mileage driven was related to travel to drilling operations, $^1/_3$ was related to travel to G&G exploration areas, and $^1/_3$ was related to travel to production locations.

REQUIRED:

a. Give any entries necessary to record depreciation on the truck for the first year that it was in service assuming the 7 leases were located on different reservoirs.

b. Give any entries necessary to record depreciation on the truck for the first year that it was in service assuming the 7 leases were on the same reservoir. The total of net wells and related E&F, including the truck above, was $900,000. Proved reserves at the end of the year were 300,000 barrels, proved developed reserves at the end of the year were 120,000, and production during the year was 30,000 barrels.

NOTES

1. Wright, Charlotte, "Environmental Accounting: Implications for the Oil and Gas Industry," *Petroleum Accounting and Financial Management Journal*, Vol. 17, No. 2, Summer, 1998, pp. 30-49.

2. Landrum, Tony, "Successful-Efforts Companies: Oil and Gas Properties—Realization Test," *Journal of Extractive Industries Accounting*, Spring, 1983, p. 101.

3. Gallun, Rebecca A., and Nichols, Linda. "Implementation of SFAS No. 121 in the Oil and Gas Industry: A Survey Update," *Petroleum Accounting and Financial Management*, Vol. 16, No. 2, Summer, 1997, pp. 25-37.

chapter seven

Full Cost Accounting (FC)

Under full cost (FC), as stated in chapter 2, both successful and unsuccessful costs incurred in the search for oil and gas are considered necessary to finding oil or gas. In other words, it is necessary to drill a certain number of dry holes as well as successful wells—and incur other nonproductive exploration costs—in order to find oil and gas reserves. According to this reasoning, both successful and unsuccessful expenditures should be capitalized and amortized over production as part of the cost of the oil and gas. Thus, directly relating costs incurred to specific reserves discovered is not relevant under full-cost accounting. In contrast, under successful-efforts accounting, only those exploration costs that can be directly related to specific reserves are capitalized.

Some would argue that FC is a departure from traditional historical cost accounting since nonproductive exploration costs, having no future economic benefit, are capitalized. Others contend that FC actually produces more meaningful financial statements. Unlike other industries, the primary assets of oil and gas companies are not the property, plant, and equipment, but the oil and gas in the ground. FC financial statements are based on the fact that unsuccessful expenditures are a necessary and unavoidable part of discovering those assets.

Specifically, under full cost, all costs associated with property acquisition, exploration, and development activities are capitalized. Consequently, G&G studies, delay rentals, and exploratory dry holes are capitalized. Even when property is impaired or abandoned, the costs of impairment or abandonment continue as part of the capitalized costs of the cost center. Any portion of general and administrative costs directly related to acquisition, exploration, and development activities, such as the internal costs of the land department, may also be capitalized. Costs related to production, general corporate overhead, and similar activities are expensed (ASR 258, *Reg SX 4-10*, 3.18[i]). General corporate overhead cost normally includes all costs incurred in maintaining the various administrative offices throughout the company unless those office costs can be identified with acquisition, exploration, and development activities. Examples of general corporate overhead include executive compensation and offices, salaries of accounting personnel and the costs of their offices, and other administrative costs not related to acquisition, exploration, and development activities.

Under full-cost accounting, the cost center for accumulating capitalized costs to be amortized is a country, *i.e.*, the United States, Canada, etc. A country also includes any offshore area under the country's legal jurisdiction. Theoretically, only one capital account per cost center is necessary to account for the activities of a full-cost company. However, because of management, tax, and regulatory requirements, detailed records similar to those of a successful-efforts company are normally kept. As discussed later in this chapter, more detailed record-keeping is also dictated by certain full-cost rules that allow exclusion of specific costs from amortization and require a ceiling test to be performed. Exclusion of certain costs from amortization also necessitates identification of unsuccessful costs, such as dry holes and impairment or abandonment of unproved properties.

The examples and homework problems in this chapter use journal entries to illustrate full-cost accounting. Even though some full-cost companies may not actually make formal journal entries for all transactions, the journal entries used in this book reflect the actual process or mechanics of full-cost accounting.

Figure 7–1 shows the accounting treatment of the four basic types of costs incurred in exploration and production activities under full cost. These four costs—acquisition, exploration, development, and production—are defined for full cost as they are for successful efforts.

The following is an example of accounting for costs using the full-cost method. In studying the example, compare the accounting treatment under full cost to that under successful efforts—which costs are capitalized versus expensed, as well as account titles used. Because the full-cost rules allow exclusion of certain costs from the amortization base, all nondrilling costs that can be associated, when incurred, with specific unproved properties are debited to an unproved property control account. Drilling costs are initially charged to wells in progress (W/P) and, if proved reserves are found, transferred to wells and related E&F. If proved reserves are not found, exploratory wells in progress are transferred to exploratory dry holes. Although different accounts are used for successful exploratory wells versus exploratory dry holes, both accounts are asset accounts under FC accounting and are written off to expense through DD&A. Development wells in progress that do not find proved reserves are transferred to wells and related E&F in the same manner as under successful efforts.

Fig. 7–1 — *Full Cost Summary*

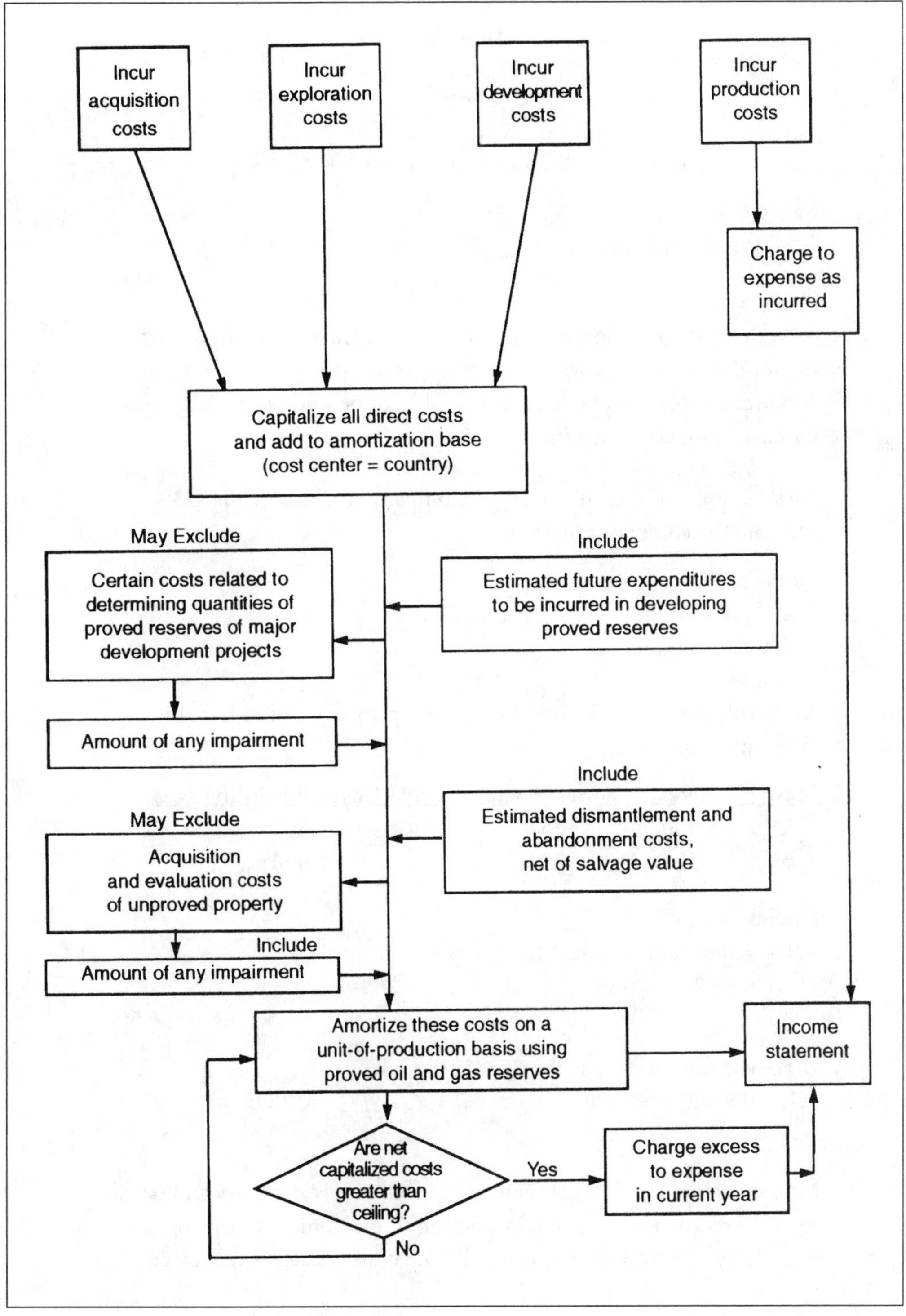

EXAMPLE

FC Entries

a. Lucky Company acquired shooting rights to 10,000 acres, paying $5,000.

Entry

G&G costs—nondirect	5,000	
Cash		5,000

Note that although these G&G costs are recorded as capitalized costs, they are not debited to unproved property control. They are instead recorded as nondirect G&G costs because they could not be associated with a specific unproved property when incurred.

b. Lucky Company then hired Davis Company to conduct the G&G work and paid the company $30,000.

Entry

G&G costs—nondirect	30,000	
Cash		30,000

c. As a result of the G&G work, Lucky Company decided to lease the following properties:

Lease	Acres	Bonus $/acre	Legal Costs, Recording Fees
A	1,000	$25	$ 500
B	1,500	20	1,000

Entries

U/P—acquisition—Lease A		
(1,000 x $25 + $500)	25,500	
Cash		25,500
U/P—acquisition—Lease B		
(1,500 x $20 + $1,000)	31,000	
Cash.		31,000

Note that although the cost center under full cost is a country and detailed records are not required, Lucky's accounting records are kept on a lease basis. When DD&A is computed, all lease costs are aggregated by country.

d. During the following year, Lucky Company incurred and paid the following items:

1) Ad valorem taxes of $1,500 on Lease A

Entry

U/P—ad valorem taxes—Lease A	1,500	
Cash		1,500

Note that these ad valorem taxes and many of the costs below are non-drilling costs that can be associated with an unproved property when incurred, and therefore, are debited to an unproved property control account.

2) Dry-hole contribution of $10,000 on Lease B

Entry

U/P—test-well contributions—Lease B	10,000	
Cash		10,000

3) Delay rentals on Leases A and B of $3,000 and $4,000, respectively

Entries

U/P—delay rental—Lease A	3,000	
Cash.		3,000
U/P—delay rental—Lease B	4,000	
Cash		4,000

4) Nonrecoverable delinquent taxes on Lease B, $2,000 (not specified in lease contract)

Entry

U/P—exploration costs—Lease B	2,000	
Cash		2,000

5) Costs of maintaining lease records for Leases A and B were $800 for the year (allocated based on relative acreage)

Entry ($800/(1,000 + 1,500) = $0.32 per acre)

U/P—maintenance costs—Lease A ($0.32 x 1,000)	320	
U/P—maintenance costs—Lease B ($0.32 x 1,500)	480	
Cash		800

Note that the cost of the lease records department should be allocated between lease acquisition and maintenance activities versus those activities that are of a general administrative nature. The costs allocated to lease acquisition and maintenance should be capitalized and the cost allocated to general administration should be expensed as part of general corporate overhead.

e. At the beginning of the next year, Lucky Company began drilling operations on both Lease A and Lease B. The company incurred the following costs:

Lease	L&WE	IDC
Lease A	$50,000	$200,000
Lease B	20,000	130,000

Entries

Lease A

W/P—L&WE	50,000	
W/P—IDC	200,000	
Cash		250,000

Lease B

W/P—L&WE	20,000	
W/P—IDC	130,000	
Cash		150,000

Drilling costs are accumulated in the W/P accounts until drilling is complete.

f. Assume the well on Lease A is successful, and additional IDC of $10,000 and additional equipment costs of $70,000 are incurred to complete the well.

Entry

W/P—L&WE	70,000	
W/P—IDC	10,000	
Cash		80,000

Entry to record completion of work on well, *i.e.*, to close out W/P accounts

Wells and related E&F—L&WE	120,000	
Wells and related E&F—IDC	210,000	
W/P—L&WE		120,000
W/P—IDC.		210,000

Entry to reclassify lease as proved

Proved property	30,320	
Unproved property		30,320

Unproved Property	
25,500	
1,500	
3,000	
320	
30,320	

The separation of property into proved and unproved and the separation of drilling costs into completed and uncompleted is not required under full-cost accounting theory. However, this distinction is useful for management needs, income tax accounting, etc.

g. Assume the well on Lease B is determined to be dry and is plugged and abandoned for an additional $3,000.

Entry to record additional costs

W/P—IDC	3,000	
Cash		3,000

Entry to record completion of work on well

Exploratory dry holes—L&WE	20,000	
Exploratory dry holes—IDC	133,000	
W/P—L&WE		20,000
W/P—IDC		133,000

Remember that under full cost, all exploration costs are capitalized whether successful or unsuccessful. However, the costs of the exploratory dry hole are transferred into an account identified as unsuccessful rather than into wells and related equipment and facilities. This transfer facilitates the use of certain procedures related to exclusion of costs from amortization. The separation of exploratory dry-hole costs is also beneficial for management decision-making purposes and for regulatory purposes.

h. Lucky Company decides to abandon Lease B.

Entry

Abandoned costs	47,480	
Unproved property.		47,480

Unproved Property	
31,000	
10,000	
4,000	
2,000	
480	
47,480	

Remember that abandoned costs or impairment costs are not expensed but remain capitalized.

Disposition of Capitalized Costs

Under the theory of full-cost accounting, all successful and unsuccessful costs associated with acquisition, exploration, and development activities are considered to be part of the cost of any oil or gas found and produced and should therefore be amortized over production. All capitalized costs within a cost center are, with certain exceptions, amortized over proved reserves using the unit-of-production basis.

If oil and gas are produced jointly, the oil and gas are converted to a common unit of measure based on relative energy content. However, oil and gas prices may be so disproportionate relative to their energy content that the unit-of-production method would result in an improper matching of the cost of oil and gas production against related revenue received. When that is the case, the unit-of-revenue is a more appropriate basis of computing amortization. (The revenue method is not allowed under SE accounting.) If the unit-of-revenue method is used, the actual selling price of the oil and gas should be used to value the production during the year, and current prices (year-end) rather than future prices should generally be used in valuing the proved reserves. Future prices should be used only when provided by contractual arrangements.

The basic formula for computing amortization using unit-of-production is essentially the same as the one used for computing amortization under successful efforts:

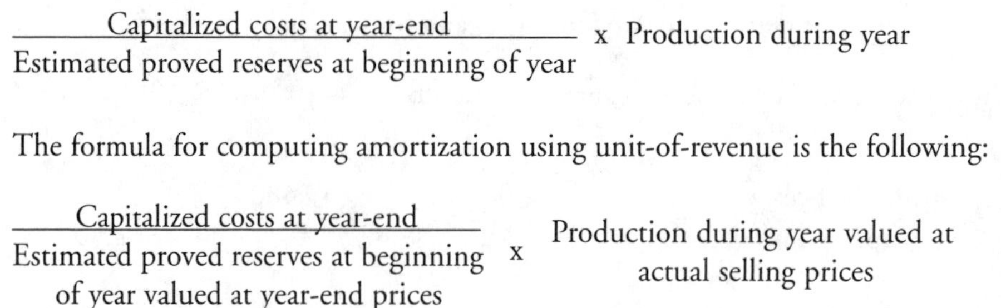

$$\frac{\text{Capitalized costs at year-end}}{\text{Estimated proved reserves at beginning of year}} \times \text{Production during year}$$

The formula for computing amortization using unit-of-revenue is the following:

$$\frac{\text{Capitalized costs at year-end}}{\text{Estimated proved reserves at beginning of year valued at year-end prices}} \times \text{Production during year valued at actual selling prices}$$

The following example illustrates both DD&A computed using unit-of-production and DD&A computed using unit-of-revenue.

EXAMPLE

Full Cost DD&A

Lucky Company has oil and gas properties located only in the United States. Data for Lucky Company as of December 31, 2006, are as follows:

Costs to be amortized	$2,700,000
Production during 2006	
Oil	5,000 bbl
Gas	12,000 Mcf
Proved reserves, 12/31/06	
Oil	80,000 bbl
Gas	300,000 Mcf
Current selling price, 12/31/06	
Oil	$20/bbl
Gas	$2.20/Mcf
Selling price during 2006	
Oil	$22/bbl
Gas	$2.00/Mcf

a. Computation of DD&A based on equivalent energy units for 2006 is as follows:

Equivalent Units
Production:

Oil		5,000
Gas	12,000/6 =	2,000 BOE
		7,000 BOE

Reserves:

Oil		80,000
Gas	300,000/6 =	50,000 BOE
		130,000 BOE

DD&A

$$\frac{\$2{,}700{,}000}{130{,}000 + 7{,}000} \times 7{,}000 = \underline{\underline{\$137{,}956}}$$

b. Computation of DD&A based on unit-of-revenue is as follows:

Unit-of-Revenue

Production:

Oil	5,000 x $22 =	$110,000
Gas	12,000 x $2.00 =	24,000
		$134,000

Reserves:

Oil	80,000 x $20 =	$1,600,000
Gas	300,000 x $2.20 =	660,000
		$2,260,000

DD&A

$$\frac{\$2,700,000}{\$2,260,000 + \$134,000} \times \$134,000 = \underline{\$151,128}$$

c. Which would be the appropriate basis in this case? One barrel of oil is approximately equivalent to 6 Mcf of gas, but the current price of one barrel of oil is $20 compared to $13.20 for 6 Mcf of gas. The selling price of one barrel of oil during the year was $22 compared to $12 for 6 Mcf of gas. This analysis indicates that the DD&A amount based on relative energy units will be different from the DD&A amount based on unit-of-revenue. Therefore, if the difference is considered to be material, unit-of-revenue would be the appropriate basis.

Entry

DD&A expense	151,128	
Accumulated DD&A		151,128

Reg. SX 4-10 states that reserves should be valued at current prices, *i.e.*, period-end prices, in computing amortization based on unit-of-revenue. However, it also states that if a significant (material) price increase occurs, the increase should be reflected in the period (quarter) following the price increase. For example, if a significant price increase occurs in the third quarter, the third-quarter amortization is computed using the old price, and the fourth-quarter computation uses the new price.

Inclusion of additional costs

Net capitalized costs, including both leasehold costs and drilling and development costs, are amortized over proved reserves under full-cost accounting. Proved reserves consist of:

1. proved developed reserves, which will be produced through existing wells and equipment
2. proved undeveloped reserves, which will be produced through future wells or through future major recompletions

Therefore, when the proved reserves are not fully developed, a portion of the proved reserves—which make up the denominator in the DD&A calculation—will be producible only as a result of relatively major future development costs. Consequently, those future development costs must be included in the costs being amortized to avoid a distortion of the DD&A rate, *i.e.,* a mismatching of costs and reserves. Estimates of future development costs should be based on current costs.

In addition, estimated future dismantlement and abandonment costs, net of salvage values, must also be included in the amortization calculation in the same manner as amortization under successful-efforts accounting (chapter 6).

EXAMPLE

DD&A: Additional Costs

Data for Lucky Company as of December 31, 2007, is as follows:

Unrecovered costs (net capitalized costs)	$2,000,000
Production during 2007	50,000 bbl
Proved reserves, January 1, 2007	600,000 bbl

Lucky Company estimates that future expenditures to develop partially developed properties (proved properties) will be $300,000. Lucky Company also estimates that future dismantlement and abandonment costs will be $400,000, with salvage of $100,000.

DD&A Calculations

Costs to be amortized:

Unrecovered costs	$2,000,000
Add: Future development costs	300,000
Add: Future dismantlement and abandonment costs	400,000
Less: Future salvage	100,000
Total costs to amortize	$2,600,000

DD&A:

$$\frac{\$2{,}600{,}000}{600{,}000} \times 50{,}000 = \underline{\underline{\$216{,}667}}$$

Exclusion of costs

Amortization for full cost is based on proved reserves; consequently, amortization of unproved properties, which by definition have no proved reserves attributable to them, can seriously distort the amortization rate. Similarly, amortization of major development projects expected to involve significant costs in order to determine the quantities of proved reserves can also seriously distort the amortization rate. In recognition of this problem, the SEC allows unproved properties or development projects entailing significant future development costs to be excluded from the amortization base. Specifically:

1. All costs directly associated with the acquisition and evaluation of unproved properties may be excluded until it is determined whether or not proved reserves can be assigned to the properties. However, exploratory dry holes must be included in the amortization base as soon as a well is determined to be dry. Further, any G&G costs that cannot be directly associated with specific unproved properties should be included in the amortization base as incurred. In practice, most companies are excluding at least a portion of these G&G costs.

2. Certain costs associated with major development projects expected to entail significant costs in order to ascertain the quantities of proved reserves attributable to the properties under development may be excluded if they have not previously been included in the amortization base. Excludable costs include both costs already incurred and future costs. The portion of common costs (*i.e.,* costs common to both the known reserves and the reserves yet to be determined) excluded should be based on a comparison of existing proved reserves to total expected proved reserves, or a comparison of the number of wells to which proved reserves have been assigned to the total number of wells expected to be drilled. Both the excluded costs and the related proved reserves should be transferred into the amortization base on a well-by-well or property-by-property basis as proved reserves are established or as impairment is determined.

The above costs may be excluded from the amortization base until proved reserves are established or until impairment (or abandonment) occurs. Any impairment is not expensed but is transferred into the amortization base and recovered, *i.e.,* written off to expense, only through DD&A. Note that if proved reserves are established, the costs must be included in amortization. Thus, even though production may not yet have begun (*e.g.,* construction of a pipeline necessary before any oil or gas is produced), the costs are amortized.

The following example illustrates the basic computations involved when costs associated with unproved properties are included and excluded from amortization:

EXAMPLE

Unproved Property Inclusion and Exclusion

Costs incurred by Lucky Company on oil and gas properties located in the United States as of December 31, 2009, were as follows:

Proved property costs	$ 40,000
Unproved property costs	70,000
Nondrilling exploration costs, proved properties	80,000
Nondrilling exploration costs, unproved properties	100,000
Drilling costs, proved properties	400,000
W/P, unproved properties	600,000
Dry holes, unproved properties	900,000
Total accumulated DD&A	(490,000)
Total capitalized costs, net of DD&A	$1,700,000

Other Data:	
Future development costs	$ 300,000
Future dismantlement costs	400,000
Proved reserves, 12/31/09, bbl	500,000
Production during year, bbl	100,000

a. DD&A assuming all possible costs are included in the amortization base:

Costs to be Amortized	
Total net capitalized costs	$1,700,000
Plus: Future development costs	300,000
Future dismantlement costs	400,000
Net costs to be amortized	$2,400,000

DD&A:

$$\frac{\$2,400,000}{500,000 + 100,000} \times 100,000 = \underline{\underline{\$400,000}}$$

b. DD&A assuming all possible costs are excluded from the amortization base:

Costs to be Amortized	
Proved property costs	$ 40,000
Nondrilling exploration costs, proved properties	80,000
Drilling costs, proved properties	400,000
Dry holes, unproved properties	900,000
Total accumulated DD&A	(490,000)
Future development costs	300,000
Future dismantlement costs	400,000
Net costs to be amortized	$1,630,000
Costs to Exclude	
Unproved property costs	$ 70,000
Nondrilling exploration costs, unproved properties	100,000
W/P, unproved properties	600,000
Total costs excluded	$ 770,000
Total costs included and excluded	$2,400,000

DD&A:

$$\frac{\$1{,}630{,}000}{500{,}000 + 100{,}000} \times 100{,}000 = \underline{\underline{\$271{,}667}}$$

An alternative approach to computing the costs to be amortized is to start with the net capitalized costs and subtract the costs that may be excluded from amortization:

Net capitalized costs	$1,700,000
Plus: Future development costs	300,000
Future dismantlement costs	400,000
Less: Unproved properties	(70,000)
Nondrilling exploration costs, U/P	(100,000)
W/P, unproved properties	(600,000)
Net costs to be amortized	$1,630,000

The following example presents the exclusion of incurred costs and future costs associated with a major development project, for which significant future costs are necessary to ascertain quantities of proved reserves attributable to the properties under development.

EXAMPLE

Major Development Project Exclusion

Lucky Company has an offshore lease on which proved reserves have been found. Preliminary estimates of proved reserves yield 200,000 barrels. Costs incurred to date relating to this lease total $3 million. To develop this lease fully and to determine the quantity of reserves, Lucky Company plans to install an offshore drilling platform from which development wells will be drilled. It is estimated that the platform and the wells will cost $8 million.

Total unrecovered costs at year-end for Lucky Company's other properties are $20 million, with proved reserves of 500,000 barrels at the beginning of the year. Production for the year was 10,000 barrels of oil.

DD&A

All costs and reserves relating to this major development project may be excluded from the DD&A calculation.

$$\frac{\$20{,}000{,}000}{500{,}000} \times 10{,}000 = \underline{\underline{\$400{,}000}}$$

Total costs excluded: $3,000,000 + $8,000,000 = $\underline{\underline{\$11{,}000{,}000}}$

To summarize the rules relating to the inclusion and exclusion of costs from the full cost amortization base:

a. All costs associated with unproved properties may be excluded except for:
   1) dry holes
   2) impairment costs
   3) abandonment costs
   4) costs if proved
   5) any G&G costs not related to specific properties
b. All costs associated with proved properties must be included except for major development projects where reserve quantities have yet to be determined
c. Future dismantlement costs net of salvage must be included
d. Future development costs must be included

Impairment

Only those unproved properties excluded from the amortization base should be impaired. The unproved properties excluded from the amortization base should be assessed at least annually. As with successful-efforts impairment, properties whose costs are individually significant must be assessed individually, while properties whose costs are not individually significant may be assessed on a group basis. Although not defined in the FC rules, the SEC suggests that, in general, individual properties or projects are individually significant if their costs exceed 10% of the net capitalized costs of the cost center.

The procedure for determining impairment for significant unproved properties, as well as the factors considered, are the same as under SE. The same method used to impair insignificant properties under successful-efforts accounting may also be used to impair insignificant properties under full-cost accounting. Insignificant properties may be grouped and assessed in one or more groups per cost center.

In contrast to successful-efforts impairment, any impairment under full cost is not expensed but is transferred into the amortization base and continues to be capitalized until recovered (*i.e.,* expensed) through DD&A. The impairment amount is recorded in an impairment account. An allowance for impairment account is also recorded and, as with successful-efforts accounting, is netted against the unproved property account to obtain the net carrying value of the unproved property—assuming the property is significant. The impairment account under full-cost accounting is an asset account that is included in the costs to be amortized. The allowance for impairment account is a contra-asset that reduces the amount of unproved property costs excluded from amortization. An example follows that illustrates impairment under full cost for significant and insignificant properties.

EXAMPLE

Impairment of Significant and Insignificant Properties

Lucky excludes all possible costs relating to unproved properties from the amortization base. Lucky's acquisition and evaluation costs related to unproved properties as of December 31, 2008, are as follows:

Significant Unproved Properties:

	Lease A	**Lease B**
Unproved property, control	$500,000	$800,000
Allowance for impairment	100,000	0

Insignificant Unproved Properties:

	Lease C	Lease D	Lease E	Total
Unproved property, control	$40,000	$60,000	$80,000	$180,000
Allowance for impairment, group				30,000

During 2009, wells on properties surrounding Lease A and Lease B are drilled and determined to be dry. Further, the primary term on Lease A is nearly over and Lucky has no firm plans to drill on Lease A. As a result, Lucky decides Lease A is impaired a total of 80% and Lease B is impaired 20%.

For insignificant unproved properties, Lucky has a policy based on historical experience of providing at year-end an allowance equal to 70% of the gross cost of the insignificant unproved properties.

Impairment for Significant Properties

Lease A: 0.80 x $500,000 = $400,000

$400,000 (total impairment) – $100,000 (impairment already recognized) = $300,000

Lease B: 0.20 x $800,000 = $160,000

Entry

Impairment for unproved properties	460,000	
Allowance for impairment—Lease A		300,000
Allowance for impairment—Lease B		160,000

Impairment for Insignificant Properties

0.70 x $180,000 = $126,000

Allowance	
30,000	← have
?	← 96,000
126,000	← want

Entry

Impairment for unproved properties—group . .	96,000	
Allowance for impairment, group		96,000

Any costs associated with major development projects excluded from the amortization base must also be evaluated for impairment. Impairment is computed as it is for unproved properties, and as on unproved properties, any impairment is transferred into the amortization base and recovered through DD&A.

Abandonment of properties

In full-cost accounting, all costs, both successful and unsuccessful, are considered necessary to finding oil or gas and consequently are considered part of the cost of oil or gas. Thus, when a property, either proved or unproved, is abandoned, the costs generally are not expensed, but remain capitalized and are recovered through DD&A. The abandonment of a property can be handled in several ways, depending upon the accounts used by a company and whether costs are excluded from the amortization base:

1. No entry is made
2. Accumulated DD&A or an abandoned cost account is charged
3. If the abandonment significantly affects the amortization per unit, however, the costs are removed and a gain or loss is recognized

In this book, abandoned properties are generally charged to abandoned costs. When a proved property is abandoned, all the related costs are generally charged to abandoned costs. An individual property will not have accumulated DD&A associated with it specifically because the cost center under FC is a country and any accumulated DD&A relates to the entire cost center and not an individual property. Thus, when a proved property is abandoned, only the proved property account is closed out and charged to abandoned costs.

EXAMPLE

Abandonment of Proved Property

Lucky Company abandoned a proved property late in 2004. Costs relating to the property are as follows:

Proved property	$ 50,000
Wells and related E&F	600,000
Exploratory dry holes	400,000

Entry to record abandonment

Abandoned costs	1,050,000	
Proved property		50,000
Wells and related E&F		600,000
Exploratory dry holes		400,000

The specific accounting entry to record abandonment of an unproved property depends on whether the property is considered to be significant or insignificant. If significant, both the unproved property account and the allowance for impairment account should be closed out, with the net balance charged to abandoned costs. If the property is insignificant, the unproved property account should be charged to the allowance account. (See next example.)

Reclassification of properties

When proved reserves are discovered on an unproved property, the property should be reclassified from unproved to proved. As with the abandonment of unproved properties, the specific accounting entry depends on whether the property is considered to be significant or insignificant. For a property assessed on an individual basis, both the unproved property account and the allowance for impairment account should be closed out, with the net carrying value transferred to a proved property account. For a property assessed on a group basis, the gross unproved property account balance should be transferred because, as with successful-efforts accounting, a net carrying value cannot be determined on a separate property basis for properties assessed on a group basis.

The following example illustrates impairment, abandonment, and reclassification for significant unproved properties:

EXAMPLE

Impairment, Abandonment, and Reclassification of Significant Unproved Properties

Lucky Company excludes all possible costs relating to unproved properties from the amortization base. Lucky's direct acquisition and evaluation costs at December 31, 2006, related to two significant, unproved properties are as follows:

	Lease A	Lease B
Unproved property, control	$100,000	$ 75,000
Wells in progress	300,000	200,000

During 2007, the wells on each lease are determined to be dry. As a result, Lucky impairs Lease A $40,000 and Lease B $25,000.

Entries

Exploratory dry holes ($300,000 + $200,000)	500,000	
W/P—Lease A		300,000
W/P—Lease B		200,000

Impairment for unproved properties ($40,000 + $25,000)	65,000	
Allowance for impairment—Lease A		40,000
Allowance for impairment—Lease B		25,000

Costs to be Amortized Relating to above Leases at 12/31/07

	Lease A	Lease B	Total
Impairment	$ 40,000	$ 25,000	$ 65,000
Dry holes	300,000	200,000	500,000
	$340,000	$225,000	$565,000

Costs that may be Excluded

	Lease A	Lease B	Total
Unproved property, control	$ 100,000	$ 75,000	$ 175,000
Less: Allowance	(40,000)	(25,000)	(65,000)
	$ 60,000	$ 50,000	$ 110,000

Total Capitalized Costs	$675,000

Note the $40,000 impairment amount for Lease A is included in the costs to be amortized, and the $40,000 allowance for impairment is shown as a reduction to the unproved property amount, reducing the amount excluded from amortization.

During 2008, Lucky abandons Lease A and drills a successful well on Lease B at a cost of $450,000.

Entries

Lease A:

Abandoned costs	60,000	
Allowance for impairment—Lease A	40,000	
Unproved property—Lease A		100,000

Note the entry to record abandonment of an impaired property under full cost is similar to the entry to record abandonment of an impaired property under successful efforts except that under full cost the abandoned amount continues to be capitalized, while under successful efforts the amount is expensed.

Lease B:

Wells and related E&F	450,000	
Cash		450,000

Proved property	50,000	
Allowance for impairment—Lease B	25,000	
Unproved property—Lease B		75,000

Note the entry to reclassify a property from unproved to proved is the same under full-cost and successful-efforts accounting.

Costs to be Amortized Relating to above Leases at 12/31/08

	Lease A	Lease B	Total
Impairment	$ 40,000	$ 25,000	$ 65,000
Dry holes	300,000	200,000	500,000
Abandoned costs	60,000		60,000
Proved property		50,000	50,000
Wells and related E&F		450,000	450,000
Total capitalized costs	$400,000	$725,000	$1,125,000

The above example illustrates that as properties are impaired, the impairment amounts are put into an asset account that is included in the amortization base. The impairment account remains untouched when the unproved property is either abandoned or is reclassified as proved. If abandoned, the unproved property and related allowance account are eliminated, with the net carrying value of the property being charged to abandoned costs, an asset account. If proved, the unproved property and allowance accounts are also eliminated, with the net carrying value transferred to proved property. It is therefore probable that the impairment asset account will have a greater balance than the contra-asset allowance for impairment account. The following example illustrates such a situation.

EXAMPLE

Impairment Account Versus Allowance Account

Lucky Company's direct acquisition and evaluation costs at December 31, 2003, related to two significant unproved properties are as follows:

	Lease A	Lease B
Unproved property, control	$100,000	$ 75,000

During 2004, wells on properties surrounding each lease are determined to be dry. As a result, Lucky impairs Lease A $60,000 and Lease B $25,000.

Entry

Impairment for unproved properties	85,000	
Allowance for impairment—Lease A		60,000
Allowance for impairment—Lease B		25,000

During 2005, Lucky abandons Lease A. Lease B remains classified as unproved, and no further impairment is made.

Entry

Abandoned costs	40,000	
Allowance for impairment—Lease A	60,000	
Unproved property—Lease A		100,000

The balance in the unproved property control account is now $75,000 and the allowance for impairment control account has a balance of $25,000. However, the impairment asset account has a balance of $85,000, $60,000 greater than the allowance for impairment account.

Costs to be Amortized

Impairment for unproved properties	$ 85,000
Abandoned costs	40,000
Total costs to be amortized	$125,000

Costs to be Excluded

Unproved property—Lease B	$ 75,000
Less: Allowance for impairment	(25,000)
Total costs to be excluded	$ 50,000

The following example illustrates impairment, abandonment, and reclassification for insignificant properties:

EXAMPLE

Impairment, Abandonment, and Reclassification of Insignificant Unproved Properties

On December 31, 2009, Lucky's unproved property account containing leases not considered individually significant had a $500,000 balance and the allowance for impairment account had a $60,000 balance. Based on historical experience, Lucky has a policy of providing at year-end an allowance equal to 60% of gross unproved properties.

Entry

Impairment for unproved properties—group . .	240,000	
Allowance for impairment, group		240,000

Calculations

500,000 x 0.60 = $300,000

Allowance		
	60,000	← have
	?	← 240,000
	300,000	← balance needed

On March 15, leases with a cost of $30,000 were abandoned.

Entry

Allowance for impairment, group	30,000	
Unproved property		30,000

On April 15, proved reserves were found on a lease with capitalized costs of $8,000.

Entry

Proved property	8,000	
Unproved property		8,000

Another method of amortizing insignificant unproved properties, in which the computation is based on layers of costs, is described in an article by Adkerson and Moore in the *Journal of Extractive Industries Accounting.*[1]

The following example illustrates a complex DD&A problem in which abandoned costs and impairment are included. Part A illustrates the computation of DD&A when all possible costs are amortized, and Part B illustrates the computation of DD&A when all possible costs are excluded from amortization. In the example, the column headed I relates to Part A of the problem where all possible costs are included in the amortization base. The column headed E relates to Part B of the problem where all possible costs are excluded from the amortization base. An I in the columns indicates that the cost must be included in amortization. An E in the columns indicates that the cost may be excluded from amortization.

EXAMPLE

Complex DD&A

Lucky Company has oil and gas properties located only in the United States. Costs for Lucky Company as of December 31, 2007, were as follows:

I	E		
I	I	Proved properties	$ 100,000
I	E	Unproved properties, acquisition costs	105,000
I	I	Wells in progress, proved properties	250,000
I	E	Wells in progress, unproved properties	550,000
I	I	Nondirect G&G costs	70,000
I	I	Direct G&G costs, proved properties	120,000
I	E	Direct G&G costs, unproved properties	210,000
I	I	Dry holes, proved properties	300,000
I	I	Dry holes, unproved properties	700,000
I	I	Delay rentals, proved properties	25,000
I	E	Delay rentals, unproved properties	75,000
I	I	Successful drilling costs	600,000
I	I	Abandoned costs of unproved properties	95,000
I	I	Total accumulated DD&A	(500,000)
I	I	Future development costs	400,000
		Net costs, plus future development costs	$3,100,000

Total capitalized costs, plus future development costs, net of DD&A	$3,100,000
Proved reserves, 12/31/07	4,240,000 bbl
Production during year	245,000 bbl

Note in the preceding data, several costs associated with proved properties such as delay rentals are incurred when the associated property is unproved. Under FC, these costs are capitalized instead of being expensed as in SE, and so are still on the books when the property becomes proved.

a. DD&A is computed as follows, assuming all possible costs are included in the amortization base:

Costs to Amortize

All of the above costs, less accumulated DD&A, would be included in computing DD&A. (The costs to be amortized are indicated above by an I in the column headed I.)

Total capitalized costs	$3,200,000
Plus: Future development costs	400,000
Less: Accumulated DD&A	(500,000)
Net costs to be amortized	$3,100,000

DD&A

$$\frac{\$3{,}100{,}000}{4{,}240{,}000 + 245{,}000} \times 245{,}000 = \underline{\$169{,}342}$$

b. DD&A is next computed, assuming that all possible costs are excluded from the amortization base. It is also assumed that the allowance for impairment for the above unproved properties is $20,000 and that unproved properties have been impaired a total of $80,000. (As discussed, when a significant unproved property that has been impaired is either abandoned or proved, the costs of the property, net of impairment, are transferred to abandoned costs or proved property. In either case, the allowance for impairment account is closed out for that property. For abandoned insignificant unproved properties, the allowance for impairment, group account would be debited. In contrast, once the impairment amount is put on the books, it becomes a part of the total capitalized costs and is not removed. Thus, it is possible for the allowance account for unproved properties to be smaller than the impairment account.)

Total Costs under New Assumptions

Net costs plus future development costs under old assumptions:	$3,100,000
Less: Allowance for impairment*	(20,000)
Plus: Impairment	80,000
Total costs, new assumptions	$3,160,000

* Contra account to unproved property, which is included at gross in the $3,100,000.

Costs to Amortize

All of the above costs associated with unproved properties except dry-hole costs, abandoned costs, and impairment would be excluded from amortization. The costs to be amortized are indicated above by an I in the column headed E. The costs indicated by an E should be excluded.

Proved properties		$ 100,000
Wells in progress, proved properties		250,000
Nondirect G&G costs		70,000
Direct G&G costs, proved properties		120,000
Dry holes, proved properties		300,000
Dry holes, unproved properties		700,000
Delay rentals, proved properties		25,000
Successful drilling costs		600,000
Abandoned costs of unproved properties		95,000
Impairment for unproved properties		80,000
Total accumulated DD&A		(500,000)
Future development costs		400,000
Net costs to be amortized		$2,240,000
Costs to Exclude		
Unproved properties	$105,000	
Less: Allowance for impairment	20,000	
Net unproved properties		$ 85,000
Wells in progress, unproved properties		550,000
Direct G&G costs, unproved properties		210,000
Delay rentals, unproved properties		75,000
Costs to exclude		$ 920,000
Total costs included and excluded		$3,160,000

DD&A

$$\frac{\$2{,}240{,}000}{4{,}240{,}000 + 245{,}000} \times 245{,}000 = \underline{\underline{\$122{,}363}}$$

Support equipment and facilities

Support equipment and facilities are defined and handled similarly under full-cost accounting and successful-efforts accounting. The depreciation and operating costs of support equipment and facilities should be allocated to exploration, development, or production as appropriate. The portion of the costs allocated to exploration and development would be capitalized; the portion allocated to production would be expensed. As with SE accounting, when support equipment and facilities serve only one cost center, the support equipment and facilities should be depreciated using the unit-of-production method. Since the cost center for FC accounting is a country, it is highly unlikely that support equipment and facilities would support more than one cost center. As a result, support equipment and facilities under FC accounting would almost always be depreciated using unit-of-production.

Although depreciation of support equipment and facilities should theoretically be allocated to exploration, development, or production as appropriate, in practice support equipment and facilities are typically capitalized directly to the cost center and amortized with other capitalized costs using unit-of-production over the life of the cost center. The depreciation on the support equipment and facilities is not separated and allocated between multiple activities; instead it is expensed as DD&A expense. This treatment may result in minor differences in the timing of expense recognition; however, these differences are not likely to be material.

DD&A under SE versus FC

A comparison of the DD&A computation for SE and FC is presented in the following example.

EXAMPLE

SE Compared to FC

Lucky Company, a new company, has only two oil and gas properties, both located in the United States and both considered significant. Lease A was acquired in 2005 and production began on Lease A on January 1, 2006. Lease B was acquired early in 2007 and is still unproved. No equipment was salvaged from any of the dry holes drilled. Data for Lucky as of 12/31/09 are given below, along with columns indicating whether each cost would be:

a. amortized (I) under successful efforts (SE)
b. amortized (I) under full cost when all possible costs are included in amortization (FC-I)
c. amortized (I) under full cost when all possible costs are excluded from amortization (FC-E)

An E indicates the cost may be excluded from amortization under FC.

	Lease A, Proved				Lease B, Unproved			
		Amortize Under (I)				Amortize Under (I)		
Item	**Cost**	**SE**	**FC-I**	**FC-E**	**Cost**	**SE**	**FC-I**	**FC-E**
Bonus, before impairment					$ 60,000		I	E
Allowance for impairment					(10,000)		NA	*
Impairment for U/P, 2008					10,000		NA	I
Bonus, net of impairment	$ 50,000	I	I	I				
Delay rentals, cumulative	25,000		I	I	20,000		I	E
Direct G&G, cumulative	90,000		I	I	70,000		I	E
W/P—IDC	450,000		I	I	300,000		I	E
W/P—L&WE	100,000		I	I	100,000		I	E
Exploratory dry holes—IDC	500,000		I	I	425,000		I	I
Exploratory dry holes—L&WE	200,000		I	I	175,000		I	I
Development dry holes—IDC	310,000	I	I	I	NA			
Development dry holes—L&WE	120,000	I	I	I	NA			
Successful wells—IDC	475,000	I	I	I	NA			
Successful wells—L&WE	225,000	I	I	I	NA			
Storage tanks, 18-year	150,000	I	I	I	NA			
Future development costs	400,000		I	I	NA			
Estimated salvage**	90,000	(I)	(I)	(I)	0			
Total, net of SV	$3,005,000				$1,150,000			

* Acquisition costs net of allowance for impairment may be excluded from amortization.

** Subtract from total costs to amortize.

Other data, Lease A	**Oil, bbl**	**Gas, Mcf**
Proved reserves, 12/31/09	200,000	600,000
Proved developed reserves, 12/31/09	80,000	240,000
Production	10,000	24,000

a. DD&A under SE is calculated in the simplest manner possible assuming gas is the dominant mineral. Assume accumulated DD&A for leaseholds is $5,000 and accumulated DD&A for wells and related E&F is $430,000.

Leasehold

$$\frac{\$50{,}000 - \$5{,}000}{600{,}000 + 24{,}000} \times 24{,}000 = \underline{\underline{\$\ 1{,}731}}$$

Well and Related E&F

Costs to be amortized: $310,000 + $120,000 +
$475,000 + $225,000 + $150,000 – $90,000
= $1,190,000

$$\frac{\$1{,}190{,}000 - \$430{,}000}{240{,}000 + 24{,}000} \times 24{,}000 = \underline{\underline{\$69{,}091}}$$

$$\text{Total DD\&A} = \underline{\underline{\$70{,}822}}$$

b. DD&A under FC is calculated using the energy basis. All costs that may be included in the amortization base are included. Assume accumulated DD&A is $930,000. Remember if Lease B, an unproved property, is included in DD&A, the lease would not have been impaired. Note that the costs to amortize from Lease A include future development costs.

	Production		Proved Reserves
Oil:		10,000		200,000
Gas:	24,000/6 =	4,000 BOE	600,000/6 =	100,000 BOE
		14,000 BOE		300,000 BOE

Costs to Amortize

$3,005,000 (Lease A) + $1,150,000 (Lease B) – $930,000 = $3,225,000

$$\frac{\$3{,}225{,}000}{300{,}000 + 14{,}000} \times 14{,}000 = \underline{\underline{\$143{,}790}}$$

c. DD&A under FC is calculated using the revenue method, assuming prices at the end of the year were $25 per barrel and $3 per Mcf, and that oil sold during the year for $28 per barrel and gas sold for $2 per Mcf. All costs that may be excluded from the amortization base are excluded. Assume accumulated DD&A is $630,000.

Production		Proved Reserves	
10,000 x $28 =	$280,000	200,000 x $25 =	$5,000,000
24,000 x $ 2 =	48,000	600,000 x $ 3 =	1,800,000
	$328,000		$6,800,000

Costs to amortize from Lease B

Net capitalized costs	$1,150,000
Less: Bonus, net of impairment	50,000
Delay rental	20,000
Direct G&G	70,000
W/P—IDC	300,000
W/P—L&WE	100,000
Costs to amortize	$ 610,000

Costs to amortize from Lease B (alternate approach)

Impairment	$ 10,000
Exploratory dry holes—IDC	425,000
Exploratory dry holes—L&WE	175,000
Costs to amortize	$ 610,000

Costs to amortize from Lease A and Lease B

$3,005,000 (Lease A) + $610,000 (Lease B) = $3,615,000

$$\frac{\$3,615,000 - \$630,000}{6,800,000 + 328,000} \times 328,000 = \$137,357$$

RESERVES IN PLACE—PURCHASE

Beginning in 1984 and 1985, because of relatively lower crude oil prices than in prior years, many companies began to purchase oil and gas reserves. The cost of purchased reserves should be added to the full-cost cost pool and the reserve barrels and Mcf added to total proved reserves when computing DD&A unless substantially different lives are involved. In other words, the costs and reserve amounts are, in most cases, treated the same for purchased reserves as for developed reserves.

INTEREST CAPITALIZATION

Interest capitalization rules relating to basic computations and definitions are essentially the same for SE and FC. As described in chapter 5, interest capitalization for successful-efforts companies is, in practice, varied and subject to wide interpretation. However, for full-cost companies the capitalization procedure is clearer. Under *FASB Interpretation No. 33* (to *SFAS No. 34*), full-cost companies capitalize interest only on assets that have been excluded from the amortization base. Once costs are included in the amortization base, they lose their separate identity and cannot be identified as being under construction. Costs in the amortization base and being amortized are considered to be completed and, therefore, interest cannot be capitalized on those assets.

Capitalized interest is computed as follows:

Average accumulated expenditures during construction x Interest rate x Construction period

Interest capitalization begins when the initial expenditure related to the asset is made during the construction period. In oil and gas activities, capitalization of interest may begin with the first drilling expenditures or the first expenditures made in preparation of drilling.

A simple example follows:

EXAMPLE

Interest Capitalization

Lucky Company has unproved property costs of $60,000 for Lease A at January 1, 2004. During 2004, an exploratory well is drilled in the amount of $240,000. A 10%, $400,000 note is outstanding during the entire year and was specifically obtained for the acquisition and drilling program related to Lease A. Work on the well began on January 1, 2004, and ended on June 30, 2004. The well was successful and the property reclassified as proved property on June 30, 2004. Lucky Company excludes all unproved property costs from the amortization base. Lucky Company should capitalize interest on the unproved property prior to reclassification, computed as follows:

Average accumulated expenditures: $\frac{\$60,000 + \$300,000}{2} = \$180,000$

Capitalize: $\$180,000 \times 0.10 \times {}^{6}/_{12}{}^{*} = \$9,000$

Entry

W/P—IDC	9,000	
Interest expense		9,000

* Note that because interest capitalization ends when the asset is included in the amortization base, the construction period for interest capitalization for 2004 is only six months.

In the above entry, W/P—IDC is debited although it is unlikely that the same amount would be considered IDC for tax. In practice, when taxes are prepared, companies generally make relevant adjustments to the accounts to accommodate tax and financial accounting differences such as this one.

Limitation on Capitalized Costs—A Ceiling

A ceiling test is required under the FC rules to ensure that net capitalized costs of full-cost companies do not exceed the underlying value of the company. In applying this test, which must be performed quarterly, a value for net capitalized costs is compared to a cost ceiling value. If the capitalized cost value exceeds the ceiling, capitalized costs must be permanently written down. Specifically, capitalized costs for each cost center, less accumulated DD&A and related deferred income taxes, must not exceed a cost ceiling, which is defined as the sum of the following:

a. The present value of future net revenues from proved reserves

Plus

b. The cost of properties not being amortized (*i.e.,* unproved properties and costs of major development projects for which the quantity of proved reserves has not yet been determined)

Plus

c. The lower of cost or market of unproved properties being amortized

Less

d. The tax on the sum of a, b, and c less the tax basis of the property

If total capitalized costs exceed the cost ceiling, the excess should be expensed and may not be reinstated in the future. This test should be performed each period after recording the current period's DD&A.

In calculating net capitalized costs, only deferred taxes related to oil and gas activities should be included. Deferred taxes are subtracted—assuming a net deferred tax liability, as is the usual case—because deferred taxes are tax amounts that have already been recognized (expensed) on the financial statements but will be paid to the IRS in the future.

In calculating the ceiling, the present value of future net revenues is calculated as follows:

Future gross revenue (estimated future production x period-end prices)
Less: Future development costs (estimated using costs at year-end)
Less: Future production costs (estimated using costs at year-end)
Future net revenue, undiscounted
Discounted at 10%
Future net revenue, discounted

The cost of properties not being amortized would include all the related costs that are excluded from amortization net of impairment. Thus, unproved property costs excluded would include not only the acquisition costs, but also related direct G&G, wells in progress, etc.

EXAMPLE

Ceiling Test

Data for Lucky Oil Company is as follows:

Present value of future gross revenues		$85,000,000
Present value of future related costs		45,000,000
Capitalized costs of proved properties		
Acquisition costs	$ 2,000,000	
W/P	13,000,000	
Wells and related E&F	15,000,000	
Exploratory dry holes	20,000,000	
Total		50,000,000
Unproved properties not being amortized		
Acquisition costs	800,000	
Test-well contributions	200,000	
G&G costs	1,000,000	
W/P	6,000,000	
Total		8,000,000
Unproved properties being amortized		
Acquisition costs	500,000	
W/P	1,500,000	
Total		2,000,000
Accumulated DD&A		5,000,000
Deferred income taxes		1,000,000
Income tax effects—books versus tax		1,000,000
Market value of unproved property being amortized		2,300,000

Calculations

Capitalized Costs Less Accumulated DD&A and Deferred Taxes	
Capitalized costs of proved properties	$50,000,000
Add: Unproved properties not being amortized	8,000,000
Add: Unproved properties being amortized	2,000,000
Less: Accumulated DD&A	(5,000,000)
Less: Deferred income taxes	(1,000,000)
	$54,000,000

Ceiling	
Present value of future net revenues	
($85,000,000 – $45,000,000)	$40,000,000
Add: Unproved properties not being amortized	8,000,000
Add: LCM of unproved properties being amortized	2,000,000
Less: Income tax effects—books versus tax	(1,000,000)
	$49,000,000

Entry

Loss on producing properties	5,000,000	
Accumulated capitalized cost reduction . . .		5,000,000

Deferred taxes

The SEC's full-cost rules were written at a time when deferred tax assets were not allowed. Since these rules were written, GAAP has been changed to require the recognition of deferred tax assets subject to a valuation allowance. Recognizing a loss due to a ceiling test write-down could generate recognition of a deferred tax asset. In applying the ceiling test, net costs, which are compared to the FC ceiling, must include related deferred taxes. However, including a deferred tax asset increases net costs and would then result in net costs again exceeding the FC ceiling. In such a situation, there are two unknowns: the amount of the write down and the amount of the deferred tax asset.

Income tax effects

The income tax effects that should be considered in computing the amount to be deducted from estimated future net revenues are:

- tax basis of oil and gas properties
- net operating loss carry-forwards

- foreign tax credit carry-forward
- investment tax credits (ITC)
- minimum taxes on tax preference items
- impact of percentage depletion

Although investment tax credits are not currently allowed by the tax code, they frequently have been in the past, and may well be allowed in the future. Therefore, ITCs are included in the discussion of the ceiling test and in the example showing the income tax effects.

Only net operating loss (NOL) carry-forwards that are directly applicable to oil and gas operations should be used in the above computation. Also, only ITC carry-forward and foreign tax credit carry-forwards clearly attributable to oil and gas operations should be used in computing the income tax effects.

The SEC rules in *Reg. SX 4-10* were modified by *Staff Accounting Bulletin (SAB) 47.* The computation of the income tax effects by using the short-cut method presented in *SAB 47* is illustrated below:

EXAMPLE

Ceiling Test Income Tax Effect

Assumptions

Capitalized costs of oil and gas assets		$2,000,000
Accumulated DD&A		(400,000)
Book basis of oil and gas assets		1,600,000
Related deferred income taxes		90,000
Net book basis to be recovered		$1,510,000
NOL carry-forward*		$ 50,000
Foreign tax credit carry-forward*		3,000
ITC-carry-forward*	$5,000	
Present value of ITC relating to future development costs	2,000	7,000
Estimated preference (minimum) tax on percentage depletion in excess of cost depletion		2,000
Tax basis of oil and gas assets		1,200,000
Present value of statutory depletion attributable to future deductions		30,000
Statutory tax rate		46%
Present value of future net revenues from proved oil and gas reserves		1,300,000

Cost of properties not being amortized			75,000
Lower of cost or estimated fair value of unproved properties included in costs being amortized			65,000

* All carry-forward amounts in this example represent amounts that are available for tax purposes and that relate to oil and gas operations.

Calculation

Present value of future net revenue			$1,300,000
Cost of properties not being amortized			75,000
Lower of cost or estimated fair value of unproved properties included in costs being amortized			65,000
Tax effects:			
Total of above items		$ 1,440,000	
Less: Tax basis of properties	$(1,200,000)		
Statutory depletion	(30,000)		
NOL carry-forward	(50,000)	(1,280,000)	
Future taxable income		160,000	
Tax rate		x 46%	
Tax payable at statutory rate		$ (73,600)	
ITC		7,000	
Foreign tax credit carryforward		3,000	
Estimated preference tax		(2,000)	
Total tax effects			(65,600)
Cost center ceiling			$1,374,400
Less: Net book basis			(1,510,000)
Required Write-off, net of tax**			$ (135,600)

** For accounting purposes, the gross write-off should be recorded to adjust both the oil and gas properties account and the related deferred income taxes.

An exemption to the cost ceiling limitation may be obtained when the cost of purchased reserves in-place exceeds the present value of estimated future net revenues from the sale of such reserves. The primary reason for paying more for reserves-in-place than the expected future benefit is that the purchaser expects oil and gas prices to escalate. Further, a value is usually placed on probable and possible reserves. In such cases, the excess cost would not be written off if an exemption is obtained from the SEC.

Fairness of the ceiling test

Costs are capitalized under the FC method without regard to future benefits. The SEC consequently felt that the capitalized costs of a full-cost company could well exceed its future net revenues and therefore mandated the full-cost ceiling test. The full-cost ceiling test is an unusual and harsh test in many respects. First, the test must be performed quarterly using end-of-period prices, but restoration of quarterly write-downs is not allowed even if prices improve before year-end. This rule is especially harsh for companies producing predominantly natural gas. Gas pricing is highly seasonal, with the lowest prices being in the summer months. Regardless, permanent write-downs must be made at quarter-end even if the average gas price for the year or the year-end price is expected to be much higher than the seasonal quarter-end price. Oil prices, while not seasonal, are highly volatile. The price of oil can be significantly different just weeks before or after period-end as compared to the period-end price. Potentially huge write-downs thus depend in part on how lucky a company is in relation to the timing of its period-end as compared to the current monetary oil price.

Second, in order to compute the cost ceiling, the amount and timing of production of proved reserves must be obtained from petroleum engineers. These reserve estimates have proven to be very imprecise. The unreliability of the reserve estimates causes the ceiling test results to be unreliable. Furthermore, only proved reserves are considered when computing the cost ceiling. Some portion of probable and possible reserves will normally be transferred to the proved category at some future date, but the ceiling test places no value on these reserves.

SFAS No. 121

As discussed in chapter 6, SE companies must now apply *SFAS No. 121,* "Accounting for the Impairment of Long-Lived Assets and for Long-Lived Assets to Be Disposed Of." Because FC companies are already required to apply the FC ceiling test, FC companies are not required to apply *SFAS No. 121* to their oil and gas properties. However, FC companies must apply *SFAS No. 121* to their long-lived non oil and gas assets. On the surface, the FC ceiling test appears to be more ominous than the test under *SFAS No. 121* in that the FC ceiling test as compared to *SFAS No. 121:*

- must be applied based on discounted net revenue versus screened based on undiscounted net revenues
- must be applied using sales prices as of period end versus expected future prices
- must be applied on an interim and annual basis versus only when circumstances indicate possible impairment exists
- must be applied using only proved reserves versus proved, probable, and even possible reserves

The FC ceiling test appears to be less stringent than the test under *SFAS No. 121* only in that it is applied at a country level as compared to presumably a field level under the *SFAS*

No. 121 impairment test. However, in practice, the effect of the FC ceiling test may be much less dire than *SFAS No. 121* because the larger cost center used for the FC ceiling test allows bad properties to be offset by good properties. In addition, in 1999, the discount rate used to apply *SFAS No. 121* was around 15 to 20% compared to the required and fixed 10% used in the FC ceiling test.

Post balance sheet events and the ceiling test

Two types of events occurring after year-end but prior to publication of the financial statements can have a major impact on the full-cost ceiling test. First, if proved reserves are discovered after year-end but before the financial statements are published, the ceiling test should be revised to include the new reserves. Since the reserves existed at balance sheet date, but were just unknown at that time, this is the type of subsequent event for which GAAP requires an adjustment to the financial statements. Second, the SEC apparently allows some changes in the price of oil or gas after year-end but before the financial statements are issued to be used in calculating the FC ceiling test. However, because a change in price is not an underlying condition as of balance sheet date, the allowed treatment by the SEC is inconsistent with GAAP. GAAP allows an adjustment to the financial statements only when the subsequent event provides evidence of a condition that existed at balance sheet date.

Reasons for Changing Methods

A 1990 survey examined the reasons why companies changed between the full-cost and successful-efforts methods during the 1985 through 1989 period.[2] The controllers of oil and gas companies switching accounting methods during this period were asked why they made the change. The most important reason stated for change to either method was to improve the appearance of the financial statements. The officers of firms changing to successful-efforts accounting stated that a major reason for the change was due to substantial losses that would need to be recognized under the full-cost method because of the full-cost ceiling test. Tax considerations and auditor recommendations were considered the least important reasons to change methods. The trend since 1986 has been for companies to switch from the full-cost accounting method to the successful-efforts method.

PROBLEMS

1. Indicate whether each of the following costs would be expensed (E) or capitalized (C) under full-cost (FC) and successful-efforts (SE) accounting.

Cost	SE	FC
a. Aerial magnetic study—an area of interest is identified		
b. Seismic studies on 20,000 acres—no land is leased		
c. Brokers' fees		
d. Bottom-hole contribution Well is productive Well is dry		
e. Dry-hole contribution		
f. Delay rental payment		
g. Cost of landmen in acquiring properties		
h. G&G to select specific drillsite		
i. Cash bonus		
j. Productive exploratory drilling Intangible costs Tangible costs		
k. Lease record maintenance on unproved properties		
l. Exploratory dry-hole drilling Intangible costs Tangible costs		
m. Shooting rights		
n. Development dry-hole drilling Intangible costs Tangible costs		
o. Production costs		

2. Given the following, compute DD&A, assuming no exclusions from the amortization base.

Unrecovered costs	$700,000
Net salvage value of properties	60,000
Estimated future development costs for proved reserves	100,000
Proved reserves, 12/31	100,000 bbl
Proved developed reserves, 12/31	75,000 bbl
Production	20,000 bbl

3. Data for Pride Oil Company for 12/31/06 are as follows:

Unrecovered costs	$700,000
Estimated future development costs for proved reserves	100,000
Estimated future dismantlement costs	200,000
Net salvage value of properties	60,000

	Oil bbl	Gas Mcf
Proved reserves, 12/31/06	100,000	300,000
Proved developed reserves, 12/31/06	75,000	120,000
Production during year	20,000	50,000
Selling price during year	$30/bbl	$1.00/Mcf
Selling price, 12/31/06	25/bbl	1.50/Mcf
Expected selling price, 2007	27/bbl	2.00/Mcf

a. Compute DD&A using a common unit of measure based on BOE.
b. Compute DD&A using the unit-of-revenue basis.
c. Which would be the appropriate basis?

4. When oil and gas are produced jointly, what basis or units of measure are allowed for FC amortization? When oil and gas are produced jointly, what basis or units of measure are allowed for SE amortization?

5. Which special costs or reserves must or may be excluded from amortization for FC? Which costs or reserves must or may be excluded for SE?

6. Which special costs or reserves must or may be included in amortization for FC? Which costs or reserves must or may be included for SE?

7. In FC and SE accounting, costs are amortized over either proved reserves (PR) or proved developed reserves (PDR). Fill in the following table to indicate which reserves (PR or

PDR) should be used to amortize the costs under each accounting method. If the costs should not be amortized, put an X to indicate no amortization.

Cost	SE	FC
a. proved property acquisition		
b. unproved property acquisition		
c. successful exploratory drilling		
d. successful development drilling		
e. unsuccessful exploratory drilling		
f. unsuccessful development drilling		
g. G&G costs		
h. W/P—P/P		

8. Data for Ramsey's U.S. properties:

I	E		
		Acquisition costs of unproved properties, net of impairment	$100,000
		Acquisition costs proved properties	150,000
		Delay rentals, unproved properties—cumulative	30,000
		Delay rentals, proved properties—cumulative	50,000
		Dry holes, proved properties	300,000
		Dry holes, unproved properties	160,000
		Abandoned costs	40,000
		Successful well costs	400,000
		W/P—unproved properties	175,000
		W/P—proved properties	125,000
		Major development project necessary to determine quantity of proved reserves	500,000
		Future development costs (proved properties)	300,000
		Accumulated DD&A	(800,000)
		Production, bbl	50,000
		Proved reserves, beginning of year, bbl	600,000

For the column headed I above, assume all possible costs are included in the amortization base. For the column headed E assume all possible costs are excluded from the amortization base. Place an I in the columns above if a cost must be included in amortization. Place an E in the columns above if the cost may be excluded from amortization.

Calculate DD&A assuming the following:

a. Inclusion of both unproved property costs and the major development project.
b. Exclusion of both unproved property costs and the major development project.
c. Inclusion of unproved property costs and exclusion of the major development project.
d. Exclusion of unproved property costs and inclusion of the major development project.
e. How would impairment for unproved properties have been handled?

9. Wildcat Oil Company began operations January 1, 2005. Transactions for the first three years include the data below. Using that data:
   a. Prepare journal entries assuming FC (ignore revenue entries and assume no exclusions from the amortization base).
   b. Prepare income statements under FC and SE for all three years, again assuming no exclusions from the FC amortization base. Ignore severance tax. Assume a 1/8 royalty interest.
   c. Recalculate DD&A assuming Wildcat is a full cost company and that Wildcat excludes all possible costs from the amortization base.
   d. Which of the journal entries given in a above would have been different if Wildcat had been excluding all possible costs from the amortization base rather than including all costs?

Data:

2005—acquired three leases, assume a ⅛ royalty interest:

Lease	Bonus	G&G Costs, Direct
A	$20,000	$30,000
B	15,000	70,000
C	30,000	50,000

2006

1) A delay rental of $2,000 is paid for Lease B.
2) A delay rental of $4,000 is paid for Lease A.
3) Drilling costs of $120,000 are paid on Lease C. Proved reserves are found and estimated to be 200,000 total gross barrels and proved developed reserves are estimated to be 70,000 total gross barrels as of December 31, 2006. During 2006, 20,000 total gross barrels of oil are produced and sold. Lifting costs were $7/bbl and the selling price was $25/bbl. (Expense lifting costs as lease operating expense.) Future development costs are estimated to be $100,000.

2007

1) Lease B is surrendered.
2) A dry hole is drilled on Lease A at a cost of $250,000. As a result, Wildcat feels that Lease A is worth only $^1/_4$ of the amount capitalized as unproved property. (Note: SE and FC impairment amounts will be different.)
3) An additional well (development) is drilled on Lease C at a cost of $300,000. Proved reserves at 12/31/07 are estimated to be 230,000 total gross barrels, and proved developed reserves are estimated to be 90,000 total gross barrels.

During 2007, 25,000 total gross barrels of oil are produced and sold for $24/bbl. Lifting costs are $8.00/bbl. Future development costs are estimated to be $150,000.

10. Aggie Oil Company began operations in 2006. Give the entries, assuming the following transactions in the first three years of operations. Calculate DD&A twice, once assuming no exclusions and once assuming all possible exclusions from the amortization base. Ignore the ceiling test for 2006, but apply it for 2007 and 2008. Calculate the ceiling test twice for each of those years, once assuming no exclusions and once assuming all possible exclusions from the amortization base. Assume all leases are located in the United States. You may combine entries. Note that Problem 11 in chapter 5 is similar to this problem. Compare these two problems for a comparison of SE and FC accounting.

Year	Transaction	Lease A	Lease B	Lease C
2006	Lease bonuses	$ 50,000	$ 40,000	$ 55,000
	G&G costs—direct	60,000	50,000	90,000
	Lease record maintenance	2,000	5,000	1,000
	Legal costs for title defense	15,000	none	10,000
	Drilling costs		Well 1, exploratory	Well 1, exploratory
	IDC	none	300,000	150,000
	Equipment	none	125,000	40,000
	Drilling results	none	Well dry	Drilling uncompleted
2007	Delay rental	4,000	none	none
	Drilling costs	Well 1, exploratory	Well 2, exploratory	Well 1
	IDC	275,000	225,000	60,000
	Equipment	50,000	50,000	40,000

	Drilling results	Drilling completed, found PR of 100,000 bbl	Drilling uncompleted	Drilling completed, found PR of 300,000 bbl
	Installed flow lines, tanks, etc.			
	Installation costs	$ 5,000	none	$ 3,000
	Purchase costs	30,000	none	45,000
	Future development costs	600,000	none	400,000
	Production during year	4,000 bbl	none	6,000 bbl
	PV of future net revenue as of 12/31/07	500,000	FMV of Lease B = $200,000	800,000
2008	Drilling costs	Well 2, development	Well 2	Well 2, development
	IDC	300,000	5,000	250,000
	Equipment	80,000		100,000
	Drilling results	Well dry	Well dry, abandon lease	Drilling completed, found PR of 200,000 bbl
	Future development costs	200,000		0
	Production during year	5,000	none	20,000
	PV of future net revenue as of 12/31/08	1,000,000	FMV of Lease B = $0	3,000,00

PR = proved reserves

11. Data as of 12/31/05 for Gusher Oil Company's U.S. properties are as follows: (This problem is similar to Problem 14 in Chapter 6.)

Unproved properties, control	$ 200,000
W/P—unproved properties	350,000
Dry exploratory wells on unproved properties	425,000
Abandoned costs of unproved properties	275,000
Proved properties	190,000
Wells and related E&F	600,000
Accumulated DD&A	375,000

Gusher's activities during 2006 were as follows:

Unproved properties acquired	$ 35,000

Delay rentals paid	$ 10,000
Test well contributions paid on unproved properties	20,000
Title exams paid on unproved properties	8,000
Title defenses paid on unproved properties	15,000
Unproved properties proved during 2006	30,000
Unproved properties abandoned	20,000
Exploratory dry hole drilled on unproved property (Total cost of well was $375,000; $100,000 was in progress at 12/31/05 and therefore, not incurred during 2006; $275,000 was incurred in 2006.)	275,000
Successful exploratory well drilled	400,000
Development dry hole drilled	300,000
Service well drilled	325,000
Tanks, separators, etc., installed	125,000
Development well partially drilled, in progress 12/31/06	140,000
Estimated future development costs	225,000

	Oil bbl	Gas Mcf
Production	100,000	500,000
Proved reserves, 12/31/06	1,020,000	5,000,000
Proved developed reserves, 12/31/06	900,000	4,700,000

a. Use T accounts to accumulate costs.
b. Calculate DD&A for 2006, assuming no cost exclusions and using a common unit of measure based on BOE.
c. Calculate DD&A for 2006, assuming all possible cost exclusions and using a common unit of measure based on BOE. In addition, assume impairment for unproved properties was $75,000 and allowance for impairment was $25,000 at 12/31/06.

12. The Jones Oil Company started its oil and gas exploration and production business in 2005. During the year 2005 and 2006, the company provided the following information relating to leases located both in the U.S. and in Canada:

Year		Lease A (U.S.)	Lease B (U.S.)	Lease C (Canada)
2005	Leasehold costs	$ 40,000	$ 30,000	
	Well Number	Well 1	Well 1	
	IDC	300,000	500,000	
	L&WE	0	350,000	
	Results	Well dry, abandon well	Proved	

	Future development costs		$200,000	
	Production: Oil Gas		4,000 bbl 20,000 Mcf	
	Estimated reserves Oil, PR — 12/31/05 Gas, PR — 12/31/05 Oil, PDR — 12/31/05 Gas, PDR — 12/31/05		 50,000 bbl 300,000 Mcf 30,000 bbl 200,000 Mcf	
	Selling Price: Oil Gas		$20/bbl $1.50/Mcf	
	Price (12/31/05): Oil Gas		$22/bbl $2.00/Mcf	
2006	Leasehold costs			$ 60,000
	Well Number IDC L&WE	Well 2 $400,000 300,000		Well 1 600,000 400,000
	Results	Proved		Proved
	Future development costs	100,000	150,000	175,000
	Production: Oil Gas	5,000 bbl 20,000 Mcf	8,000 bbl 30,000 Mcf	8,000 bbl
	Estimated reserves Oil, PR — 12/31/06 Gas, PR — 12/31/06 Oil, PDR — 12/31/06 Gas, PDR — 12/31/06	 40,000 bbl 250,000 Mcf 30,000 bbl 200,000 Mcf	 40,000 bbl 280,000 Mcf 28,000 bbl 180,000 Mcf	 80,000 bbl 30,000 bbl

a. Record the above information for both years. Ignore revenue entries.
b. Compute and record DD&A for both years. Assume the revenue method may be ignored in the second year. If there is not a significant difference between the revenue basis and energy basis in the first year (less than $150,000), use the energy basis (equivalent Mcf). Assume that all possible costs are included in DD&A.
c. Compute DD&A using a common unit of measure based on Mcf assuming Jones Oil Company used SE accounting instead of FC accounting.

13. Indicate by a C whether the costs given below should be capitalized. Indicate by an I if the costs must be included in computing amortization or by an E if the cost may be excluded from amortization under FC. Indicate by an I if the costs should be amortized under SE.

Cost	Costs to Capitalize		Costs to Amortize		
				FC	
	SE	FC	SE	P/P	U/P
a. Bonus on unproved property					
b. Bottom hole contribution					
c. Ad valorem taxes, paid on U/P					
d. Legal costs for title defense, U/P					
e. Nondirect G&G, cumulative					
f. Delay rentals, cumulative					
g. W/P					
h. Dry exploratory wells					
i. Successful exploratory wells					
j. Successful development wells					
k. Dry development wells					
l. Abandoned costs					
m. Impairment of unproved properties					
n. Future development costs					
o. Estimated dismantlement costs					

14. The Ebert Oil Company provides the following information for the year ended December 31, 2007:

PV of future gross revenues	$60,000,000
PV of future related costs	15,000,000
Capitalized costs of proved properties	50,000,000
Unproved properties not being amortized	11,000,000
Accumulated DD&A	(4,000,000)

REQUIRED:

a. Prepare a ceiling test and an entry, if necessary, for the write-off of capitalized costs. Ignore income taxes.

b. Assuming the PV of future gross revenues is $70,000,000, repeat item *a*'s requirements.

15. Cruser Oil, a full-cost company, incurs the following costs during 2006:

G&G costs—nondirect	$25,000
Acquisition costs:	
Lease A	26,000
Lease B	32,000
G&G costs—direct to Lease A	15,000
Ad valorem taxes on Lease B	1,200
Dry-hole contribution on land adjacent to Lease B	40,000
Nonrecoverable delinquent taxes on Lease A paid at the time of acquiring the lease	2,100
Cost of maintaining lease records:	
Lease A	500
Lease B	600

During 2007, the following costs were incurred:

Delay rentals were paid on Leases A & B of $2,000 and $3,000 respectively.

In the latter half of 2007, drilling operations were commenced on both leases, and costs were incurred as follows:

	Lease A	**Lease B**
G&G costs to locate wellsite	$ 5,000	$ 6,000
Contractor's fee:		
IDC	325,000	350,000
L&WE	110,000	125,000
	Lease A	**Lease B**
Flowlines:		
Equipment	3,000	5,000
Labor to install	2,000	3,000
Tank Battery	15,000	17,000

REQUIRED: Record the above transactions.

16. The Hard Luck Oil Company incurs unproved property (Lease A) costs of $60,000 on April 1, 2005. An 8% loan is obtained on April 1, 2005, for $500,000 to finance a drilling program. Hard Luck started a well on Lease A on June 1, 2005, and the well is still in progress at 12/31/05. Drilling costs to 12/31/05 are $300,000. The company excludes all possible costs from the cost pool.

REQUIRED: Compute the interest capitalization amount and record the interest.

17. Determine the amount of the total amortization cost base, assuming (1) no exclusions from the amortization base, and (2) all possible costs are excluded from the amortization base.

	(1) Amount—no exclusions	(2) Amount—all exclusions
Unproved property—Cost $50,000, amount impaired $20,000 (if excluded)		
Unproved property—abandoned—cost $22,000, no impairment		
Unproved property—Found proved reserves—cost $18,000, no impairment		
Unproved property purchased—cost $40,000		
Wells in progress on Unproved Property—$100,000		
Well completed on U/P—dry-hole cost $220,000		
Total	?	?

18. Hays Oil Company has the following account balances at 12/31/05:

Lease A:	Unproved property	$ 20,000
Lease B:	Proved property	30,000
	Wells and related E&F—IDC	250,000
	Wells and related E&F—L&WE	150,000

The above properties are abandoned in 2006. Record the abandonment entry.

19. Wildcat Oil Company has the following account balances at 12/31/05:

Unproved property—Lease A	$ 60,000
Unproved property—group	220,000
Allowance for impairment, group	12,000

At 12/31/05, Lease A is considered to be 40% impaired. Wildcat Company's estimated abandonment rate of insignificant unproved properties is 60% (*i.e.*, the impairment rate is 60%).

Prepare entries to record impairment.

20. Use the same facts as Problem 19 and prepare entries using the following independent assumptions:

    a. Lease A is abandoned in 2006.
    b. Lease A is proved in 2006.
    c. Insignificant Lease Y, with a cost of $3,000 is abandoned.
    d. Insignificant Lease X, with a cost of $5,000 is proved.

21. Roberts Oil Company has the following information at 12/31/05:

Capitalized costs, plus future development costs	$2,000,000
Accumulated DD&A	800,000
Proved reserves—Oil	200,000 bbl
—Gas	800,000 Mcf

Production in 2005:

Oil	15,000 bbl	Sold at average price of $20 per bbl
Gas	70,000 Mcf	Sold at average price of 1.25 per Mcf

The current prices at 12/31/05 are:

Oil	$22.00 bbl
Gas	1.50 Mcf

Compute DD&A using the unit-of-revenue method.

22. Basic Company started operations on 1/1/05. At 12/31/05, the company owned the following leases in Canada:

Lease A — Proved property	$ 75,000
— Wells and related E&F	500,000
Lease B — Proved property	90,000
— Wells and related E&F	800,000
Lease C — Unproved property	60,000
W/P—Lease C	125,000
Proved Reserves — Oil	200,000 bbl
— Gas	900,000 Mcf
Production — Oil	10,000 bbl
— Gas	50,000 Mcf

The production was sold at $24/bbl and $1.50/Mcf. Current prices at 12/31/05 are $25/bbl and $2.00/Mcf.

Compute DD&A for Canada as follows assuming:

a. No exclusions from the amortization base, and using unit-of-production converted to a common unit of measure based on energy (equivalent Mcf).
b. No exclusions from the amortization base, and using unit-of-revenue method.
c. All possible costs are excluded from amortization, and using a common unit of measure based on energy (equivalent Mcf).
d. All possible costs are excluded from amortization, and using unit-of-revenue method.

23. Data for Lucky is as follows for all U.S. properties:

Acquisition costs of U/P, net of impairment	$ 300,000
Nondrilling evaluation costs of U/P—direct	400,000
Abandoned costs	100,000
Dry holes on U/P	600,000
Capitalized costs of P/P	2,400,000
Costs of major development project (costs not included above)	800,000
Accumulated DD&A (assume same for Part A and Part B)	700,000
Deferred income taxes	500,000
PV of future gross revenue	4,100,000
PV of future related costs	1,400,000
Income tax effects—books versus tax	500,000
LCM of unproved properties	550,000

a. Apply the FC ceiling test and record any entries necessary, assuming that all possible costs are excluded from amortization.
b. Apply the FC ceiling test and record any entries necessary, assuming that all possible costs are being amortized.

24. Big John Oil Company, located in Southern California, has been operating for three years. Big John uses full-cost accounting and excludes all possible costs from the amortization base. The following account balances are as of the end of 2010:

Insignificant Leases		
	Cost	Total Impairment
Lease A	$ 50,000	
Lease B	60,000	
Lease C	30,000	
	$140,000	$84,000

Significant Leases		
	Cost	Impairment
Lease D	$400,000	$100,000
Lease E	300,000	200,000
Lease F	600,000	400,000

REQUIRED: Give any entries necessary for the following events and transactions.

a. Lease A was abandoned on March 20, 2011.
b. Lease B was proved on May 13.
c. Insignificant leases costing $80,000 were acquired during 2011.
d. Lease E was abandoned early in July.
e. Lease F was proved October 21.
f. At December 31, Big John decides that Lease D should be impaired an additional $150,000.
g. At December 31, Big John decides to continue its policy of maintaining an allowance for impairment equal to 60% of unproved insignificant leases.

25. Gusher Oil Company began operations on 1/5/2001 and has acquired only two properties. The two properties, which are both considered significant, are located in different states. Lease B was proved on 1/1/2003. Costs incurred from 1/5/2001 through 12/31/2003 are as follows:

	Lease A	**Lease B**	**Unallocated**
Seismic studies, nondirect			$ 70,000
Bonus	$ 50,000	$ 60,000	
Title exams	10,000	5,000	
G&G costs, direct	90,000	80,000	
Test well contributions	15,000	18,000	
Insurance	2,000	3,000	
Exploratory dry holes—IDC	220,000	250,000	
Exploratory dry holes—L&WE	30,000	40,000	
Exploratory W/P—IDC	100,000	180,000	
Exploratory W/P—L&WE	22,000	19,000	
Wells and related E&F—IDC	—	700,000	
Wells and related E&F—L&WE	—	260,000	
Tanks and separators	—	110,000	
Lease operating costs	—	134,000	
Total	$539,000	$1,859,000	$ 70,000

Additional data		
Future development costs	—	$ 500,000
Proved reserves, 1/1/2003	—	600,000 bbl
Proved developed reserves, 1/1/2003	—	200,000 bbl
Production during 2003	—	40,000 bbl

Other information:
The company also owns a building that it purchased 1/1/2001 at a cost of $500,000. The building houses the corporate headquarters and has an estimated life of 20 years (ignore salvage). The operations conducted in the building are general in nature and are not directly attributable to any specific exploration, development, or production activities. Since the building is not related to exploration, development, or production, it is depreciated using straight-line depreciation for financial accounting.

REQUIRED:

a. Give the entry to record DD&A for 2003 under FC accounting assuming all possible costs are excluded from amortization.
b. Give the entry to record DD&A for 2003 under SE accounting.

26. Bonnel Company uses the full-cost method. Recently, the company acquired a truck costing $60,000 with an estimated life of five years (ignore salvage value). The foreman drives the truck to oversee operations on seven leases all in the same general geographical area but on multiple reservoirs. The foreman keeps a log of his mileage in order to determine how the truck is utilized. Analysis of the log indicated that 1/3 of the mileage driven was related to travel to drilling operations, ⅓ was related to travel to G&G exploration areas, and ⅓ was related to travel to production locations.

    Total net capitalized costs, including the truck above, are $2,300,000. Proved reserves at the beginning of the year were 300,000 barrels and production during the year was 30,000 barrels. Bonnel Company amortizes all possible costs.

    REQUIRED:
    Give any entries necessary to record depreciation on the truck for the first year that it was in service.

NOTES

1. Adkerson, Richard C., and Moore, David L., "Calculating Full Cost Depreciation, Depletion, and Amortization Under the New SEC Rules," *Journal of Extractive Industries Accounting,* Spring, 1984, pp. 9-27.

2. Nichols, Linda M., "Survey of Reasons for Accounting Method Changes," *Petroleum Accounting and Financial Management Journal,* Fall/Winter, 1990, pp. 127–135.

chapter eight

Accounting *for* Production Activities

This chapter discusses the accounting treatment of costs incurred in producing oil and gas. Accounting for these costs is essentially the same whether the company is a successful-efforts company or a full-cost company.

After a well has been completed, and flow lines, heater treaters, separators, storage tanks, etc., have been installed, production activities begin. Production activities involve lifting the oil and gas to the surface and then gathering, treating, processing, and storing the oil and gas (*SFAS No. 19*, par. 23). Costs incurred in these activities are called production costs, lease operating costs, or lifting costs and are defined under both SE and FC accounting as follows:

> *Production costs are those costs incurred to operate and maintain an enterprise's wells and related equipment and facilities, including depreciation and applicable operating costs of support equipment and facilities (par. 26) and other costs of operating and maintaining those wells and related equipment and facilities. They become part of the cost of oil and gas produced. Examples of production costs (sometimes called lifting costs) are:*
>
> *a. Costs of labor to operate the wells and related equipment and facilities.*
> *b. Repairs and maintenance.*
> *c. Materials, supplies, and fuel consumed and services utilized in operating the wells and related equipment and facilities.*

d. Property taxes and insurance applicable to proved properties and wells and related equipment and facilities.

*e. Severance taxes.**

Depreciation, depletion, and amortization of capitalized acquisition, exploration, and development costs also become part of the cost of oil and gas produced along with production (lifting) costs identified in paragraph 24. (SFAS No. 19, par. 24, 25*)*

---

* Tax levied by states on the volume or value of oil or gas produced and sold or consumed. This tax is usually referred to as either a severance tax or a production tax.

Accounting Treatment

Production costs are expensed as incurred by companies using successful-efforts accounting and as well as by those using full-cost accounting.

EXAMPLE

Production Costs

Lucky Company paid wages of $1,500 to employees engaged in operating the wells and equipment solely on Lease A.

Entry

Lease operating expense	1,500	
Cash		1,500

The definition of production costs for both SE and FC states that production costs become part of the cost of the oil and gas produced. If these costs were actually treated as part of the cost of the oil and gas produced, the costs would be inventoried as "finished goods" until sold, at which point the costs would become "cost of goods sold." In practice, however, as seen in the above example, production costs are expensed as incurred by both SE and FC companies. In fact, the SEC full-cost rules (*ASR No. 258*) specifically state that "All costs relating to production activities, including workover costs incurred solely to maintain or increase levels of production from an existing completion interval, shall be charged to expense as incurred." (See Fig. 8–1 for the accounting treatment of production costs.)

If the costs of production are not inventoried, an issue arises regarding the recording of any inventory of oil or gas on hand at year-end. Although most oil and gas is sold on or soon

Fig. 8–1 — *Production Costs*

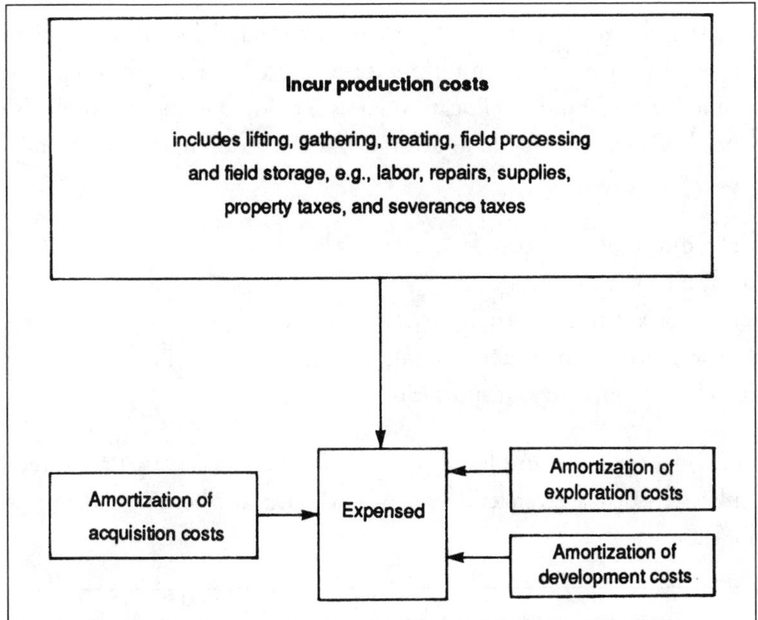

after the date it is produced, it is possible for a company to have crude oil inventory in tanks or gas inventory in certain storage facilities. However, for the most part, the quantity of oil or gas inventory on hand at the end of the period is immaterial. Additionally where such inventory exists, the amount of inventory on hand at period-end does not generally vary much from period to period. As a result, many firms historically have elected not to recognize oil and gas inventory on their financial statements.

A 1999 survey confirmed that only a slight majority of SE companies with oil and gas inventory recognize inventory on their financial statements. Of those companies recording inventory, most value the inventory at the average or posted sales price of the oil or gas. Only about 30% of the companies recording inventory—about 15% of all companies responding to the survey—value the oil and gas based on production costs. None of the FC companies responding to the survey record crude oil inventory; 20% record gas inventory.[1]

As with acquisition, exploration, and development costs, production costs must be accumulated by cost center. Therefore, a company using successful-efforts accumulates costs by an individual lease, reservoir, or field. A company using full-cost accounting accumulates costs by country. However, a full-cost company (as well as a successful-efforts company) also normally accumulates costs on a smaller cost center such as a lease or even on an individual well basis because of tax, regulatory, contractual obligations, and management requirements. Thus, even though the accounting method itself does not require allocation of costs to a well or lease, a company typically allocates costs to individual wells or leases.

Production costs can be divided into those that are directly attributable to a specific well or lease and those that must be assigned to the well or lease through some method of allocation. For costs such as repairs to a specific well, or wages and benefits of an employee working solely on one lease, accumulation of production costs by lease is straight forward: the costs are simply charged directly to the well or lease involved. Allocable costs such as the cost of a saltwater disposal facility serving multiple leases must be allocated to each well or lease on some reasonable basis. Common allocation bases include the number of wells or number of barrels produced. Other reasonable allocation bases for different types of allocable costs include:

1. Number of direct labor hours
2 Amount of direct labor costs
3. Number of miles driven for transportation and hauling
4. Gallons of water used for waterflooding
5. Miles traveled by boat for transportation

Table 8–1 provides several examples of production-related costs that are directly attributable to a particular well or lease versus those costs which must be assigned to a well or lease through cost allocation.

Directly Attributable Costs	Allocable Costs
a. Direct materials, supplies, and fuel—wells and leases involved identified in the invoices	a. Field offices and facilities serving several leases
b. Direct labor (pumpers), gaugers, etc.—employees who work on one lease only or who designate hours worked on certain wells or leases	b. Salaries and fringe benefits of field or operations center supervisors of several leases or fields
c. Contract labor or services for oxidizing, refracturing, scrubbing, etc.—invoices indicate wells	c. Depreciation—support facilities, gathering systems, treatment systems—several leases involved
d. Repairs and maintenance that can be traced to individual wells and leases	d. Transportation and hauling—several leases involved
e. Property taxes and insurance traceable from tax receipts or property descriptions on insurance policies	e. Operating costs of saltwater disposal system—several leases involved
f. Production or severance taxes—reports to the state identify these taxes to specific leases	f. Boat and fuel expenses, offshore operations—several leases involved
g. Operating costs of waterflooding system—only one lease involved	g. Operating costs of waterflooding system—several leases involved

Table 8–1 — *Directly Attributable versus Allocable Production Costs*

EXAMPLE

Cost Allocation

Expenses for Field Office A of Lucky Oil Company amounted to $10,000 for the month of May. The field office has supervision over the following leases and wells:

Lease	Number of Wells	Barrels of Oil Produced
A	1	1,000
B	3	500
C	4	2,000
D	2	1,500
Total	10	5,000

a. If the field office expense is allocated on the basis of the number of barrels produced, each lease would be charged the following amount:

Lease	Computations	Amount
A	($10,000 x 1,000 bbl/5,000 bbl)	$ 2,000
B	($10,000 x 500/5,000)	1,000
C	($10,000 x 2,000/5,000)	4,000
D	($10,000 x 1,500/5,000)	3,000
Total		$10,000

b. If the field office expense is allocated on the basis of the number of wells supervised, each lease would be charged the following amount:

Lease	Computations	Amount
A	($10,000 x 1 well/10 wells)	$ 1,000
B	($10,000 x 3/10)	3,000
C	($10,000 x 4/10)	4,000
D	($10,000 x 2/10)	2,000
Total		$10,000

INDIVIDUAL PRODUCTION COSTS

Labor costs to operate wells and related equipment include salaries, wages, and employee benefits of first-level supervisors and field employees who work on the leases. First-level supervisors generally supervise employees and contract laborers who are directly employed on a property in maintaining and operating the wells and equipment. First-level supervisor costs are directly related to the operation of the leases and are typically charged to the leases on the basis of detailed time records that show the amount of time worked on each property. If detailed time records are not kept, the costs are apportioned on some equitable basis. Field employees perform routine maintenance and provide supervision of the measurement process. Field employees include pumpers, gaugers, roustabouts, etc. If field work covers more than one lease or well, detailed time sheets should be maintained for field employees so their salaries and benefits can be charged to the individual leases or wells served.

Labor costs include **employee benefits**, which are charged to leases along with other labor costs. Generally, companies estimate their total employee benefits and direct labor costs, and then compute the ratio of employee benefits to direct labor costs. For example, a company may estimate that employee benefits will equal 25% of direct labor costs. Then, as direct labor costs are charged to leases, an additional 25% is charged representing employee benefits.

Repairs and maintenance include normal repairs and maintenance, certain workover operations, and certain recompletions. Normal repair and maintenance may be necessary for all lease equipment such as tanks and flow lines, lease roads, and buildings on the lease. Examples of repair and maintenance include replacing a defective meter, repairing a broken flow line, and lubricating production equipment. Repair and maintenance cost is expensed unless the useful life of the asset is materially lengthened or the productivity of the asset is materially increased. Invoices should indicate which wells or leases to charge; consequently, these costs are directly attributable to particular wells or leases.

Often, repair and maintenance costs, plus costs of operation of the wells and equipment may be incurred on a contract basis, with an independent contractor providing this pumping service. Pumping service includes such items as routine maintenance, meter reading, and gauging operations. The invoice from the independent contractor for pumping service should identify the well or lease to be charged and provide the data for the charges to particular properties.

Production costs also include many types of workovers. As discussed in chapter 5, **workover operations** may include repairing sucker rods, tubing, Christmas trees and leaks, cleaning out sand-filled perforations, acidizing, etc. If the workover cost is incurred for the purpose of stimulating or restoring production in the same producing horizon, it should be expensed as a production cost. If the workover or recompletion cost is incurred when deepening a well or plugging back to a shallower horizon to produce reserves from a new formation or to increase reserves, the costs are considered exploratory or development drilling costs rather than production. Consequently, the costs should be capitalized or expensed, depending upon the financial method of accounting being used, the classification of the well, and the outcome of the drilling. A recompletion may also involve recompleting in a producing zone or in a zone that produced previously. This type of recompletion is expensed as a production cost because its purpose is restoring or increasing production without an increase in proved reserves.

Costs of materials, **supplies**, and **fuel** used in well operations include materials and supplies used in routine repair and maintenance as well as fuel used to operate the production equipment—separators, heater-treaters, etc. These costs are considered production costs and can usually be identified with a particular lease.

Property taxes and insurance on proved properties are also lease operating costs. **Property taxes** are ad valorem taxes levied by governmental agencies. Property taxes on proved properties are typically significant in amount; property taxes assessed on unproved properties are generally insignificant in amount and are considered exploration costs rather than production costs (chapter 3). Types of **insurance** coverage on proved property would include general liability, worker's compensation, and fire and casualty. Both property taxes and insurance are easily identified with a particular property.

Shut-in payments normally must be paid to the royalty owner when a well that is capable of producing is shut-in. Companies using either successful-efforts accounting or full-cost accounting would normally expense shut-in payments as production costs if they were not recoverable from future production. When shut-in payments are recoverable from future production, a receivable from the royalty owner is recorded for the amount of the payments by both full-cost and successful-efforts companies.

Depreciation and **operating costs** of support equipment and facilities should theoretically be allocated to exploration, development, or production as appropriate. The portion allocated to production is expensed as a cost of production. Depreciation of support equipment and facilities is discussed in chapters 6 and 7.

Overhead costs, such as administrative costs, that may not be directly related to oil and gas acquisition, exploration, development, and production activities are also expensed as incurred. This type of overhead may include home office expenses such as officers' salaries, legal fees, accounting, etc. Overhead costs not directly related to oil and gas producing activities are not production costs and generally are not allocated to individual leases or wells for financial accounting and reporting purposes.

Secondary and tertiary recovery

Installation of secondary and tertiary recovery systems frequently requires large expenditures, *e.g.*, the cost of drilling multiple injection wells, converting producing wells to injection wells, and the cost of purchasing injection equipment. The expenditures for the wells and equipment are considered part of the development phase and are therefore capitalized and amortized using the unit-of-production method. The costs of any chemical injectants that are utilized are commonly accounted for using one of two methods. First, the injectants may be considered depreciable property and recorded as wells and related equipment and facilities. The costs are then amortized along with other wells and related equipment costs for the cost center. Second, the costs of injectants may be charged directly to production expense.

Routine maintenance and operating costs of secondary and tertiary recovery systems are considered production costs and are expensed as incurred.

The following example illustrates drilling and operating a secondary recovery system:

EXAMPLE

Secondary Recovery System

Lucky Company incurred $3,000,000 in drilling and equipping a water-flood secondary recovery system. In the month following completion of the installation, $4,000 was incurred for supplies and fuel to operate the system. In addition, the company incurred a $10,000 cost for water that was purchased and injected into the formation.

Entries

Secondary recovery system	3,000,000	
Notes payable		3,000,000
Lease operating expense ($4,000 + $10,000) . .	14,000	
Cash		14,000

Gathering systems

A gathering system begins with pipelines that move the oil and gas produced from individual wells to a central point where the oil and gas are separated and the **basic sediment and water** (BS&W) is removed. After treatment, the oil and gas is then transferred either to tanks or into a pipeline to be transported to a refinery or gas processing plant. An oil-gathering system includes equipment such as oil and gas separators, heater treaters, and gathering tanks. A gas-gathering system includes equipment such as compressors and dehydrators. The term gathering system is also used to refer to just the network of flowlines.

Attendant costs incurred for gathering-system installation are development costs subject to DD&A. The operating costs, however, are considered production costs and are expensed as incurred.

Saltwater disposal systems

Saltwater, an unwanted by-product, is normally produced along with oil and gas. The saltwater, which must be disposed of, may be reinjected back into the formation as a means of disposal that does not harm the environment and that helps to maintain natural reservoir pressures. A saltwater disposal system gathers the saltwater, treats it to remove chemicals and corrosive materials, and then reinjects it into the earth. The cost of the system is capitalized and depreciated using either the unit-of-production method—if the system serves a single field—or any other acceptable method if the system serves multiple fields. If, as would normally be the case, the disposal system covers several leases, the operating costs must be allocated to the individual properties served.

In situations where the wells are producing approximately equal volumes of saltwater, the relative number of wells connected to the disposal system is a commonly used allocation basis. When the individual wells or properties are producing significantly different volumes of saltwater, the saltwater coming in from each lease will be metered and relative saltwater throughput will be used as the allocation basis. If only one lease is served by a saltwater disposal system, the operating costs are charged to lease operating expense for that lease.

Tubular goods

Another capital expenditure that gives rise to production costs involves tubular goods. In the oil and gas industry, the term **tubular goods** refers to casing and tubing. When a well or facility is originally drilled or constructed, the purchase and installation costs of tubular goods are capitalized. Subsequent repair and replacement of tubular goods are considered production costs and are expensed as incurred. In general, the price of tubular goods includes the invoice amount plus transportation costs.

The following example presents the purchase and installation of tubing for replacement of damaged tubing:

EXAMPLE

Tubular Goods

Lucky Company obtained a new string of tubing to replace damaged tubing in a producing well on Lease A. The net cost of the new tubing was $100,000, and transportation costs were $10,000. Installation costs of the new tubing were $20,000. Record the expenditures.

Entry

Lease operating expense	130,000	
Cash		130,000

Severance taxes

Severance tax is a tax levied by state governments on oil and gas production. Severance taxes, also called **production taxes**, are typically a specified percentage of the selling price of oil or gas. If not a specified percentage of the selling price, severance taxes are based on the quantity of oil or gas sold, *e.g.*, $2 per barrel.

While production costs are generally recorded as expenses when incurred, severance taxes, which are usually based on the selling price of the product, are incurred in conjunction with the sale of oil or gas. As a result, this type of production cost is recorded when the related revenues are recorded. Recording severance taxes is illustrated in chapter 9.

Production Costs Statements

Each month, primarily for internal management purposes, operating statements similar to an income statement are prepared for each property. These statements are referred to as production costs statements. The statements normally include income and expense but may show expense only. Current-month along with year-to-date figures for revenues and expenses are usually reported.

Joint Interest Operations

When more than one company owns an undivided working interest in a property, the companies must enter into a joint operating agreement (JOA). One of the companies is designated the *operator*, typically the party with the largest interest. In production activities, the operator typically pays all of the direct costs related to operating the property and then bills the nonoperators for their proportionate share of the cost. Indirect costs, including corporate and administrative overhead incurred by the operator are recovered through overhead charges to the nonoperators, as provided for in the JOA. (This topic is discussed in detail in chapter 11.)

Now that the costs involved in drilling, completing, and operating a well are known and understood, attention can be turned to the decision of whether to complete a well and whether a well or property is ultimately profitable.

Decision to Complete a Well

In making a decision whether to complete a well, the incremental costs to complete the well should be compared with the future net cash flows expected to be received from the sale of oil or gas produced from the well. This comparison entails estimating the following items:

1. Quantity of oil or gas recoverable from the reservoir
2. Timing of future production of the oil or gas

3. Future selling price of the oil or gas
4. Future production costs of the oil or gas, including severance taxes
5. Completion costs
6. Cost of capital

If the expected future net revenue is greater than the expected completion costs, the well is usually completed.

Estimation of total recoverable reserves is an inexact science, especially for the first well discovering oil or gas in a field. A study done by PricewaterhouseCoopers found that reserve estimates, when a field is first discovered, are inaccurate by at least ± 50%, and it may be only after five years of production that reserve estimates can be made within ± 20% (1979).[2]

Reserve estimates are usually made by a reservoir engineer, who may be an employee of the company or an independent contractor. Some factors or characteristics taken into account by the reservoir engineer in preparing a reserve estimate are as follows:

1. Size of the reservoir
2. Porosity and permeability of the reservoir
3. Pressure and temperature in the reservoir
4. Oil, gas, and water contained in the reservoir pores

In addition to estimating the total recoverable reserves, the reserve engineer also must estimate the *timing* of production of the oil and gas reserves. Reserve recovery timing depends on such factors as the characteristics of the reservoir, product demand, and government regulations. Product demand is especially important for natural gas because gas is difficult and expensive to store.

Another important consideration affecting the decision to complete a well is the future product price and the cost of producing the product. Both items must be estimated before net revenue can be determined. Estimating the future price of the product is a difficult and imprecise process. The price of oil or gas is affected by product supply and demand probably more than any other factor. Governmental intervention, whether domestic or foreign, also affects the price and is usually impossible to predict. On the other hand, lifting costs can generally be estimated with a relatively high degree of accuracy. Severance taxes are part of production costs and must also be estimated.

Finally, completion costs must be estimated and compared to future net cash inflows. Completion costs can typically be estimated relatively easily and accurately. An example follows showing the comparison of net cash inflows to completion costs necessary in the decision of whether to complete a well. Costs already incurred, *i.e.*, sunk costs are not relevant to that decision.

EXAMPLE

Completion Decision

Lucky Company's data follow:

Proved property cost (acquisition cost) . . .	$ 40,000
Drilling cost incurred to date	200,000
Estimated completion cost	150,000
Estimated selling price per bbl	20
Estimated lifting costs per bbl	4
State severance tax	5%
Working interest share of revenue	90%
Royalty interest percentage	10%

REQUIRED: Should the well be completed assuming the following total production?

Production

Case A: 7,500 bbl
Case B: 15,000 bbl
Case C: 30,000 bbl

Computations

	Case A	Case B	Case C
Total revenue (bbl x $20)	$150,000	$300,000	$600,000
Less: RI's revenue (10%)	(15,000)	(30,000)	(60,000)
Revenue to WI	135,000	270,000	540,000
Less: Severance tax			
(5% x revenue to WI)	(6,750)	(13,500)	(27,000)
Net revenue before lifting costs	128,250	256,500	513,000
Less: Lifting costs ($4 x bbl)	(30,000)	(60,000)	(120,000)
Net revenue to WI owner	$ 98,250	$196,500	$393,000

Case A: The well should not be completed. The net revenue to the WI owner is only $98,250, and estimated completion costs are $150,000. Reserve estimates for new production are often higher than actual; therefore, the actual net cash inflow may be even less.

Case B: Further analysis of reserve estimates may be needed. The net cash inflow is projected to be $196,500 as compared with completion costs of $150,000. If the reserve estimate is 30% off on the downside, completion costs would not be recovered.

Case C: The well should be completed. The net revenue is $393,000 and completion costs are $150,000.

In the above example, the timing of reserve recovery was not addressed and the time value of money was not discussed. The estimated future net cash inflows should be discounted to present value, using an appropriate interest rate. Discounting is especially important in this type of analysis because many of the costs are up front, while the revenues are received over time. For example, in Case B above, the estimated future net cash inflows are only slightly more than the completion costs. If the net cash inflows are received over a period of time, the completion costs will in all likelihood exceed the discounted future net cash inflows.

Also in the preceding example, the cost of production facilities was ignored. In fields where these facilities are not already in place, the cost of the production facilities would have to be considered in the completion decision. In some cases, it might be necessary to drill additional exploratory wells to determine the extent of the reserves and whether those reserves justify the facilities.

Profitability of a well or a property

Even though a decision is made to complete a well because future net revenue is expected to be greater than completion costs, the well may still not be profitable. In order to be profitable, the net revenue from the well must exceed not only completion costs but all other costs as well. The costs incurred before deciding to complete a well, such as drilling costs, etc., are sunk costs and do not enter into the completion decision. These costs are relevant, however, in determining whether the well has ultimately been profitable.

The data from the preceding example, except for changes in reserve and production amounts, are used in the two following similar examples. These examples illustrate an analysis to determine the ultimate profitability of a well and an analysis to determine the ultimate profitability of a property.

EXAMPLE

Profitability of a Well

The following information applies to a well drilled and produced by Lucky Company:

Drilling cost	$200,000
Completion cost	150,000
Selling price per bbl	20
Lifting costs per bbl	4
State severance tax	5%
Working interest share of revenue	90%
Royalty interest	10%

REQUIRED: Determine whether the well was profitable.

Production

Case A: 750 bbl per month for 30 months
Case B: 1,200 bbl per month for 30 months

Computations

	Case A	Case B
Total revenue per month	$ 15,000	$ 24,000
Less: RI's revenue	(1,500)	(2,400)
Revenue to WI	13,500	21,600
Less: Severance tax	(675)	(1,080)
Net revenue before lifting costs	12,825	20,520
Less: Lifting costs	(3,000)	(4,800)
Net revenue to WI owner per month	$ 9,825	$ 15,720
Total net revenue (revenue x 30 mos.)	$294,750	$471,600

Costs to Recover:	Estimated drilling cost	$200,000
	Estimated completion cost	150,000
		$350,000

Case A: The well was not profitable—$294,750 net revenue, compared to $350,000 total costs.

Case B: The well was profitable—$471,600 net revenue, compared to $350,000 total costs.

An important point to note is that if a well is completed, it is classified as successful. But, even though classified as successful, the well may still be unprofitable. Thus, even though one exploratory well in approximately three wells is successful, far fewer wells are profitable.

EXAMPLE

Profitability of a Property

Lucky Company provided the following information:

Property cost (acquisition cost)	$ 40,000
Drilling cost	200,000
Completion cost	150,000
Selling price per bbl	20
Lifting costs per bbl	4
State severance tax	5%
Working interest share of revenue	90%
Royalty interest	10%

REQUIRED: Would an investment made in this property have been successful, assuming management or the investor wanted a 36-month payout?

Production

Case A: 750 bbl per month
Case B: 1,200 bbl per month

Computations

	Case A	Case B
Total revenue per month	$15,000	$24,000
Less: RI's revenue	(1,500)	(2,400)
Revenue to WI	13,500	21,600
Less: Severance tax	(675)	(1,080)
Net revenue before lifting costs	12,825	20,520
Less: Lifting costs	(3,000)	(4,800)
Net revenue to WI owner per month	$ 9,825	$15,720

Costs to Recover:	Proved property cost	$ 40,000
	Estimated drilling cost	200,000
	Estimated completion cost	150,000
		$390,000

Payout:

Case A: $\frac{\$390{,}000}{\$9{,}825} = \underline{\underline{39.69 \text{ months}}}$

Case B: $\frac{\$390{,}000}{\$15{,}720} = \underline{\underline{24.81 \text{ months}}}$

Case A: No, the investment would not have been successful—Payout is 39.69 months

Case B: Yes, the investment would have been successful—Payout is 24.81 months

PROBLEMS

1. Define and discuss the following:
   tubular goods
   gathering systems

2. Indicate whether each of the following items is a directly attributable operating expense (D), an allocable operating expense (A), or a capital expenditure (C):

   ____ Repairs that can be traced to individual wells
   ____ Cost of pumping unit
   ____ Installation of pumping unit
   ____ Ordinary repair parts for pumping unit
   ____ Repair of pumping unit
   ____ Lease road maintenance
   ____ Wages for employees who work only on one lease
   ____ Wages for employees who designate hours worked on certain wells or leases
   ____ Wages for employees who work on multiple leases, detailed records not kept
   ____ Severance tax
   ____ District office expenses
   ____ Fuel for leases, invoices indicate particular lease
   ____ Depreciation of truck used on one lease
   ____ Depreciation of truck used on multiple leases, detailed mileage records not kept
   ____ Property taxes
   ____ Contract labor for refracturing, invoices indicate wells
   ____ Workover for the purpose of restoring production
   ____ Workover/recompletion for the purpose of obtaining new production
   ____ Depreciation of district office facilities

3. District office expenses were $48,000 for July 2008. The district office supervised the following leases:

Lease	Number of Wells	Oil Produced, bbl
A	6	5,000
B	2	1,000
C	4	2,000

   a. Record lease operating expense, assuming allocation based on the number of wells.
   b. Record lease operating expense, assuming allocation based on the barrels of oil produced.

4. Hard Luck Oil Company has a waterflooding system for a reservoir in Texas. Costs of operating the system for the month of June were $20,400.
   a. Give the entry to record operating expense, assuming the reservoir underlies only Lease A.
   b. Assume the reservoir underlies the following three leases:

Lease	Producing Wells	Injection Wells	Volume of Water
A	2	1	200
B	5	2	450
C	6	3	625

   Give the entry to record operating expense, assuming allocation based on the volume of water.

5. Ebert Oil Company incurs the following costs relative to a gathering system:
   a. Purchase and installation costs of separators and compressors $200,000
   b. Operational costs for the system 10,000

   Give the entry to record the above costs, assuming the gathering system serves only one lease.

6. Miller Oil Company purchased new tubing and casing to replace damaged tubular goods in a well. The new tubular goods were installed in a producing well. The cost of the items were as follows:
   a. Tubing and casing $80,000
   b. Installation of tubular goods 15,000
   c. Loading, hauling and unloading charges 10,000

   Give any entries necessary.

7. What factors are important in determining whether to plug or complete a well?

8. Davenport Energy Company data in connection with Lease A is as follows:

Property cost (acquisition cost)	$ 30,000
Drilling cost (one well)	240,000
Estimated completion cost	200,000
Estimated selling price per bbl	25
Estimated lifting costs per bbl	5
State severance tax	6%
Working interest share of revenue	87.5%
Royalty interest	12.5%

REQUIRED: Should the well be completed, assuming the following total production? Discuss your answer.
Case A: 10,000 bbl
Case B: 20,000 bbl
Case C: 30,000 bbl

9. Assuming the same data as given in Problem 8, was the well in each case profitable? Discuss your answer.

10. Assume the same data as given in Problem 8, except the company expects the following production:
Case A: 600 bbl per month
Case B: 1,000 bbl per month

REQUIRED:
a. Determine the number of months needed for payout.
b. Would an investment in this property be considered successful if an investor wanted a 30-month payout?

11. A saltwater disposal system is added to Lease A's gathering system at a cost of $150,000. The expense for the month of May, 2006 is $15,000. Record the acquisition cost and the monthly expense assuming the following:
Case A: The disposal system serves several wells on two different leases. There are five wells on Lease A and 10 wells on Lease B.
Case B: The disposal system serves only the wells on Lease A.

12. Ebert Oil Company pays the following amounts during June 2006, relating to producing leases:

Supplies for Lease A	$ 300
Fuel for Lease A	800
Labor cost for pumpers and gaugers—Lease A	2,000

Fringe benefits of pumpers and gaugers who work on several leases	$ 400
Salaries and fringe benefits of regional supervisors	10,000
Contract labor for refracturing well on Lease A	1,000
Property tax—Lease A	1,500
Transportation cost for several leases	1,200

Record the above transactions.

13. Miller Oil Company has the following data in connection with Lease A:

Proved property cost	$ 40,000
Drilling cost	200,000
Estimated completion cost	150,000
Estimated selling price per bbl	22
Estimated lifting costs per bbl	6
State severance tax rate	5%
Royalty interest	$^3/_{16}$

Miller Oil Company is sole owner of the working interest.

If reserves are determined to be 20,000 barrels, is the well profitable? Would the well be profitable if reserves are 30,000 barrels? How many barrels of total production would it take to recover drilling and completion costs?

14. Hard Luck Oil Company estimates the following costs to acquire, drill, and complete a well on Lease A:

Acquisition costs	$ 50,000
Drilling and completion costs	400,000
Selling price of oil per bbl	22
Lifting costs per bbl	4
State severance tax rate	5%
Royalty interest	20%

Would the investment be profitable if proved reserves are:

a. 20,000 barrels?
b. 30,000 barrels?
c. 40,000 barrels?

15. Knight Company has the following expenditures in May 2006:

Lease A—Well #2: Cleaning and reacidizing formation	$10,000
Lease B—Well #3: Testing, perforating, and completion at 10,000 feet. Current production is at 12,000 feet.	
IDC	60,000
Equipment from inventory	10,000

Record the above transactions.

16. Good Luck Oil Company drilled a successful gas well. Although capable of production, the well was shut-in awaiting the completion of a pipeline. Shut-in royalty payments of $300 per month were paid during the months of June through August. Assume the following independent cases:
    a. The shut-in royalty payments were recoverable from future production. Give the entry to record payment of the shut-in royalty for June.
    b. The shut-in royalty payments were not recoverable from future production. Give the entry to record payment of the shut-in royalty for June.

17. Tyler Company operates the Salt Creek Field. Along with oil this field produces large quantities of saltwater. The saltwater is highly corrosive and, as a result, the downhole production equipment is subject to frequent replacement. During the first half of 2004, Tyler must replace the tubulars in wells A, B, and C. Wells X, Y, and Z (newly drilled wells) are in the process of being completed.

    During April, Tyler receives a shipment containing 500 joints of production tubing. The invoice totals $500,000 plus $3,000 for transportation and hauling. The production tubing is installed in the following wells:

Well	Production Tubing
A	100
B	50
C	75
X	100
Y	125
Z	50

    Record the purchase and installation of the production tubing.

Notes

1. PricewaterhouseCoopers, *Survey of U.S. Petroleum Accounting Practices*, Denton, Texas: Institute of Petroleum Accounting, 1999, p. 83-84.

2. Pricewaterhouse. *Reserve Recognition Accounting: An Evaluation of its Viability and Application,* New York: PriceWaterhouse, 1979.

chapter nine

Accounting *for* Revenue from Oil & Gas Sales

Accounting for production activities was examined in the last chapter. In this chapter, accounting for revenue from the sale of oil and gas is discussed. As with production costs, revenue accounting is the same for both full-cost accounting and successful-efforts accounting. Before discussing revenue accounting, the process of measuring and selling oil and gas is briefly discussed. Definitions of common terminology related to the measurement and sale of oil and gas are presented in the following glossary:

API gravity. The measure of the density of oil, expressed in degrees. The lighter the oil, the greater the gravity and the higher the price.

Associated gas. Natural gas that overlies and is in contact with crude oil in the reservoir but not in solution with the oil. Commonly known as *gas-cap gas.* See Figure 9–3 and definitions of dissolved gas and nonassociated gas.

Barrel. A unit of measure of crude oil equal to 42 U.S. gallons.

BS&W. Basic sediment and water, *i.e.*, impurities in suspension with crude oil.

Btu. British thermal unit; the amount of heat required to raise the temperature of one pound of water by one degree Fahrenheit.

Casinghead gas. Natural gas that is produced with crude oil from oil wells.

City gate. System inlet of the local distribution company to which gas is delivered from a natural gas pipeline company.

Commingled gas. Gas from two or more sources combined into one stream of gas.

Condensate. A hydrocarbon liquid that condenses from a gaseous state to a liquid state when produced.

Crude oil. Unrefined liquid petroleum (hydrocarbons).

Dissolved gas. Natural gas in solution with the crude oil in the reservoir. Also referred to as *solution gas.*

Field facility. An installation designed to process natural gas from two or more leases in order to control the quality of gas sold to the market. Usually separates condensate from the natural gas and may extract other products such as propane and butane.

Gauging. Measuring the level of oil in a tank.

Gas balancing agreement. An agreement that specifies how any gas production imbalance will be balanced between the different parties.

Gas settlement statement. A form used to record the amount of natural gas transferred to the pipeline.

Heater-treater. A vessel that heats the well fluids to separate the water, sediments, and any remaining gas from the oil to raise the oil to a quality acceptable for sale.

LACT unit. Lease automatic custody transfer unit that measures the flow of oil from the lease to a pipeline and determines the API gravity, BS&W, and temperature.

Local distribution company (LDC). A company that purchases gas and resells the gas to the public, or to other end-users.

Mcf. The standard unit for measuring natural gas—a thousand cubic feet.

MMcf. One million cubic feet of natural gas.

Natural gas. A mixture of hydrocarbons occurring in the gaseous state.

Natural gas processing plant. An installation in which the natural gas liquids are separated from the natural gas and/or the natural gas liquids are separated into liquid products such as butane and propane.

Nonassociated gas. Natural gas in a reservoir that does not contain any oil (See Fig. 9–3).

Posted field price. The announced or published price that some purchasers in a particular area will pay for crude oil.

Processed gas. Natural gas processed in a natural gas processing plant in order to remove the liquid hydrocarbons.

Psia. Pressure per square inch absolute.

Run ticket. A document showing the amount of oil run from a lease tank into a connecting pipeline or tank truck.

Scrubber. A vessel used to separate entrained liquids or solids from gas.

Separator. A vessel used to separate well fluids into gases and liquids.

Shrinkage. The reduction in the volume of natural gas due to the removal of impurities such as liquids, sulfur, and water vapor.

Spot sales. Short term sale of oil or gas in the spot market, without a long-term contract.

Tank battery. Two or more tanks on a lease that can hold oil production from wells prior to sale.

Tank strapping. Measuring an oil tank so that a tank table can be prepared that shows the volume of oil in the tank at any given height.

Thief. A brass, aluminum, or glass cylinder that is lowered into an oil storage tank to obtain an oil sample.

Measurement and Sale of Oil and Gas

Oil and gas must be measured either when produced or when sold. Understanding the measurement process is important to the accountant because the amount of revenue recorded is based on information obtained during this process. The following sections discuss the measurement and sale of oil and gas.

Oil measurement and sale

Fluids produced from an oil well normally contain oil, gas, and water. Before the oil can be sold, well fluids must be separated, treated, and measured. Flow lines transfer the well fluids from individual wells to a centralized separating, treating, and/or storage facility. At the centralized facility, the fluid enters separators that remove the gas from the liquid (oil and water). The liquid is then treated in heater-treater units to remove the water and other impurities from the oil. The water and other impurities, called **BS&W** or basic sediment and water, must be below an established limit in oil sold. The limit is usually 1%.

The treated crude is then stored in a **tank battery** on the lease. Typically, a tank battery has storage capacity equal to three to seven days of production. Each tank in a tank battery is **strapped**, which means that the dimensions of a tank are determined and the volume of oil (in barrels) that a tank can hold is calculated at height intervals, usually $^1/_4$ inch apart. The oil enters a tank through an opening at the top of the tank and is sold, *i.e.*, delivered or run to a tank truck, barge, or pipeline through an opening near the bottom of the tank. The oil delivered is measured by **gauging** the height of the oil in the tank both before and after delivery of the oil. A tank gauger normally determines the height of the oil in the tank in a nonautomated system by lowering a steel tape with a plumb bob on the end into the tank. When the tape touches the tank bottom, it is withdrawn, and the gauger notes the oil level on the tape.

The unit of measure of oil is a **barrel**, which is 42 U.S. gallons at standard conditions, net of any BS&W. Generally, standard conditions are a temperature of 60°F and API gravity at 60°F. The volume of oil sold is converted to standard conditions by using the characteristics of the oil sold, *i.e.*, the API gravity, temperature, and BS&W. Thus, in addition to measuring the amount of oil delivered, the oil must also be tested to determine its characteristics. A container known as a **thief** is used to obtain samples of the oil at various levels in the tank, and these samples are then analyzed to determine BS&W content, the gravity of the oil, and the temperature. In addition, the temperature of the oil in the tank is taken both before and after the run. These two temperature readings are used to correct the volume of the fluid in the tank before and after the run to a standard condition of 60°F. The temperature of the oil from the oil sample is used to correct the observed gravity at the observed temperature to the standard API gravity at 60°F. The BS&W factor is used to convert gross barrels of oil delivered to barrels net of BS&W content, at standard API gravity.

The BS&W, API gravity, temperature, the number of barrels of oil removed from a tank, and the calculations converting the barrels to standard conditions, net of BS&W are recorded on a run ticket at the time of delivery of the oil (Fig. 9–1). The run ticket is the source document for recording the amount of oil sold.

When the oil is sold, the pipeline gauger, observed by the lease pumper-gauger, measures the oil removed and determines its characteristics. When the amount of oil removed from the tank and its characteristics have been recorded on the run ticket, both the pipeline gauger and the lease pumper-gauger sign the ticket.

Another method of measuring oil delivered is used when the oil is delivered or transferred automatically into a pipeline. In this case, measurement of the oil is performed by a **lease automatic custody transfer (LACT) unit**. A LACT unit is an unattended unit that automatically transfers oil from the lease to the pipeline, takes samples, and collects and records production/accounting data, *i.e.*, volume measurement, temperature, and BS&W content. While the use of LACT units appears to be much more efficient than manual measurement,

COMPANY

ACCT NO

RUN TICKET NO

DIST NO

DATE

OPERATOR

LEASE NAME	LEASE NO.

WELL NO.		FT.	INCHES	TEMP	BBLS
	1ST MEAS.				
TANK NO.	2ND MEAS.				

	GRAVITY			
TANK SIZE	OBSERVED	TEMP.	TRUE	PRICE
BS &W	OFF←SEAL NO.'S→ON			AMOUNT

THIEFING	BEFORE	AFTER	CALCULATIONS
FT – IN BS & WATER			

REMARKS

1S T MEAS	DATE	GAUGER
	TIME A.M P.M.	WELL OWNER'S WITNESS
2ND MEAS	DATE	GAUGER
	TIME A.M P.M.	WELL OWNER'S WITNESS

This ticket covers all claims for allowance. The oil represented by this ticket was received and run as property of

Fig. 9–1 — *Typical Run Ticket*

Source: The run tickets presented in this chapter and in the homework answers are based on a pipeline run ticket prepared and printed by Kraftbilt, Tulsa, Oklahoma, 1-800-331-7290.

only about 50% of the producing leases in the U.S. are equipped with LACT units. LACT units are most common in large fields or reservoirs with numerous producing wells.

If a tank battery is to collect crude oil from several wells, the wells must be connected to a gathering system. A gathering system is a network of flowlines going from the various wells to the central tank battery. Since the tank battery collects crude oil from several wells, the quantity of crude oil produced must be allocated back to the various wells. The amount of production per well is determined by using a routine productivity test. Each well is periodically put on test by diverting the flow from the production separator to a test separator that determines the production rates of oil, gas, and water of the well being tested. Once this information is obtained, each well's percentage of production can be calculated. From these percentages, actual production of the commingling facility can be allocated to each well. (See further discussion later in this chapter.)

Run ticket calculation

The data entered on a run ticket is used to determine the volume of oil released from a storage tank or flowed through a meter. The process used to complete a run ticket and determine the net volume from a tank run is described below and is illustrated with numbers from a completed run ticket (Fig. 9–2):

a. At the time a tank is installed, it is strapped and assigned a unique number, in this case number 1. This number, which is placed on the run ticket, indicates the strapping table that corresponds with the volumetric characteristics of the tank. This table will be used to convert the feet and inches measurement on the run ticket into a run volume from the tank expressed in barrels.
b. The level of oil in the tank and the temperature of the oil are measured at the beginning and at the end of the run. In this example, the beginning level was 13'0" and the ending level was 1'4". Each of these measurements, which are in feet and inches, are converted to a volume in barrels using the appropriate tank table (Table 9–1). In this example, the level of oil in the tank converted to barrels at the beginning and at the end of the run is 180.83 barrels and 18.60 barrels, respectively. The difference between these two numbers is the measured volume of the run.
c. A thief is used to get a sample of the oil. From this sample, the temperature (74°), the BS&W (0.8%), and the observed gravity (32°) are determined. The volume of oil sold must be adjusted to a standard temperature of 60°, net of any BS&W. The observed gravity must also be adjusted to a true gravity based on the observed temperature of the sample.
d. A true gravity is determined using the observed gravity (32°) and observed temperature (74°) of the oil during the run. The true gravity, which is the observed gravity corrected for temperature, is determined using a gravity temperature correction table given in Table 9–2. The true gravity for the oil run is 31° API.

COMPANY

RUN TICKET NO

ACCT NO 492

DIST NO 14 DATE 5/1/2005

OPERATOR
Lucky Company

LEASE NAME	LEASE NO.
Ebert	5

WELL NO.		FT.	INCHES	TEMP	BBLS	
	1ST MEAS.	13	0	80	180	83
TANK NO. 1	2ND MEAS.	01	4	75	18	60
	GRAVITY				159	44* = Net Barrels
TANK SIZE	OBSERVED		TEMP.	TRUE	PRICE	
210	32.0		74	31	$25	00
BS &W	OFF←SEAL NO.'S→ON				AMOUNT	
8/10%					$3986	00

THIEFING	BEFORE		AFTER	
FT - IN BS & WATER				

REMARKS

CALCULATIONS

*180.83 x .9910 = 179.20
18.60 x .9932 = 18.47
160.73
x .992
159.44

1ST MEAS	DATE	GAUGER
	TIME A.M P.M.	WELL OWNER'S WITNESS
2ND MEAS	DATE	GAUGER
	TIME A.M P.M.	WELL OWNER'S WITNESS

This ticket covers all claims for allowance. The oil represented by this ticket was received and run as property of

* Net Barrels

Fig. 9–2 – *Typical Run Ticket Data*

Source: The run tickets presented in this chapter and in the homework answers are based on a pipeline run ticket prepared and printed by Kraftbilt, Tulsa, Oklahoma, 1-800-331-7290.

TANK NO. 1   DISTRICT ____

OWNER ____   COUNTY ____

LEASE ____   STATE ____

LEASE NO. ____   STRAPPER ____

	.00	Ft. 1	13.95	Ft. 2	27.90	Ft. 3	41.85	Ft. 4	55.76	Ft. 5	69.66	Ft. 6	83.56	Ft. 7	97.45
¼	.29	¼	14.24	¼	28.19	¼	42.14	¼	56.05	¼	69.95	¼	83.85	¼	97.74
½	.58	½	14.53	½	28.48	½	42.43	½	56.34	½	70.24	½	84.14	½	98.03
¾	.87	¾	14.82	¾	28.77	¾	42.72	¾	56.63	¾	70.53	¾	84.43	¾	98.32
1	1.16	1	15.11	1	29.06	1	43.01	1	56.92	1	70.82	1	84.72	1	98.61
¼	1.45	¼	15.40	¼	29.35	¼	43.30	¼	57.21	¼	71.11	¼	85.01	¼	98.90
½	1.74	½	15.69	½	29.64	½	43.59	½	57.50	½	71.40	½	85.30	½	99.19
¾	2.03	¾	15.98	¾	29.94	¾	43.88	¾	57.79	¾	71.69	¾	85.58	¾	99.48
2	2.32	2	16.27	2	30.23	2	44.17	2	58.08	2	71.98	2	85.87	2	99.77
¼	2.61	¼	16.56	¼	30.52	¼	44.46	¼	58.36	¼	72.27	¼	86.16	¼	100.06
½	2.90	½	16.85	½	30.81	½	44.75	½	58.65	½	72.56	½	86.45	½	100.35
¾	3.19	¾	17.15	¾	31.10	¾	45.04	¾	58.94	¾	72.85	¾	86.74	¾	100.64
3	3.48	3	17.44	3	31.39	3	45.33	3	59.23	3	73.14	3	87.03	3	100.93
¼	3.77	¼	17.73	¼	31.68	¼	45.62	¼	59.52	¼	73.43	¼	87.32	¼	101.22
½	4.06	½	18.02	½	31.97	½	45.91	½	59.81	½	73.72	½	87.61	½	101.51
¾	4.36	¾	18.31	¾	32.26	¾	46.20	¾	60.10	¾	74.00	¾	87.90	¾	101.80
4	4.65	4	18.60	4	32.55	4	46.49	4	60.39	4	74.29	4	88.19	4	102.09
¼	4.94	¼	18.89	¼	32.84	¼	46.78	¼	60.68	¼	74.58	¼	88.48	¼	102.38
½	5.23	½	19.18	½	33.13	½	47.07	½	60.97	½	74.87	½	88.77	½	102.67
¾	5.52	¾	19.47	¾	33.42	¾	47.36	¾	61.26	¾	75.16	¾	89.06	¾	102.95
5	5.81	5	19.76	5	33.71	5	47.65	5	61.55	5	75.45	5	89.35	5	103.24
¼	6.10	¼	20.05	¼	34.00	¼	47.94	¼	61.84	¼	75.74	¼	89.64	¼	103.53
½	6.39	½	20.34	½	34.30	½	48.23	½	62.13	½	76.03	½	89.93	½	103.82
¾	6.68	¾	20.63	¾	34.59	¾	48.52	¾	62.42	¾	76.32	¾	90.22	¾	104.11
6	6.97	6	20.92	6	34.88	6	48.81	6	62.71	6	76.61	6	90.51	6	104.40
¼	7.26	¼	21.21	¼	35.17	¼	49.09	¼	63.00	¼	76.90	¼	90.80	¼	104.69
½	7.55	½	21.51	½	35.46	½	49.38	½	63.29	½	77.19	½	91.09	½	104.98
¾	7.84	¾	21.80	¾	35.75	¾	49.67	¾	63.58	¾	77.48	¾	91.37	¾	105.27
7	8.13	7	22.09	7	36.04	7	49.96	7	63.87	7	77.77	7	91.66	7	105.56
¼	8.42	¼	22.38	¼	36.33	¼	50.25	¼	64.16	¼	78.06	¼	91.95	¼	105.85
½	8.72	½	22.67	½	36.62	½	50.54	½	64.45	½	78.35	½	92.24	½	106.14
¾	9.01	¾	22.96	¾	36.91	¾	50.83	¾	64.74	¾	78.64	¾	92.53	¾	106.43
8	9.30	8	23.25	8	37.20	8	51.12	8	65.03	8	78.93	8	92.82	8	106.72
¼	9.59	¼	23.54	¼	37.49	¼	51.41	¼	65.32	¼	79.22	¼	93.11	¼	107.01
½	9.88	½	23.83	½	37.78	½	51.70	½	65.61	½	79.51	½	93.40	½	107.30
¾	10.17	¾	24.12	¾	38.07	¾	51.99	¾	65.90	¾	79.79	¾	93.69	¾	107.59
9	10.46	9	24.41	9	38.36	9	52.28	9	66.19	9	80.08	9	93.98	9	107.88
¼	10.75	¼	24.70	¼	38.66	¼	52.57	¼	66.48	¼	80.37	¼	94.27	¼	108.17
½	11.04	½	24.99	½	38.95	½	52.86	½	66.77	½	80.66	½	94.56	½	108.46
¾	11.33	¾	25.28	¾	39.24	¾	53.15	¾	67.06	¾	80.95	¾	94.85	¾	108.74
10	11.62	10	25.57	10	39.53	10	53.44	10	67.34	10	81.24	10	95.14	10	109.03
¼	11.91	¼	25.87	¼	39.82	¼	53.73	¼	67.63	¼	81.53	¼	95.43	¼	109.32
½	12.20	½	26.16	½	40.11	½	54.02	½	67.92	½	81.82	½	95.72	½	109.61
¾	12.49	¾	26.45	¾	40.40	¾	54.31	¾	68.21	¾	82.11	¾	96.01	¾	109.90
11	12.78	11	26.74	11	40.69	11	54.60	11	68.50	11	82.40	11	96.30	11	110.19
¼	13.08	¼	27.03	¼	40.98	¼	54.89	¼	68.79	¼	82.69	¼	96.59	¼	110.48
½	13.37	½	27.32	½	41.27	½	55.18	½	69.08	½	82.98	½	96.88	½	110.77
¾	13.66	¾	27.61	¾	41.56	¾	55.47	¾	69.37	¾	83.27	¾	97.16	¾	111.06

Table 9–1 — *Strapping Table*

e. The number of barrels in the tank at the beginning of the run (180.83 bbl) is adjusted to standard temperature by referring to Table 9–3. Using the true gravity and temperature of the oil at the beginning of the run (80°), a correction factor of 0.9910 is obtained from the table. The number of barrels in the tank at the end of the run is also corrected for temperature. The second correction factor for the number of barrels in the tank at the end of the run, also taken from Table 9–3 using the true gravity and temperature of the oil at the end of the run (75°), is 0.9932. The factors from the table are multiplied times the volumes of oil before

¼" INCREMENTS

144 X .29068
240 X .28968
720 X .28950
208.62720

Computed by
R. K. McFARLAND

Ft.		Ft.		Ft.		Ft.		Ft.		Ft.		Ft.		Ft.	
8	111.35	9	125.25	10	139.14	11	153.04	12	166.93	13	180.83	14	194.73	15	208.6
¼	111.64	¼	125.54	¼	139.43	¼	153.33	¼	167.22	¼	181.12	¼	195.02	¼	
½	111.93	½	125.83	½	139.72	½	153.62	½	167.51	½	181.41	½	195.31	½	
¾	112.22	¾	126.11	¾	140.01	¾	153.91	¾	167.80	¾	181.70	¾	195.59	¾	
1	112.51	1	126.40	1	140.30	1	154.20	1	168.09	1	181.99	1	195.88	1	
¼	112.80	¼	126.69	¼	140.59	¼	154.49	¼	168.38	¼	182.28	¼	196.17	¼	
½	113.09	½	126.98	½	140.88	½	154.78	½	168.67	½	182.57	½	196.46	½	
¾	113.38	¾	127.27	¾	141.17	¾	155.06	¾	168.96	¾	182.86	¾	196.75	¾	
2	113.67	2	127.56	2	141.46	2	155.35	2	169.25	2	183.15	2	197.04	2	
¼	113.96	¼	127.85	¼	141.75	¼	155.64	¼	169.54	¼	183.44	¼	197.33	¼	
½	114.25	½	128.14	½	142.04	½	155.93	½	169.83	½	183.73	½	197.62	½	
¾	114.53	¾	128.43	¾	142.33	¾	156.22	¾	170.12	¾	184.01	¾	197.91	¾	
3	114.82	3	128.72	3	142.62	3	156.51	3	170.41	3	184.30	3	198.20	3	
¼	115.11	¼	129.01	¼	142.91	¼	156.80	¼	170.70	¼	184.59	¼	198.49	¼	
½	115.40	½	129.30	½	143.20	½	157.09	½	170.99	½	184.88	½	198.78	½	
¾	115.69	¾	129.59	¾	143.48	¾	157.38	¾	171.28	¾	185.17	¾	199.07	¾	
4	115.98	4	129.88	4	143.77	4	157.67	4	171.57	4	185.46	4	199.36	4	
¼	116.27	¼	130.17	¼	144.06	¼	157.96	¼	171.86	¼	185.75	¼	199.65	¼	
½	116.56	½	130.46	½	144.35	½	158.25	½	172.15	½	186.04	½	199.94	½	
¾	116.85	¾	130.75	¾	144.64	¾	158.54	¾	172.43	¾	186.33	¾	200.23	¾	
5	117.14	5	131.04	5	144.93	5	158.83	5	172.72	5	186.62	5	200.52	5	
¼	117.43	¼	131.33	¼	145.22	¼	159.12	¼	173.01	¼	186.91	¼	200.81	¼	
½	117.72	½	131.62	½	145.51	½	159.41	½	173.30	½	187.20	½	201.10	½	
¾	118.01	¾	131.90	¾	145.80	¾	159.70	¾	173.59	¾	187.49	¾	201.38	¾	
6	118.30	6	132.19	6	146.09	6	159.99	6	173.88	6	187.78	6	201.67	6	
¼	118.59	¼	132.48	¼	146.38	¼	160.28	¼	174.17	¼	188.07	¼	201.96	¼	
½	118.88	½	132.77	½	146.67	½	160.57	½	174.46	½	188.36	½	202.25	½	
¾	119.17	¾	133.06	¾	146.96	¾	160.85	¾	174.75	¾	188.65	¾	202.54	¾	
7	119.46	7	133.35	7	147.25	7	161.14	7	175.04	7	188.94	7	202.83	7	
¼	119.75	¼	133.64	¼	147.54	¼	161.43	¼	175.33	¼	189.23	¼	203.12	¼	
½	120.04	½	133.93	½	147.83	½	161.72	½	175.62	½	189.52	½	203.41	½	
¾	120.32	¾	134.22	¾	148.12	¾	162.01	¾	175.91	¾	189.80	¾	203.70	¾	
8	120.61	8	134.51	8	148.41	8	162.30	8	176.20	8	190.09	8	203.99	8	
¼	120.90	¼	134.80	¼	148.70	¼	162.59	¼	176.49	¼	190.38	¼	204.28	¼	
½	121.19	½	135.09	½	148.99	½	162.88	½	176.78	½	190.67	½	204.57	½	
¾	121.48	¾	135.38	¾	149.27	¾	163.17	¾	177.07	¾	190.96	¾	204.86	¾	
9	121.77	9	135.67	9	149.56	9	163.46	9	177.36	9	191.25	9	205.15	9	
¼	122.06	¼	135.96	¼	149.85	¼	163.75	¼	177.65	¼	191.54	¼	205.44	¼	
½	122.35	½	136.25	½	150.14	½	164.04	½	177.94	½	191.83	½	205.73	½	
¾	122.64	¾	136.54	¾	150.43	¾	164.33	¾	178.22	¾	192.12	¾	206.02	¾	
10	122.93	10	136.83	10	150.72	10	164.62	10	178.51	10	192.41	10	206.31	10	
¼	123.22	¼	137.12	¼	151.01	¼	164.91	¼	178.80	¼	192.70	¼	206.60	¼	
½	123.51	½	137.41	½	151.30	½	165.20	½	179.09	½	192.99	½	206.89	½	
¾	123.80	¾	137.69	¾	151.59	¾	165.49	¾	179.38	¾	193.28	¾	207.17	¾	
11	124.09	11	137.98	11	151.88	11	165.78	11	179.67	11	193.57	11	207.46	11	
¼	124.38	¼	138.27	¼	152.17	¼	166.07	¼	179.96	¼	193.86	¼	207.75	¼	
½	124.67	½	138.56	½	152.46	½	166.36	½	180.25	½	194.15	½	208.04	½	
¾	124.96	¾	138.85	¾	152.75	¾	166.64	¾	180.54	¾	194.44	¾	208.33	¾	

Table 9–1 cont'd — *Strapping Table*

and after the run to get the number of barrels adjusted to standard temperature, 179.20 and 18.47 respectively. The adjusted number of barrels at the beginning of the run and at the end of the run are then subtracted to get gross measured volume, 160.73 barrels.

f. The measured volume, now corrected for gravity and temperature, is multiplied by the correction factor for BS&W (1.00 – 0.008) to determine the net barrels (159.44 barrels). The net barrels are multiplied by the selling price per barrel to compute gross revenue from the run.

TEMP. F	30.0	30.5	31.0	31.5	32.0	32.5	33.0	33.5	34.0	34.5	35.0	TEMP. F
				API GRAVITY AT OBSERVED TEMPERATURE CORRESPONDING API GRAVITY AT 60 F								
60.0	30.0	30.5	31.0	31.5	32.0	32.5	33.0	33.5	34.0	34.5	35.0	60.0
60.5	30.0	30.5	31.0	31.5	32.0	32.5	33.0	33.5	34.0	34.5	35.0	60.5
61.0	29.9	30.4	30.9	31.4	31.9	32.4	32.9	33.4	33.9	34.4	34.9	61.0
61.5	29.9	30.4	30.9	31.4	31.9	32.4	32.9	33.4	33.9	34.4	34.9	61.5
62.0	29.9	30.4	30.9	31.4	31.9	32.4	32.9	33.3	33.8	34.3	34.8	62.0
62.5	29.8	30.3	30.8	31.3	31.8	32.3	32.8	33.3	33.8	34.3	34.8	62.5
63.0	29.8	30.3	30.8	31.3	31.8	32.3	32.8	33.3	33.8	34.3	34.8	63.0
63.5	29.8	30.3	30.8	31.3	31.7	32.2	32.7	33.2	33.7	34.2	34.7	63.5
64.0	29.7	30.2	30.7	31.2	31.7	32.2	32.7	33.2	33.7	34.2	34.7	64.0
64.5	29.7	30.2	30.7	31.2	31.7	32.2	32.7	33.2	33.7	34.2	34.7	64.5
65.0	29.7	30.2	30.6	31.1	31.6	32.1	32.6	33.1	33.6	34.1	34.6	65.0
65.5	29.6	30.1	30.6	31.1	31.6	32.1	32.6	33.1	33.6	34.1	34.6	65.5
66.0	29.6	30.1	30.6	31.1	31.6	32.1	32.6	33.1	33.6	34.0	34.5	66.0
66.5	29.5	30.0	30.5	31.0	31.5	32.0	32.5	33.0	33.5	34.0	34.5	66.5
67.0	29.5	30.0	30.5	31.0	31.5	32.0	32.5	33.0	33.5	34.0	34.5	67.0
67.5	29.5	30.0	30.5	31.0	31.5	32.0	32.4	32.9	33.4	33.9	34.4	67.5
68.0	29.4	29.9	30.4	30.9	31.4	31.9	32.4	32.9	33.4	33.9	34.4	68.0
68.5	29.4	29.9	30.4	30.9	31.4	31.9	32.4	32.9	33.4	33.9	34.4	68.5
69.0	29.4	29.9	30.4	30.9	31.4	31.8	32.3	32.8	33.3	33.8	34.3	69.0
69.5	29.3	29.8	30.3	30.8	31.3	31.8	32.3	32.8	33.3	33.8	34.3	69.5
70.0	29.3	29.8	30.3	30.8	31.3	31.8	32.3	32.8	33.3	33.7	34.2	70.0
70.5	29.3	29.8	30.3	30.8	31.2	31.7	32.2	32.7	33.2	33.7	34.2	70.5
71.0	29.2	29.7	30.2	30.7	31.2	31.7	32.2	32.7	33.2	33.7	34.2	71.0
71.5	29.2	29.7	30.2	30.7	31.2	31.7	32.2	32.7	33.1	33.6	34.1	71.5
72.0	29.2	29.7	30.2	30.6	31.1	31.6	32.1	32.6	33.1	33.6	34.1	72.0
72.5	29.1	29.6	30.1	30.6	31.1	31.6	32.1	32.6	33.1	33.6	34.1	72.5
73.0	29.1	29.6	30.1	30.6	31.1	31.6	32.0	32.5	33.0	33.5	34.0	73.0
73.5	29.1	29.6	30.0	30.5	31.0	31.5	32.0	32.5	33.0	33.5	34.0	73.5
74.0	29.0	29.5	30.0	30.5	31.0	31.5	32.0	32.5	33.0	33.4	33.9	74.0
74.5	29.0	29.5	30.0	30.5	31.0	31.5	31.9	32.4	32.9	33.4	33.9	74.5
75.0	29.0	29.5	29.9	30.4	30.9	31.4	31.9	32.4	32.9	33.4	33.9	75.0

* DENOTES EXTRAPOLATED VALUE

API GRAVITY = 30.0 TO 35.0

Table 9–2 — *Generalized Crude Oils API Correction to 60°F*

API GRAVITY AT 60 F

FACTOR FOR CORRECTING VOLUME TO 60 F

TEMP. F	30.0	30.5	31.0	31.5	32.0	32.5	33.0	33.5	34.0	34.5	35.0	TEMP F
75.0	0.9933	0.9933	0.9932	0.9932	0.9931	0.9931	0.9931	0.9930	0.9930	0.9929	0.9929	75.0
75.5	0.9931	0.9930	0.9930	0.9930	0.9929	0.9929	0.9928	0.9928	0.9927	0.9927	0.9926	75.5
76.0	0.9929	0.9928	0.9928	0.9927	0.9927	0.9926	0.9926	0.9925	0.9925	0.9925	0.9924	76.0
76.5	0.9926	0.9926	0.9925	0.9925	0.9925	0.9924	0.9924	0.9923	0.9923	0.9922	0.9922	76.5
77.0	0.9924	0.9924	0.9923	0.9923	0.9922	0.9922	0.9921	0.9921	0.9920	0.9920	0.9919	77.0
77.5	0.9922	0.9921	0.9921	0.9920	0.9920	0.9919	0.9919	0.9918	0.9918	0.9917	0.9917	77.5
78.0	0.9920	0.9919	0.9919	0.9918	0.9918	0.9917	0.9917	0.9916	0.9916	0.9915	0.9915	78.0
78.5	0.9917	0.9917	0.9916	0.9916	0.9915	0.9915	0.9914	0.9914	0.9913	0.9913	0.9912	78.5
79.0	0.9915	0.9915	0.9914	0.9914	0.9913	0.9913	0.9912	0.9911	0.9911	0.9910	0.9910	79.0
79.5	0.9913	0.9912	0.9912	0.9911	0.9911	0.9910	0.9910	0.9909	0.9909	0.9908	0.9907	79.5
80.0	0.9911	0.9910	0.9910	0.9909	0.9908	0.9908	0.9907	0.9907	0.9906	0.9906	0.9905	80.0
80.5	0.9908	0.9908	0.9907	0.9907	0.9906	0.9906	0.9905	0.9904	0.9904	0.9903	0.9903	80.5
81.0	0.9906	0.9906	0.9905	0.9904	0.9904	0.9903	0.9903	0.9902	0.9902	0.9901	0.9900	81.0
81.5	0.9904	0.9903	0.9903	0.9902	0.9902	0.9901	0.9900	0.9900	0.9899	0.9899	0.9898	81.5
82.0	0.9902	0.9901	0.9901	0.9900	0.9899	0.9899	0.9898	0.9897	0.9897	0.9896	0.9896	82.0
82.5	0.9900	0.9899	0.9898	0.9898	0.9897	0.9896	0.9896	0.9895	0.9894	0.9894	0.9893	82.5
83.0	0.9897	0.9897	0.9896	0.9895	0.9895	0.9894	0.9893	0.9893	0.9892	0.9891	0.9891	83.0
83.5	0.9895	0.9894	0.9894	0.9893	0.9892	0.9892	0.9891	0.9890	0.9890	0.9889	0.9888	83.5
84.0	0.9893	0.9892	0.9891	0.9891	0.9890	0.9889	0.9889	0.9888	0.9887	0.9887	0.9886	84.0
84.5	0.9891	0.9890	0.9889	0.9889	0.9888	0.9887	0.9886	0.9886	0.9885	0.9884	0.9884	84.5
85.0	0.9888	0.9888	0.9887	0.9886	0.9886	0.9885	0.9884	0.9883	0.9883	0.9882	0.9881	85.0
85.5	0.9886	0.9885	0.9885	0.9884	0.9883	0.9883	0.9882	0.9881	0.9880	0.9880	0.9879	85.5
86.0	0.9884	0.9883	0.9882	0.9882	0.9881	0.9880	0.9879	0.9879	0.9878	0.9877	0.9877	86.0
86.5	0.9882	0.9881	0.9880	0.9879	0.9879	0.9878	0.9877	0.9876	0.9876	0.9875	0.9874	86.5
87.0	0.9879	0.9879	0.9878	0.9877	0.9876	0.9876	0.9875	0.9874	0.9873	0.9873	0.9872	87.0
87.5	0.9877	0.9876	0.9876	0.9875	0.9874	0.9873	0.9873	0.9872	0.9871	0.9870	0.9869	87.5
88.0	0.9875	0.9874	0.9873	0.9873	0.9872	0.9871	0.9870	0.9869	0.9869	0.9868	0.9867	88.0
88.5	0.9873	0.9872	0.9871	0.9870	0.9869	0.9869	0.9868	0.9867	0.9866	0.9865	0.9865	88.5
89.0	0.9870	0.9870	0.9869	0.9868	0.9867	0.9866	0.9866	0.9865	0.9864	0.9863	0.9862	89.0
89.5	0.9868	0.9867	0.9867	0.9866	0.9865	0.9864	0.9863	0.9862	0.9862	0.9861	0.9860	89.5
90.0	0.9866	0.9865	0.9864	0.9863	0.9863	0.9862	0.9861	0.9860	0.9859	0.9858	0.9857	90.0

* DENOTES EXTRAPOLATED VALUE

API GRAVITY = 30.0 TO 35.0

Table 9–3 — *Generalized Crude Oils Volume Correction to 60°F*

Gas measurement and sale

Gas, unlike oil, is not stored on the lease after production but is instead transferred immediately to a pipeline. (Fig. 9–3 diagrams the flow of gas from the wellhead to the point of consumption.) Gas measurement is more complicated than oil measurement because of different state and federal measurement standards and because of the physical characteristics of natural gas. Gas volume varies significantly with changes in pressure and temperature. Therefore, it is especially important that measured volumes of gas be converted to volumes at a standard set of conditions. Generally, standard conditions are a temperature of 60°F and a pressure at 14.65 psia. Federal standard conditions are a temperature of 60°F and a pressure at 14.73 psia for onshore leases and 15.025 psia for offshore leases.

Gas volume is measured in thousand cubic feet, *i.e.*, Mcf adjusted to standard conditions. Several instruments are used for measuring natural gas. One of the most important and widely used instruments is the orifice meter. An orifice meter records the flow of gas through pressure differentials. The data are recorded and are then converted to standard cubic feet by using conversion factors based upon the size of the line, the size of the orifice, the flowing temperature, the flowing pressure, the specific gravity of the gas, and the differential pressure.

Other measuring devices are used, such as mass flowmeters (normally at refineries), turbine meters, positive-displacement meters, and electronic flow computers (used in connection with an orifice or turbine, and giving instantaneous and continuous metered volumes). For further details concerning these methods, see *COPAS Bulletin No. 7.*

Sales of natural gas (whether processed or nonprocessed) are normally reported on a gas settlement statement. The form of the statement varies by purchaser, but the necessary information should include purchaser identification, lease identity, producer identification, quantity delivered from lease, month of production, Btu factor, gross or net value due the lease, lease production taxes, value due the producer, and pressure base(s). This information is necessary whether the sale is directly from the lease or from a central facility. If a central facility is used, gas allocation to individual wells must be made.

Standard division order

Due to the fact that U.S. laws allow private ownership of oil and gas reserves, leases located in the U.S. frequently have numerous interest owners. It is the operator's responsibility to assure that all owners are identified and compensated. The legal document setting up each interest owner's share of production is called a division order. A division order must be executed in order for the operator or the purchaser to legally collect the revenue from the sale of the oil and gas and to pay the correct owners their correct share of production or the proceeds from production. Thus, a division order must be prepared and signed by all interest owners prior to the sale of oil or gas from a particular lease. The information from the division order is set up in a division order database in the operator's accounting system so that the appropriate accounting entries can be made and payment distributed to the appropriate parties. As

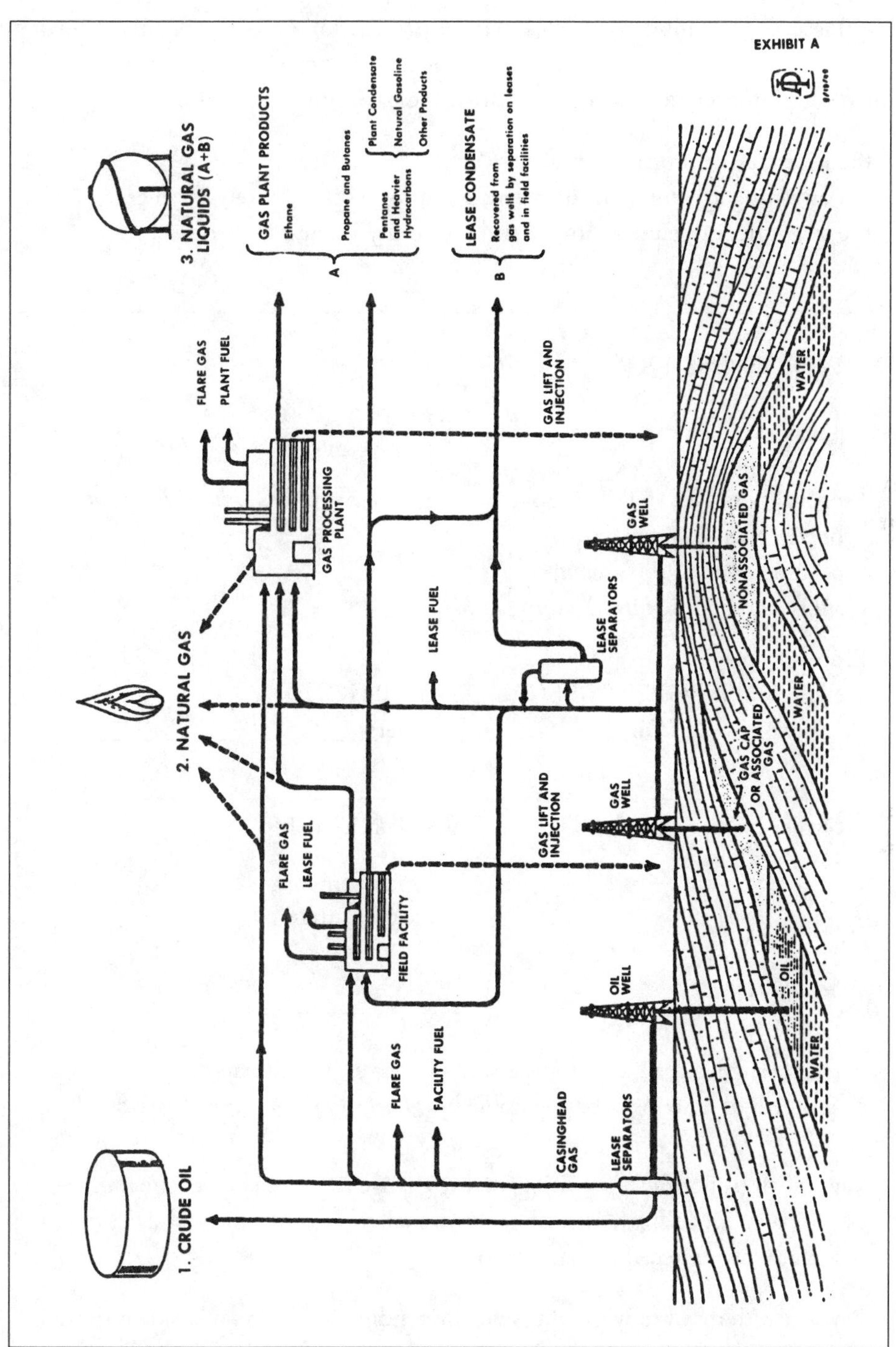

Fig. 9–3 — *Flow of Gas*

Source: COPAS Bulletin No. 7, April 1993, "Gas Accounting Manual," Page I-4. Reprinted with permission of the American Petroleum Institute and COPAS.

shown in Figure 9–4, exhibit A, which is part of the division order, is prepared, showing the interest owners' names, addresses, and interests owned, *i.e.*, ownership decimals.

The ownership decimals in a division order are based on:

- the lease contract for royalty owners
- the operating agreement or unit agreement for working interest owners
- the assignment document for overriding royalty interest owners

DIVISION ORDER*

To: Date: *November 20, 2000*

Property Number: *5474–3456–3* Effective Date: *Date of First Sale*
Property Name: *Lomax Heirs*
Operator: *Lucky Oil Company*
County and State: *Harris County, Texas*

Property
Description: [PROPERTY LEGAL DESCRIPTION]
Production: _*X*_ Oil _*X*_ Gas ___ Other: ________________

Owner Name and Address:	OWNER NUMBER
	Type of Interest:
	Decimal Interest:
See Exhibit A	

The undersigned certifies the ownership of their decimal interest in production or proceeds as described above payable by _*Lucky Oil Company*_ (Payor).
(Company Name)

Payor shall be notified, in writing, of any change in ownership, decimal interest, or payment address. All such changes shall be effective the first day of the month following receipt of such notice.

Payor is authorized to withhold payment pending resolution of a title dispute or adverse claim asserted regarding the interest in production claimed herein by the undersigned. The undersigned agrees to indemnify and reimburse

Fig. 9–4 — *Division Order*

Payor any amount attributable to an interest to which the undersigned is not entitled. Payor may accrue proceeds until the total amount equals *$25.00* , or pay _____ whichever occurs first, or as required by applicable state statute.

This Division Order does not amend any lease or operating agreement between the undersigned and the lessee or operator or any other contracts for the purchase of oil or gas.

In addition to the terms and conditions of this Division Order, the undersigned and Payor may have certain statutory rights under the laws of the state in which the property is located.

Special Clauses:

Owner(s) Signature(s): ______________ ______________
Owner(s) Tax I.D. Number(s): ______________ ______________

Owner Daytime Telephone #: ______________ ______________
Owner FAX Telephone #: ______________ ______________

Federal Law requires you to furnish your Social Security or Taxpayer Identification Number. Failure to comply will result in 31% tax withholding and will not be refundable by Payor.

Exhibit A

Property Number: *5474–3456–3* Effective Date: *Date of First Sale*
Property Name: *Lomax Heirs*
Operator: *Lucky Oil Company*
County and State: *Harris County, Texas*

Property
Description: [PROPERTY LEGAL DESCRIPTION]
Production: _X_ Oil _X_ Gas ___ Other: ____________________

Fig. 9–4 cont'd — *Division Order*

OWNER NUMBER	OWNER NAME	OWNER ADDRESS	TYPE OF INTEREST	DECIMAL INTEREST
243662	Lucky Oil Company		WI	0.365625
345927	Big John Oil Company		WI	0.162500
298732	Wildcat Oil Company		WI	0.121875
675233	Mabel Lomax		RI	0.093750
723122	E.B. Lomax		RI	0.031250
723123	Bruce Lomax		RI	0.031250
723123	Kay Jerin Lomax		RI	0.031250
823345	James Larkin		ORI	0.081250
823345	Janice Larkin		ORI	0.040625
823346	Estelle Bradt		ORI	0.040625

* The above division order is the National Association of Division Order Analysts (NADOA) Model Form Division Order (Adopted 9/95).

Fig. 9–4 cont'd — *Division Order*

DETERMINATION OF REVENUE

Revenue from the sale of oil or gas produced from a property must be divided among all owners of an economic interest in that property according to the ownership interests shown on the division order. The following example illustrates the computation of the share of revenue for all the economic interests in a property. In contrast to revenue, operating costs are divided only among the working interest owners. Consequently, a working interest's percentage of revenue is usually not the same as the working interest's percentage of costs. The following example also illustrates what portion of the costs each interest bears:

EXAMPLE

Revenue and Cost Allocation to Ownership Interests

a. Mr. Z owns the mineral rights in some land in Texas. He leases this land to Lucky Company, reserving a $^1/_8$ royalty. Each owner's share of the first year's production of 40,000 barrels is as follows.

Royalty Interest, Mr. Z:	$^1/_8$ RI
	$^1/_8$ x 40,000 = 5,000 bbl
	(0% of the costs)
Working Interest, Lucky Company:	100% of the WI ($^7/_8$ revenue interest)
	$^7/_8$ x 40,000 = 35,000 bbl
	(100% of the costs)

b. Assume that on January 1 of year two, Lucky Company assigned to Mr. Over a $^1/_5$ ORI (overriding royalty interest) and that production in year two was 80,000 barrels. Note that an interest created out of the WI is stated in terms of that WI's revenue share at the time the interest was created. Consequently, the ORI's share of revenue is $^1/_5$ x $^7/_8$.

Royalty Interest, Mr. Z:	$^1/_8$ RI
	$^1/_8$ x 80,000 = 10,000 bbl
	(0% of the costs)
ORI, Mr. Over:	$^1/_5$ x WI share = $^1/_5$ x $^7/_8$ ORI
	$^1/_5$ x $^7/_8$ x 80,000 = 14,000 bbl
	(0% of the costs)
WI, Lucky Company:	Revenue Interest:
	$^7/_8$ x $^4/_5$*
	$^7/_8$ x $^4/_5$ x 80,000 = 56,000 bbl
	(100% of the costs)

c. Assume that on January 1 of year three, Lucky Company assigned to Joint Company a joint working interest of 25% and that the production is again 80,000 barrels.

Royalty Interest, Mr. Z:	$^1/_8$ RI
	$^1/_8$ x 80,000 = 10,000 bbl
	(0% of the costs)

ORI, Mr. Over:	$\frac{1}{5} \times \frac{7}{8}$ ORI $\frac{1}{5} \times \frac{7}{8} \times 80{,}000 = 14{,}000$ bbl (0% of the costs)
WI, Lucky Company:	$(\frac{7}{8} \times \frac{4}{5}) \times 75\%$ Joint WI $(\frac{7}{8} \times \frac{4}{5}) \times 75\% \times 80{,}000 = 42{,}000$ bbl (75% of the costs)
WI, Joint Company:	$(\frac{7}{8} \times \frac{4}{5}) \times 25\%$ Joint WI $(\frac{7}{8} \times \frac{4}{5}) \times 25\% \times 80{,}000 = 14{,}000$ bbl (25% of the costs)

---

* The WI share of revenue is what is left after paying all the nonworking interests. To compute the amount of the working interest's share of revenue, start with the original whole interest (1) and subtract the royalty interest ($\frac{1}{8}$) and the ORI ($\frac{1}{5} \times \frac{7}{8}$) as follows:

$$\text{WI} = 1 - \frac{1}{8} - \frac{1}{5} \times \frac{7}{8}$$

The WI share of revenue can then be reduced as follows:

$$\begin{aligned} \text{WI} &= \frac{7}{8} - \frac{1}{5} \times \frac{7}{8} \\ &= \frac{7}{8}(1 - \frac{1}{5}) \quad \text{(factoring out the } \frac{7}{8}) \\ &= \frac{7}{8} \times \frac{4}{5} \end{aligned}$$

Note that as opposed to the DD&A calculations shown in chapters 6 and 7, total production is used in the distribution of revenue versus only the working interest's share of production.

In the example above, ownership interests were shown as fractions because in the authors' opinion, fractions more effectively illustrate how the owners' interests are computed. In a division order, however, the interests are stated as decimals. In determining division order decimals, fractional ownership is converted to decimal ownership. The decimal calculation is comparable with the preceding example where fractional ownership is shown. Either approach may be used in determining the gross revenue applicable to each owner.

EXAMPLE

Decimal Ownership

Lucky Company has production on a lease with the following ownership interest:

RI—$^3/_{16}$ RI
ORI—$^1/_5$ of WI
PPI—20% of WI, created after the ORI was created
25% WI—Lost Cause Company
75% WI—Lucky Company

Originally, Lucky Company had 100% of the working interest. Later an ORI and then a PPI were carved out. Several years after the ORI and PPI were created, Lucky conveyed 25% of its working interest to Lost Cause Company. Based on these facts, ownership interests as shown on a division order are as follows:

Total Interest	1.000000	
RI, $^3/_{16}$	<u>0.187500</u>	
Remainder		0.812500
ORI, $^1/_5$ x 0.8125	0.162500	
Remainder		0.650000
PPI, 0.20 x 0.65	0.130000	
Remainder		0.520000
Lost Cause Company, 0.25 x 0.52	0.130000	
Lucky Company, 0.75 x 0.52	0.390000	

The division order would show the owner's interests as follows:

RI	0.187500
ORI	0.162500
PPI	0.130000
Lost Cause Company	0.130000
Lucky Company	<u>0.390000</u>
	1.000000

Each owner's portion of the total revenue is determined by multiplying the decimal amounts by total revenue. For example, if total revenue were $56,000, the royalty owner would receive $56,000 x 0.187500 or $10,500.

Unitizations

As discussed in chapter 1, two or more properties are often combined for a more efficient, economical operation. The parties voluntarily agree to unitize or, alternatively, the regulatory authority overseeing the area may require that the parties unitize. Typically, when properties are unitized, a unitization agreement is negotiated among all of the parties. In most cases, the unitization agreement effectively supercedes the lease agreement. In order to be effective, the agreement to unitize normally requires approval of a pre-agreed upon percentage of *all* of the interest parties (*i.e.*, 85% of all WI, RI, and ORI owners in the unitized area). If approved, all of the interests in all of the unitized properties are redetermined based upon whatever sharing factors are stipulated in the unitization agreement.

Sharing factors, called participation factors, typically are based on such elements as acreage contributed, pay sand or sand thickness contributed, or net recoverable barrels of oil-in-place contributed. The redetermined interests are used to share revenues and costs. All of the working interest owners typically share a common obligation to all of the royalty owners. The following simple example illustrates this process.

EXAMPLE

Unitization—Determination of Interest

Assume that the following three leases are being combined to form the South End Unit:

Lease 1

WI: 100% Lucky Company

RI: $^{1}/_{16}$ Mable Smith

$^{1}/_{16}$ James Jones

Acreage contributed: 100 acres

Estimated net recoverable bbl: 50,000 bbl

Lease 2

WI: 50% Texas Company

50% ArkTex Company

RI: $^{1}/_{8}$ Fred Luna

Acreage contributed: 500 acres

Estimated net recoverable bbl: 100,000 bbl

Lease 3
WI: 100% Frank Company
RI: 1/6 Anna Bell
Acreage contributed: 200 acres
Estimated net recoverable bbl: 10,000 bbl

Participation Factors Based on Acreage Contributed

	WI	RI
Lucky Company	100%(100/800) = .125	
Texas Company	50%(500/800) = .3125	
ArkTex Company	50%(500/800) = .3125	
Frank Company	100%(200/800) = .25	
Mable Smith		1/16 (100/800) = .0078
James Jones		1/16 (100/800) = .0078
Fred Luna		1/8 (500/800) = .0781
Anna Bell		1/6 (200/800) = .0417

Note: Recall that the working interest is stated in terms of how the costs are to be shared while the royalty interest is stated in terms of its share of gross revenue. In order to calculate the working interest owners' share of gross revenue, the working interest owners' participation factors would be multiplied by 1 minus the royalty on the property that they contributed. For example, for Lucky Company, the working interest after the unitization is .125 so Lucky will be responsible for 12.5% of the costs on the unit. Lucky's share of revenue is 10.94% (.125 x 7/8 = .1094).

Participation Factors Based on
Estimated Net Recoverable Barrels Contributed

	WI	RI
Lucky Company	100%(50/160) = .3125	
Texas Company	50%(100/160) = .3125	
ArkTex Company	50%(100/160) = .3125	
Frank Company	100%(10/160) = .0625	
Mable Smith		1/16 (50/160) = .0195
James Jones		1/16 (50/160) = .0195
Fred Luna		1/8 (100/160) = .0781
Anna Bell		1/6 (10/160) = .0104

Participation factors may also be based on multiple elements that are combined in a weighted average calculation to redetermine interests. This calculation is illustrated in the example below:

EXAMPLE

Participation Factors Based on Multiple Elements

Using the information from the previous example, assume that the parties agree to weight acreage contributed by a 45% factor and net recoverable barrels by 55%.

Participation Factors Based on Acreage Contributed (45%) and Estimated Net Recoverable Barrels Contributed (55%)

WI:			
Lucky Company	100%[(100/800 x .45) + (50/160 x .55)]	=	.2281
Texas Company	50%[(500/800 x .45) + (100/160 x .55)]	=	.3125
ArkTex Company	50%[(500/800 x .45) + (100/160 x .55)]	=	.3125
Frank Company	100%[(200/800 x .45) + (10/160 x .55)]	=	.1469
RI:			
Mable Smith	1/16 [(100/800 x .45) + (50/160 x .55)]	=	.0143
James Jones	1/16 [(100/800 x .45) + (50/160 x .55)]	=	.0143
Fred Luna	1/8 [(500/800 x .45) + (100/160 x .55)]	=	.0781
Anna Bell	1/6 [(200/800 x .45) + (10/160 x .55)]	=	.0245

Several working interest owners with properties producing from the same formation may wish to develop their properties based on an overall plan for the area. Rather than actually unitizing, the parties may enter into a **cooperative unit.** In a cooperative unit, the working interests share in the total costs and production from the unit. The royalty owners, however, receive a royalty based on the production from their own leases.

Payment of Oil and Gas Revenues

Who is responsible for paying severance taxes? Who is responsible for paying the royalty owners and other owners of economic interests their *pro rata* share of lease revenue? The answers depend on legal requirements and contractual provisions.

Frequently, the purchaser assumes the responsibility of collecting and paying severance tax on both oil and gas production. In doing so, the purchaser is acting as a conduit, *i.e.*, it collects the taxes owed by the economic interest owners of the property by simply deducting the taxes owed from revenue due the economic interest owners, and then it sends the severance tax amount to the state.

In the past, the purchaser of *oil* has frequently assumed the obligation of paying to each owner of an economic interest that owner's share of revenue, less applicable severance taxes. If the purchaser does not assume this responsibility, the lease operator normally distributes the applicable amounts to the other economic interest owners. The purchaser of *gas* usually remits 100% of the proceeds, less severance taxes, to the lease operator. The lease operator ordinarily then assumes the responsibility of paying the other economic interest owners their share of revenue, net of severance tax.

If the purchaser of either oil or gas does not assume either responsibility, *i.e.*, the responsibility of paying severance taxes and each economic interest owner his or her share of revenue net of taxes, then the purchaser remits 100% of the proceeds to the operator. The operator ordinarily then assumes both responsibilities. For simplicity, the purchaser in most of the examples in this chapter assumes either both responsibilities or neither responsibility.

Recording Oil Revenue

Gross revenue from oil sales is determined by multiplying the net amount of oil delivered by the price for that particular grade of oil. The net amount of oil delivered is recorded on a **run ticket**, which serves as the source document for determining the quantity of oil sold. The price of oil is dependent in part on the quality of the oil, measured in terms of gravity. **Gravity**, as used here, refers to degrees of gravity on the American Petroleum Institute (API) scale. On this scale, the greater the density of the petroleum, the lower the API degree. Generally, the higher the API gravity, the greater the value of the oil.

In the past, the price of the oil has generally either been given as a posted field price or negotiated as a spot price. A **posted field price** is the price purchasers are willing to pay for oil in the area. This price, which is published in a crude oil price bulletin, is based on factors such as the API gravity, location of the oil, grade, market center prices, actual negotiated transactions in the field, the NYMEX oil futures market and other available information. A **spot price** refers to a short-term price negotiated between the buyer and the seller in the spot market, without a long-term contract.

Once the sales price has been determined and the oil sold, the accounting entries are made. The following example illustrates recording revenue from the sale of oil for both the working interest owner and the royalty interest owner. In this example, as well as in the remaining examples in the chapter, it is assumed that the severance tax, as is the usual case, is a specified percentage of the selling price.

Part *a* of the first example presents an analysis of each economic interest owner's portion of gross revenue, severance tax, and revenue net of tax. This analysis provides the amounts to be debited or credited in the revenue entries made by the various interest owners.

In part *b*, it is assumed that the purchaser distributes severance tax to the state and remits the net revenue amounts to the various interest owners. Therefore, the working interest owner records a receivable for its share of revenue after tax due from the purchaser. The working interest owner also records its share of severance tax expense and credits oil sales (revenue) for its share of gross revenue. The amounts in the analysis presented in part *a* or the equivalent calculations given in parenthesis after each debit and credit may be used to obtain the amounts to debit and credit.

In part *c* of the example, it is assumed that the purchaser remits 100% of the proceeds to the working interest owner. The WI owner remits the severance tax owed by all the economic interest owners to the state and remits the revenue net of taxes to all economic interest owners—in this case, just the RI owner—for their share of revenue net of tax. Again, the analysis prepared in part *a* or the calculations given in parentheses may be used to determine the amounts to be debited or credited. In this case, The WI owner records the gross revenue amount as a receivable and, as in part *b*, records its own share of tax expense and revenue. A liability to the state for production taxes is recognized by the WI owner for both its share of production taxes and the RI owner's share. The WI owner also records a payable to the RI owner for revenue due the RI owner net of production tax.

Note that, regardless of who bears the responsibility for paying the severance tax and paying the economic interest owners their share of revenue, *i.e.*, part *b* compared to part *c*, the working interest owner (or any other economic interest owner) records the same amount of revenue and expenses.

Part *d* gives the entry for the royalty interest owner. The entry for the royalty interest owner would be the same under any situation because the royalty interest owner always receives only his share of revenue after tax, *i.e.*, his portion of revenue net of severance tax. The royalty owner records his net receivable, tax expense, and revenue.

EXAMPLE

Oil Revenue

Lucky Company owns 100% of the working interest in the Gilbert Lease in Texas. Gilbert Jones owns a $\frac{1}{8}$ royalty interest in the lease. During February, 4,000 barrels of oil (after correction for temperature, gravity, and BS&W) were produced and sold. Assume the price for oil is $16.00/bbl and the production tax in Texas is 5%.

a. **Analysis:**

	a	b	c
Revenue Interest	**Gross**	**Severance Tax**	**Net of Severance Tax**
	(4,000 x $16 x interest)	(5% x a)	(a – b)
WI ($^7/_8$)	$56,000	$2,800	$53,200
RI ($^1/_8$)	8,000	400	7,600
Total ($^8/_8$)	$64,000	$3,200	$60,800

b. The purchaser assumes the responsibility of distributing severance taxes and royalty income.

Entry by Lucky Company (WI)—Amounts may be taken from the above analysis or may be calculated as shown below:

A/R—purchaser (4,000 x $16 x 0.95 x $^7/_8$)	53,200	
Production tax expense (4,000 x $16 x 0.05 x $^7/_8$)	2,800	
Oil sales (4,000 x $16 x $^7/_8$)		56,000

c. Lucky Company assumes the responsibility and receives 100% of the proceeds.

Entry by Lucky Company (WI)—Amounts may be taken from the analysis in *a* above or from the calculations in parentheses.

A/R—purchaser (WI + RI) (4,000 x $16) . .	64,000	
Production tax expense (WI) (4,000 x $16 x 0.05 x $^7/_8$)	2,800	
Production tax payable (WI + RI) (4,000 x $16 x 0.05)		3,200
Royalty payable (net of severance tax) (4,000 x $16 x 0.95 x $^1/_8$)		7,600
Oil sales (WI) (4,000 x $16 x $^7/_8$)		56,000

d. The royalty interest owner's entry under either *b* or *c* above.

Entry by Gilbert Jones (RI)

A/R (4,000 x $16 x 0.95 x 1/8)	7,600	
Production tax expense (4,000 x $16 x 0.05 x 1/8)	400	
Oil sales (4,000 x $16 x 1/8)		8,000

The following example illustrates a situation in which the oil is sold to a refinery owned 100% by the working interest owner. The working interest owner's entry to record the sale in this situation differs only from the previous example in that an intercompany A/R and an intercompany revenue account must be recognized. The purchaser, as distinguished from the producer, *i.e.*, the working interest owner, differentiates between intercompany oil purchases and outside oil purchases.

EXAMPLE

Intercompany Sales

Assume the same facts as in the above example except that Lucky Company sold the oil to a refinery owned 100% by Lucky Company.

Entry for producer:

A/R—intercompany (WI + RI)	64,000	
Production tax expense	2,800	
Production tax payable		3,200
Royalty payable (after taxes)		7,600
Oil sales—intercompany (WI)		56,000

Entry for purchaser:

Crude oil purchases—intercompany (4,000 x $16 x 7/8)	56,000	
Crude oil purchases—outside (RI) (4,000 x $16 x 1/8)	8,000	
A/P—intercompany		64,000

EXAMPLE

Multiple Nonworking Interests

Lucky Company has production on a lease with the following ownership interests:

RI—$^1/_8$ RI
ORI—$^1/_5$ of total WI
Joint WI Interest: Ewing Company (25%) and Lucky Company (75%)

During February, 4,000 barrels of oil (after correction for temperature, gravity, and BS&W) were produced and sold. Assume the price for oil is $16.00/bbl and the production tax in Texas is 5%.

a. **Analysis:**

Revenue Interest	a Gross	b Severance Tax	c Net of Severance Tax
	(4,000 x $16 x interest)	(5% x a)	(a – b)
WI—Lucky ($^7/_8$ x $^4/_5$)75%	$33,600	$1,680	$31,920
WI—Ewing ($^7/_8$ x $^4/_5$)25%	11,200	560	10,640
ORI ($^1/_5$ x $^7/_8$)	11,200	560	10,640
RI ($^1/_8$)	8,000	400	7,600
Total ($^8/_8$)	$64,000	$3,200	$60,800

b. The purchaser assumes the responsibility of distributing severance taxes and royalty income.

Entry by Lucky, the Operator—Amounts may be taken from the above analysis or may be calculated as shown below:

A/R—purchaser (Lucky) (4,000 x $16 x 0.95 x $^7/_8$ x $^4/_5$ x 75%)	31,920	
Production tax expense (Lucky) (4,000 x $16 x 0.05 x $^7/_8$ x $^4/_5$ x 75%)	1,680	
Oil sales (Lucky) (4,000 x $16 x $^7/_8$ x $^4/_5$ x 75%)		33,600

c. Lucky Company assumes the responsibility and receives 100% of the proceeds.

Entry by Lucky—Amounts may be taken from the analysis in *a* above or from the calculations in parentheses.

A/R—purchaser (WI + RI + ORI) (4,000 x $16)	64,000	
Production tax expense (WI) (4,000 x $16 x 0.05 x $^7/_8$ x $^4/_5$ x 75%)	1,680	
Production tax payable (WI + RI + ORI) (4,000 x $16 x 0.05)		3,200
Royalty payable (4,000 x $16 x 0.95 x $^1/_8$)		7,600
ORI payable (4,000 x $16 x 0.95 x $^1/_5$ x $^7/_8$)		10,640
Ewing payable (4,000 x $16 x 0.95 x $^7/_8$ x $^4/_5$ x 25%)		10,640
Oil sales (Lucky) (4,000 x $16 x $^7/_8$ x $^4/_5$ x 75%)		33,600

d. The royalty interest owner's entry under either *b* or *c* above.

Entry by RI

A/R (4,000 x $16 x 0.95 x $^1/_8$)	7,600	
Production tax expense (4,000 x $16 x 0.05 x $^1/_8$)	400	
Oil sales (4,000 x $16 x $^1/_8$)		8,000

e. The overriding royalty interest owner's entry under either *b* or *c* above.

Entry by ORI

A/R (4,000 x $16 x 0.95 x $^1/_5$ x $^7/_8$)	10,640	
Production tax expense (4,000 x $16 x 0.05 x $^1/_5$ x $^7/_8$)	560	
Oil sales (4,000 x $16 x $^1/_5$ x $^7/_8$)		11,200

f. Ewing's entry under either *b* or *c* above.

Entry by JWI		
A/R (4,000 x \$16 x 0.95 x $^7/_8$ x $^4/_5$ x 25%)	10,640	
Production tax expense		
(4,000 x \$16 x 0.05 x $^7/_8$ x $^4/_5$ x 25%) . . .	560	
Oil sales (4,000 x \$16 x $^7/_8$ x $^4/_5$ x 25%)		11,200

Frequently, a relatively small quantity of the oil or gas produced on a lease is used on that lease to fuel generators, boilers, pumps, or even to spread on lease roads. When oil or gas is used on the lease from which it was produced, the financial accounting impact is typically neutral. Neither revenue nor expense is recognized because any revenue or expense would normally offset. Royalty interest owners generally do not receive royalty payments on oil or gas produced and used on the same lease because most lease agreements have a "free fuel" clause. The payment of severance tax on oil or gas produced and used on the same lease depends upon state law.

Since operators typically operate many leases in close proximity to one another, it is possible that oil or gas produced on one lease might be used on a nearby lease with differing ownership. In that case, revenue, royalty, and severance taxes would be recorded by the producing lease while operating expense is recorded for the lease using the product. In the following example, oil produced on Lease A is transferred to Lease B to be used on Lease B in lease operations. The working interest ownership in both leases is the same but the royalty ownership is different. The WI owner treats this transaction as a sale of the oil from Lease A and as an operating expense on Lease B. The value assigned to the oil would normally correspond to the price at which the oil produced in the area could have been sold.

EXAMPLE

Oil Used off Lease by Operator

Lucky Company operates Leases A and B. The company uses 10 barrels of oil obtained from Lease A as fuel on Lease B. Assume the oil is priced on both leases at \$20/bbl, that production taxes are 5%, and that the royalty interest on Lease A is a $^1/_8$ royalty.

a. **Analysis:**

Revenue Interest	**a** **Gross**	**b** **Severance Tax**	**c** **Net of Severance Tax**
	(10 x $20)	(5% x a)	(a – b)
WI ($^7/_8$)	$175	$ 8.75	$166.25
RI ($^1/_8$)	25	1.25	23.75
Total ($^8/_8$)	$200	$10.00	$190.00

Entry to record transfer of 10 barrels of oil

Lease operating expense—Lease B (10 x $20)	200.00	
Production tax expense—Lease A (10 x $20 x 5% x $^7/_8$).	8.75	
Production tax payable (WI + RI) (10 x $20 x 5%)		10.00
Royalty payable—Lease A (10 x $20 x 95% x $^1/_8$)		23.75
Oil sales—Lease A (10 x $20 x $^7/_8$)		175.00

Crude Oil Exchanges

Producing companies often exchange crude oil in order to operate more efficiently. For example, crude oil production in one area may be exchanged for crude in another location closer to the purchaser so that the oil does not have to be physically transported. Alternately, future production could be exchanged for an immediate delivery of crude oil in order to meet refinery demands.

Crude oil exchanges are often tracked and accounted for using volumetric information. Under the **inventory method**, the number of barrels received in an exchange is debited to a volume-based physical inventory account with a corresponding credit to exchange inventory. The reverse entry would be made when the oil received is paid back with a future delivery.

A common alternate method of recording exchanges is based on the monetary value of the oil exchanged. The **gross purchase and sales method** treats each exchange receipt as a purchase and each exchange delivery as a sale, with corresponding entries made to accounts payable and accounts receivable, respectively.

Regardless of which basic method is used, differentials will usually exist because the monetary values of the barrels exchanged will normally not be equal due to differences in quality, crude type, or gravity. These differentials are settled in cash and are recorded as a revenue by the company receiving the payment and as an expense by the company making the payment.

Recording Gas Revenue

Prior to 1985, most sales of natural gas were made on long-term contracts, often for terms of 10 or more years. In contrast, most gas sales today are made on the spot market on a short-term basis. **Spot sales** of gas involve the sale of gas at current market prices by the producer to pipelines or their affiliates. Spot sales of gas also include **direct sales** that are made by the producer to the end-user or local distribution company. In direct sales, the gas is transported by a third-party pipeline company, which charges a transportation fee. Spot sales also include sales to **marketers** or middlemen who buy and resell gas. Brokers may also be used. **Brokers**, instead of taking title to the gas, find a buyer for the gas and charge a brokerage fee for their services. This change to selling gas in the spot market has led to significant seasonal price volatility, which is expected to continue in the foreseeable future.

In the domestic market and many foreign markets, the volume of gas sold is measured in Mcf (1,000 cubic feet). The selling price of gas, however, is usually based on the heating value of the gas, as measured in MMBtu's (1,000,000 British Thermal Units: A Btu is the quantity of energy necessary to raise one pound of water by 1° Fahrenheit). Because of this difference in measurement units, it is important to know how to convert gas measurements to MMBtu's and to Mcf's. (Gas volumes are determined by metering the gas. The heating value, on the other hand, is determined through sampling and analyzing the gas in the pipeline.)

EXAMPLE

MMBtu's versus Mcf's

Lucky Company sold 500 Mcf of gas with a heat content of 1.040 MMBtu/Mcf. The selling price of the gas was $2.10 per MMBtu.

To determine MMBtu's for gas sold:
500 Mcf x 1.040 MMBtu/Mcf = 520 MMBtu
(Note that the units of Mcf's cancel giving a result in units of MMBtu's.)

To determine total sales price:
520 MMBtu x $2.10/MMBtu = $1,092
(Note that the units of MMBtu's cancel giving a result in units of dollars.)

To determine unit sales price per Mcf:
$2.10/MMBtu x 1.040 MMBtu/Mcf = $2.184 per Mcf
(Note that the units of MMBtu's cancel giving a result in units of dollars per Mcf.)

The temperature standard for gas is typically 60°F but the pressure standard varies. Both the volume of gas in Mcf and the heating content of the gas in MMBtu per Mcf must be at the same standard pressure (psia). If they are not at the same standard pressure, one of the measurements must be converted to the pressure of the other measurement. The formulas for converting from one pressure to another are as follows:

Converting Volume:

$$\text{Mcf @ current psia} \times \frac{\text{current psia}}{\text{desired psia}} = \text{Mcf @ desired psia}$$

The calculation above is based on the ideal gas law.

Converting Btu:

$$\text{MMBtu/Mcf @ current psia} \times \frac{\text{desired psia}}{\text{current psia}} = \text{MMBtu/Mcf @ desired psia}$$

In the calculation above, the desired psia is divided by the current psia rather than the current psia being divided by the desired psia because the calculation involves MMBtu *per* Mcf.

EXAMPLE

Pressure Base

Lucky Company sold 500 Mcf of gas at 14.65 psia with a heat content of 1.040 MMBtu/Mcf at 14.73 psia. The selling price of the gas was $2.00 per MMBtu.

To convert Mcf @ 14.65 psia to a standard pressure of 14.73 psia:

$$500 \text{ Mcf} \times \frac{14.65}{14.73} = \underline{\underline{497.28445 \text{ Mcf @ } 14.73 \text{ psia}}}$$

To determine MMBtu's for gas sold:

$$497.28445 \text{ Mcf} \times 1.040 \text{ MMBtu/Mcf} = \underline{\underline{517.1758 \text{ MMBtu}}}$$

Alternatively to convert MMBtu/Mcf @ 14.73 psia to a standard pressure of 14.65 psia:

$$1.040 \text{ MMBtu/Mcf} \times \frac{14.65}{14.73} = \underline{\underline{1.03435 \text{ MMBtu/Mcf @ } 14.65}}$$

To determine MMBtu's for gas sold:

$$500 \text{ Mcf} \times 1.03435 \text{ MMBtu/Mcf} = \underline{\underline{517.175 \text{ MMBtu}}}$$

To determine sales price of gas sold:

517.175 MMBtu x \$2.00/MMBtu = \$1,034.35

Although most sales contracts are written based on MMBtu's, most accounting records and internal reports are based on gas volumes and prices in Mcf's. Therefore, any subsequent examples illustrating gas sales utilize Mcf's versus MMBtu's.

Once the price of natural gas is determined and a gas settlement statement is received from the purchaser, the accounting entries are made. Accounting for gas revenue is similar to accounting for oil revenue except for the following:

a. No gas is stored on the lease prior to sale.
b. The measurement of gas is more complex than the measurement of oil.
c. Distribution of the proceeds to the economic interest owners is usually handled by the operator rather than the purchaser.

An illustration of gas revenue accounting follows:

EXAMPLE

Gas Revenue

Lucky Company owns a 100% working interest in a lease on which there is a $^1/_8$ royalty interest. During one month Lucky produced and sold 300 Mcf of gas at \$2.00/Mcf. Assume production taxes are 5%.

a. **Analysis:**

	a	**b**	**c**
Revenue Interest	**Gross**	**Severance Tax**	**Net of Severance Tax**
	(300 x \$2 x interest)	(5% x a)	(a – b)
WI ($^7/_8$)	\$525	\$26.25	\$498.75
RI ($^1/_8$)	75	3.75	71.25
Total ($^8/_8$)	\$600	\$30.00	\$570.00

b. The operator assumes the responsibility of distributing royalty and the purchaser assumes the responsibility of paying severance taxes.

Entry by Lucky Company

A/R (300 x \$2 x 95%)	570.00	
Production tax expense (300 x \$2 x .05 x $^7/_8$)	26.25	
Royalty payable (300 x \$2 x 0.95 x $^1/_8$)		71.25
Gas sales (300 x \$2 x $^7/_8$)		525.00

c. The operator assumes the responsibility of distributing royalty income and severance taxes.

Entry by Lucky Company

A/R (300 x \$2)	600.00	
Production tax expense (300 x \$2 x .05 x $^7/_8$)	26.25	
Production tax payable (WI + RI) (300 x \$2 x 5%)		30.00
Royalty payable (300 x \$2 x 0.95 x $^1/_8$)		71.25
Gas sales (300 x \$2 x $^7/_8$)		525.00

d. The royalty interest owner's entry under either *b* or *c* above:

Entry by RI

A/R (300 x \$2 x 0.95 x $^1/_8$)	71.25	
Severance tax expense (300 x \$2 x 0.05 x $^1/_8$)	3.75	
Gas sales (300 x \$2 x $^1/_8$)		75.00

Gas, like oil, is also often used for lease operation. The gas may be used as fuel for operating lease equipment such as pumping units, heater-treaters, and compressors or as fuel for the lessor's residence or other activities. Such gas is measured by metering if the quantities involved are substantial; otherwise, the volume used is normally estimated based on engineering estimates of fuel used by the various pieces of equipment. The value assigned to the gas is generally equivalent to the value at which the gas could have been sold. If gas produced on one lease is used on another lease owned by the same company, the accounting entry is similar to the entry in the previous example and is also similar to the oil revenue entry for the same situation.

EXAMPLE

Gas Used off Lease by Operator

Lucky Company owns 100% of the working interests on Leases A and B. Lucky Company used 300 Mcf of gas produced on Lease A and valued at \$2 per Mcf to operate lease equipment on Lease B. Assume production taxes were 5% and a $^1/_8$ royalty interest on Lease A.

Entry—analysis is the same as in the above example

Lease operating expense—Lease B	600.00	
Production tax expense—Lease A		
(300 x $2 x .05 x $^7/_8$)	26.25	
Production tax payable		
(RI + WI)(300 x $2 x .05)		30.00
Royalty payable—Lease A		
(300 x $2 x .95 x $^1/_8$)		71.25
Gas sales—Lease A (300 x $2 x $^7/_8$)		525.00

Gas may also be used in lease operation for gas injection or gas lift. In a **gas injection** operation, gas is injected into a formation to maintain pressure or for secondary recovery. Reproduced injected gas cannot usually be distinguished from the original formation gas. Therefore, some assumption must be made concerning the recovery of gas, *i.e.*, whether the produced gas is original formation gas or injected gas (see *COPAS Bulletin No. 7* for further details). **Gas-lift gas** is injected into the wellbore to help lift the oil to the surface. Gas-lift gas, unlike injected gas, returns immediately to the mouth of the well without entering the reservoir. Normally the sales price for recovered gas-lift gas is lower since, when it is reproduced, its pressure has diminished. Because gas going into a pipeline must either be at a prescribed pressure or at the same pressure as the gas already in the pipeline, if the pressure of the recovered gas-lift gas is too low, compression is necessary.

EXAMPLE

Gas Injection or Gas Lift

If Lucky Company had used the gas in the preceding example for injection or for gas lift, the entry to record the use of the gas would have been the same as described above. When the gas is recovered and sold, the entry would be the following (assuming 100% recovery and the same selling price):

Entry

A/R—purchaser	600	
Lease operating expense—Lease B		600

If gas is produced and used on the same lease, royalty payments and production taxes may not have to be paid. The liability for royalty payments and production taxes would be determined from the lease agreement and state laws. However, most leases provide that gas may be used on the same lease free from the obligation of payment of royalty. If state laws do not levy production taxes on such gas, then an accounting entry would not have to be made to record the production and use of gas on the same lease. If an entry is made, it would be as follows:

Entry		
Lease operating expense	XXX	
Gas sales		XXX

If the gas is used for injection or gas lift on the same lease from which it was produced, usually no accounting entry is made at the time of production and use, and usually no production tax or royalties are payable on such gas. When the gas is recovered and sold, the regular accounting entry would be made to record the sale of the gas.

Vented or flared gas

Under normal circumstances, gas flaring is prohibited because of environmental considerations. Generally, permission to vent or flare gas must be obtained from the appropriate regulatory agency. A request to flare or vent gas may arise when the gas is of low quality or low pressure or when sale of the gas is not economically feasible (*i.e.*, an oil well with no gas pipelines available). If venting or flaring is allowed, normally no value is assigned to vented or flared gas and revenue accounts are not credited. In most states, this type of gas is not subject to severance or production taxes. Vented or flared gas is not usually measured, but the volume is estimated.

Nonprocessed natural gas

Gas may be of marketable quality when produced and, therefore, not need any processing. In this case the gas may be sold directly to a gas pipeline, local distribution company, or end user. (All of the gas sales examples in this chapter apply to nonprocessed natural gas that is of marketable quality when produced.)

Natural gas processing

Often gas is not of marketable quality when produced because of impurities and liquids. Since gas may not usually be disposed of by flaring, the gas must be processed to make it marketable. Such processing involves removing impurities and saleable liquids and compressing the gas to meet pipeline pressure specifications. Typically the gas is either sold to a processing plant or is processed by a plant for a fee. In either case, accounting entries to record the sale are similar to the accounting discussed above. (For more information, see *COPAS Bulletin No. 7*.)

The demand for gas is highly seasonal. As a result, if wells are to produce at capacity during the low demand summer months, the excess gas must be placed in underground storage facilities for use in the higher demand winter months. Royalties usually must be paid to royalty interest owners when the gas is severed from the lease, even if the gas is being stored by the producer for sale at a later time.

Take-or-pay provision

Some old gas sales/purchase contracts, most often those entered into with a pipeline, contain a take-or-pay clause. This clause requires the purchaser to take a specified minimum volume of gas each month. If the minimum volume of gas is not taken, the purchaser must pay for the deficiency gas, even though the gas is not taken. However, the purchaser generally may offset the deficiency payment against any future purchases in excess of the specified volume. The producer records the deficiency payment as a deferred credit until make-up gas is taken by the purchaser, as follows:

Entry

Cash	XXX	
Deferred credit		XXX

When the purchaser buys more than the specified minimum amount of gas from the producer, the producer recognizes revenue and reduces the deferred credit. If the price of gas has increased since the deficiency arose, most contracts require the purchaser to pay an additional amount.

EXAMPLE

Take-or-Pay

Lucky Company has a contract with Southwest Pipeline in which Southwest has agreed to purchase 20,000 Mcf of gas per month at $2.00 per Mcf. Southwest must pay for the gas, even if it chooses not to take delivery of the full amount. In August, Southwest took delivery of 16,000 Mcf produced from a lease with a $^1/_8$ royalty and in which Lucky owns 100% of the WI. Production taxes are 5%. Lucky assumes the responsibility of distributing royalty income and paying production taxes.

Entries

To record the sale of 16,000 Mcf:		
Cash (16,000 x \$2.00)	32,000	
Production tax expense		
(16,000 x \$2.00 x 5% x $^7/_8$)	1,400	
Royalty payable		
(16,000 x \$2.00 x $^1/_8$ x 0.95)		3,800
Production tax payable		
(16,000 x \$2.00 x 5%)		1,600
Gas sales (16,000 x \$2.00 x $^7/_8$)		28,000
To record the deferred credit:		
Cash (4,000 x \$2.00)	8,000	
Deferred revenue		8,000

In September, Southwest took delivery of 24,000 Mcf.

Entries

To record the current month's sale:		
Cash (20,000 x \$2.00)	40,000	
Production tax expense		
(20,000 x \$2.00 x 5% x $^7/_8$)	1,750	
Royalty payable		
(20,000 x \$2.00 x $^1/_8$ x 0.95)		4,750
Production tax payable		
(20,000 x \$2.00 x 5%)		2,000
Gas sales (20,000 x \$2.00 x $^7/_8$)		35,000
To record the make-up:		
Deferred revenue (4,000 x \$2.00)	8,000	
Production tax expense		
(4,000 x \$2.00 x 5% x $^7/_8$)	350	
Royalty payable (4,000 x \$2.00 x $^1/_8$ x 0.95)		950
Production tax payable (4,000 x \$2.00 x 5%)		400
Gas sales (4,000 x \$2.00 x 7/8)		7,000

Very few contracts containing a take-or-pay provision still exist. Moreover, contracts containing such a provision are not currently being written. (For more information, see *COPAS Bulletin No. 7.*)

Timing of Revenue Recognition

Theoretically, revenue from oil and gas sales should be recognized when the sale is made rather than when the oil or gas is produced. Legally, title normally passes when physical control of the oil or gas is transferred from one party to another. Therefore, theoretically, revenue should be recognized at the point of delivery.

Revenue from oil

In practice, revenue from oil sales is typically recognized either when the oil is produced or when the oil is sold. When oil revenue is recognized at the time of sale, the revenue is recognized based on pipeline runs or run tickets. Revenue recognition at the time of the sale is the basis used for all the previous examples in this chapter.

EXAMPLE

Revenue Recognition at Time of Sale

Lucky Company produces 3,000 barrels in the month of October. The sales price is $22. Assume the purchaser of the oil will pay the royalty owner and severance taxes. Assume that Lucky owns 100% of the WI, the royalty interest is $^1/_8$ and the state severance tax is 8%.

Entry at time of sale

A/R—purchaser (3,000 x $22 x 0.92 x $^7/_8$)	53,130	
Production tax expense (3,000 x $22 x 0.08 x $^7/_8$) . . .	4,620	
Oil sales (3,000 x $22 x $^7/_8$)		57,750

Commonly, companies that recognize revenue at the point of sale disregard inventory in lease tanks; therefore, no entry is made at the time of production. If the inventory is recorded, it is often recorded at year-end at the market selling price versus the cost of producing the oil. (See chapter 8 for more discussion of oil and gas inventory.)

If, on the other hand, oil revenue is recognized when the oil is produced, the revenue is determined by multiplying the number of barrels produced times the expected sales price. Inventory is also then recognized and carried in an inventory account on the books of the producer until the crude oil is sold.

EXAMPLE

Revenue Recognition at Time of Production

Using the same data as in the previous example, the entry to record the production and sale of the oil would be the following.

a. Entry at time of production is as follows:

Entry

Crude oil inventory (3,000 x \$22 x 7/8) . . .	57,750	
Oil sales		57,750

b. The production is sold in November. Entry at the time of the sale is as follows:

Entry

A/R—purchaser (3,000 x \$22 x 0.92 x 7/8)	53,130	
Production tax expense (3,000 x \$22 x 0.08 x 7/8)	4,620	
Crude oil inventory		57,750

Any price difference between the expected price at the time of production and actual sales price at the time of sale would be treated as an adjustment to oil revenue.

Revenue from gas

Since gas cannot be stored on the lease, nonprocessed gas is generally not produced unless there is a buyer for the gas. Traditionally under the old long-term contracts, the delivery point for nonprocessed gas sales was at the wellhead, and therefore gas sales were recognized essentially concurrently with production, based on pipeline runs. (All gas revenue examples in this chapter are on this basis.) With spot market sales, title of the gas often passes at an extended delivery point such as at *city gate* rather than at wellhead. In these cases, the pipeline is transporting the gas to a delivery point versus purchasing it at the wellhead. Accounting for gas sold in the spot market is complicated by the extended delivery points. The extended delivery points result in transportation charges, pipeline losses, and the need to recognize gas in the pipeline as inventory.

Several methods of accounting for spot sales are currently used. The example below illustrates (a) the traditional accounting where the gas is sold to a pipeline at the wellhead and (b) a spot sale in which the gas is sold to an industrial plant, and the pipeline only transports the gas instead of purchasing the gas.

EXAMPLE

Gas Revenue Recognition

Delivery point wellhead (pipeline is purchaser)

Lucky Company produces and sells 20,000 Mcf of gas to a pipeline for $1.50 per Mcf. Lucky owns a 100% WI in the lease with a $1/5$ royalty interest. Ignore severance taxes.

Entry to record sale

A/R (20,000 x $1.50)	30,000	
Royalty payable (20,000 x $1.50 x $1/5$)		6,000
Gas sales (20,000 x $1.50 x $4/5$)		24,000

Delivery point gas inlet at municipal gas company (pipeline transports gas)

Lucky Company also produces 30,000 Mcf of gas from the same lease to be transported by a pipeline to the local municipal gas company. The gas is sold at $2.00 per Mcf. The transportation charge is $0.40 per Mcf. Lucky elects not to pass any of the transportation charge on to the royalty owner.

Entry to record gas delivered to pipeline

Inventory (30,000 x $2.00)	60,000	
Royalty payable (30,000 x $2.00 x $1/5$)		12,000
Deferred revenue (30,000 x $2.00 x $4/5$)		48,000

Entry to record delivery of gas to municipal gas company

A/R (30,000 x $2.00)	60,000	
Deferred revenue	48,000	
Transportation expense (30,000 x $0.40)	12,000	
A/P transporter (30,000 x $0.40)		12,000
Inventory		60,000
Gas sales (30,000 x $2.00 x $4/5$)		48,000

Some small producers defer recording both oil and gas revenue until a check and accompanying statement are received from the purchaser. This practice, referred to as recording revenue at the time of settlement, does not adhere to the principles of accrual accounting.

However, as long as production and prices are relatively stable from accounting period to accounting period, it does not result in a material distortion of revenue and therefore its use is allowed.

Revenue Reporting to Interest Owners

Royalty owners and other interest owners are normally paid monthly, based on their revenue percentage as specified in the division order. The payments are made by either the operator or the purchaser, depending upon who has assumed that responsibility. The statement is typically a check-stub type that contains the information needed by the interest owners for their tax records and for other record-keeping purposes. Information provided on the monthly check stub includes the number of barrels or Mcfs sold, the average price, the owner's decimal ownership, and the applicable state severance tax withheld.

Additional Topics

The following sections discuss miscellaneous revenue accounting topics:

Gas imbalances

One of the more challenging issues in revenue accounting is gas imbalances. Imbalances can be of two types: **producer gas imbalances** or **pipeline gas imbalances**. Producer gas imbalances occur when there are two or more working interest owners in a single property and one working interest owner takes more than his proportionate share of the gas produced from the property. Pipeline gas imbalances occur between a producer and a pipeline company. For example, a producer agrees to sell a quantity of gas to an end-user. The end-user typically has no means to store the excess gas that results when their actual demanded quantity is less than the contracted quantity and no reserve to use when their actual demanded quantity is greater than the contracted quantity. In these cases the buyer takes what it needs from the pipeline. If the take is more or less than the amount of gas that the producer put into the pipeline, a pipeline imbalance results.

Producer gas imbalances

Producer gas imbalances occur for a variety of reasons, but with the advent of spot sales and direct sales, the imbalances have become more complex. Currently, each working interest owner on a property may market its gas to one or more purchasers, which includes marketers, brokers, end-users, and other producers. This creates multiple gas purchasers. In some instances it is not possible to deliver the gas simultaneously to all purchasers, thus one purchaser will take all the production for a period of time, after which another purchaser will take all the production, and so forth. Only the working interest owner who sold the gas to the purchaser currently taking the gas is paid by that purchaser for the gas received. Thus, at any point in time, an imbalance situation exists. This imbalance is magnified due to the fact that because natural gas wells rarely produce at a steady rate, the volumes delivered to the purchasers are seldom at exactly the average quoted production rate. Imbalances may also occur because some of the working interest owners on a property may be unwilling or unable, because of market conditions, lack of connection to wellhead, etc. to market their gas currently. The working interest owners on a property usually sign a gas balancing agreement, which specifies the rights and duties of each party when a gas imbalance situation exists. Each gas balancing agreement is typically modified to conform to the particular situation. An example of a gas balancing agreement taken from *COPAS Bulletin No. 4* is presented in Figure 9–5.

EXHIBIT____________

COPAS — 1973 Recommended by the Council of Petroleum Accountants Societies of North America.

GAS BALANCING AGREEMENT FOR GAS WELL PRODUCTION

Attached to and made a part of the Operating Agreement between ......
..
..
..

The parties to the Operating Agreement to which this Gas Balancing Agreement is attached own the working interest in the gas rights underlying the Joint Property covered by such agreement and are entitled to share in the percentages as stated in the Operating Agreement.

In accordance with the terms of the Operating Agreement, each party thereto has the right to take its share of gas produced from the Unit Area and market the same. In the event any of the parties hereto does not market its share of gas or has contracted to sell its share of gas produced from the Joint Property to a purchaser which, at any time while this agreement is in effect fails to take the share of gas attributable to the interest of such party, the terms of this storage agreement shall automatically become effective.

1.

During any period or periods when the market of a party is not sufficient to take that party's full share of the gas produced from the Joint Property, or its purchaser is unable to take its share of gas produced from the Joint Property, the other party or parties shall be entitled to produce from said Joint Property (and take or deliver to a purchaser), each month, all or a part of that portion of the allowable gas production assigned to such Joint Property by the regulatory body having jurisdiction. That party shall be entitled to take and deliver to its or their purchaser all of such gas production, provided; however, that no party shall be entitled to take or deliver to a purchaser gas production in excess of __________% (percent) of its share of the allowable gas production assigned thereto by the regulatory body having jurisdiction, unless that party has gas in storage. All parties hereto shall share in and own the liquid hydrocarbons recovered from such gas by primary separation equipment in accordance with their respective interests and subject to the terms of the above described Operating Agreement.

2.

Each party unable to market its share of the gas produced, and taking less than its full share of the gas produced shall be credited with gas in storage equal to its share of the gas produced under this agreement, less such party's share of the gas taken, gas used in Joint Property operations, vented, or lost. Each party taking gas shall furnish the Joint Property operator a monthly statement of gas taken. The operator of the Joint Property will maintain a running account of the gas balance between the parties hereto and will furnish each party monthly statements showing the total quantity of gas produced, the amount thereof used in Joint Property operations, vented or lost, and the total quantity of gas delivered to markets. Measurement of gas for over and under production shall be accomplished by use of (sales) (lease) meters, and lease measurement shall be in accordance with AGA requirements.

Fig. 9–5 — *Gas Balancing Agreement for Gas Well Production*

3.

After written notice to the operator, any party may at any time begin taking or delivering to its purchaser its full share of the gas produced from said Joint Property (less any used in Joint operations, vented, or lost). To allow for the recovery of gas in storage and to balance the gas account of the parties in accordance with their respective interests, a party with gas in storage shall be entitled to take or deliver to a purchaser its full share of gas produced from said Joint Property (less any used in Joint operations, vented or lost) plus an amount determined by multiplying percent (%) of the interest of the party or parties without gas in storage by a fraction, the numerator of which is the interest in the Joint Property of such party with gas in storage and the denominator of which is the total percentage interest in the Joint Property of all parties with gas in storage.

4.

Nothing herein shall be construed to deny any party the right, from time to time to produce and take or deliver to its purchaser its full share of the allowable gas production to meet the deliverability tests required by its purchaser. Each party, shall at all times, use its best efforts to regulate its takes and deliveries from said Joint Property so that said Joint Property will not be shut-in for over producing the allowable assigned thereto by the regulatory body having jurisdiction.

5.

During the terms of this agreement, while gas is being produced from the Joint Property, each party shall make settlement with its own respective royalty owners (and the term "royalty owners" shall include owners of royalties, overriding royalties, production payments and similar interests), based on such royalty owner's respective interests in the Joint Property and on total volumes of gas produced, saved and taken or delivered to purchasers. Each party hereto agrees to hold each other party harmless from any and all claims for royalty payments asserted by royalty owners to whom each party is accountable.

6.

Each party producing and taking or delivering gas to its purchaser shall pay, or cause to be paid, all production taxes due on such gas.

7.

In the event production of gas from said Joint Property shall be discontinued before the gas account is balanced, a complete balancing will be made between the parties for gas remaining in storage. In making such settlement, the party or parties with gas remaining in storage will be paid by the other party or parties a sum of money equal to that which said other party or parties received, less applicable taxes heretofore paid, for the latest delivery of a volume of gas equal to that for which settlement is made. The operator shall be responsible for determining the final accounting of the underproduction and overproduction and the amounts due to be paid to, or by, each party.

8.

This agreement shall remain in force and effect as long as the operating agreement, to which it is attached, and shall inure to the benefit of and be binding upon the parties hereto, their successors, legal representatives and assigns.

9.

Nothing herein shall change or affect each party's obligations to pay its proportionate share of all costs and liabilities incurred in Joint operations, as its share thereof is set forth in the above-described Operating Agreement.

10.

The provisions of this agreement shall be applied to each well and/or each formation separately as if each well and/or formation was a separate well and covered by separate but identical agreements.

Fig. 9–5 cont'd — *Gas Balancing Agreement for Gas Well Production*

Source: COPAS, 1973. Recommended by the Council of Petroleum Accountants Societies of North America. COPAS Bulletin No. 4, "COPAS Forms," January 1974, pp. 33-34. COPAS bulletins may be ordered from Krafbilt, Tulsa, Oklahoma, 1-800-331-7290.

In a producer gas imbalance situation, the operator typically maintains records or reports in which allocated and delivered volumes are tracked. This enables the operator to request adjustments of future deliveries of gas to each purchaser because of deliveries being over or short in the past.

The industry has developed several methods of dealing with gas imbalances, called **production balancing**. Under one method, an "in-kind" balancing, the underproduced party may give notice to take up to an agreed upon quantity of the overproduced party's share of gas until balanced. Some agreements also allow the operator to periodically require a cash balancing. If balancing is not possible, *e.g.*, the reservoir has been depleted, the operating or gas balancing agreement may provide for a cash settlement under both the sales and entitlement method.

For financial accounting purposes, there are two methods of recognizing revenues—the sales method and the entitlement method. Both are generally accepted accounting methods. Using the sales method, the working interest owners recognize their actual sales of gas regardless of the amount of production they are entitled to for the period. The sales method assumes that any production sold by a working interest owner comes from that party's share of the total reserves-in-place. Thus, whatever quantity is sold in any given period is revenue for that party. No receivables, payables, or unearned revenue are recorded unless a working interest owner's aggregate sales from the property exceed its share of the total reserves-in-place. If such a situation arises, the parties would likely choose to cash balance or, in some instances, the overdelivered partner might choose to negotiate to buy out the underdelivered party's share.

Using the entitlement method, each owner recognizes revenue based on its ownership share of total gas actually produced during the period, regardless of which owner actually sells and receives payment for the gas. Thus, the overdelivered owner would recognize unearned revenue for the amount received in excess of its entitled share. The underdelivered owner would recognize a receivable and revenue for the amount of its share of production sold by another owner for which it was not paid. While both the sales and entitlements methods are widely used, the sales method is the more popular of the two.

Companies must determine their reserves for the property in a manner that is consistent with their revenue recognition method. For example, a company using the sales method would reduce its estimated reserves for the quantity of gas it actually sells during the period. A company using the entitlements method would reduce its reserves by the amount of the share of production it was entitled regardless of whether the quantity it actually sold was more or less than the entitled quantity.

The following example illustrates an in-kind production balancing arrangement and the sales and entitlement methods of recording revenues:

EXAMPLE

Producer Gas Imbalances

Shepherd Field is jointly owned by Lucky Company (60%) who acts as field operator, and Garza Company (40%). There is a $1/8$ royalty. The $1/8$ royalty is shared proportionally by Lucky and Garza. The two working interest owners agree that Lucky's purchaser will take gas produced in April and Garza's purchaser will take gas produced in May. Gas allocations will be equalized in June. Ignore severance taxes.

Production and gas prices were as follows:

	Production	Price
April	300,000 Mcf	$2.25/Mcf
May	280,000 Mcf	2.25/Mcf
June	200,000 Mcf	2.25/Mcf

Each working interest owner receives payment only for gas delivered to his purchaser(s). The following Gas Balance Report prepared by Lucky Company summarizes the production deliveries and equalization of gas for April through June:

Gas Balance Report
(Mcf)

	April	May	Cumulative	June	Cumulative Total
Total Production:	300,000	280,000	580,000	200,000	780,000
Allocated shares based on WI% (including royalty to be paid):					
Lucky Company (60%)	180,000	168,000	348,000	120,000	468,000
Garza Company (40%)	120,000	112,000	232,000	80,000	312,000
	300,000	280,000	580,000	200,000	780,000
Deliveries taken by:					
Lucky Company's Purchaser	300,000	—	300,000	168,000	468,000
Garza Company's Purchaser	—	280,000	280,000	32,000	312,000
	300,000	280,000	580,000	200,000	780,000
Overdelivered (Underdelivered):					
Lucky Company	120,000	(168,000)	(48,000)	48,000	0
Garza Company	(120,000)	168,000	48,000	(48,000)	0

Sales Method:
The journal entries made by each company for the three-month period are shown below assuming that both companies use the sales method:

APRIL

Lucky Company

Accounts receivable (300,000 x \$2.25)	675,000	
Royalty payable (300,000 x \$2.25 x 1/8) . . .		84,375
Gas Revenue (300,000 x \$2.25 x 7/8)		590,625

Garza Company

No entry

MAY

Lucky Company

No entry

Garza Company

Accounts receivable (280,000 x \$2.25)	630,000	
Royalty payable (280,000 x \$2.25 x 1/8) . . .		78,750
Gas Revenue (280,000 x \$2.25 x 7/8)		551,250

JUNE

Lucky Company

Accounts receivable (168,000 x \$2.25)	378,000	
Royalty payable (168,000 x \$2.25 x 1/8) . . .		47,250
Gas Revenue (168,000 x \$2.25 x 7/8)		330,750

Garza Company

Accounts receivable (32,000 x \$2.25)	72,000	
Royalty payable (32,000 x \$2.25 x 1/8)		9,000
Gas Revenue (32,000 x \$2.25 x 7/8)		63,000

Entitlement Method:

The journal entries made by each company for the three-month period are shown below assuming that both companies use the entitlement method for both revenue and royalty.

APRIL

Lucky Company

Accounts receivable (300,000 x \$2.25)	675,000	
Royalty payable (180,000 x \$2.25 x 1/8) . . .		50,625
Unearned revenue—overdelivered (120,000 x \$2.25)		270,000
Gas Revenue (180,000 x \$2.25 x 7/8)		354,375

Garza Company

Accounts receivable—underdelivered		
(120,000 x $2.25)	270,000	
Royalty payable (120,000 x $2.25 x $^1/_8$) . . .		33,750
Gas revenue (120,000 x $2.25 x $^7/_8$)		236,250

MAY

Lucky Company

Unearned revenue—overdelivered		
(120,000 x $2.25)	270,000	
Accounts receivable—underdelivered		
(48,000* x $2.25)	108,000	
Royalty payable (168,000 x $2.25 x $^1/_8$) . . .		47,250
Gas revenue (168,000 x $2.25 x $^7/_8$)		330,750

* Lucky's share of production to date based on his working interest percentage is 348,000 Mcf. The actual amount taken to date by Lucky's purchaser is 300,000 Mcf.

Garza Company

Accounts receivable (280,000 x $2.25)	630,000	
Royalty payable (112,000 x $2.25 x $^1/_8$) . . .		31,500
Accounts receivable—underdelivered		
(120,000 x $2.25)		270,000
Unearned revenue—overdelivered		
(48,000* x $2.25)		108,000
Gas revenue (112,000 x $2.25 x $^7/_8$)		220,500

* Garza's share of production to date based on his working interest percentage is 232,000 Mcf. The actual amount taken to date by Garza's purchaser is 280,000 Mcf.

JUNE

Lucky Company

Accounts receivable (168,000* x $2.25)	378,000	
Royalty payable (120,000 x $2.25 x $^1/_8$) . . .		33,750
Accounts receivable—underdelivered		108,000
Gas revenue (120,000 x $2.25 x $^7/_8$)		236,250

* The 168,000 barrels equals Lucky's 60% share of June's 200,000 barrels of production plus the 48,000 underdelivered barrels.

Garza Company

Accounts receivable (32,000* x \$2.25)	72,000	
Unearned revenue—overdelivered	108,000	
Royalty payable (80,000 x \$2.25 x 1/8) . . .		22,500
Gas revenue (80,000 x \$2.25 x 7/8)		157,500

* The 32,000 barrels equals Garza's 40% share of June's 200,000 barrels of production less the 48,000 overdelivered barrels.

In addition to recognizing revenue on gas sales, companies must also pay royalties and severance taxes. While lease agreements typically do not specify whether the sales or entitlements method is to be used in the calculation of royalties, most companies—even those using the entitlements method for financial accounting purposes—use the sales method for determining royalties. Most states require the sales method for determination of severance taxes.

Pipeline gas imbalances

In order for the various interest owners in a lease to market their own gas there must be communication between each working interest owner (producer), the operator, the pipeline company, and each producer's customer. The operator must advise each producer of the anticipated gas production from each well during the next month. Each producer negotiates with its purchaser for the quantity of gas the purchaser will require during the next month. Then through a process of **nomination** and **confirmation**, the operator agrees to put a quantity of gas into the pipeline on behalf of the producer, and the purchaser agrees to take that quantity out of the pipeline during the next month.

The actual amount of gas production rarely equals the total of the confirmed gas nominations. The parties involved should execute an agreement that includes a method to determine (a) the number of Mcf of actual production to be allocated to each party, and (b) each producer's resulting overdeliveries and underdeliveries to the pipeline. Two methods commonly used to determine the imbalance position of the parties are allocation of production volumes based on *confirmed nominations* and allocation based on *entitlements.* These two methods are similar to the sales and entitlement methods illustrated and discussed in the producer gas imbalance situation. Regardless of which method is used, a monthly production volume allocation statement is prepared.

Financial accounting for pipeline gas imbalances involves recording overdeliveries and underdeliveries to the pipeline company as receivables or payables. Once overdeliveries and underdeliveries are determined through allocation, the issue of valuing the receivables and payables must be resolved. Since each producer is selling its gas to its own various customers, multiple gas sales prices will exist. Which price should be used to value the imbalance? The FASB Emerging Issues Task Force (EITF) and the SEC have studied this issue and generally agree that in valuing the imbalances, receivables should be valued at the lowest and payables valued at the highest of the possible sales prices (see EITF Issue No. 90-22).

The following example demonstrates accounting for pipeline imbalances.

EXAMPLE

Pipeline Gas Imbalances

Lehman Field is jointly owned by Lucky Company who acts as field operator, and Conroe Company. Each company owns 50% of the WI. Lucky estimates that gas production during August will be 50,000 Mcf. Lucky Company makes confirmed nominations of 40,000 Mcf and Conroe Company makes confirmed nominations of 10,000 Mcf to a pipeline company. Actual August production totals 80,000 Mcf. Assume that the appropriate price is $2 per Mcf. The pipeline company has paid Lucky and Conroe for the confirmed nominations only; therefore, Lucky and Conroe must recognize receivables for the gas sold in excess of the confirmed nominations.

Allocation using the confirmed nomination method and entitlements method would be:

Confirmed Nomination Method:

	Lucky Company	Conroe Company
Actual production	80,000 Mcf	80,000 Mcf
Nomination ratio	x 40/50 = 80%	x 10/50 = 20%
Allocated actual production	64,000 Mcf	16,000 Mcf
Less: confirmed nomination	(40,000) Mcf	(10,000) Mcf
Imbalance receivable from pipeline	24,000 Mcf	6,000 Mcf
$2.00/Mcf	x $ 2	x $ 2
Pipeline Imbalance Receivable	$48,000	$12,000
Allocated actual production	64,000 Mcf	16,000 Mcf
$2.00/Mcf	x 2	x 2
Revenue recognized	$128,000	$32,000

Entitlement Method:

	Lucky Company	Conroe Company
Actual production	80,000 Mcf	80,000 Mcf
Ownership interest	x 0.50	x 0.50
Allocated actual production	40,000 Mcf	40,000 Mcf
Less: confirmed nomination	(40,000) Mcf	(10,000) Mcf
Imbalance receivable from pipeline	0 Mcf	30,000 Mcf
$2.00/Mcf	x $ 2	x $ 2
Pipeline Imbalance Receivable	$ 0	$60,000
Allocated actual production	40,000 Mcf	40,000 Mcf
$2.00/Mcf	x 2	x 2
Revenue recognized	$80,000	$80,000

The conditions that result in pipeline gas imbalances also result in producer gas imbalances. For example, in the illustration above, Lucky Company sold more and Conroe Company sold less than their respective 50% share of gas production. Thus, in addition to accounting for the pipeline gas imbalance, it would be necessary for the companies to also account for the resulting producer gas imbalances.

Allocation of oil and gas

Production from different wells or leases is often commingled and measured at the point of delivery. Total production sold may need to be allocated back or reconciled to the individual well or lease production in order to comply with tax requirements and management needs or because of different ownership of the properties involved.

The method for determining production quantities from individual wells depends upon whether the well is metered individually. Each gas well/completion is usually metered individually, as is generally required by state law. Oil wells are not required to be metered individually and, as a result, usually are not metered individually but instead are put on test periodically to determine or estimate the total amount of oil, gas, and water produced from the well.

A well **put on test** means that the stream of fluid from the well is diverted from the production separator to a test separator. The test separator measures the amount of oil, gas, and water produced from the well for 24 hours. The test results are used to estimate the production from that well for a period of time—usually a month. When all wells have been tested, the test results are used to allocate total production to the various wells or properties. This allocation method is similar to the relative sales value method used for many other accounting purposes.

The following example illustrates a one-stage allocation of oil in which production from one lease is allocated back or reconciled to the individual well production, based on a 24-hour test rate of production for the individual wells.

EXAMPLE

Allocation Back to Well

Assume the following data for Lease A:

Well	24-hr Test, bbl	Days Produced	Total Production Sold
1	20	30	2,900 barrels—measured at
2	35	26	point of delivery (production
3	50	24	from 3 wells commingled)

Allocation

Well	24-hr Test, bbl	Days Produced	Theoretical Production*		Ratio**		Actual Production
1	20	30	600	x	1.07	=	642
2	35	26	910	x	1.07	=	974
3	50	24	1,200	x	1.07	=	1,284
			2,710	x	1.07	=	2,900

An equivalent computational approach more commonly used in accounting when allocating a lump sum, referred to as the relative sales method, is as follows:

Well 1: $2{,}900 \times \frac{600}{600 + 910 + 1{,}200} = 642$ bbl

Well 2: $2{,}900 \times \frac{910}{600 + 910 + 1{,}200} = 974$ bbl

Well 3: $2{,}900 \times \frac{1{,}200}{600 + 910 + 1{,}200} = 1{,}284$ bbl

* 24-hr Test x Days Produced = Theoretical Production

** 2,900/2,710 = 1.07

The same basic principles of allocation apply when production from more than one lease has been commingled. This usually occurs when one central processing system has been installed as the central delivery point for two or more leases. The processing system will consist of the equipment necessary for processing the oil and gas from the wells. Included would typically be a separator, heater-treater, gas scrubber (to remove liquids entrained with the gas), test separator, tanks, and a LACT unit.

When the production from more than one lease is being commingled, generally:

- the production from all the leases will be measured at the central tank battery
- the production from each lease will be measured
- the production from each well will be measured or estimated

In this case, at least a two-stage allocation would be involved, starting with the disposition point of the oil or gas and flowing toward the source of the oil or gas. The following example illustrates a two-stage allocation of oil in which production from two leases is delivered to a central tank battery.

EXAMPLE

Two-Stage Allocation

The production from each well on Lease A and Lease B is estimated based on a 24-hour test. The production from each well on each lease is then commingled and the commingled flow measured before leaving the individual lease. The production from both leases is commingled and delivered to a central tank battery. Assume the following data for Lease A:

Well	24-hr Test, bbl	Days Produced	Theoretical Production
1	20	30	600
2	35	26	910
3	50	24	1,200
			2,710

Assume the following data for Lease B:

Well	24-hr Test, bbl	Days Produced	Theoretical Production
1	25	28	700
2	40	29	1,160
			1,860

Measured production from Lease A is 2,450 barrels and 1,663 barrels from Lease B. After treatment at the central tank battery, 4,000 barrels are sold.

Allocation

Step 1: Allocate sales from tank battery to each lease.

Ratio of sales to total production from leases = 4,000/4,113 = .9725

Lease	Measured Production		Ratio		Allocated Sales
A	2,450	x	0.9725	=	2,383
B	1,663	x	0.9725	=	1,617
	4,113				4,000

Step 2: Allocate sales from each lease to individual wells.

Lease A: Ratio of sales allocated to Lease A to total theoretical production from wells on Lease A = 2,383/2,710 = .8793

Well	24-hr Test, bbl	Days Produced	Theoretical Production		Ratio		Allocated Sales
1	20	30	600	x	0.8793	=	528
2	35	26	910	x	0.8793	=	800
3	50	24	1,200	x	0.8793	=	1,055
			2,710	x	0.8793	=	2,383

Lease B: Ratio of sales allocated to Lease B to total theoretical production from wells on Lease B = 1,617/1,860 = 0.8694

Well	24-hr Test, bbl	Days Produced	Theoretical Production		Ratio		Allocated Sales
1	25	28	700	x	0.8694	=	609
2	40	29	1,160	x	0.8694	=	1,008
			1,860	x	0.8694	=	1,617

The exact method of allocation depends upon the points at which the production is measured or metered.

The allocation procedure may be complicated when the quality of the oil varies. The value of a barrel of oil is partially dependent on the quality of the oil, specifically the API gravity of the oil, which can vary from lease to lease. Consequently, when significant differences in gravity between the various leases exist, the allocation procedure illustrated above must be modified to determine sales revenue due each lease from the quantities of oil produced and sold from each lease. (For further discussion, see *COPAS Bulletin No. 17.*)

While conceptually the allocation of gas is the same as allocation of oil, the actual allocation process is much more complicated. (For further discussion, see *COPAS Bulletin No. 7.*)

MINIMUM ROYALTY—AN ADVANCE REVENUE TO ROYALTY OWNERS

Minimum royalties usually arise from an agreement in which the lessee agrees to pay a stated minimum amount of money to the royalty owner whenever the actual production revenue to which the royalty owner is entitled is less than the minimum amount. This payment may be paid prior to production or during the production phase. Minimum royalty payments may or may not be recoverable from future production. If nonrecoverable, the payments usually are expensed. If recoverable, the payments should be held in a suspense account until recovered. If the payments have not been recovered by the time the lease terminates, they should be expensed.

EXAMPLE

Minimum Royalty

a. During 2006 before production was established, Lucky Company paid minimum royalty payments of $100 a month for five months. The minimum royalty payments were recoverable out of future royalty payments. Give the entry to record the minimum royalty payments paid.

Entry

Minimum royalty suspense	500	
Cash		500

b. Production was established late in 2006. The royalty payment payable from the first month of production was $250. Give the entry to record payment of the royalty.

Entry

Royalty payable	250	
Minimum royalty suspense		150
Cash		100

Note that the royalty owner is still entitled to a minimum of $100 per month and thus only $150 of the suspense may be recovered. The royalty payable account originated in the entry to record revenue.

c. The royalty payment from the second month of production was $400. Give the entry to record the payment of the royalty.

Entry

Royalty payable	400	
Minimum royalty suspense		300
Cash		100

d. Disaster struck and the lease was abandoned. Give any entry necessary related to the minimum royalty.

Entry

Expense	50	
Minimum royalty suspense		50

e. Instead, assume in *a* above that the minimum royalty payments were not recoverable. Give the entry to record the minimum royalty payments.

Entry

Minimum royalty expense	500	
Cash		500

Oil and gas revenue accounting is a complicated topic. This chapter discusses only the fundamental points. For more guidance on some of these issues, see *COPAS Bulletins No. 17* and *No. 7.*

PROBLEMS

1. Why is the temperature of oil or gas important when measuring the volume of oil or gas sold?

2. Define the following:
   BS&W
   gravity
   tank battery
   LACT unit
   Mcf
   minimum royalty
   MMBtu

3. Define the following:
    run ticket
    settlement statement
    thief
    gauging
    strapping

4. Mr. Davis owns the mineral rights in some land in Texas. He leases the land to Aggie Oil Company, reserving a $1/5$ royalty. During 2006, Aggie Oil Company makes the following assignments:
    a. To Mr. Jones, an ORI of $1/7$ of Aggie's interest.
    b. To Mr. Brown, a PPI of 30,000 barrels of oil to be paid out of $1/4$ of the WI's share of production (*i.e.*, Mr. Brown gets $1/4$ of this production until he receives a total of 30,000 barrels).
    c. To Mr. Smith, a joint working interest of $1/3$ after giving consideration to all the above assignments.

    Assuming production is 45,000 barrels, calculate each owner's share.

5. Blow Out Oil Company owns 100% of the WI in Lease A. Lease A is burdened with a $1/6$ royalty. During the month of June, 12,000 barrels of oil were produced and sold. Assume the selling price of the oil was $24/bbl and the production tax was 5%. Give the entry required to record the sale of the oil for each of the following:
    a. The purchaser assumes responsibility for distributing taxes and royalty income.
    b. Blow Out Oil Company assumes the responsibility.

6. Ebert Oil Company sold 1,000 Mcf of gas at $3.00/Mcf. The lease provides a $1/5$ RI. The WI owner receives 100 percent of the revenues (net of 5% severance tax) and then distributes the amount due to the RI owner. Give the entry by Ebert to record the sale of the gas.

7. Deep Hole Oil Company used 100 Mcf of gas obtained from Lease A and valued at $2.10/Mcf for gas injection on Lease B. Assume production taxes are 5% and the RI on Lease A is a $1/6$ RI.
    a. Give the entry necessary to record the transfer of the gas.
    b. Give the entry assuming 100 percent of the gas is recovered and sold at $2.10/Mcf.

8. Fortunate Oil Company sold or used the gas produced on Lease A during January as follows:
    a. 300 Mcf used as fuel to operate lease equipment.
    b. 800 Mcf sold to R Company at $3/Mcf.

Assume a $^1/_7$ RI and a production tax of 5%, and assume that the lease agreement has a free fuel clause, but production taxes still have to be paid according to state law. The selling price of gas is currently \$3/Mcf. Give the entries necessary to record the gas sold or used, assuming the operator distributes taxes and royalty.

9. Big John Oil Company purchased 200 barrels of oil from JD Operator. The gross value of the oil was \$5,000. The severance tax rate was 4 percent. Give the entry to record revenue for JD, assuming Big John disbursed the royalty and remitted all taxes and assuming a division order as follows:

Property No. 35	**Interest**
JD Company	0.875
Mabel Royalty Owner	0.125

10. Gusher Oil Company has a working interest in a property. In addition to the $^1/_5$ royalty, Gusher agreed to pay the royalty owner a minimum royalty of \$400 per month. Gas production on the lease began in the third month after the lease contract was signed. Total sales revenue during the third month and the next two months was \$6,000 each month. The severance tax rate was 10%. Give the revenue and minimum royalty entries for the first five months, assuming Gusher takes the responsibility of distributing taxes and royalty and assuming:
    a. The minimum royalty payments were not recoverable
    b. The minimum royalty payments were recoverable

11. Complete the run ticket on p. 341 and give the entry to record the sale of the oil at \$30/bbl assuming a severance tax rate of 5% and a $^1/_5$ RI. Use the tables given in the chapter. The tank number is 1, the observed temperature is 73° and the observed gravity is 33°. The first tank measurement at the beginning of the run is 14'4" at a temperature of 77° and the second tank measurement at the end of the run is 1'6" at 78°. The BS&W content is .002.

12. Information from a run ticket shows that 1,000 net barrels of oil with an API gravity of 36° were sold. The selling price is based on a contract price of \$22 per barrel, adjusted downward 4¢ for each degree of gravity less than 40. Compute the selling price for the 1,000 barrels.

COMPANY

ACCT NO

RUN TICKET NO

DIST NO

DATE

OPERATOR

LEASE NAME	LEASE NO.

WELL NO.		FT.	INCHES	TEMP	BBLS
	1ST MEAS.				
TANK NO.	2ND MEAS.				
	GRAVITY				
TANK SIZE	OBSERVED	TEMP.	TRUE		PRICE
BS &W	OFF←SEAL NO.'S→ON				AMOUNT

THIEFING	BEFORE	AFTER	CALCULATIONS
FT - IN BS & WATER			
REMARKS			

1ST MEAS	DATE	GAUGER
	TIME A.M P.M.	WELL OWNER'S WITNESS
2ND MEAS	DATE	GAUGER
	TIME A.M P.M.	WELL OWNER'S WITNESS

This ticket covers all claims for allowance. The oil represented by this ticket was received and run as property of

13. Determine the barrels of production allocated to each well given the following data:

Well	24-hr. Test, bbl	Days Produced	Total Production Sold
1	40	28	3,300 barrels—measured at
2	50	30	point of delivery (production
3	35	24	from 3 wells commingled)

Round the ratio to three decimal places and round barrels to the nearest whole barrel.

14. Cameron Oil Company produced 2,000 barrels of oil in June 2007. The expected selling price was \$20 per barrel. The purchaser pays the severance taxes and the royalty interest owner and remits the remainder to Cameron Oil. The royalty interest is $^1/_5$ and the severance tax rate is 10%.
    a. Prepare the entry for Cameron Oil Company and the RI owner, assuming the oil was sold in June 2007 at the expected selling price.
    b. Prepare entries for Cameron Oil Company assuming that revenue is recognized as produced based on the expected selling price and the oil produced in June was sold in July 2007 for \$22 per barrel.

15. Whitmire Oil Company owns a working interest in the Carpenter Lease in Texas. The lease is burdened with a $^3/_{16}$ royalty interest. During February, 3,000 bbl of oil are sold at \$22/bbl to a refinery owned 100 percent by Whitmire Oil. Assume the severance tax rate is 5% in Texas.

    Prepare entries for the producer and the refinery purchaser.

16. Cameron Oil Company operates Leases X and Y. Cameron Oil transfers 50 barrels of oil from Lease X to Lease Y to be used as fuel on Lease Y. The current spot oil price is \$22/bbl and the severance tax rate is 5%. Cameron owns 100% working interests in Lease X and Lease Y. Lease X has a $^1/_8$ royalty interest and Lease Y a $^3/_{16}$ royalty interest.

    Prepare an entry for the transfer of the oil.

17. Stephens Oil Company produces 2,000 bbl of oil in June that is sold in July. The posted field price and the actual selling price is \$22/bbl. The severance tax rate is 5%. The purchaser of the oil will pay the severance tax to the state and also will pay the royalty interest owner. The royalty interest is $^1/_8$.

    Prepare entries assuming Stephens Oil recognizes revenue (a) at time of sale, and (b) at time of production based on the posted field price.

18. Knight Oil Company has the following transactions in 2006.
    a. Minimum royalty payments of $200 per month are paid during the months of January through March. The minimum royalty payments are recoverable from future royalty payments.
    b. Production was sold in April 2006, and the royalty payable in April is $300.
    c. The royalty payment in May is $500.
    d. The well quit producing in June and the lease was abandoned.

    REQUIRED:
    1. Prepare entries for the above transactions.
    2. Prepare entries assuming the royalty payments are not recoverable.

19. Mr. Dube owns some mineral rights in Texas that he leases to Seagull Oil Company, reserving a $^1/_8$ royalty interest. During 2006, Seagull Oil made the following assignments:
    a. To Mr. Hall, an ORI of $^1/_6$.
    b. To Mr. Irwin, a PPI of 10,000 barrels of oil to be paid out of $^1/_5$ of the working interest's share of production.
    c. To Mr. Brown, a joint working interest of 40% after giving consideration to the above assignments.

    REQUIRED:
    a. Prepare the decimals to be used in the division order.
    b. Assuming production of 12,000 barrels of oil, calculate the number of barrels each owner would receive.

20. Joyner Oil Company sells 10,000 Mcf of gas at $1.50/Mcf. The lease provides for a $^1/_6$ RI, and the WI owner has distributed an ORI of $^1/_{10}$. The severance tax rate is 7%.

    Record the entries for the RI owner, ORI owner, and WI owner, assuming:
    a. The purchaser assumes responsibility for distributing taxes and royalty income.
    b. Joyner Oil assumes the responsibility for distributing taxes and royalty income.

21. Gusher Oil Company's production for Leases A and B is gathered into a common system and sold. Total Sales for the month are 6,562 barrels. Assume the following data for Leases A and B:

	Well	**24 hour Test, bbl**	**Days Produced**
Lease A:	1	40	24
	2	30	30
	3	50	28
Lease B:			
	1	50	30
	2	80	24

Measured production from Lease A is 3,300 barrels and 3,500 barrels from Lease B.

REQUIRED:

a. Allocate production to Leases A and B.

b. Allocate amounts per lease determined in part *a* to the wells. Round ratios to four decimal places.

22. Dube Company's production from each well on Lease C and Lease D is estimated based on a 24-hour test. Oil produced from each well on each lease is commingled and measured before leaving each lease. The oil produced from each lease is then commingled and delivered to a central tank battery. Assume the following data for Lease C:

Well	24 hour Test, bbl	Days Produced
1	30	28
2	40	30

Assume the following data for Lease D:

Well	24-hr Test, bbl	Days Produced
1	20	27
2	30	26
3	35	30

Measured production from Lease C is 2,000 barrels and 2,280 barrels from Lease D. After treatment at the central tank battery, 4,100 barrels are sold.

REQUIRED: Allocate the 4,100 barrels sold to each lease and then to each well in a two-stage allocation. Round ratios to three decimal places.

23. Cameron Company and Adams Company own 70% and 30% of the WI of the Dowling Field, respectively. There is a $\frac{1}{8}$ royalty owner. The $\frac{1}{8}$ royalty is shared proportionally by Cameron and Adams. Cameron and Adams agree that Cameron's purchaser will take March's gas production and Adam's purchaser will take April's production. Gas allocations will be equalized in May. Ignore severance taxes.

Production and gas prices were as follows:

	Production	Price
March	300,000 Mcf	$2.00/Mcf
April	200,000 Mcf	2.00/Mcf
May	220,000 Mcf	1.80/Mcf

Each working interest owner receives payment only for gas delivered to his purchaser.

REQUIRED:

a. Prepare a gas balance report.
b. Prepare entries for the three-month period for both parties assuming both companies use the entitlement method for both revenue and royalty.

24. Ramsey Company has a 100% WI in some property in Texas. The property is burdened with a $^1/_8$ royalty. Ramsey produces and sells a total of 130,000 Mcf of gas from the property during July. Of the 130,000 Mcf, 50,000 Mcf of gas is sold to a pipeline for $2.00/Mcf. Ramsey sells the remaining 80,000 Mcf of gas to the local gas company for $2.00/Mcf. The 80,000 Mcf of gas, which must be transported to the local gas company, will be transported by Isaac Pipeline Company. Transportation charges are $0.26 per Mcf.

    REQUIRED: Ignore severance taxes. Prepare entries to record the sales by Ramsey Company.

25. Two leases in far West Texas are being combined to form the West End Unit.

    Lease 1

WI: 100% Glenn Company	
RI: $^1/_7$ Mike Merino	
$^1/_7$ John Hill	
Acreage contribute	300 acres
Estimated net recoverable bbls:	150,000 bbls

    Lease 2

WI: 75% Core Company	
25% Format Company	
RI: $^1/_8$ E.B. Lomax	
Acreage contributed:	200 acres
Estimated net recoverable bbls:	250,000 bbls

    REQUIRED:

    a. Determine the participation factors for each party assuming the participation factors are based on the acreage contributed.
    b. Determine the participation factors for each party assuming the participation factors are based on the net recoverable barrels contributed.
    c. Determine the participation factors for each party assuming the participation factors are based on the acreage contributed weighted 40% and the net recoverable barrels weighted 60%.

26. Fossil Oil Company has production on a lease in Louisiana with the following ownership interest:

RI—$^1/_5$ RI
ORI—$^1/_{16}$ of $^4/_5$ of gross production
Joint WI: Lomax Company (40%) and Fossil Company (60%)

During April, 10,000 barrels of oil (after correction for temperature, gravity, and BS&W) were produced and sold. Assume the price for oil is $20.00/bbl and the severance tax rate in Louisiana is 5%.

REQUIRED:

a. Prepare the journal entry for Fossil Oil to record the sale of the oil given the purchaser assumes the responsibility of distributing severance taxes and royalty income.
b. Prepare the journal entry for Fossil Oil to record the sale of the oil given Fossil Company assumes the responsibility and receives 100% of the proceeds.
c. Prepare the journal entry for the royalty interest owner to record the sale of the oil.
d. Prepare the journal entry for the overriding royalty interest owner to record the sale of the oil.
e. Prepare the journal entry for Lomax to record the sale of the oil.

27. Churchwell Company, a successful-efforts company located in California, sold 2,500 Mcf of gas with a heat content of 1.030 MMBtu/Mcf. The selling price of the gas was $2.00 per MMBtu.

REQUIRED:

a. Determine the MMBtu's for the gas sold.
b. Determine the total sales price.
c. Determine the unit sales price per Mcf.

28. During September, 2010, Fortunate Oil Company sold 2,000 Mcf of gas at 14.65 psia with a heat content of 1.030 MMBtu/Mcf at 14.73 psia. The selling price of the gas was $2.20 per MMBtu.

REQUIRED:

a. Convert Mcf to a standard pressure of 14.73 psia and determine the MMBtu's for the gas sold.
b. Convert MMBtu/Mcf to a standard pressure of 14.65 psia and determine the MMBtu for the gas sold.
c. Determine the sales price of the gas sold.

29. Fossil Field is jointly owned by Allen Company (70% WI) who acts as field operator, and Garza Company (30% WI). There is a $^1/_6$ royalty. The $^1/_6$ royalty is shared proportionally by Allen and Garza. The two working interest owners have agreed that Allen's purchaser will take gas produced in July and Garza's purchaser will take gas produced in

August. Gas allocations will be equalized in September. Assume each working interest owner receives payment only for gas delivered to his purchaser(s). Ignore severance taxes.

Production and gas prices were as follows:

	Production	**Price**
July	100,000 Mcf	$2.00/Mcf
August	120,000 Mcf	2.00/Mcf
September	190,000 Mcf	2.00/Mcf

REQUIRED:

a. Prepare the Gas Balance Report for Allen Company to summarize the production deliveries and equalization of gas for July through September.
b. Prepare the journal entries for each company during the three-month period assuming that both companies use the sales method for both revenue and royalty.
c. Prepare the journal entries for each company during the three-month period assuming that both companies use the entitlement method for both revenue and royalty.

30. Sherwood Field, located in East Texas, is jointly owned by Kelly Company (60%) and Tiger Company (40%). Kelly, who is the operator, estimates that gas production during July will be 40,000 Mcf. Kelly Company makes confirmed nominations of 30,000 Mcf and Tiger Company makes confirmed nominations of 10,000 Mcf to a pipeline company. Actual July production totals 50,000 Mcf. Assume that the appropriate price is $1.80 per Mcf.

    REQUIRED: Determine the number of Mcf of actual production to be allocated to each party and each producer's resulting over- and underdeliveries to the pipeline using:

    a. the confirmed nominations method, and
    b. the entitlements method.

chapter ten

Basic Oil & Gas Tax Accounting

Tax accounting for oil and gas producing activities differs in many respects from financial accounting. The purpose of tax accounting is to gather the necessary information for the efficient preparation of income tax returns according to rules and regulations in the Internal Revenue Code and Regulations. This chapter introduces basic oil and gas tax accounting, which is applicable to all entities, large or small, integrated or independent. Exceptions relating to independent producers, such as percentage depletion, and differences applying only to integrated companies, such as 30% holdback of IDC, are noted and examined. Transactions that are discussed include those involving nondrilling costs, acquisition costs, drilling costs, completion costs, operational costs, and the disposition of capitalized costs.

This chapter is intended to provide only a basic overview of taxation for oil and gas exploration and production activities. It is not intended to provide the reader with an authoritative discussion of tax issues.

Lessee's Transactions

The lessee's transactions are discussed in the following sections. The lessor's transactions are discussed later in the chapter.

Nondrilling costs

This section discusses the following nondrilling costs: geological and geophysical (G&G) costs, delay rentals, ad valorem taxes, legal costs, dry-hole contributions, bottom-hole contributions, and shut-in royalties.

Generally, the first costs incurred in an oil and gas operation are G&G costs. G&G costs are considered inherently capital in nature and must be added to the tax basis of property acquired or retained. As a first step, a reconnaissance G&G survey is often made on a broad area called a **project area**. The survey is done to locate portions of the project area having the greatest possibility of containing oil and gas. These promising portions are labeled areas of interest. **Interest areas** are defined as noncontiguous project portions that warrant detailed G&G (primarily subsurface) studies. (Noncontiguous areas are those areas that are not adjoining or in direct contact with each other except when sharing a common corner.)

All costs of the project area are initially allocated to a single project. If **interest areas** warranting further evaluation are identified, the initial reconnaissance costs of the project area are allocated equally to the areas of interest regardless of their relative sizes. If only one area of interest is identified, all reconnaissance costs are allocated to that area of interest. If no areas of interest are identified, the reconnaissance costs are deductible.

After areas of interest are identified, any costs incurred to further evaluate an area are capitalized to that area of interest. If further evaluation results in property interests being acquired within an area of interest, the allocated reconnaissance costs and the cost of detail work performed within the area of interest are allocated to any property taken on a comparative acreage basis. There is no deduction if only a part of an area of interest is acquired. If the detailed survey does not result in property acquisition, the accumulated costs may be written off in the year in which the company decides that it is completely and permanently not interested in pursuing property acquisition in that area.

EXAMPLE

G&G Costs

Lucky Company incurs G&G costs of $6,000 for a reconnaissance survey of project area 1102. The survey indicates three areas of interest. Detailed surveys are made on the areas of interest at the following costs:

Interest Area A	640 acres	$25,000
Interest Area B	335 acres	20,000
Interest Area C	640 acres	35,000

After evaluating the surveys, Lucky Company acquired the following acreage:

Interest Area A	One lease of 320 acres
Interest Area B	One lease of 160 acres
Interest Area C	No leases acquired

Allocation of G&G Cost in Determination of Tax Basis Per Acre

	Area A	Area B	Area C
Allocation of reconnaissance costs ($6,000/3)	$ 2,000	$ 2,000	$ 2,000
Detailed survey costs	25,000	20,000	35,000
Tax basis of area of interest	$27,000	$22,000	$37,000
Acreage acquired (acres)	320	160	0
Tax loss—immediately deductible			$37,000
Tax basis per property	27,000	22,000	NA
Tax basis per acre	$84.375	$137.50	NA

Assume *instead* that the 320 acres acquired in Interest Area A included the following leases:

Lease 1	120 acres	100% WI
Lease 2	120 acres	25% WI
Lease 3	80 acres	25% WI

Net Acreage Acquired in the Interest Area

Lease 1	120 acres x 1	=	120
Lease 2	120 acres x $^1/_4$	=	30
Lease 3	80 acres x $^1/_4$	=	20
Total net acreage			170

Tax Basis Per Acre for the Leases Acquired in Area A

$27,000/170 acres = $158.82

Allocation of Costs of Interest Area A

Lease 1	120 acres x $158.82	=	$19,059
Lease 2	30 acres x $158.82	=	4,765
Lease 3	20 acres x $158.82	=	3,176
Total			$27,000

Shooting rights, which are the rights to access a property, should be accounted for in the same way as G&G costs. Consequently, the cost of shooting rights should ultimately be capitalized if a lease is obtained. If no lease is obtained, these rights become deductible as indicated above.

A typical lease contract requires the lessee to commence drilling operations within one year. If the lessee does not commence such operations within one year, a delay rental payment must be made. The tax law allows the lessee to either capitalize delay rentals or deduct them in the period incurred. The election must be made on a year-by-year, lease-by-lease basis. Except for companies with significant net operating losses, delay rental payments are almost universally expensed as incurred. For the lessor, any delay rentals payments received must be treated as ordinary income not subject to depletion.

Ad valorem taxes paid by the lessee on his working interest in an unproved property are ordinary and necessary business expenses, and are deductible in the period incurred. Ad valorem taxes paid by the lessee for the lessor can generally be recovered from future royalty payments to the lessor, and are neither an expense nor revenue to the lessee. If the ad valorem tax amount is recoverable, a receivable is established for the payment.

Legal costs incurred to defend or examine title to a property are considered capital expenditures and should be capitalized as part of the basis of the underlying property. Maintenance costs of lease and land records, however, appear to be ordinary and necessary business expenses.

Bottom-hole contributions for successful wells are considered to be acquisition costs or retention costs and are capitalized as incurred. These costs are deemed to be payment for information relating to the acquisition of adjoining acreage or retention of an interest, and are capitalized and added to the basis of the underlying property. Bottom-hole contributions for dry holes and dry-hole contributions also should be capitalized.

Shut-in royalty payments may have three basic characteristics:

- the payments are not recoverable from future production
- failure to make payments terminates the lease
- the payments are recoverable from future production

When shut-in payments have one of the first two characteristics above, the payments are considered to be delay rentals and are accounted for as described previously. Those shut-in payments recoverable from future production are treated as a receivable until recovered from production.

Acquisition costs

Acquisition costs paid by the lessee should be capitalized and added to the basis of the property acquired. The major acquisition cost is the lease bonus paid by the lessee to the lessor. Other acquisition costs that should be capitalized include broker fees, legal fees, filing fees, and the cost of title examinations. Acquisition costs are recovered through cost depletion or percentage depletion. Certain other costs, although not typical acquisition costs, are treated as acquisition costs and are capitalized and depleted along with normal acquisition costs. These costs include capitalized G&G, shooting rights, test-well contributions, and capitalized delay rental payments, if any. The lessor treats any bonus received as ordinary income.

EXAMPLE

Acquisition Costs

Lucky Company incurred $10,000 of G&G reconnaissance costs in North Texas that resulted in one area of interest being identified. Detailed survey costs totaling $15,000 were subsequently incurred in the area resulting in two leases being acquired. Lucky paid a lease bonus of $50 an acre to Sam Jones in return for a 150 acre lease and $45 an acre to Manuel Ramirez for a 250 acre lease. Legal fees, landman commissions, filing fees, and title examination fees were incurred in the amount of $2,000 on the Jones lease and $1,500 on the Ramirez lease.

The Tax Basis of Each Lease

	Jones Lease	Ramirez Lease
Capitalized G&G = $10,000 + $15,000		
Acre-by-acre allocation:		
$25,000/400 acres = $62.50/acre		
$62.50 x 150	$ 9,375	
$62.50 x 250		$15,625
Lease bonus:		
$50 x 150	7,500	
$45 x 250		11,250
Other fees	2,000	1,500
Tax basis	$18,875	$28,375

When a producing property is subleased (*i.e.*, the current WI owner conveys the working interest in the property to another party and retains a nonoperating interest), the **grantee** or **sublessee** (*i.e.*, the new WI owner) of the producing property must apportion the consideration paid between the leasehold and equipment. The equipment received should be allocated its fair market value, with the remainder being allocated to the leasehold.

Producing property that is subleased generally provides for a retained overriding royalty interest by the **grantor** or **lessee** (*i.e.*, the original WI owner). Therefore, when a producing property is subleased to another party by the lessee, the consideration received by the grantor (lessee) is considered as payment for equipment. When the cash consideration received is greater than the depreciable basis of the equipment, the excess is considered as a lease bonus received, *i.e.*, treated as ordinary income. When the cash consideration received is less than the basis of the equipment, the excess equipment basis should be transferred to the depletable basis of the interest retained and depleted along with that interest.

EXAMPLE

Sublease

Lucky Company owns a producing lease with leasehold costs of $5,000, accumulated depletion of $1,000, well equipment of $300,000, and accumulated depreciation of $75,000. The lease is subleased to Sam Jones for $400,000, and Lucky retains a 3/8 ORI in the property. At the date of sublease, the equipment had a fair market value of $325,000.

Tax Treatment of the Sublease by Lucky Company

Tax basis of property:	
Cost	$ 5,000
Less: Accumulated depletion	(1,000)
Basis	$ 4,000

Tax basis of equipment:	
Cost	$300,000
Less: Accumulated depreciation	(75,000)
Basis	$225,000

Since the cash received ($400,000) is greater than the depreciable basis of the equipment ($225,000), the excess ($175,000) is treated as ordinary income as shown below.

Payment for equipment:	
Sales price	$400,000
Less: Basis	(225,000)
Taxable income	$175,000

Lucky's basis of the proved property (versus equipment) is transferred to the basis of the retained ORI.

Basis of retained ORI	$ 4,000

Tax Treatment of the Sublease by Sam Jones

The FMV of the equipment $325,000 is allocated to equipment and the remainder, $75,000, to the leasehold.

Tax basis of equipment (FMV)	$ 325,000
Tax basis of leasehold	75,000
Total payment	$ 400,000

Drilling operations

As will be seen, the tax treatment of costs related to drilling operations depends upon whether the producer is an independent producer or an integrated producer. An **integrated producer** is an entity that has refining or retailing activities in addition to producing activities. In general, a refiner engages in refining when the refinery runs exceed 50,000 barrels on any one day during the taxable year. A retailer sells more than $5,000,000 annually of oil, gas, or any resulting product through retail outlets connected to or controlled by the taxpayer. For more information, see Code Sec. 613A(d).

During drilling operations, the most important consideration is the proper separation of drilling costs between (1) intangible drilling and development costs (IDC), and (2) lease and well equipment. This importance is due to the ability, if the proper election is made, to expense as incurred the majority of IDC for an integrated producer or 100% of IDC for an independent producer. Lease and well equipment, in contrast, must be capitalized and recovered through depreciation. Two separate elections may be involved in expensing or capitalizing IDC. If an election is made in the first tax return filed by the taxpayer to expense IDC, then all IDC incurred by the taxpayer may be expensed in the year incurred, except for integrated producers who must capitalize 30% of IDC incurred on productive wells. Even if a taxpayer fails to elect to expense IDC, a second election may be made to currently expense IDC of dry holes.

The 30% capitalized by integrated producers must be amortized over 60 months, beginning with the month in which the costs are paid or incurred. The total amount amortized for the year is deducted currently by an integrated producer in addition to the remaining 70% of the related IDC expenditures. Note that 30% of the IDC for a well in progress must also be capitalized and amortized. If the well is determined to be dry, any remaining unamortized IDC would be expensed in the year in which the well is plugged and abandoned.

Equipment costs, which must be capitalized, generally include all tangible property costs incurred both before and after the Christmas tree, plus intangible costs incurred after the Christmas tree. Intangible costs incurred prior to and including the installation of the Christmas tree are IDC amounts.

IDC is defined as any expenditure that in itself does not have a salvage value and is *incident to and necessary for the drilling of wells and the preparation of wells for the production of oil and gas (U.S. Treas. Reg. Section 1.612-4[a])*. Costs incurred relating to the following activities are considered to be IDC:

a. Agreements and negotiations in obtaining an operator for the well
b. Agreements and negotiations with drilling contractors for bids on drilling
c. Survey and seismic work for locating a wellsite
d. Constructing a road to the drilling location
e. Dirt work on location—cellar, pits, and drilling pack
f. Transporting the rig to location and rig-up costs
g. Water, fuel, and other items necessary for drilling wells
h. Setting deadmen (anchors in the ground used to stabilize the drilling rig)
i. Technical services rendered during the drilling activities by engineers, geologists, and fluid technicians
j. Logging and drillstem test services
k. Swabbing, fracturing, and acidizing
l. Rental equipment for oil storage during testing
m. Removing rig from location, trucking, bulldozers, and labor
n. Dirt work for cleanup of drillsite
o. Cementing and installing surface casing
p. Transportation of casing and tubing from supply point
q. Perforating casing and electrical logging
r. Saltwater, freshwater, and gas injection wells drilled solely for pressurization or flooding of producing zone (intangible costs only)
s. Water supply wells if the water is to be used for drilling or secondary recovery (intangible costs only)

As shown by this list, IDC is incurred for *wages, fuel, repairs, hauling, supplies, etc.,* in preparing to drill wells (clearing ground, making roads, digging mud pits, etc.); in drilling, shooting, and cleaning wells; in constructing derricks; and in completing the well by installing casing and the Christmas tree.

Equipment costs

Lease and well equipment installed or constructed in connection with the drilling, completion, and production of a well must be capitalized. These costs normally are recovered through depreciation. Some typical lease and well equipment costs follow:

a. Surface casing— even though permanently cemented and unsalvageable
b. Well casing
c. Tubing

d. Transportation of casing and tubing from manufacturer to supply point
e. Stabilizers, guide shoes, centralizers, and other downhole equipment
f. Wellhead—Christmas tree
g. Saltwater disposal equipment and necessary pipelines, including cost of drilling disposal wells
h. Pump jack, treaters, separators
i. Recycling equipment, including necessary flow lines
j. Dirt-moving necessary for the tank battery and operation roads
k. Digging, refilling, and backhoe work for installing flow lines from the well to the tank battery
l. Installation and labor costs for the tank battery, flow lines, pump jacks, separators, and similar items
m. Construction of a turnaround pad at the tank battery, with additional overflow pits

The following example illustrates the tax treatment of IDC and equipment costs.

EXAMPLE

IDC versus Equipment

Lucky Company incurred the following costs in connection with the drilling and completion of a well during the first three months of 2008. The company had elected to expense IDC as incurred. The IDC costs are indicated by an X in the column headed IDC, and the equipment costs are indicated by an X in the column headed Eq. Lucky's IDC deduction and the basis of Lucky's equipment are shown below assuming:

a. Lucky is an independent producer and
b. Lucky is an integrated company.

IDC	Eq.		
X		Legal and negotiating costs for hiring an operator and drilling contractor	$ 10,000
X		G&G costs in selection of the wellsite	15,000
X		Construction of roads to the drilling site ...	12,000
X		Water well and fuel costs during drilling ...	9,000
X		Drilling contractor costs on a footage and day-rate basis	316,000
X		Costs incurred in connection with technical services (engineers, geologists, logging, swabbing, testing)	75,000

IDC	Eq.		
X		Dirt work costs for cleaning up the drillsite	$ 2,000
X		Cement for casing	2,000
	X	Well casing	20,000
X		Transportation of the well casing from the supply point	500
	X	Equipment necessary for completing the well (including wellhead equipment, treaters, separators, flow lines)	200,000
X		Labor costs for installing the casing and wellhead equipment	4,000
	X	Labor to install the treaters, separators, and flow lines	12,000

a. If Lucky is an independent producer:

The deduction for IDC and the basis of lease and well equipment are equal to the sum of their respective costs (as itemized above).

IDC deduction =

$10,000 + $15,000 + $12,000 + $9,000 + $316,000
+ $75,000 + $2,000 + $2,000 + $500 + $4,000 = $445,500

Basis of lease and well equipment =

$20,000 + $200,000 + $12,000 = $232,000

b. If Lucky Company is an integrated producer:

IDC deduction:

70% of current IDC expenditures (0.70 x $445,500) = $311,850

Capitalized IDC:

0.30 x $445,500 = $133,650 deductible over the next 60 months

Basis of lease and well equipment:

$20,000 + $200,000 + $12,000 = $232,000

The capitalized IDC is amortized on a straight-line basis over 60 months at a rate of ($133,650/60) = $2,227.50 per month beginning with the month in which the IDC was incurred or paid. In the event that the well is determined to be dry, any remaining unamortized IDC would be expensed immediately.

The following is a more detailed example of the required treatment for IDC incurred by an integrated producer.

EXAMPLE

Capitalized IDC of Integrated Producer

Lucky Company, an integrated producer, incurs the following IDC costs for the years indicated. All IDC costs marked with an asterisk (*) relate to dry-hole IDC. Assume no IDC was incurred prior to 2006.

Date Incurred	Amount	Date Incurred	Amount	Date Incurred	Amount
5/11/06	$100,000*	3/14/07	$300,000	6/12/08	$180,000*
9/11/06	200,000*	7/22/07	360,000	10/14/08	400,000

Amount of IDC Deductible in each of the Following Years

2006

IDC incurred in 2006:	$100,000 x 1.00 =	$100,000
	$200,000 x 1.00 =	200,000
		$300,000

2007

IDC incurred in 2007:	$300,000 x 0.70 =	$210,000
	$300,000 x 0.30 x $^{10}/_{60}$ =	15,000
	$360,000 x 0.70 =	252,000
	$360,000 x 0.30 x $^{6}/_{60}$ =	10,800
IDC incurred in 2006:	100% expensed in 2006 =	0
		$487,800

2008

IDC incurred in 2008:	$180,000 x 1.00 =	$180,000
	$400,000 x 0.70 =	280,000
	$400,000 x 0.30 x $^{3}/_{60}$ =	6,000
IDC incurred in 2007:	$300,000 x 0.30 x $^{12}/_{60}$ =	18,000
	$360,000 x 0.30 x $^{12}/_{60}$ =	21,600
IDC incurred in 2006:	100% expensed in 2006 =	0
		$505,600

Assume 2008 was the last year in which IDC was incurred. The amount of IDC deductible in 2013 would be computed as follows:

2013

All IDC incurred prior to 2008 would be fully expensed by 2013
IDC incurred in 2008:

100% of \$180,000 expensed in 2008
\$400,000 x 0.70 expensed in 2008
\$400,000 x 0.30 x $^{9}/_{60}$ = \$18,000 expensed in 2013

The number of months of amortization per year for the capitalized IDC relating to the \$400,000 can be diagramed as follows:

Year:	2008	2009	2010	2011	2012	2013	Total months amortized
# of Months:	3	12	12	12	12	9	60

Production operations

Revenue arising from the production and sale of oil and gas products is ordinary income. The portion of revenue from production paid to the royalty interest (RI) owner is sometimes referred to as a royalty expense. This payment to the RI owner is actually a division of revenue and not included in gross income of the working interest owner. The royalty is taxable income to the royalty owner.

Ordinary and necessary business expenses incurred in lifting the oil and gas and treating it for sale are deductible as necessary and reasonable operating expenditures. Examples of these costs are pumpers' and gaugers' salaries, the cost of travel to and from the lease, repairs, minor workovers, fuel costs, maintenance costs, depreciation, and depletion.

Depreciation and depletion, which are two important expenses connected with lifting and treating the oil and gas, are discussed in the following section.

Cost disposition

Leasehold costs are recovered through cost depletion or possibly through percentage depletion. (Percentage depletion, which is currently available only to independent producers and royalty owners and only if larger than cost depletion, is discussed later in this chapter.) Leasehold costs include such costs as lease bonuses, title examinations, and legal fees in drafting contracts. Other capitalized costs that are treated as leasehold costs and depleted include G&G costs, test well contributions, elected capitalized IDC (versus the 30% IDC required to be capitalized on productive wells by integrated producers), and capitalized delay rentals, if any. Cost depletion is based on activities connected with a particular property. Generally, a property is defined as either a single mineral interest or multiple mineral interests that are obtained at the same time and under the same lease contract. Operating and nonoperating interests are classified as separate property units.

Cost depletion is computed as follows:

$$B \times \frac{Y}{Z + Y}$$

where:

B = basis at end of period (unrecovered costs at year-end)
Z = reserves at year-end
Y = barrels or Mcf sold during year

Reserves at year-end for the property unit are generally proved reserves rather than proved developed reserves. Proved reserves are used because the costs being depleted usually represent leasehold costs. (The IRS is currently contending that the reserves used in computing depletion should include not only proved reserves but also probable reserves.) Note that although the cost depletion formula is very similar to the unit-of-production formula used for SE and FC, barrels or Mcf sold are used in the tax formula rather than barrels or Mcf produced.

Depreciable property placed in service after December 31, 1986, is recovered using the depreciation methods delineated in the Tax Reform Act of 1986 and the 1993 Revenue Reconcilation Act. Depreciable property consists of tangible assets, including the intangible costs of installing such assets, if after the Christmas tree. For the assets commonly used in oil and gas exploration activities, the law currently provides for a 7-year life for tangible personal property and a 39-year life for nonresidential real property.

Lease and well equipment of a producer, which is typically seven-year property, includes the following: exploration equipment, offshore drilling equipment and platforms, facilities to store oil and gas, etc. Lease and well equipment is depreciated over a seven-year period initially using double declining (200%) balance and later switching to straight-line at such time as to maximize the deduction. A half-year convention modified by a mid-quarter convention is specified. Under the half-year convention, assets are considered to have been placed in service at mid-year regardless of when the particular assets were actually placed in service. However, if more than 40% of the assets were placed in service during the last quarter of the year, the mid-quarter convention applies. Under the mid-quarter convention, assets purchased during a quarter are considered to have been placed in service at mid-quarter, regardless of when the assets were actually placed in service during that quarter. As a result of the half-year convention, the recovery life in effect spans eight years for lease and well equipment. Lease and well equipment of a producer placed in service in a foreign location is typically depreciated over 12 to 14 years.

EXAMPLE

Depreciation of Seven-Year Property

On May 2, 2007, Lucky Company placed in service lease and well equipment costing $20,000.

Year	Depreciation	Year	Depreciation
2007	$2,857*	2011	$1,785 SL***
2008	4,898**	2012	1,785 SL
2009	3,499	2013	1,785 SL
2010	2,499	2014	892 SL

* ($20,000 x 1/7 x 2)/2

** ($20,000 – $2,857) x 1/7 x 2

*** $6,247/3.5 years

The taxpayer may elect to depreciate a class of property using straight-line depreciation. This election is irrevocable and applies to all property in that class placed in service during that taxable year. The taxpayer may also elect to exclude a property from the specified IRS depreciation provisions and depreciate the property using the unit-of-production method if the taxpayer can determine the oil and gas reserves applicable to the property.

Nonresidential real property, *i.e.*, commercial real estate property, includes those equipment items permanently attached to the land or having the characteristic of a building, such as a gas processing plant. Under the 1993 Revenue Reconciliation Act, nonresidential real property must be depreciated over 39 years using the mid-month convention. Under the mid-month convention, property placed in service (or disposed of) during any month is treated as being placed in service (or disposed of) on the mid-point of the month.

Salvage value is ignored for assets being depreciated using both double declining balance and straight-line depreciation. The following example illustrates depreciation for both 7-year and 39-year property and depletion.

EXAMPLE

Depreciation and Depletion

Lucky Company, an independent producer, owns the WI in a lease in Nueces County, Texas. The following information concerning the lease is for the year ended December 31, 2007. Assume Lucky's cost depletion is larger than percentage depletion and therefore Lucky's depletion expense is based on cost depletion. Also assume IDC was expensed in the year incurred. All reserve, production, and sales data apply only to Lucky Company.

Well equipment cost (end of second full year of life).....	$100,000
Leasehold cost and capitalized G&G	18,000
Capitalized test-well contribution	2,000
Building (18-year life, placed in service 1/1/06)	135,000
Accumulated depreciation, well equipment (1/1/07).....	14,286
Accumulated depreciation, building (1/1/07)	3,317
Accumulated depletion (1/1/07)	1,200
Percentage depletion for 2007 (assumed)	1,600
Proved reserves 12/31/07..........................	36,000 bbl
Production for the year	5,000 bbl
Sales for the year...............................	4,000 bbl

Computation of Depreciation for 2007

Well equipment: Year 1: $(\$100{,}000 \times {}^{2}/_{7})/2 = \$14{,}286$

Year 2: $(\$100{,}000 - \$14{,}286) \times {}^{2}/_{7} = \underline{\underline{\$24{,}490}}$

Building: Year 1: $\$135{,}000/39 \times {}^{11.5}/_{12} = \$3{,}317$ first year*

Year 2: $\$135{,}000/39 = \underline{\underline{\$3{,}462}}$ second year

* Since the building was placed in service in January 2006, and given that the mid-month convention applies, depreciation for only 11.5 months out of 12 months would be recognized in the first year.

Computation of Cost Depletion for 2007

$$(\$18{,}000 + \$2{,}000 - \$1{,}200) \times \frac{4{,}000}{36{,}000 + 4{,}000} = \underline{\underline{\$1{,}880}}$$

Losses from unproductive property

For the lessee, losses from unproductive property may be taken for tax purposes in the following situations:

- Abandonment of unproved property
- Abandonment of wells and related equipment and facilities when the well or reservoir is depleted but the lease is not abandoned
- Abandonment of leasehold, wells, and related equipment and facilities when the reservoir is depleted and the lease is abandoned
- Drilling exploratory or developmental dry holes

In general, losses may be deducted in the year in which the property is deemed worthless. Property is deemed worthless if the title is relinquished or if the property is abandoned. Abandonment or title relinquishment is considered to be a closed and completed transaction, thereby proving worthlessness.

Determination of worthlessness may also be made by examining identifiable events. The most common identifiable events that indicate worthlessness are the following:

- Depletion of oil and gas
- Drilling dry hole(s) on leased property
- Drilling dry hole(s) on adjacent property

Once a property is deemed worthless, the deductible amount is determined as follows:

- Abandonment of unproved property: The deductible loss amount is the acquisition cost of the property, including other capitalized costs such as G&G costs and test-well contributions.
- Abandonment of wells and related equipment and facilities when a well or reservoir is depleted but the lease is not abandoned: The amount deductible is the undepreciated value of the equipment (*i.e.,* book value), plus the unamortized portion of any capitalized IDC, less salvage value. However, if the company uses a system whereby the depreciable assets are combined into one main account referred to as mass asset accounting, no deduction is usually allowed for the abandonment.
- Abandonment of leasehold, wells, and related equipment and facilities when the reservoir is depleted and the lease is abandoned: The deductible amount is the total unrecovered cost of the equipment, leasehold, and IDC, less salvage value.
- Drilling exploratory and developmental dry holes: The deductible amount for an independent producer is the total of IDC and equipment, less salvage value. The deductible amount for an integrated producer is also the total of IDC and equipment, less salvage value, without a reduction for the 30% IDC holdback.

The following two examples present abandonment transactions for an independent producer and an integrated producer.

EXAMPLE

Independent Producer Abandonment

Transactions for the Lucky Company, an independent producer, in 2009 follow:

a. Abandoned unproved Lease A—acquisition costs totaling $4,000.
b. Abandoned Wells 1 and 2 on the Jones lease. The Jones lease was not abandoned because Well 3 is producing in another formation.

Costs Incurred on the Jones Lease

	Jones Lease	Well 1	Well 2	Well 3
Acquisition costs	$5,000			
IDC (deducted in year drilled)		$320,000	$400,000	$380,000
Equipment costs		160,000	180,000	200,000
Accumulated depletion	4,000			
Accumulated depreciation		120,000	150,000	90,000

The salvageable equipment located on Wells 1 and 2 was transferred to a warehouse owned by Lucky Company. The FMV of the transferred equipment was $50,000. Assume each well was depreciated individually.

c. Use the same facts as item b, except that Well 3 and the lease were also abandoned.
d. A dry hole costing $275,000 was drilled on unproved Lease B. The IDC cost was $240,000 and equipment was $35,000. (No equipment was salvaged.)

Lucky Company's tax losses for each of these situations are determined below.

Tax Losses

a. Lease A—abandonment loss		$ 4,000
Remaining basis of Lease A		$ 0
b. Cost of equipment on Well 1	$ 160,000	
Less: Accumulated depreciation	(120,000)	
Equipment Basis Well 1		$ 40,000

Cost of equipment on Well 2	180,000	
Less: Accumulated depreciation	(150,000)	
Equipment Basis Well 2		30,000
Less: Salvage value of equipment		(50,000)
Abandonment loss on well equipment		$ 20,000

c. Cost of equipment on Well 1	$ 160,000	
Less: Accumulated depreciation	(120,000)	
Equipment Basis Well 1		$ 40,000
Cost of equipment on Well 2	180,000	
Less: Accumulated depreciation	(150,000)	
Equipment Basis Well 2		30,000
Cost of equipment on Well 3	200,000	
Less: Accumulated depreciation	(90,000)	
Equipment Basis Well 3		110,000
Less: Salvage value of equipment		(50,000)
Abandonment loss on well equipment		130,000
Leasehold	5,000	
Less: Accumulated depletion	(4,000)	
Abandonment loss on leasehold		1,000
Total abandonment loss		$131,000

d. IDC (deductible as drilled)	$ 240,000	
Non-salvageable equipment	35,000	
Total tax loss on dry hole (assuming the lease itself is retained)		$275,000

EXAMPLE

Integrated Producer Abandonment

Transactions for the Lucky Company, an integrated producer, in 2009 are as follows:

a. Abandoned unproved Lease A—acquisition costs totaling $4,000.
b. Abandoned Wells 1 and 2 on the Jones lease. The Jones lease was not abandoned because Well 3 is producing in another formation.

Costs Incurred on the Jones Lease

	Jones Lease	Well 1	Well 2	Well 3
Acquisition costs	$5,000			
IDC (deducted in year drilled)		$320,000	$400,000	$380,000
Capitalized IDC (30%)		137,000	171,000	163,000
Equipment costs		160,000	180,000	200,000
Accumulated depletion	4,000			
Accumulated depreciation		120,000	150,000	90,000
Accumulated IDC amortization (assumed)		137,000	57,000	41,000

The salvageable equipment located on Wells 1 and 2 was transferred to a warehouse owned by Lucky Company. The FMV of the transferred equipment was $50,000. Assume each well was depreciated individually.

c. Use the same facts as in item b, except that Well 3 and the lease were also abandoned.
d. A dry hole costing $275,000 was drilled on unproved Lease B. The IDC cost was $240,000 and equipment was $35,000. (No equipment was salvaged.)

Lucky Company's tax losses for each of these situations are determined below:

Tax Losses

a. Lease A—abandonment loss		$ 4,000
Remaining basis of Lease A		$ 0
b. Cost of equipment on Well 1	$ 160,000	
Less: Accumulated depreciation	(120,000)	
Equipment Basis Well 1		$ 40,000
Capitalized IDC	137,000	
Less: IDC written off	(137,000)	
Net IDC remaining		0
Cost of equipment on Well 2	180,000	
Less: Accumulated depreciation	(150,000)	
Equipment Basis Well 2		30,000
Capitalized IDC	171,000	
Less: IDC written off	(57,000)	
Net IDC remaining		114,000
Less: Salvage value of equipment		(50,000)
Net abandonment loss on equipment and IDC		$134,000

c. Cost of equipment on Well 1	$ 160,000	
Less: Accumulated depreciation	(120,000)	
Equipment Basis Well 1		$ 40,000
Capitalized IDC	137,000	
Less: IDC written off	(137,000)	
Net IDC remaining		0
Cost of equipment on Well 2	180,000	
Less: Accumulated depreciation	(150,000)	
Equipment Basis Well 2		30,000
Capitalized IDC	171,000	
Less: IDC written off	(57,000)	
Net IDC remaining		114,000
Cost of equipment on Well 3	200,000	
Less: Accumulated depreciation	(90,000)	
Equipment Basis Well 3		110,000
Capitalized IDC	163,000	
Less: IDC written off	(41,000)	
Net IDC remaining		122,000
Less: Salvage value of equipment		(50,000)
Net abandonment loss on equipment and IDC		366,000
Leasehold	5,000	
Less: Accumulated depletion	(4,000)	
Abandonment loss on leasehold		1,000
Total abandonment loss		$367,000

d. As well is drilled:

IDC deduction	$240,000 x 70% =	$ 168,000	
IDC capitalized	$240,000 x 30% =	$ 72,000	
Equipment capitalized		$ 35,000	

When well is determined to be dry*:

IDC (deduct remaining capitalized IDC)	$ 72,000	
Non-salvageable equipment	35,000	
Total tax loss on dry hole (assuming the lease itself is retained)		$107,000
Total deductions relating to well ($168,000 + $107,000)		$275,000

* Capitalized IDC is amortized over 60 months beginning with the month incurred. This calculation assumes the well was drilled and determined to be dry all within one month.

Note that for income tax purposes, the designation of a well as exploratory or development has no meaning. The only considerations are whether the costs are IDC versus equipment and whether the well is dry or successful.

The following comprehensive example illustrates tax accounting for both independent and integrated producers.

EXAMPLE

Comprehensive Example

During 2008

Lucky Company, an integrated company, has decided to survey a project area in Deaf Smith County, Texas. A G&G team was hired to perform a reconnaissance survey of the eastern half of the county for $30,000. After the survey was complete and a thorough study was made of the seismic information, three interest areas were identified and labeled areas A, B, and C. The G&G team made a detailed survey of each of the interest areas. Interest Area A's survey, covering 640 acres, cost $24,000; Interest Area B's survey, covering 320 acres, cost $15,000; Interest Area C's survey, covering 160 acres, cost $10,000. After a detailed evaluation, Lucky Company acquired the following leases:

Interest Area A: None
Interest Area B: Richards—320 acres; 100% WI; $1/8$ RI; bonus, $75 per acre; other acquisition costs, $2,500; delay rental, $2 per acre
Interest Area C: Rylander—80 acres; 100% WI; $3/16$ RI; bonus, $50 per acre; other costs, $1,500; delay rental, $1 per acre

Tax Effects of Transactions Occurring During 2008

Basis of Areas of Interest

	Area A (640 acres)	Area B (320 acres)	Area C (160 acres)
Allocation of reconnaissance costs ($30,000/3)	$10,000	$10,000	$10,000
Detailed survey costs	24,000	15,000	10,000
Capitalized basis of areas	$34,000	$25,000	$20,000

Disposition of Capitalized Area of Interest Costs

Area A

No interest, therefore, entire basis of $34,000 is currently deductible.

Area B

Acquired Richards Lease (320 acres). Costs capitalized to Area B added to basis of Richards Lease:

Capitalized costs of Area B	$25,000
Bonus (320 x $75)	24,000
Other acquisition costs	2,500
Basis of Richards Lease	$51,500

Area C

Acquired Rylander Lease (80 acres). Costs capitalized to Area C added to basis of Rylander Lease:

Capitalized costs of Area C	$20,000
Bonus (80 x $50)	4,000
Other acquisition costs	1,500
Basis of Rylander Lease	$25,500

During 2009

Lucky Company made the following payments in 2009:

Delay rental—Richards lease, $640, paid March 2009

Dry-hole contribution of $5,000 in connection with the Richards lease

Bottom-hole contribution in connection with the Rylander lease, $10,000 (well was dry)

Drilling costs of Rylander No. 1:

IDC (assume incurred September 16)	$302,000
Equipment costs (assume incurred during September)	212,000

The well was completed in late September 2009, and 1,600 total gross barrels of oil were produced and sold during the year. Estimated total gross reserves at year-end were 30,400 bbl. The oil was sold for $20/bbl in 2009. Lifting costs are $5/bbl. The lessee distributes the royalty to Rylander. The severance tax rate is 7%. Because Lucky Company is an integrated producer, the company must use cost depletion.

Lucky Company's taxable income and expenses in 2009 and the tax basis of these two properties are determined below.

Tax Effects of Transactions Occurring During 2009

Gross income (1,600 x $^{13}/_{16}$ x $20)		$ 26,000
Deductions for both leases:		
Delay Rental	$ 640	
IDC (0.70 x 302,000)	211,400	
Capitalized IDC (see below*)	6,040	
Severance taxes (26,000 x 0.07)	1,820	
Lifting costs (1,600 x $5)	8,000	
Depletion (see below**)	1,775	
Depreciation (see below***)	30,286	
Total		259,961
Net tax loss on properties		$(233,961)

* IDC:

Capitalized IDC—Rylander No. 1 ($302,000 x 0.30)	$90,600
2009 write-off ($90,600 x $^{4}/_{60}$)	(6,040)
Remaining balance to be written over next 56 months	$84,560

** Depletion:

	Richards	**Rylander**
1/1/09 Basis	$51,500	$25,500
Test well contributions	5,000	10,000
Balance subject to depletion	$56,500	$35,500
2009 Depletion	0	(1,775)
12/31/09 Basis	$56,500	$33,725

$$\frac{1{,}600 \times {}^{13}/_{16}}{30{,}400 \times {}^{13}/_{16} + 1{,}600 \times {}^{13}/_{16}} \times \$35{,}500 = \$1{,}775$$

***Depreciation:

Capitalized cost of tangible equipment	$212,000
2009 write-off ($212,000 x $^{2}/_{7}$ x $^{1}/_{2}$)	(30,286)
Undepreciated balance	$181,714

If Lucky had been an independent producer, all of the income and expenses would have been identical with the exception of IDC. In that case, the entire $302,000 of IDC would have been deductible during 2008.

Table 10–1 presents a comparison of capitalizing or expensing selected expenditures under SE, FC, and tax accounting. The abbreviations used are *E* (Expense), *C* (Capitalize), and *NA* (Not Applicable). The tax treatments of expenditures by the lessor, which are included in the table below, are discussed in more detail later in the chapter.

			Tax	
Item	SE	FC	Lessor	Lessee
1. Broad G&G studies—an area of interest is identified	E	C	NA	C
2. Broad G&G studies—an area of interest is not identified	E	C	NA	E
3. Seismic studies on 2,000 acres—land is leased	E	C	NA	C
4. Seismic studies on 2,000 acres—no land is leased	E	C	NA	E
5. Bottom-hole contribution—well successful	E	C	NA	C
6. Dry-hole contribution	E	C	NA	C
7. Bonus	C	C	Income	C
8. Delay rental payments	E	C	Income	E(C)
9. Minor Workover	E	E	NA	E
10. Water injection well (intangible costs)	C	C	NA	E(C)/IDC*
11. Option to lease	C	C	Income	C
12. Water disposal well (intangible costs)	C	C	NA	C
13. Successful exploratory drilling				
a. Intangible costs	C	C	NA	E(C)*
b. Tangible costs	C	C	NA	C
14. Exploratory drilling—dry hole				
a. Intangible costs	E	C	NA	E(C)**
b. Tangible costs(net of salvage)	E	C	NA	E
15. Successful development drilling				
a. Intangible costs	C	C	NA	E(C)*
b. tangible costs	C	C	NA	C
16. Development drilling-dry hole				
a. Intangible costs	C	C	NA	E(C)*
b. Tangible costs(net of salvage)	C	C	NA	E
17. Deepening a development well to unexplored depths—dry hole				
a. Intangible costs	E	C	NA	E(C)/IDC**
b. Tangible costs(net of salvage)	E	C	NA	E

Table 10–1 — ***Summary of Accounting Treatment of Costs—SE vs. FC vs. Tax***

Item	SE	FC	Tax	
			Lessor	Lessee
18. Abandonment of well on producing lease (well had been depreciated separately for tax purposes)	***	***	NA	E
19. Abandonment of lease (no other nearby leases)	E	***	NA	E
20. Production costs	E	E	NA	E
21. Water supply well (intangible costs)				
a. Water to be used for development drilling	C	C	NA	E(C)/IDC*
b. Water to be used for secondary recovery	C	C	NA	E(C)/IDC*
22. Exploratory stratigraphic test well—dry (intangible costs)	E	C	NA	E(C)/IDC**

* Depends upon election to expense and whether independent or integrated producer.
** Depends upon election to expense.
*** Do not recognize a loss, instead charge accumulated DD&A or abandoned costs.

Table 10–1 cont'd — *Summary of Accounting Treatment of Costs—SE vs. FC vs. Tax*

The next example compares the computation of DD&A under tax accounting, SE accounting, and FC accounting.

EXAMPLE

DD&A under Three Accounting Methods

Lucky Company began operations in September 2007. By December 2009, Lucky had acquired only two properties, both located in the United States and both considered significant. Costs incurred on those properties during the first years of operations are given below net of accumulated DD&A. During the first years of operations, Lucky drilled two dry holes and one successful well on Lease A and one dry hole on Lease B. No equipment was salvaged from the dry holes. Production on Lease A began on January 1, 2008. As a result of the dry hole on Lease B, Lucky impaired Lease B $10,000 on December 31, 2008, for SE accounting purposes. Early in August 2009, Lucky incurred $30,000 of IDC on Lease A (in W/P—IDC at 12/31/09) and $150,000 on Lease B (in W/P—IDC at 12/31/09). All reserve, production, and sales data given below apply only to Lucky Company.

Costs Incurred 9/07 through 12/09

	Lease A	Lease B
Related G&G costs	$ 50,000	$ 80,000
Bonus	60,000	50,000
Impairment (SE accounting)		(10,000)
Delay rentals, cumulative	0	5,000
W/P—IDC (at 12/31/09)	30,000	150,000
Dry exploratory wells—IDC	200,000	250,000
Dry exploratory wells—L&WE	70,000	75,000
Successful exploratory well—IDC	300,000	—
Successful exploratory well—L&WE	100,000	—
Dry development wells—IDC	350,000	—
Dry development wells—L&WE	120,000	—
Other data:		
Salvage value of wells	45,000	—
Future development costs	400,000	—
Production	10,000 bbl	—
Sales	9,000 bbl	—
Proved reserves, 12/31/09	500,000 bbl	—
Proved developed reserves, 12/31/09	100,000 bbl	—

Lucky also placed in service on 1/1/08 a building that cost $108,000 and has a life of 20 years with no salvage value. The building houses the corporate headquarters that supports oil and gas operations in the United States and non oil and gas operations in Canada. The operations conducted in the building are general in nature and are not directly attributable to any specific exploration, development, or production activity. Since the building is not directly related to exploration, development, or production and supports activities in more than one cost center, it is depreciated using straight-line depreciation for financial accounting.

a. DD&A for 2009 for financial accounting purposes assuming Lucky uses successful-efforts accounting is computed below.

Proved Property, Lease A only

$$\$60{,}000 \times \frac{10{,}000}{500{,}000 + 10{,}000} = \underline{\underline{\$1{,}176}}$$

Wells and Related E&F, Lease A

Total wells = \$300,000 + \$100,000 + \$350,000 + \$120,000 – \$45,000 = \$825,000

$$\$825{,}000 \times \frac{10{,}000}{100{,}000 + 10{,}000} = \underline{\underline{\$75{,}000}}$$

Depreciation Building (supporting general and administrative functions only)

$$\frac{\$108{,}000 - \$0}{20} = \underline{\underline{\$5{,}400}}$$

b. DD&A for 2009 for financial accounting purposes assuming Lucky uses full-cost accounting and assuming inclusion of all possible costs in the amortization base is shown below. The revenue method is ignored.

Total costs to amortize = all costs from both Lease A (minus salvage) and Lease B (impairment not included because Lease B would not have been impaired due to its inclusion in amortization) and plus future development costs
= \$1,235,000 (Lease A) + \$610,000 (Lease B) + \$400,000 (development costs)
= \$2,245,000

$$\$2{,}245{,}000 \times \frac{10{,}000}{500{,}000 + 10{,}000} = \underline{\underline{\$44{,}020}}$$

Depreciation Building (supporting general and administrative functions only)

$$\frac{\$108{,}000 - \$0}{20} = \underline{\underline{\$5{,}400}}$$

c. DD&A for 2009 for tax reporting purposes assuming that Lucky is an independent producer is shown below. Percentage depletion is ignored.

Depletion, Lease A only

$$(\$50{,}000 + \$60{,}000) \times \frac{9{,}000}{500{,}000 + 9{,}000} = \underline{\underline{\$1{,}945}}$$

Depreciation in Second Year

L&WE from successful exploratory well (assume $100,000 is gross cost):

($100,000 – $14,286*) x 2/7 = $24,490

* Depreciation for the first year: $100,000 x 2/7 x 1/2 = $14,286

Building (assume $108,000 is gross cost):

$108,000/39 = $2,769/year (except first and last years)

d. DD&A for 2009 for tax reporting purposes assuming Lucky is an integrated producer is shown below.

Depletion and depreciation would be unchanged. However, 30% of all the IDC for successful wells and wells in progress must be capitalized and amortized over 60 months.

Total IDC incurred for successful wells and wells in progress:

Prior to 2009 (but within 60 months)	$300,000
During 2009	180,000
	$480,000

Total IDC Capitalized:

Prior to 2009	$300,000 x 0.30 =	$ 90,000
During 2009	180,000 x 0.30 =	54,000
		$144,000

Total IDC amortized during 2009:

90,000/60 months x 12 months months =	$ 18,000
54,000/60 months x 5 months months =	4,500
	$ 22,500

Percentage depletion

Percentage depletion is one of the two methods used for income tax purposes in determining the depletion amount to be deducted in arriving at taxable income. Percentage depletion, when allowed and if greater than cost depletion, must be taken in lieu of cost depletion. Generally, percentage depletion is allowed only for royalty interest owners and independent producers of oil and gas. Companies with integrated functional areas, *i.e.*, exploring, producing, refining, and distributing, are not allowed to take percentage depletion.

Percentage depletion is computed by multiplying the gross revenues from a property by a percentage established by federal statutes. The depletion percentage is currently 15% for nonmarginal wells. The amount of oil or gas subject to depletion cannot exceed an average daily production of 1,000 barrels of oil or 6,000 Mcf of gas. The amount of percentage depletion allowed cannot be greater than 100% of net income from a particular property. Further, percentage depletion may not exceed 65% of the taxpayer's taxable income before depletion from all sources. Amounts disallowed as a result of the 65% limitation may be carried over.

Percentage depletion allowed is deducted from the capitalized leasehold costs and elected capitalized IDC (if any). Percentage depletion may be taken even if all capitalized costs have been recovered. In contrast, cost depletion is limited to the amount of capitalized costs.

Prior to October 12, 1990, a transferee of a proved oil or gas property was denied percentage depletion. The Revenue Act of 1990 repealed this prohibition on percentage depletion for transferees. Therefore, a person receiving a proved property can take percentage depletion if the other percentage depletion rules are met.

EXAMPLE

Percentage Depletion

Lucky Company is an independent producer. In 2006, the company had an average production of 400 barrels a day from Lease A. The average selling price for the oil was $20/bbl. Net income was $985,000 from Lease A, and taxable income was the same amount.

Computation of Percentage Depletion

	Lease A
Gross amount of oil runs (400 x 365 x $20)	$2,920,000
Tentative % depletion (gross revenue x rate: $2,920,000 x 0.15)	438,000
Net income limitation (100% x NI)	985,000
Tentative allowable deduction	438,000
Taxable income limitation (0.65 x taxable income: 0.65 x $985,000)	640,250
Percentage depletion	$ 438,000*

* This amount must be compared with cost depletion—the greater amount would be the actual depletion expense taken.

In the above example, only one lease was used to illustrate percentage depletion. When more than one lease is involved, the computation of the net income and the application of the depletable quantity limitation and the 65% limitation become complicated. Allocation of various expenses must be made between leases before net income can be determined, and the maximum percentage depletion allowable under the 65% tax income limitation must be allocated and possibly reallocated between properties. The following example briefly illustrates the 100% and 65% limitations. (For a more detailed discussion of these rules, see Klingstedt, Brock, and Mark, PDI, *Oil and Gas Taxation*, 1989.)[1]

EXAMPLE

Percentage Depletion: 100% and 65% Limitations

Lucky Company owns three properties (classified as individual properties for tax purposes) at year-end, 2007. Data for the three properties are presented below:

	Properties			
	A	**B**	**C**	**Total**
Total gross revenues subject to % depletion	$400,000	$500,000	$300,000	$1,200,000
Net (loss) income before depletion—2007	50,000	100,000	(30,000)	120,000

Calculation of Allowable Percentage Depletion

	A	B	C	Total
Gross revenues x 15%	$ 60,000	$ 75,000	$ 45,000	$ 180,000
100% Net income limitation	50,000	100,000	(30,000)	120,000
Allowable % depletion subject to 100% limitation	50,000	75,000	0	125,000
Total taxable income before depletion x 65% ($120,000 x 0.65)				78,000
Comparing $125,000 with $78,000				
Percentage depletion				$ 78,000

Property

The term property is important in the following areas of oil and gas tax accounting:

- G&G costs capitalized and allocated to a property unit
- Capitalized IDC
- Cost and percentage depletion
- Gain or loss on disposal of property
- Recapture of IDC as ordinary income
- Tax preference items

Property means each separate interest owned by the taxpayer in each mineral deposit in each tract or parcel of land (*U.S. Treas. Reg.* Section 1.614-1[b]). Each economic interest is generally regarded as a separate property, unless it were acquired at the same time, is the same kind of interest, is from the same assignor, and is contiguous. Interests generally are classified as operating (working interest) or nonoperating (royalty or ORI, production payment, or net profits interest). Nonoperating interests may, with the consent of the commissioner, be aggregated.

Separate tract or parcel relates to the physical scope of land. If two parcels of land are not contiguous, then even if they are purchased at the same time from the same assignor, two separate properties are created.

Property differentiation can be very important in the recapture of IDC and depletion. As will be seen in the next section, separation of properties into small property units may be advantageous for IDC recapture.

Recapture of IDC and depletion

All IDC that has been expensed and any depletion expense that reduced the basis of the property must be recaptured as ordinary income versus a capital gain to the extent a property is sold at a gain. (As there is currently no difference between the capital gains rate versus the ordinary income rate for a business, this section does not have any tax implications for a business unless the business has a capital loss.) Specifically, the taxpayer is required to recapture the lower of the following amounts:

- The total of any IDC expensed as incurred, IDC expensed through amortization, and any depletion expense that reduced the adjusted basis of the property
- The amount of the gain on the sale

EXAMPLE

Recapturable IDC and Depletion Given

Mineral deposit A has recapturable IDC and depletion of $400,000 and mineral deposit B has recapturable IDC of $200,000. Both properties together are sold for $1,000,000. The selling price is allocated $100,000 to A and $900,000 to B. Both deposits have a zero basis.

Assuming the same property—Ordinary income is the lower of $600,000 versus $1,000,000:

Section 1231 Gain (capital gain)	$400,000
Ordinary income	600,000

Assuming separate properties

Property A—Ordinary income is the lower of $400,000 versus $100,000:

Section 1231 Gain	$ 0
Ordinary income	100,000

Property B—Ordinary income is the lower of $200,000 versus $900,000:

Section 1231 Gain	$700,000
Ordinary income	200,000

The following example illustrates the computation of the amount of IDC and depletion to recapture as ordinary income for both an integrated and independent company:

EXAMPLE

Recapturable IDC and Depletion Not Given

On January 1, 2007, Lucky Company bought an undeveloped lease for $200,000. Early in January 2007, Lucky Company incurred IDC of $450,000. Reserves of 75,000 barrels were discovered, and 30,000 barrels were produced and sold. Gross income from production was $1,000,000. On January 3, 2008, Lucky Company sold the property for $420,000. All reserve, production, and sales data apply only to Lucky Company.

a. The income tax effects of the sale assuming an independent producer and percentage depletion of 15% are determined below. The 100% and 65% limitations are ignored.

Cost Depletion

$$\$200{,}000 \times \frac{30{,}000}{75{,}000} = \underline{\underline{\$80{,}000}}$$

Percentage Depletion

0.15 x $1,000,000 = $150,000

Depletion expense for 2007 is $150,000.

Amount of IDC and Depletion Recapturable

$150,000 (depletion) + $450,000 (IDC) = $600,000

The tax basis of the property is $50,000 ($200,000 – $150,000); therefore, the gain on the sale is $370,000 ($420,000 – $50,000). Ordinary income is the lower of $600,000 versus $370,000. Therefore, the entire gain on the sale is ordinary income.

b. The income tax effects for the sale are determined below assuming the same facts as shown in the above example except that Lucky Company is an integrated producer. (An integrated producer cannot take percentage depletion, and 30% of any IDC must be capitalized and amortized over 60 months.)

Actual Cost Depletion Expense

$$\$200{,}000 \times \frac{30{,}000}{75{,}000} = \$80{,}000$$

IDC Expensed to Date

$450,000 x 0.70 =	$315,000
($450,000 x 0.30) x $^{12}/_{60}$ =	27,000
	$342,000

Amount of IDC and Depletion Recapturable

$80,000 (depletion) + $342,000 (IDC) = $422,000

Amount of Gain

Basis in property:	
Leasehold	$200,000
Less: Cost depletion	(80,000)
Leasehold basis	120,000
Plus: Capitalized IDC ($135,000 – $27,000)	108,000
	$228,000

Gain = $420,000 (Selling price) – $228,000 (Basis in property) = $192,000

Ordinary income is the lower of $422,000 versus $192,000. Therefore, the entire gain on the sale is ordinary income.

The recapture provision illustrated by the preceding example applies to the disposition of all properties placed in service after December 31, 1986.

The example is unrealistically simple because the property was owned for only one year. If the property had been owned for more than one year, the amount of recapturable IDC would have had to be calculated for each year the property was owned. In addition, in the above example, equipment was ignored. Generally in a transaction where IDC must be recaptured, equipment would also be sold. The selling price should be allocated between the equipment, the leasehold, and the IDC, based on relative fair market values. In practice, the equipment is treated as being sold at net book value, which results in any gain being allocated to the leasehold and IDC.

Enhanced oil recovery credit

All taxpayers (independent and integrated producers) can take a 15% credit for certain costs that are incurred related to an enhanced recovery project. Costs must be incurred in connection with a qualified enhanced oil recovery project. Three types of costs qualify for the credit. These costs are IDC, equipment, and qualified injectant expenses. The 15% credit rate is multiplied times the qualified cost amount. However, this credit amount must be modified if the reference price of crude oil rises above $28 a barrel.

LESSOR'S TRANSACTIONS

The previous discussion has centered upon the determination of taxable income by the lessee. Attention is now turned to the income tax accounting for transactions involving the lessor.

Acquisition costs

The lessor should capitalize any acquisition costs incurred prior to leasing the property to an oil company and deplete the costs capitalized. If the mineral rights were obtained separately from the surface rights, the amount capitalized should be the acquisition cost of the mineral rights. If the property were purchased in fee, the purchase price should be allocated between the mineral and surface rights.

Provided the lessor is a royalty owner—or an independent producer—either cost depletion or percentage depletion may be taken on acquisition cost amounts, with the actual depletion expense being the larger of the two. Percentage depletion, unlike cost depletion, is allowed even after the cost has been recovered. The lessor may, in many situations, have zero or a small dollar amount of acquisition costs. Therefore, cost depletion is not a viable option for many lessors.

EXAMPLE

Lessor's Acquisition Costs

Mr. Lucky purchased some mineral rights for $40,000. Lucky later leased the property and received a $^1/_5$ royalty. The year-end total gross proved reserve estimate is 45,000 barrels and total gross sales during the year were 5,000 barrels.

The tax treatment for the acquisition cost of the mineral rights and cost depletion for Lucky are determined below. Percentage depletion is ignored.

Tax Treatment

Depletion calculation:

Royalty interest production ($^1/_5$ x 5,000)	=	1,000 bbl
Royalty interest reserves ($^1/_5$ x 45,000)	=	9,000 bbl

Cost depletion:

$$\frac{1{,}000}{9{,}000 + 1{,}000} \times \$40{,}000 = \underline{\underline{\$4{,}000}}$$

Mineral rights acquisition cost:

Tax basis of property at beginning of year	$40,000
Less: Cost depletion	(4,000)
Tax basis of property at end of year	$36,000

Revenue

Revenue received by a lessor as his share of production is a royalty and is treated as ordinary income. Percentage depletion is calculated by taking 15% of the revenue from a property. For a royalty interest owner, this translates into 15% of the royalty from a property. Lease bonuses paid to the lessor are also considered to be in the nature of a royalty amount paid in advance—an advance royalty. However, even though considered a royalty, effective August 16, 1986, advance royalty payments, lease bonuses or any other amount payable without regard to production from a property no longer qualify for percentage depletion by royalty owners or independent producers.

Delay rentals, ad valorem taxes paid by the lessee for the lessor, minimum and shut-in royalties, and options to lease are all considered to be ordinary income to the lessor. Delay rentals received and ad valorem taxes paid by the lessee are not considered royalties and therefore cannot be reduced by percentage depletion. Minimum and shut-in royalties are advance royalties; however, they also are not subject to percentage depletion. Percentage depletion can be taken only on royalties from actual production.

EXAMPLE

Lessor's Revenue

Mabel Lomax received the following benefits in 2007 in connection with Lease A, from which no oil or gas was produced during 2007:

Received delay rental	$2,000
Received notice that lessee paid Mabel's ad valorem taxes in the amount of	1,500
Received shut-in royalty for well drilled in 2007	5,000
Royalties from actual production	0

Income tax treatment of transactions

Ordinary income not subject to percentage depletion

Delay rental	$2,000
Ad valorem taxes paid by lessee	1,500
Shut in royalty	5,000
Total	$8,500

Income subject to percentage depletion	$0
Depletion deduction	$0

PROBLEMS

1. Casing Oil Company incurs G&G costs of $21,000 for Project Area 12. Three areas of interest are identified. Detailed G&G is conducted on the areas of interest at the following costs:

Interest Area	Cost
1	$55,000
2	60,000
3	70,000

As a result of the detailed G&G studies, the following leases were acquired:

Interest Area	Lease	Acres	Interest
1	A	1,000	100% of WI
	B	3,000	50% of WI
	C	2,000	25% of WI
2	D	5,000	100% of WI
3	None	None	None

Determine the tax basis of any assets and the amount of any tax deductions.

2. Basic Oil Company incurred intangible costs during 20XA related to the following:

	Amount	Date Incurred
Dry exploratory well	$100,000	May 1
Dry development well	200,000	May 1
Successful exploratory well	250,000	May 1
Water injection well	270,000	Aug 1
Deepening a well (successful)	80,000	Aug 1
Workover, minor	50,000	Aug 1

a. Assuming Basic is an independent producer, how much IDC can it deduct for 20XA?

b. How much IDC could Basic deduct as an integrated producer?

3. On January 1, 20XA, Core Oil Company bought a developed lease for $300,000. During 20XA, Core Oil Company incurred $600,000 of IDC. Reserves of 400,000 barrels were discovered, and 100,000 barrels were produced and sold. Gross income from production was $2,000,000. On January 4, 20XB, the company sold the property for $700,000. All reserve, production, and sales data apply only to Core Oil Company.

a. Determine the amount of ordinary income and the amount of any capital gain assuming Core Oil is an integrated producer and that part of the IDC is capitalized as required. Assume that all of the IDC was incurred on April 1, 20XA.

b. Determine the amount of ordinary income and the amount of any capital gain, assuming an independent producer, percentage depletion of 15% and that all the IDC was expensed as incurred. Ignore the 100% and 65% limitations.

4. During 20XA, Gravity Oil Company incurred G&G costs of $20,000 for Project Area 15. Two areas of interest were identified. Detailed seismic studies were conducted on the areas of interests at the following costs:

Interest Area	Cost
1	$60,000
2	40,000

As a result of the detailed seismic studies, the following leases were obtained:

Interest Area	Lease	Acres	Interest	Bonus Paid
1	None	None	-	-
2	A	2,000	20% of WI	$50/acre
	B	4,000	100% of WI	$30/acre

During 20XB, Gravity Oil Company made the following payments:

	Lease A	Lease B
Delay rental, paid March 20XB	$10,000	
Bottom-hole contribution (well was successful)		$ 25,000
Dry-hole contribution	20,000	
Lessor's ad valorem taxes (nonrecoverable)	2,000	
Drilling costs:		
IDC (assume incurred January 1, 20XB)		300,000
Equipment		80,000

The well on Lease B was completed early in January 20XC and was successful. Gravity Oil Company's share of production from the well was 10,000 barrels of oil. All 10,000 barrels of oil were sold during 20XC. Gravity's share of estimated reserves at year-end was 300,000 barrels. The selling price of the oil was $20 per barrel and lifting costs were $50,000. Lease A was abandoned in March 20XC and Lease B was abandoned early in January 20XD. No oil was produced during 20XD.

a. Determine the tax effects for the above transactions in each year, assuming Gravity is an independent producer. Ignore percentage depletion, but remember DD&A.

b. Determine any tax effects that would be different if Gravity were an integrated producer rather than an independent producer.

5. Universal Oil Company, an independent producer, began operations in June 20XA. During the first 2½ years of operation, Universal acquired only two U.S. properties, which were noncontiguous. Costs incurred on those properties during that 2½ years are given below, net of accumulated DD&A. Universal drilled three dry holes and one successful well. No equipment was salvaged from the dry holes. Production on the successful well started on January 1, 20XB. During 20XC, Universal incurred total IDC of \$100,000 on Lease A (in W/P-IDC at 12/31/20XC) and \$200,000 on Lease B (in W/P-IDC at 12/31/20XC). All 20XC IDC on both leases was incurred early in May. All reserve, production, and sales data below apply only to Universal Oil Company.

Costs incurred 6/20XA through 12/20XC:

	Lease A		Lease B
Related G&G costs	\$ 20,000		\$ 40,000
Bonus	30,000		50,000
Delay rentals, cumulative	10,000		5,000
W/P—IDC (at 12/31/20XC)	100,000		200,000
Dry exploratory wells—IDC	250,000		300,000
Dry exploratory wells—L&WE	50,000		75,000
Successful exploratory well IDC	400,000		-
Successful exploratory well—L&WE	150,000		-
Dry development wells—IDC	200,000		-
Dry development wells—L&WE	100,000		-
Other data			
Future development costs	300,000		
Production	20,000	bbl	-
Sales	18,000	bbl	-
Proved reserves, 12/31/20XC	600,000	bbl	-
Proved developed reserves, 12/31/20XC	200,000	bbl	-

Universal also placed in service on 1/1/20XB a building that cost \$117,000 and has a life of 25 years with a salvage value of \$7,000. The building houses the corporate headquarters that supports oil and gas operations in the United States and non oil and gas operations in Mexico. The operations conducted in the building are general in nature and are not directly attributable to any specific exploration, development, or production activity. Since the building is not directly related to exploration, development, or production and supports activities in more than one cost center, it is depreciated using straight-line depreciation for financial accounting.

a. Compute DD&A for 20XC for the following accounting methods assuming that Universal is an independent producer:
   1) SE
   2) FC, assuming inclusion of all possible costs in the amortization base. Ignore the revenue method
   3) Tax: ignore percentage depletion and assume that the cost of the equipment and the building are gross costs

b. Calculate DD&A for tax, assuming that Universal had been an integrated producer instead.

6. How does the formula for tax depletion differ from the formula for SE or FC depletion?

7. How does amortization of leasehold costs differ under the three methods? Include in your answer a discussion of the reserves used (PR or PDR), which costs are amortized, and the cost center used (*i.e.*, are property costs amortized separately or by some type of grouping?).

8. How does amortization of drilling costs differ under the three methods? Include in your answer a discussion of the method used, including which reserves are used, if any, which drilling costs are amortized—dry versus successful, exploratory well versus development well, completed versus uncompleted, IDC versus L&WE—and the cost center used (*i.e.*, are drilling costs amortized separately or by some grouping?).

9. How does amortization of lease equipment such as storage tanks and separators differ under the three methods? Include in your answer a discussion of the method used, including which reserves are used, if any, and the cost center used.

10. Hard Times Oil Company, an integrated producer, has an unproved property with acquisition and capitalized G&G costs of $35,000. Hard Times also has a proved property with the following costs:

	Lease	Well 1	Well 2	Well 3	Total
Acquisition costs	$40,000				$40,000
Capitalized IDC costs		$50,000	$280,000	$170,000	500,000
Equipment costs		100,000	500,000	300,000	900,000
Accumulated depletion	(8,000)				(8,000)
Accumulated amortization		(13,000)	(110,000)	(57,000)	(180,000)
Accumulated depreciation	______	(25,000)	(200,000)	(100,000)	(325,000)
Net	$32,000	$112,000	$470,000	$313,000	$927,000

Determine the amount of the tax loss in each of the following situations:

a. Hard Times drilled a dry hole on the unproved property costing $250,000 for IDC and $60,000 for equipment. Equipment worth $10,000 was salvaged.
b. As a result of the dry hole, Hard Times decided to abandon the unproved property.
c. Hard Times abandoned Well 1 on the proved property. Wells 2 and 3 are still producing. Assume that the wells had been depreciated separately.
d. Assume that instead of *c*, Hard Times abandoned the entire lease and that equipment worth $27,000 was salvaged.

11. Indicate which items are to be capitalized (C), expensed (E), and part capitalized and part expensed (C/E) for successful-efforts, full-cost, and tax accounting. Assume the maximum tax deductions are taken.

Item	**SE**	**FC**	**Tax Lessee**
Broad G&G studies—an area of interest is identified			
Broad G&G studies—an area of interest is not identified			
Seismic studies on 2,000 acres—land is leased			
Seismic studies on 2,000 acres—no land is leased			
Bottom-hole contribution—well successful			
Dry-hole contribution			
Bonus			
Delay rental payments			
Minor workover			
Water injection well (intangible costs)			
Option to lease			
Water disposal well (intangible costs)			
Wells in progress—proved property a. Intangible costs b. Tangible costs			
Wells in progress—unproved property a. Intangible costs b. Tangible costs			

Item	SE	FC	Tax Lessee
Successful exploratory drilling a. Intangible costs b. Tangible costs			
Exploratory drilling—dry hole a. Intangible costs b. Tangible costs (net of salvage)			
Successful development drilling a. Intangible costs b. Tangible costs			
Development drilling—dry hole a. Intangible costs b. Tangible costs (net of salvage)			
Deepening a development well to unexplored depths—dry hole a. Intangible costs b. Tangible costs (net of salvage))			
Abandonment of well on producing lease (well had been depreciated separately for tax purposes)			
Abandonment of lease (no other nearby leases)			
Production costs			
Water supply well (intangible costs) a. Water to be used for development drilling b. Water to be used for secondary recovery			
Exploratory stratigraphic test well—dry (intangible costs)			

12. On March 1, 20XA, Jerry Barnes purchases mineral rights for $30,000. On June 1, 20XA, he leases the mineral rights to Brown Oil Company retaining a $^1/_5$ royalty interest. Brown Oil Company pays Barnes a lease bonus of $10,000. On June 1, 20XB, a delay rental of $1,000 is received by Barnes. Oil royalties of $8,000 are paid to Barnes in 20XC. Proved reserves at 12/31/20XC are 20,000 barrels, and production and sales for the year are 3,000 barrels. The reserve, production, and sales data apply only to Jerry Barnes.

    Determine the tax basis of any assets owned by Jerry Barnes and the amount of any tax revenues reported and any tax deductions taken by Jerry Barnes in each of the three years.

13. Jones Oil Company paid the following amounts in 20XD:

Shut-in royalty payments (not recoverable)	$1,200
Shut-in royalty payments (failure to make payments terminates lease)	4,000

Shut-in royalty payments (recoverable from future production)	2,400

Determine the tax basis of any assets and the amount of any tax deductions.

14. Davenport Energy Company, an independent producer, has the following account balances at 1/1/20XA:

Unproved property—Lease A	$ 10,000
Equipment—Lease B	150,000
Acquisition costs—Lease B	8,000
Accumulated depletion	5,000
Accumulated depreciation	110,000

Determine the amount of the tax loss on the following dates:

a. On March 1, 20XA, the unproved property is abandoned.
b. On April 2, 20XA Lease B is abandoned with salvageable equipment in the amount of $12,000.

15. Aggie Oil Company has the following information:

Leasehold costs	$ 6,000
Accumulated depletion	1,000
L&WE	200,000
Accumulated depreciation	75,000

The lease is subleased to Acme Oil Company for $300,000, and Aggie retains an $^1/_{16}$ ORI. At the date of the sublease, the FMV of the equipment is 180,000.

REQUIRED: Determine the tax basis of Aggie's and Acme's assets and the amount of any tax revenue.

16. During 20XB, Dixie Company incurs the following costs relating to Lease A, a producing property:

Supplies for Lease A	$ 300
Labor cost for pumpers and gaugers—Lease A	800
Ad valorem tax on Lease A	1,600
Maintenance cost of lease records	1,500
Refracturing cost—Lease A (workover)	2,000
Transportation of personnel to wells—Lease A	800

REQUIRED: Determine the tax basis of any assets and the amount of any tax deductions.

17. Aggie Oil Company, an independent producer, has average production from Lease A of 100 barrels per day in 20XA from Lease A. The average selling price of oil in 20XA is $23 a barrel. Net income from Lease A in 20XA is $325,000 and taxable income of the company is $800,000.

    Compute percentage depletion.

18. Define the following:

    a. IDC
    b. elected capitalized IDC
    c. sublease

19. Dowling Company, an integrated producer, incurs IDC costs in the following years as indicated. The IDC marked with an asterisk (*) relate to dry hole IDC.

Date Incurred	Amount	Date Incurred	Amount	Date Incurred	Amount	Date Incurred	Amount
3/12/XA	$200,000	4/10/XB	$280,000*	2/1/XC	$600,000	3/4/XD	$240,000
7/18/XA	300,000*	11/2/XB	320,000*	12/2/XC	150,000	9/14/XD	180,000*

REQUIRED: Compute the amount that may be deducted for IDC in the years 20XA, 20XB, 20XC, and 20XD.

20. Isaac Company owns and operates four oil and gas properties that are classified for tax purposes as four separate properties. Data for the four properties are presented below:

Properties	1	2	3	4
Total gross revenues subject to % depletion	$200,000	$100,000	$60,000	$300,000
Net (loss) income before depletion	40,000	10,000	(5,000)	70,000

REQUIRED: Compute the amount of percentage depletion that could be deducted on Isaac's tax return.

21. Aggie Oil Company owns only one lease in the United States, Lease Q. The following information for Lease Q, which is burdened with a $\frac{1}{6}$ royalty, is as of 12/31/20XD. All reserve, production, and sales data apply only to Aggie Oil Company.

Acquisition costs	$ 40,000
Test-well contributions	12,000
G&G, direct.	70,000
W/P—IDC (incurred 6/24/20XD)	80,000
W/P—L&WE	6,000
Well—L&WE (end of 2nd year)	250,000
Wells—IDC (incurred in 20XB and all productive)	600,000
Future development costs	400,000
Proved reserves, 12/31/20XD	300,000
Proved developed reserves, 12/31/20XD	100,000
Production	18,000
Sales.	15,000

Additional data: Aggie also placed in service on 8/1/20XB a building that cost $140,000 and has a life of 10 years with a salvage value of $9,000. The building houses the corporate headquarters that supports oil and gas operations in the United States and non oil and gas operations in Canada. The operations conducted in the building are general in nature and are not directly attributable to any specific exploration, development, or production activity. Since the building is not directly related to exploration, development, or production and supports activities in more than one cost center, it is depreciated using straight-line depreciation for financial accounting.

Compute DD&A for 20XD for the following accounting methods.

a. Tax: assuming Aggie is an independent producer (ignore percentage depletion) and accumulated depletion is $10,000.
b. Tax: assuming Aggie is an integrated producer and accumulated depletion is $10,000.
c. SE: assuming accumulated DD&A - P/P is $5,000 and accumulated DD&A - wells is $100,000.
d. FC: assuming exclusion of all possible costs from the amortization base. Assume accumulated DD&A is $400,000.

NOTES

1. Klingstedt, John P., Brock, Horace R., Mark, Richard S., *Oil and Gas Taxation*, Denton, Texas: Professional Development Institute, 1989.

chapter eleven

Joint Interest Accounting

Joint Operations

Oil and gas exploration and production operations are typically high risk, high cost activities. In order to spread the cost and risk, oil and gas properties—especially those requiring high capital expenditures—typically are jointly owned by two or more working interest owners. In some instances, companies jointly acquire the working interest from the outset. In other cases, the joint operation may not come about until later. For example, if a number of companies own working interests in several small leases in an area, it may make economic sense for the companies to jointly operate all of the leases as one joint property. In other instances, where leases in an area have produced essentially all of the reserves that are economically recoverable without engineering help, secondary or tertiary recovery may provide a means to produce additional reserves. In many instances, secondary or tertiary production may only be possible if the leases are unitized and operated as a single property. Finally, due to conservation and economic conditions, the local state or federal government may require the companies operate the properties jointly.

If separate properties are combined before reserves are discovered, the combination of the properties is typically referred to as a pooling. If separate properties are combined after the development of all or some of the properties, the combination of the properties is typically referred to as a unitization. Poolings and unitizations are discussed in more detail in chapters

1 and 12. Joint working interests may also come about as a result of any number of different types of sharing arrangements. These types of arrangements are discussed in chapter 12.

Joint operations may be undertaken in three legal forms:

- joint venture of undivided interests
- legal partnership
- jointly owned corporation

Joint ventures of undivided interests are by far the most common form of joint operation. In an undivided interest, the parties share the interest in an entire lease. For example, a company having a 50% undivided interest in a 640 acre lease owns a 50% interest in the entire 640 acres. This differs from a situation where a company owns a 100% interest in 320 acres carved out of a 640 acre lease. The former is an undivided interest and the latter is a divided interest.

In a joint venture of undivided interests, companies may acquire an undivided interest in a property when the property is initially acquired. Alternatively, an undivided interest may be created at a later date if the companies pool, unitize, or otherwise join their properties together in such a manner that all of the parties have an undivided interest in the entire property. Companies having divided interests before a unitization or pooling will have an undivided interest after the properties are pooled or unitized. Additionally, a company may also acquire an undivided interest in a property through any number of different types of sharing arrangements (chapter 12).

Legal partnerships are much less common than joint ventures. When companies join together to form a legal partnership they normally do so under the laws of a particular state. Typically, companies join together in a partnership formed for the exploration and production of a particular project. Legal partnerships are frequently utilized to achieve certain income tax or legal objectives.

Jointly owned corporations are very rare domestically. Internationally, most operations are conducted as joint ventures; however, there may be some instances where jointly owned corporations are formed. The laws of certain foreign countries may call for oil and gas operations to be carried out only by locally incorporated companies. In order to comply with the local laws, two or more companies may find it necessary to set up a company in that country. The companies establishing the foreign company usually own all of the stock in the new corporation.

Joint venture contracts

There are several types of contracts that may be encountered in joint operations. Examples of typical contracts include unit agreements, pooling agreements, and exploration agreements. The most commonly encountered agreement is the joint operating agreement (JOA). This agreement exists among working interest owners who own undivided interests in a joint property. Through negotiation, the working interest owners designate one of the owners—typically the one with the largest interest percentage—as the operator. The other working interest owners are designated as nonoperators. The operator manages the day-to-day operations of the property, but all the working interest owners participate in the property by voting on any major decisions affecting the property. The operator, as normally required by the operating agreement, generally pays all of the costs of exploration, development, and operation of the joint interest property, and then sends a joint interest billing to the nonoperators for their proportionate share of the costs. In some instances, particularly in international operations, the operator may estimate the amount of cash that will be required to operate the property each month and request a cash advance from the nonoperators for their portion of the expected cash outlays. The operator then pays the respective vendors. At the end of the month, the operator sends a settlement statement to the nonoperators detailing how cash was actually spent.

The operator may act for the nonoperators in marketing the oil and gas or, depending on the particular situation, each of the working interest owners may contract individually to market its own share of production. If the operator markets the oil or gas, the revenue is generally distributed to the working interest owners at the end of each month, according to their respective percentage ownership interests, and the royalties are paid to the royalty owners. If each working interest owner separately markets its respective share of oil and gas, then each working interest owner is responsible for paying the royalty on the quantity of oil and gas sold by that particular working interest owner during the month.

THE JOINT OPERATING AGREEMENT

When a joint interest situation is created, the parties involved, *i.e.*, the operator and nonoperators, generally execute an operating agreement. In domestic operations the typical form used for the JOA is the *AAPL Form 610*, which is available from the American Association of Petroleum Landmen (AAPL) based in Fort Worth, Texas.

The JOA spells out how the property is to be operated. One important part of the JOA is the accounting procedure. Both the JOA itself and the accounting procedure are of significant importance to the accountant. The main provisions of a typical JOA and accounting procedure are discussed following the sample JOA (Fig. 11-1). It is important to remember that, as with any negotiated contract, the parties are free to include any provisions and language they believe to be important. Therefore, the discussion that follows relates to the typical case. It is always important to thoroughly analyze any contract to be sure the specific provisions the parties have agreed upon are understood.

Fig. 11-1 — *Operating Agreement*

A.A.P.L. FORM 610 - 1989

MODEL FORM OPERATING AGREEMENT

OPERATING AGREEMENT

DATED

____________, 19 ____,

OPERATOR ____

CONTRACT AREA ____

____

____

____

____

COUNTY OR PARISH OF ____________________, **STATE OF** ____________________

OPERATING AGREEMENT

THIS AGREEMENT, entered into by and between ______________________________,
hereinafter designated and referred to as "Operator," and the signatory party or parties other than Operator, sometimes hereinafter referred to individually as "Non-Operator," and collectively as "Non-Operators."

WITNESSETH:

WHEREAS, the parties to this agreement are owners of Oil and Gas Leases and/or Oil and Gas Interests in the land identified in Exhibit "A," and the parties hereto have reached an agreement to explore and develop these Leases and/or Oil and Gas Interests for the production of Oil and Gas to the extent and as hereinafter provided,

NOW, THEREFORE, it is agreed as follows:

ARTICLE I.
DEFINITIONS

As used in this agreement, the following words and terms shall have the meanings here ascribed to them:

A. The term "AFE" shall mean an Authority for Expenditure prepared by a party to this agreement for the purpose of estimating the costs to be incurred in conducting an operation hereunder.

B. The term "Completion" or "Complete" shall mean a single operation intended to complete a well as a producer of Oil and Gas in one or more Zones, including, but not limited to, the setting of production casing, perforating, well stimulation and production testing conducted in such operation.

C. The term "Contract Area" shall mean all of the lands, Oil and Gas Leases and/or Oil and Gas Interests intended to be developed and operated for Oil and Gas purposes under this agreement. Such lands, Oil and Gas Leases and Oil and Gas Interests are described in Exhibit "A."

D. The term "Deepen" shall mean a single operation whereby a well is drilled to an objective Zone below the deepest Zone in which the well was previously drilled, or below the Deepest Zone proposed in the associated AFE, whichever is the lesser.

E. The terms "Drilling Party" and "Consenting Party" shall mean a party who agrees to join in and pay its share of the cost of any operation conducted under the provisions of this agreement.

F. The term "Drilling Unit" shall mean the area fixed for the drilling of one well by order or rule of any state or federal body having authority. If a Drilling Unit is not fixed by any such rule or order, a Drilling Unit shall be the drilling unit as established by the pattern of drilling in the Contract Area unless fixed by express agreement of the Drilling Parties.

G. The term "Drillsite" shall mean the Oil and Gas Lease or Oil and Gas Interest on which a proposed well is to be located.

H. The term "Initial Well" shall mean the well required to be drilled by the parties hereto as provided in Article VI.A.

I. The term "Non-Consent Well" shall mean a well in which less than all parties have conducted an operation as provided in Article VI.B.2.

J. The terms "Non-Drilling Party" and "Non-Consenting Party" shall mean a party who elects not to participate in a proposed operation.

K. The term "Oil and Gas" shall mean oil, gas, casinghead gas, gas condensate, and/or all other liquid or gaseous hydrocarbons and other marketable substances produced therewith, unless an intent to limit the inclusiveness of this term is specifically stated.

L. The term "Oil and Gas Interests" or "Interests" shall mean unleased fee and mineral interests in Oil and Gas in tracts of land lying within the Contract Area which are owned by parties to this agreement.

M. The terms "Oil and Gas Lease," "Lease" and "Leasehold" shall mean the oil and gas leases or interests therein covering tracts of land lying within the Contract Area which are owned by the parties to this agreement.

N. The term "Plug Back" shall mean a single operation whereby a deeper Zone is abandoned in order to attempt a Completion in a shallower Zone.

O. The term "Recompletion" or "Recomplete" shall mean an operation whereby a Completion in one Zone is abandoned in order to attempt a Completion in a different Zone within the existing wellbore.

P. The term "Rework" shall mean an operation conducted in the wellbore of a well after it is Completed to secure, restore, or improve production in a Zone which is currently open to production in the wellbore. Such operations include, but are not limited to, well stimulation operations but exclude any routine repair or maintenance work or drilling, Sidetracking, Deepening, Completing, Recompleting, or Plugging Back of a well.

Q. The term "Sidetrack" shall mean the directional control and intentional deviation of a well from vertical so as to change the bottom hole location unless done to straighten the hole or to drill around junk in the hole to overcome other mechanical difficulties.

R. The term "Zone" shall mean a stratum of earth containing or thought to contain a common accumulation of Oil and Gas separately producible from any other common accumulation of Oil and Gas.

Unless the context otherwise clearly indicates, words used in the singular include the plural, the word "person" includes natural and artificial persons, the plural includes the singular, and any gender includes the masculine, feminine, and neuter.

ARTICLE II.
EXHIBITS

The following exhibits, as indicated below and attached hereto, are incorporated in and made a part hereof:

_______ A. Exhibit "A," shall include the following information:
- (1) Description of lands subject to this agreement,
- (2) Restrictions, if any, as to depths, formations, or substances,
- (3) Parties to agreement with addresses and telephone numbers for notice purposes,
- (4) Percentages or fractional interests of parties to this agreement,
- (5) Oil and Gas Leases and/or Oil and Gas Interests subject to this agreement,
- (6) Burdens on production.

_______ B. Exhibit "B," Form of Lease.
_______ C. Exhibit "C," Accounting Procedure.
_______ D. Exhibit "D," Insurance.
_______ E. Exhibit "E," Gas Balancing Agreement.
_______ F. Exhibit "F," Non-Discrimination and Certification of Non-Segregated Facilities.
_______ G. Exhibit "G," Tax Partnership.
_______ H. Other: ____

If any provision of any exhibit, except Exhibits "E," "F" and "G," is inconsistent with any provision contained in the body of this agreement, the provisions in the body of this agreement shall prevail.

ARTICLE III.
INTERESTS OF PARTIES

A. Oil and Gas Interests:

If any party owns an Oil and Gas Interest in the Contract Area, that Interest shall be treated for all purposes of this agreement and during the term hereof as if it were covered by the form of Oil and Gas Lease attached hereto as Exhibit "B," and the owner thereof shall be deemed to own both royalty interest in such lease and the interest of the lessee thereunder.

B. Interests of Parties in Costs and Production:

Unless changed by other provisions, all costs and liabilities incurred in operations under this agreement shall be borne and paid, and all equipment and materials acquired in operations on the Contract Area shall be owned, by the parties as their interests are set forth in Exhibit "A." In the same manner, the parties shall also own all production of Oil and Gas from the Contract Area subject, however, to the payment of royalties and other burdens on production as described hereafter.

Regardless of which party has contributed any Oil and Gas Lease or Oil and Gas Interest on which royalty or other burdens may be payable and except as otherwise expressly provided in this agreement, each party shall pay or deliver, or cause to be paid or delivered, all burdens on its share of the production from the Contract Area up to, but not in excess of, ______________________________ and shall indemnify, defend and hold the other parties free from any liability therefor. Except as otherwise expressly provided in this agreement, if any party has contributed hereto any Lease or Interest which is burdened with any royalty, overriding royalty, production payment or other burden on production in excess of the amounts stipulated above, such party so burdened shall assume and alone bear all such excess obligations and shall indemnify, defend and hold the other parties hereto harmless from any and all claims attributable to such excess burden. However, so long as the Drilling Unit for the productive Zone(s) is identical with the Contract Area, each party shall pay or deliver, or cause to be paid or delivered, all burdens on production from the Contract Area due under the terms of the Oil and Gas Lease(s) which such party has contributed to this agreement, and shall indemnify, defend and hold the other parties free from any liability therefor.

No party shall ever be responsible, on a price basis higher than the price received by such party, to any other party's lessor or royalty owner, and if such other party's lessor or royalty owner should demand and receive settlement on a higher price basis, the party contributing the affected Lease shall bear the additional royalty burden attributable to such higher price.

Nothing contained in this Article III.B. shall be deemed an assignment or cross-assignment of interests covered hereby, and in the event two or more parties contribute to this agreement jointly owned Leases, the parties' undivided interests in said Leaseholds shall be deemed separate leasehold interests for the purposes of this agreement.

C. Subsequently Created Interests:

If any party has contributed hereto a Lease or Interest that is burdened with an assignment of production given as security for the payment of money, or if, after the date of this agreement, any party creates an overriding royalty, production payment, net profits interest, assignment of production or other burden payable out of production attributable to its working interest hereunder, such burden shall be deemed a "Subsequently Created Interest." Further, if any party has contributed hereto a Lease or Interest burdened with an overriding royalty, production payment, net profits interest, or other burden payable out of production created prior to the date of this agreement, and such burden is not shown on Exhibit "A," such burden also shall be deemed a Subsequently Created Interest to the extent such burden causes the burdens on such party's Lease or Interest to exceed the amount stipulated in Article III.B. above.

The party whose interest is burdened with the Subsequently Created Interest (the "Burdened Party") shall assume and alone bear, pay and discharge the Subsequently Created Interest and shall indemnify, defend and hold harmless the other parties from and against any liability therefor. Further, if the Burdened Party fails to pay, when due, its share of expenses chargeable hereunder, all provisions of Article VII.B. shall be enforceable against the Subsequently Created Interest in the same manner as they are enforceable against the working interest of the Burdened Party. If the Burdened Party is required under this agreement to assign or relinquish to any other party, or parties, all or a portion of its working interest and/or the production attributable thereto, said other party, or parties, shall receive said assignment and/or production free and clear of said Subsequently Created Interest, and the Burdened Party shall indemnify, defend and hold harmless said other party, or parties, from any and all claims and demands for payment asserted by owners of the Subsequently Created Interest.

ARTICLE IV.
TITLES

A. Title Examination:

Title examination shall be made on the Drillsite of any proposed well prior to commencement of drilling operations and, if a majority in interest of the Drilling Parties so request or Operator so elects, title examination shall be made on the entire Drilling Unit, or maximum anticipated Drilling Unit, of the well. The opinion will include the ownership of the working interest, minerals, royalty, overriding royalty and production payments under the applicable Leases. Each party contributing Leases and/or Oil and Gas Interests to be included in the Drillsite or Drilling Unit, if appropriate, shall furnish to Operator all abstracts (including federal lease status reports), title opinions, title papers and curative material in its possession free of charge. All such information not in the possession of or made available to Operator by the parties, but necessary for the examination of the title, shall be obtained by Operator. Operator shall cause title to be examined by attorneys on its staff or by outside attorneys. Copies of all title opinions shall be furnished to each Drilling Party. Costs incurred by Operator in procuring abstracts, fees paid outside attorneys for title examination (including preliminary, supplemental, shut-in royalty opinions and division order title opinions) and other direct charges as provided in Exhibit "C" shall be borne by the Drilling Parties in the proportion that the interest of each Drilling Party bears to the total interest of all Drilling Parties as such interests appear in Exhibit "A." Operator shall make no charge for services rendered by its staff attorneys or other personnel in the performance of the above functions.

Each party shall be responsible for securing curative matter and pooling amendments or agreements required in connection with Leases or Oil and Gas Interests contributed by such party. Operator shall be responsible for the preparation and recording of pooling designations or declarations and communitization agreements as well as the conduct of hearings before governmental agencies for the securing of spacing or pooling orders or any other orders necessary or appropriate to the conduct of operations hereunder. This shall not prevent any party from appearing on its own behalf at such hearings. Costs incurred by Operator, including fees paid to outside attorneys, which are associated with hearings before governmental agencies, and which costs are necessary and proper for the activities contemplated under this agreement, shall be direct charges to the joint account and shall not be covered by the administrative overhead charges as provided in Exhibit "C."

Operator shall make no charge for services rendered by its staff attorneys or other personnel in the performance of the above functions.

No well shall be drilled on the Contract Area until after (1) the title to the Drillsite or Drilling Unit, if appropriate, has been examined as above provided, and (2) the title has been approved by the examining attorney or title has been accepted by all of the Drilling Parties in such well.

B. Loss or Failure of Title:

1. Failure of Title: Should any Oil and Gas Interest or Oil and Gas Lease be lost through failure of title, which results in a reduction of interest from that shown on Exhibit "A," the party credited with contributing the affected Lease or Interest (including, if applicable, a successor in interest to such party) shall have ninety (90) days from final determination of title failure to acquire a new lease or other instrument curing the entirety of the title failure, which acquisition will not be subject to Article VIII.B., and failing to do so, this agreement, nevertheless, shall continue in force as to all remaining Oil and Gas Leases and Interests; and,

(a) The party credited with contributing the Oil and Gas Lease or Interest affected by the title failure (including, if applicable, a successor in interest to such party) shall bear alone the entire loss and it shall not be entitled to recover from Operator or the other parties any development or operating costs which it may have previously paid or incurred, but there shall be no additional liability on its part to the other parties hereto by reason of such title failure;

(b) There shall be no retroactive adjustment of expenses incurred or revenues received from the operation of the Lease or Interest which has failed, but the interests of the parties contained on Exhibit "A" shall be revised on an acreage basis, as of the time it is determined finally that title failure has occurred, so that the interest of the party whose Lease or Interest is affected by the title failure will thereafter be reduced in the Contract Area by the amount of the Lease or Interest failed;

(c) If the proportionate interest of the other parties hereto in any producing well previously drilled on the Contract Area is increased by reason of the title failure, the party who bore the costs incurred in connection with such well attributable to the Lease or Interest which has failed shall receive the proceeds attributable to the increase in such interest (less costs and burdens attributable thereto) until it has been reimbursed for unrecovered costs paid by it in connection with such well attributable to such failed Lease or Interest;

(d) Should any person not a party to this agreement, who is determined to be the owner of any Lease or Interest which has failed, pay in any manner any part of the cost of operation, development, or equipment, such amount shall be paid to the party or parties who bore the costs which are so refunded;

(e) Any liability to account to a person not a party to this agreement for prior production of Oil and Gas which arises by reason of title failure shall be borne severally by each party (including a predecessor to a current party) who received production for which such accounting is required based on the amount of such production received, and each such party shall severally indemnify, defend and hold harmless all other parties hereto for any such liability to account;

(f) No charge shall be made to the joint account for legal expenses, fees or salaries in connection with the defense of the Lease or Interest claimed to have failed, but if the party contributing such Lease or Interest hereto elects to defend its title it shall bear all expenses in connection therewith; and

(g) If any party is given credit on Exhibit "A" to a Lease or Interest which is limited solely to ownership of an interest in the wellbore of any well or wells and the production therefrom, such party's absence of interest in the remainder of the Contract Area shall be considered a Failure of Title as to such remaining Contract Area unless that absence of interest is reflected on Exhibit "A."

2. Loss by Non-Payment or Erroneous Payment of Amount Due: If, through mistake or oversight, any rental, shut-in well payment, minimum royalty or royalty payment, or other payment necessary to maintain all or a portion of an Oil and Gas Lease or Interest is not paid or is erroneously paid, and as a result a Lease or Interest terminates, there shall be no monetary liability against the party who failed to make such payment. Unless the party who failed to make the required payment secures a new Lease or Interest covering the same interest within ninety (90) days from the discovery of the failure to make proper payment, which acquisition will not be subject to Article VIII.B., the interests of the parties reflected on Exhibit "A" shall be revised on an acreage basis, effective as of the date of termination of the Lease or Interest involved, and the party who failed to make proper payment will no longer be credited with an interest in the Contract Area on account of ownership of the Lease or Interest which has terminated. If the party who failed to make the required payment shall not have been fully reimbursed, at the time of the loss, from the proceeds of the sale of Oil and Gas attributable to the lost Lease or Interest,

calculated on an acreage basis, for the development and operating costs previously paid on account of such Lease or Interest, it shall be reimbursed for unrecovered actual costs previously paid by it (but not for its share of the cost of any dry hole previously drilled or wells previously abandoned) from so much of the following as is necessary to effect reimbursement:

(a) Proceeds of Oil and Gas produced prior to termination of the Lease or Interest, less operating expenses and lease burdens chargeable hereunder to the person who failed to make payment, previously accrued to the credit of the lost Lease or Interest, on an acreage basis, up to the amount of unrecovered costs;

(b) Proceeds of Oil and Gas, less operating expenses and lease burdens chargeable hereunder to the person who failed to make payment, up to the amount of unrecovered costs attributable to that portion of Oil and Gas thereafter produced and marketed (excluding production from any wells thereafter drilled) which, in the absence of such Lease or Interest termination, would be attributable to the lost Lease or Interest on an acreage basis and which as a result of such Lease or Interest termination is credited to other parties, the proceeds of said portion of the Oil and Gas to be contributed by the other parties in proportion to their respective interests reflected on Exhibit "A"; and,

(c) Any monies, up to the amount of unrecovered costs, that may be paid by any party who is, or becomes, the owner of the Lease or Interest lost, for the privilege of participating in the Contract Area or becoming a party to this agreement.

3. Other Losses: All losses of Leases or Interests committed to this agreement, other than those set forth in Articles IV.B.1. and IV.B.2. above, shall be joint losses and shall be borne by all parties in proportion to their interests shown on Exhibit "A." This shall include but not be limited to the loss of any Lease or Interest through failure to develop or because express or implied covenants have not been performed (other than performance which requires only the payment of money), and the loss of any Lease by expiration at the end of its primary term if it is not renewed or extended. There shall be no readjustment of interests in the remaining portion of the Contract Area on account of any joint loss.

4. Curing Title: In the event of a Failure of Title under Article IV.B.1. or a loss of title under Article IV.B.2. above, any Lease or Interest acquired by any party hereto (other than the party whose interest has failed or was lost) during the ninety (90) day period provided by Article IV.B.1. and Article IV.B.2. above covering all or a portion of the interest that has failed or was lost shall be offered at cost to the party whose interest has failed or was lost, and the provisions of Article VIII.B. shall not apply to such acquisition.

ARTICLE V.
OPERATOR

A. Designation and Responsibilities of Operator:

________________________ shall be the Operator of the Contract Area, and shall conduct and direct and have full control of all operations on the Contract Area as permitted and required by, and within the limits of this agreement. In its performance of services hereunder for the Non-Operators, Operator shall be an independent contractor not subject to the control or direction of the Non-Operators except as to the type of operation to be undertaken in accordance with the election procedures contained in this agreement. Operator shall not be deemed, or hold itself out as, the agent of the Non-Operators with authority to bind them to any obligation or liability assumed or incurred by Operator as to any third party. Operator shall conduct its activities under this agreement as a reasonable prudent operator, in a good and workmanlike manner, with due diligence and dispatch, in accordance with good oilfield practice, and in compliance with applicable law and regulation, but in no event shall it have any liability as Operator to the other parties for losses sustained or liabilities incurred except such as may result from gross negligence or willful misconduct.

B. Resignation or Removal of Operator and Selection of Successor:

1. Resignation or Removal of Operator: Operator may resign at any time by giving written notice thereof to Non-Operators. If Operator terminates its legal existence, no longer owns an interest hereunder in the Contract Area, or is no longer capable of serving as Operator, Operator shall be deemed to have resigned without any action by Non-Operators, except the selection of a successor. Operator may be removed only for good cause by the affirmative vote of Non-Operators owning a majority interest based on ownership as shown on Exhibit "A" remaining after excluding the voting interest of Operator; such vote shall not be deemed effective until a written notice has been delivered to the Operator by a Non-Operator detailing the alleged default and Operator has failed to cure the default within thirty (30) days from its receipt of the notice or, if the default concerns an operation then being conducted, within forty-eight (48) hours of its receipt of the notice. For purposes hereof, "good cause" shall mean not only gross negligence or willful misconduct but also the material breach of or inability to meet the standards of operation contained in Article V.A. or material failure or inability to perform its obligations under this agreement.

Subject to Article VII.D.1., such resignation or removal shall not become effective until 7:00 o'clock A.M. on the first day of the calendar month following the expiration of ninety (90) days after the giving of notice of resignation by Operator or action by the Non-Operators to remove Operator, unless a successor Operator has been selected and assumes the duties of Operator at an earlier date. Operator, after effective date of resignation or removal, shall be bound by the terms hereof as a Non-Operator. A change of a corporate name or structure of Operator or transfer of Operator's interest to any single subsidiary, parent or successor corporation shall not be the basis for removal of Operator.

2. Selection of Successor Operator: Upon the resignation or removal of Operator under any provision of this agreement, a successor Operator shall be selected by the parties. The successor Operator shall be selected from the parties owning an interest in the Contract Area at the time such successor Operator is selected. The successor Operator shall be selected by the affirmative vote of two (2) or more parties owning a majority interest based on ownership as shown on Exhibit "A"; provided, however, if an Operator which has been removed or is deemed to have resigned fails to vote or votes only to succeed itself, the successor Operator shall be selected by the affirmative vote of the party or parties owning a majority interest based on ownership as shown on Exhibit "A" remaining after excluding the voting interest of the Operator that was removed or resigned. The former Operator shall promptly deliver to the successor Operator all records and data relating to the operations conducted by the former Operator to the extent such records and data are not already in the possession of the successor operator. Any cost of obtaining or copying the former Operator's records and data shall be charged to the joint account.

3. Effect of Bankruptcy: If Operator becomes insolvent, bankrupt or is placed in receivership, it shall be deemed to have resigned without any action by Non-Operators, except the selection of a successor. If a petition for relief under the federal bankruptcy laws is filed by or against Operator, and the removal of Operator is prevented by the federal bankruptcy court, all Non-Operators and Operator shall comprise an interim operating committee to serve until Operator has elected to reject or assume this agreement pursuant to the Bankruptcy Code, and an election to reject this agreement by Operator as a debtor in possession, or by a trustee in bankruptcy, shall be deemed a resignation as Operator without any action by Non-Operators, except the selection of a successor. During the period of time the operating committee controls operations, all actions shall require the approval of two (2) or more parties owning a majority interest based on ownership as shown on Exhibit "A." In the event there are only two (2) parties to this agreement, during the period of time the operating committee controls operations, a third party acceptable to Operator, Non-Operator and the federal bankruptcy court shall be selected as a member of the operating committee, and all actions shall require the approval of two (2) members of the operating committee without regard for their interest in the Contract Area based on Exhibit "A."

C. Employees and Contractors:

The number of employees or contractors used by Operator in conducting operations hereunder, their selection, and the hours of labor and the compensation for services performed shall be determined by Operator, and all such employees or contractors shall be the employees or contractors of Operator.

D. Rights and Duties of Operator:

1. Competitive Rates and Use of Affiliates: All wells drilled on the Contract Area shall be drilled on a competitive contract basis at the usual rates prevailing in the area. If it so desires, Operator may employ its own tools and equipment in the drilling of wells, but its charges therefor shall not exceed the prevailing rates in the area and the rate of such charges shall be agreed upon by the parties in writing before drilling operations are commenced, and such work shall be performed by Operator under the same terms and conditions as are customary and usual in the area in contracts of independent contractors who are doing work of a similar nature. All work performed or materials supplied by affiliates or related parties of Operator shall be performed or supplied at competitive rates, pursuant to written agreement, and in accordance with customs and standards prevailing in the industry.

2. Discharge of Joint Account Obligations: Except as herein otherwise specifically provided, Operator shall promptly pay and discharge expenses incurred in the development and operation of the Contract Area pursuant to this agreement and shall charge each of the parties hereto with their respective proportionate shares upon the expense basis provided in Exhibit "C." Operator shall keep an accurate record of the joint account hereunder, showing expenses incurred and charges and credits made and received.

3. Protection from Liens: Operator shall pay, or cause to be paid, as and when they become due and payable, all accounts of contractors and suppliers and wages and salaries for services rendered or performed, and for materials supplied on, to or in respect of the Contract Area or any operations for the joint account thereof, and shall keep the Contract Area free from

liens and encumbrances resulting therefrom except for those resulting from a bona fide dispute as to services rendered or materials supplied.

4. Custody of Funds: Operator shall hold for the account of the Non-Operators any funds of the Non-Operators advanced or paid to the Operator, either for the conduct of operations hereunder or as a result of the sale of production from the Contract Area, and such funds shall remain the funds of the Non-Operators on whose account they are advanced or paid until used for their intended purpose or otherwise delivered to the Non-Operators or applied toward the payment of debts as provided in Article VII.B. Nothing in this paragraph shall be construed to establish a fiduciary relationship between Operator and Non-Operators for any purpose other than to account for Non-Operator funds as herein specifically provided. Nothing in this paragraph shall require the maintenance by Operator of separate accounts for the funds of Non-Operators unless the parties otherwise specifically agree.

5. Access to Contract Area and Records: Operator shall, except as otherwise provided herein, permit each Non-Operator or its duly authorized representative, at the Non-Operator's sole risk and cost, full and free access at all reasonable times to all operations of every kind and character being conducted for the joint account on the Contract Area and to the records of operations conducted thereon or production therefrom, including Operator's books and records relating thereto. Such access rights shall not be exercised in a manner interfering with Operator's conduct of an operation hereunder and shall not obligate Operator to furnish any geologic or geophysical data of an interpretive nature unless the cost of preparation of such interpretive data was charged to the joint account. Operator will furnish to each Non-Operator upon request copies of any and all reports and information obtained by Operator in connection with production and related items, including, without limitation, meter and chart reports, production purchaser statements, run tickets and monthly gauge reports, but excluding purchase contracts and pricing information to the extent not applicable to the production of the Non-Operator seeking the information. Any audit of Operator's records relating to amounts expended and the appropriateness of such expenditures shall be conducted in accordance with the audit protocol specified in Exhibit "C."

6. Filing and Furnishing Governmental Reports: Operator will file, and upon written request promptly furnish copies to each requesting Non-Operator not in default of its payment obligations, all operational notices, reports or applications required to be filed by local, State, Federal or Indian agencies or authorities having jurisdiction over operations hereunder. Each Non-Operator shall provide to Operator on a timely basis all information necessary to Operator to make such filings.

7. Drilling and Testing Operations: The following provisions shall apply to each well drilled hereunder, including but not limited to the Initial Well:

(a) Operator will promptly advise Non-Operators of the date on which the well is spudded, or the date on which drilling operations are commenced.

(b) Operator will send to Non-Operators such reports, test results and notices regarding the progress of operations on the well as the Non-Operators shall reasonably request, including, but not limited to, daily drilling reports, completion reports, and well logs.

(c) Opeator shall adequately test all Zones encountered which may reasonably be expected to be capable of producing Oil and Gas in paying quantities as a result of examination of the electric log or any other logs or cores or tests conducted hereunder.

8. Cost Estimates. Upon request of any Consenting Party, Operator shall furnish estimates of current and cumulative costs incurred for the joint account at reasonable intervals during the conduct of any operation pursuant to this agreement. Operator shall not be held liable for errors in such estimates so long as the estimates are made in good faith.

9. Insurance: At all times while operations are conducted hereunder, Operator shall comply with the workers compensation law of the state where the operations are being conducted; provided, however, that Operator may be a self-insurer for liability under said compensation laws in which event the only charge that shall be made to the joint account shall be as provided in Exhibit "C." Operator shall also carry or provide insurance for the benefit of the joint account of the parties as outlined in Exhibit "D" attached hereto and made a part hereof. Operator shall require all contractors engaged in work on or for the Contract Area to comply with the workers compensation law of the state where the operations are being conducted and to maintain such other insurance as Operator may require.

In the event automobile liability insurance is specified in said Exhibit "D," or subsequently receives the approval of the parties, no direct charge shall be made by Operator for premiums paid for such insurance for Operator's automotive equipment.

ARTICLE VI.
DRILLING AND DEVELOPMENT

A. Initial Well:

On or before the ________ day of ________________ , 19 ______ , Operator shall commence the drilling of the Initial Well at the following location:

and shall thereafter continue the drilling of the well with due diligence to

The drilling of the Initial Well and the participation therein by all parties is obligatory, subject to Article VI.C.1. as to participation in Completion operations and Article VI.F. as to termination of operations and Article XI as to occurrence of force majeure.

B. Subsequent Operations:

1. Proposed Operations: If any party hereto should desire to drill any well on the Contract Area other than the Initial Well, or if any party should desire to Rework, Sidetrack, Deepen, Recomplete or Plug Back a dry hole or a well no longer capable of producing in paying quantities in which such party has not otherwise relinquished its interest in the proposed objective Zone under this agreement, the party desiring to drill, Rework, Sidetrack, Deepen, Recomplete or Plug Back such a well shall give written notice of the proposed operation to the parties who have not otherwise relinquished their interest in such objective Zone under this agreement and to all other parties in the case of a proposal for Sidetracking or Deepening, specifying the work to be performed, the location, proposed depth, objective Zone and the estimated cost of the operation. The parties to whom such a notice is delivered shall have thirty (30) days after receipt of the notice within which to notify the party proposing to do the work whether they elect to participate in the cost of the proposed operation. If a drilling rig is on location, notice of a proposal to Rework, Sidetrack, Recomplete, Plug Back or Deepen may be given by telephone and the response period shall be limited to forty-eight (48) hours, exclusive of Saturday, Sunday and legal holidays. Failure of a party to whom such notice is delivered to reply within the period above fixed shall constitute an election by that party not to participate in the cost of the proposed operation. Any proposal by a party to conduct an operation conflicting with the operation initially proposed shall be delivered to all parties within the time and in the manner provided in Article VI.B.6.

If all parties to whom such notice is delivered elect to participate in such a proposed operation, the parties shall be contractually committed to participate therein provided such operations are commenced within the time period hereafter set forth, and Operator shall, no later than ninety (90) days after expiration of the notice period of thirty (30) days (or as promptly as practicable after the expiration of the forty-eight (48) hour period when a drilling rig is on location, as the case may be), actually commence the proposed operation and thereafter complete it with due diligence at the risk and expense of the parties participating therein; provided, however, said commencement date may be extended upon written notice of same by Operator to the other parties, for a period of up to thirty (30) additional days if, in the sole opinion of Operator, such additional time is reasonably necessary to obtain permits from governmental authorities, surface rights (including rights-of-way) or appropriate drilling equipment, or to complete title examination or curative matter required for title approval or acceptance. If the actual operation has not been commenced within the time provided (including any extension thereof as specifically permitted herein or in the force majeure provisions of Article XI) and if any party hereto still desires to conduct said operation, written notice proposing same must be resubmitted to the other parties in accordance herewith as if no prior proposal had been made. Those parties that did not participate in the drilling of a well for which a proposal to Deepen or Sidetrack is made hereunder shall, if such parties desire to participate in the proposed Deepening or Sidetracking operation, reimburse the Drilling Parties in accordance with Article VI.B.4. in the event of a Deepening operation and in accordance with Article VI.B.5. in the event of a Sidetracking operation.

2. Operations by Less Than All Parties:

(a) Determination of Participation. If any party to whom such notice is delivered as provided in Article VI.B.1. or VI.C.1. (Option No. 2) elects not to participate in the proposed operation, then, in order to be entitled to the benefits of this Article, the party or parties giving the notice and such other parties as shall elect to participate in the operation shall, no later than ninety (90) days after the expiration of the notice period of thirty (30) days (or as promptly as practicable after the expiration of the forty-eight (48) hour period when a drilling rig is on location, as the case may be) actually commence the proposed operation and complete it with due diligence. Operator shall perform all work for the account of the Consenting Parties; provided, however, if no drilling rig or other equipment is on location, and if Operator is a Non-Consenting Party, the Consenting Parties shall either: (i) request Operator to perform the work required by such proposed operation for the account of the Consenting Parties, or (ii) designate one of the Consenting Parties as Operator to perform such work. The rights and duties granted to and imposed upon the Operator under this agreement are granted to and imposed upon the party designated as Operator for an operation in which the original Operator is a Non-Consenting Party. Consenting Parties, when conducting operations on the Contract Area pursuant to this Article VI.B.2., shall comply with all terms and conditions of this agreement.

If less than all parties approve any proposed operation, the proposing party, immediately after the expiration of the applicable notice period, shall advise all Parties of the total interest of the parties approving such operation and its recommendation as to whether the Consenting Parties should proceed with the operation as proposed. Each Consenting Party, within forty-eight (48) hours (exclusive of Saturday, Sunday and legal holidays) after delivery of such notice, shall advise the proposing party of its desire to (i) limit participation to such party's interest as shown on Exhibit "A" or (ii) carry only its proportionate part (determined by dividing such party's interest in the Contract Area by the interests of all Consenting Parties in the Contract Area) of Non-Consenting Parties' interests, or (iii) carry its proportionate part (determined as provided in (ii)) of Non-Consenting Parties' interests together with all or a portion of its proportionate part of any Non-Consenting Parties' interests that any Consenting Party did not elect to take. Any interest of Non-Consenting Parties that is not carried by a Consenting Party shall be deemed to be carried by the party proposing the operation if such party does not withdraw its proposal. Failure to advise the proposing party within the time required shall be deemed an election under (i) . In the event a drilling rig is on location, notice may be given by telephone, and the time permitted for such a response shall not exceed a total of forty-eight (48) hours (exclusive of Saturday, Sunday and legal holidays). The proposing party, at its election, may withdraw such proposal if there is less than 100% participation and shall notify all parties of such decision within ten (10) days, or within twenty-four (24) hours if a drilling rig is on location, following expiration of the applicable response period. If 100% subscription to the proposed operation is obtained, the proposing party shall promptly notify the Consenting Parties of their proportionate interests in the operation and the party serving as Operator shall commence such operation within the period provided in Article VI.B.1., subject to the same extension right as provided therein.

(b) Relinquishment of Interest for Non-Participation. The entire cost and risk of conducting such operations shall be borne by the Consenting Parties in the proportions they have elected to bear same under the terms of the preceding paragraph. Consenting Parties shall keep the leasehold estates involved in such operations free and clear of all liens and encumbrances of every kind created by or arising from the operations of the Consenting Parties. If such an operation results in a dry hole, then subject to Articles VI.B.6. and VI.E.3., the Consenting Parties shall plug and abandon the well and restore the surface location at their sole cost, risk and expense; provided, however, that those Non-Consenting Parties that participated in the drilling, Deepening or Sidetracking of the well shall remain liable for, and shall pay, their proportionate shares of the cost of plugging and abandoning the well and restoring the surface location insofar only as those costs were not increased by the subsequent operations of the Consenting Parties. If any well drilled, Reworked, Sidetracked, Deepened, Recompleted or Plugged Back under the provisions of this Article results in a well capable of producing Oil and/or Gas in paying quantities, the Consenting Parties shall Complete and equip the well to produce at their sole cost and risk, and the well shall then be turned over to Operator (if the Operator did not conduct the operation) and shall be operated by it at the expense and for the account of the Consenting Parties. Upon commencement of operations for the drilling, Reworking, Sidetracking, Recompleting, Deepening or Plugging Back of any such well by Consenting Parties in accordance with the provisions of this Article, each Non-Consenting Party shall be deemed to have relinquished to Consenting Parties, and the Consenting Parties shall own and be entitled to receive, in proportion to their respective interests, all of such Non-Consenting Party's interest in the well and share of production therefrom or, in the case of a Reworking, Sidetracking,

Deepening, Recompleting or Plugging Back, or a Completion pursuant to Article VI.C.1. Option No. 2, all of such Non-Consenting Party's interest in the production obtained from the operation in which the Non-Consenting Party did not elect to participate. Such relinquishment shall be effective until the proceeds of the sale of such share, calculated at the well, or market value thereof if such share is not sold (after deducting applicable ad valorem, production, severance, and excise taxes, royalty, overriding royalty and other interests not excepted by Article III.C. payable out of or measured by the production from such well accruing with respect to such interest until it reverts), shall equal the total of the following:

(i) ______ % of each such Non-Consenting Party's share of the cost of any newly acquired surface equipment beyond the wellhead connections (including but not limited to stock tanks, separators, treaters, pumping equipment and piping), plus 100% of each such Non-Consenting Party's share of the cost of operation of the well commencing with first production and continuing until each such Non-Consenting Party's relinquished interest shall revert to it under other provisions of this Article, it being agreed that each Non-Consenting Party's share of such costs and equipment will be that interest which would have been chargeable to such Non-Consenting Party had it participated in the well from the beginning of the operations; and

(ii) ______ % of (a) that portion of the costs and expenses of drilling, Reworking, Sidetracking, Deepening, Plugging Back, testing, Completing, and Recompleting, after deducting any cash contributions received under Article VIII.C., and of (b) that portion of the cost of newly acquired equipment in the well (to and including the wellhead connections), which would have been chargeable to such Non-Consenting Party if it had participated therein.

Notwithstanding anything to the contrary in this Article VI.B., if the well does not reach the deepest objective Zone described in the notice proposing the well for reasons other than the encountering of granite or practically impenetrable substance or other condition in the hole rendering further operations impracticable, Operator shall give notice thereof to each Non-Consenting Party who submitted or voted for an alternative proposal under Article VI.B.6. to drill the well to a shallower Zone than the deepest objective Zone proposed in the notice under which the well was drilled, and each such Non-Consenting Party shall have the option to participate in the initial proposed Completion of the well by paying its share of the cost of drilling the well to its actual depth, calculated in the manner provided in Article VI.B.4. (a). If any such Non-Consenting Party does not elect to participate in the first Completion proposed for such well, the relinquishment provisions of this Article VI.B.2. (b) shall apply to such party's interest.

(c) Reworking, Recompleting or Plugging Back. An election not to participate in the drilling, Sidetracking or Deepening of a well shall be deemed an election not to participate in any Reworking or Plugging Back operation proposed in such a well, or portion thereof, to which the initial non-consent election applied that is conducted at any time prior to full recovery by the Consenting Parties of the Non-Consenting Party's recoupment amount. Similarly, an election not to participate in the Completing or Recompleting of a well shall be deemed an election not to participate in any Reworking operation proposed in such a well, or portion thereof, to which the initial non-consent election applied that is conducted at any time prior to full recovery by the Consenting Parties of the Non-Consenting Party's recoupment amount. Any such Reworking, Recompleting or Plugging Back operation conducted during the recoupment period shall be deemed part of the cost of operation of said well and there shall be added to the sums to be recouped by the Consenting Parties ______ % of that portion of the costs of the Reworking, Recompleting or Plugging Back operation which would have been chargeable to such Non-Consenting Party had it participated therein. If such a Reworking, Recompleting or Plugging Back operation is proposed during such recoupment period, the provisions of this Article VI.B. shall be applicable as between said Consenting Parties in said well.

(d) Recoupment Matters. During the period of time Consenting Parties are entitled to receive Non-Consenting Party's share of production, or the proceeds therefrom, Consenting Parties shall be responsible for the payment of all ad valorem, production, severance, excise, gathering and other taxes, and all royalty, overriding royalty and other burdens applicable to Non-Consenting Party's share of production not excepted by Article III.C.

In the case of any Reworking, Sidetracking, Plugging Back, Recompleting or Deepening operation, the Consenting Parties shall be permitted to use, free of cost, all casing, tubing and other equipment in the well, but the ownership of all such equipment shall remain unchanged; and upon abandonment of a well after such Reworking, Sidetracking, Plugging Back, Recompleting or Deepening, the Consenting Parties shall account for all such equipment to the owners thereof, with each party receiving its proportionate part in kind or in value, less cost of salvage.

Within ninety (90) days after the completion of any operation under this Article, the party conducting the operations for the Consenting Parties shall furnish each Non-Consenting Party with an inventory of the equipment in and connected to the well, and an itemized statement of the cost of drilling, Sidetracking, Deepening, Plugging Back, testing, Completing, Recompleting, and equipping the well for production; or, at its option, the operating party, in lieu of an itemized statement of such costs of operation, may submit a detailed statement of monthly billings. Each month thereafter, during the time the Consenting Parties are being reimbursed as provided above, the party conducting the operations for the Consenting Parties shall furnish the Non-Consenting Parties with an itemized statement of all costs and liabilities incurred in the operation of the well, together with a statement of the quantity of Oil and Gas produced from it and the amount of proceeds realized from the sale of the well's working interest production during the preceding month. In determining the quantity of Oil and Gas produced during any month, Consenting Parties shall use industry accepted methods such as but not limited to metering or periodic well tests. Any amount realized from the sale or other disposition of equipment newly acquired in connection with any such operation which would have been owned by a Non-Consenting Party had it participated therein shall be credited against the total unreturned costs of the work done and of the equipment purchased in determining when the interest of such Non-Consenting Party shall revert to it as above provided; and if there is a credit balance, it shall be paid to such Non-Consenting Party.

If and when the Consenting Parties recover from a Non-Consenting Party's relinquished interest the amounts provided for above, the relinquished interests of such Non-Consenting Party shall automatically revert to it as of 7:00 a.m. on the day following the day on which such recoupment occurs, and, from and after such reversion, such Non-Consenting Party shall own the same interest in such well, the material and equipment in or pertaining thereto, and the production therefrom as such Non-Consenting Party would have been entitled to had it participated in the drilling, Sidetracking, Reworking, Deepening, Recompleting or Plugging Back of said well. Thereafter, such Non-Consenting Party shall be charged with and shall pay its proportionate part of the further costs of the operation of said well in accordance with the terms of this agreement and Exhibit "C" attached hereto.

3. Stand-By Costs: When a well which has been drilled or Deepened has reached its authorized depth and all tests have been completed and the results thereof furnished to the parties, or when operations on the well have been otherwise terminated pursuant to Article VI.F., stand-by costs incurred pending response to a party's notice proposing a Reworking, Sidetracking, Deepening, Recompleting, Plugging Back or Completing operation in such a well (including the period required under Article VI.B.6. to resolve competing proposals) shall be charged and borne as part of the drilling or Deepening operation just completed. Stand-by costs subsequent to all parties responding, or expiration of the response time permitted, whichever first occurs, and prior to agreement as to the participating interests of all Consenting Parties pursuant to the terms of the second grammatical paragraph of Article VI.B.2. (a), shall be charged to and borne as part of the proposed operation, but if the proposal is subsequently withdrawn because of insufficient participation, such stand-by costs shall be allocated between the Consenting Parties in the proportion each Consenting Party's interest as shown on Exhibit "A" bears to the total interest as shown on Exhibit "A" of all Consenting Parties.

In the event that notice for a Sidetracking operation is given while the drilling rig to be utilized is on location, any party may request and receive up to five (5) additional days after expiration of the forty-eight hour response period specified in Article VI.B.1. within which to respond by paying for all stand-by costs and other costs incurred during such extended response period; Operator may require such party to pay the estimated stand-by time in advance as a condition to extending the response period. If more than one party elects to take such additional time to respond to the notice, standby costs shall be allocated between the parties taking additional time to respond on a day-to-day basis in the proportion each electing party's interest as shown on Exhibit "A" bears to the total interest as shown on Exhibit "A" of all the electing parties.

4. Deepening: If less than all the parties elect to participate in a drilling, Sidetracking, or Deepening operation proposed pursuant to Article VI.B.1., the interest relinquished by the Non-Consenting Parties to the Consenting Parties under Article VI.B.2. shall relate only and be limited to the lesser of (i) the total depth actually drilled or (ii) the objective depth or Zone of which the parties were given notice under Article VI.B.1. ("Initial Objective"). Such well shall not be Deepened beyond the Initial Objective without first complying with this Article to afford the Non-Consenting Parties the opportunity to participate in the Deepening operation.

In the event any Consenting Party desires to drill or Deepen a Non-Consent Well to a depth below the Initial Objective, such party shall give notice thereof, complying with the requirements of Article VI.B.1., to all parties (including Non-

Consenting Parties). Thereupon, Articles VI.B.1. and 2. shall apply and all parties receiving such notice shall have the right to participate or not participate in the Deepening of such well pursuant to said Articles VI.B.1. and 2. If a Deepening operation is approved pursuant to such provisions, and if any Non-Consenting Party elects to participate in the Deepening operation, such Non-Consenting party shall pay or make reimbursement (as the case may be) of the following costs and expenses:

(a) If the proposal to Deepen is made prior to the Completion of such well as a well capable of producing in paying quantities, such Non-Consenting Party shall pay (or reimburse Consenting Parties for, as the case may be) that share of costs and expenses incurred in connection with the drilling of said well from the surface to the Initial Objective which Non-Consenting Party would have paid had such Non-Consenting Party agreed to participate therein, plus the Non-Consenting Party's share of the cost of Deepening and of participating in any further operations on the well in accordance with the other provisions of this Agreement; provided, however, all costs for testing and Completion or attempted Completion of the well incurred by Consenting Parties prior to the point of actual operations to Deepen beyond the Initial Objective shall be for the sole account of Consenting Parties.

(b) If the proposal is made for a Non-Consent Well that has been previously Completed as a well capable of producing in paying quantities, but is no longer capable of producing in paying quantities, such Non-Consenting Party shall pay (or reimburse Consenting Parties for, as the case may be) its proportionate share of all costs of drilling, Completing, and equipping said well from the surface to the Initial Objective, calculated in the manner provided in paragraph (a) above, less those costs recouped by the Consenting Parties from the sale of production from the well. The Non-Consenting Party shall also pay its proportionate share of all costs of re-entering said well. The Non-Consenting Parties' proportionate part (based on the percentage of such well Non-Consenting Party would have owned had it previously participated in such Non-Consent Well) of the costs of salvable materials and equipment remaining in the hole and salvable surface equipment used in connection with such well shall be determined in accordance with Exhibit "C." If the Consenting Parties have recouped the cost of drilling, Completing, and equipping the well at the time such Deepening operation is conducted, then a Non-Consenting Party may participate in the Deepening of the well with no payment for costs incurred prior to re-entering the well for Deepening.

The foregoing shall not imply a right of any Consenting Party to propose any Deepening for a Non-Consent Well prior to the drilling of such well to its Initial Objective without the consent of the other Consenting Parties as provided in Article VI.F.

5. <u>Sidetracking:</u> Any party having the right to participate in a proposed Sidetracking operation that does not own an interest in the affected wellbore at the time of the notice shall, upon electing to participate, tender to the wellbore owners its proportionate share (equal to its interest in the Sidetracking operation) of the value of that portion of the existing wellbore to be utilized as follows:

(a) If the proposal is for Sidetracking an existing dry hole, reimbursement shall be on the basis of the actual costs incurred in the initial drilling of the well down to the depth at which the Sidetracking operation is initiated.

(b) If the proposal is for Sidetracking a well which has previously produced, reimbursement shall be on the basis of such party's proportionate share of drilling and equipping costs incurred in the initial drilling of the well down to the depth at which the Sidetracking operation is conducted, calculated in the manner described in Article VI.B.4(b) above. Such party's proportionate share of the cost of the well's salvable materials and equipment down to the depth at which the Sidetracking operation is initiated shall be determined in accordance with the provisions of Exhibit "C."

6. <u>Order of Preference of Operations.</u> Except as otherwise specifically provided in this agreement, if any party desires to propose the conduct of an operation that conflicts with a proposal that has been made by a party under this Article VI, such party shall have fifteen (15) days from delivery of the initial proposal, in the case of a proposal to drill a well or to perform an operation on a well where no drilling rig is on location, or twenty-four (24) hours, exclusive of Saturday, Sunday and legal holidays, from delivery of the initial proposal, if a drilling rig is on location for the well on which such operation is to be conducted, to deliver to all parties entitled to participate in the proposed operation such party's alternative proposal, such alternate proposal to contain the same information required to be included in the initial proposal. Each party receiving such proposals shall elect by delivery of notice to Operator within five (5) days after expiration of the proposal period, or within twenty-four (24) hours (exclusive of Saturday, Sunday and legal holidays) if a drilling rig is on location for the well that is the subject of the proposals, to participate in one of the competing proposals. Any party not electing within the time required shall be deemed not to have voted. The proposal receiving the vote of parties owning the largest aggregate percentage interest of the parties voting shall have priority over all other competing proposals; in the case of a tie vote, the

initial proposal shall prevail. Operator shall deliver notice of such result to all parties entitled to participate in the operation within five (5) days after expiration of the election period (or within twenty-four (24) hours, exclusive of Saturday, Sunday and legal holidays, if a drilling rig is on location). Each party shall then have two (2) days (or twenty-four (24) hours if a rig is on location) from receipt of such notice to elect by delivery of notice to Operator to participate in such operation or to relinquish interest in the affected well pursuant to the provisions of Article VI.B.2.; failure by a party to deliver notice within such period shall be deemed an election not to participate in the prevailing proposal.

7. Conformity to Spacing Pattern. Notwithstanding the provisions of this Article VI.B.2., it is agreed that no wells shall be proposed to be drilled to or Completed in or produced from a Zone from which a well located elsewhere on the Contract Area is producing, unless such well conforms to the then-existing well spacing pattern for such Zone.

8. Paying Wells. No party shall conduct any Reworking, Deepening, Plugging Back, Completion, Recompletion, or Sidetracking operation under this agreement with respect to any well then capable of producing in paying quantities except with the consent of all parties that have not relinquished interests in the well at the time of such operation.

C. Completion of Wells; Reworking and Plugging Back:

1. Completion: Without the consent of all parties, no well shall be drilled, Deepened or Sidetracked, except any well drilled, Deepened or Sidetracked pursuant to the provisions of Article VI.B.2. of this agreement. Consent to the drilling, Deepening or Sidetracking shall include:

☐ Option No. 1: All necessary expenditures for the drilling, Deepening or Sidetracking, testing, Completing and equipping of the well, including necessary tankage and/or surface facilities.

☐ Option No. 2: All necessary expenditures for the drilling, Deepening or Sidetracking and testing of the well. When such well has reached its authorized depth, and all logs, cores and other tests have been completed, and the results thereof furnished to the parties, Operator shall give immediate notice to the Non-Operators having the right to participate in a Completion attempt whether or not Operator recommends attempting to Complete the well, together with Operator's AFE for Completion costs if not previously provided. The parties receiving such notice shall have forty-eight (48) hours (exclusive of Saturday, Sunday and legal holidays) in which to elect by delivery of notice to Operator to participate in a recommended Completion attempt or to make a Completion proposal with an accompanying AFE. Operator shall deliver any such Completion proposal, or any Completion proposal conflicting with Operator's proposal, to the other parties entitled to participate in such Completion in accordance with the procedures specified in Article VI.B.6. Election to participate in a Completion attempt shall include consent to all necessary expenditures for the Completing and equipping of such well, including necessary tankage and/or surface facilities but excluding any stimulation operation not contained on the Completion AFE. Failure of any party receiving such notice to reply within the period above fixed shall constitute an election by that party not to participate in the cost of the Completion attempt; provided, that Article VI.B.6. shall control in the case of conflicting Completion proposals. If one or more, but less than all of the parties, elect to attempt a Completion, the provisions of Article VI.B.2. hereof (the phrase "Reworking, Sidetracking, Deepening, Recompleting or Plugging Back" as contained in Article VI.B.2. shall be deemed to include "Completing") shall apply to the operations thereafter conducted by less than all parties; provided, however, that Article VI.B.2 shall apply separately to each separate Completion or Recompletion attempt undertaken hereunder, and an election to become a Non-Consenting Party as to one Completion or Recompletion attempt shall not prevent a party from becoming a Consenting Party in subsequent Completion or Recompletion attempts regardless whether the Consenting Parties as to earlier Completions or Recompletions have recouped their costs pursuant to Article VI.B.2.; provided further, that any recoupment of costs by a Consenting Party shall be made solely from the production attributable to the Zone in which the Completion attempt is made. Election by a previous Non-Consenting Party to participate in a subsequent Completion or Recompletion attempt shall require such party to pay its proportionate share of the cost of salvable materials and equipment installed in the well pursuant to the previous Completion or Recompletion attempt, insofar and only insofar as such materials and equipment benefit the Zone in which such party participates in a Completion attempt.

2. Rework, Recomplete or Plug Back: No well shall be Reworked, Recompleted or Plugged Back except a well Reworked, Recompleted, or Plugged Back pursuant to the provisions of Article VI.B.2. of this agreement. Consent to the Reworking,

Recompleting or Plugging Back of a well shall include all necessary expenditures in conducting such operations and Completing and equipping of said well, including necessary tankage and/or surface facilities.

D. Other Operations:

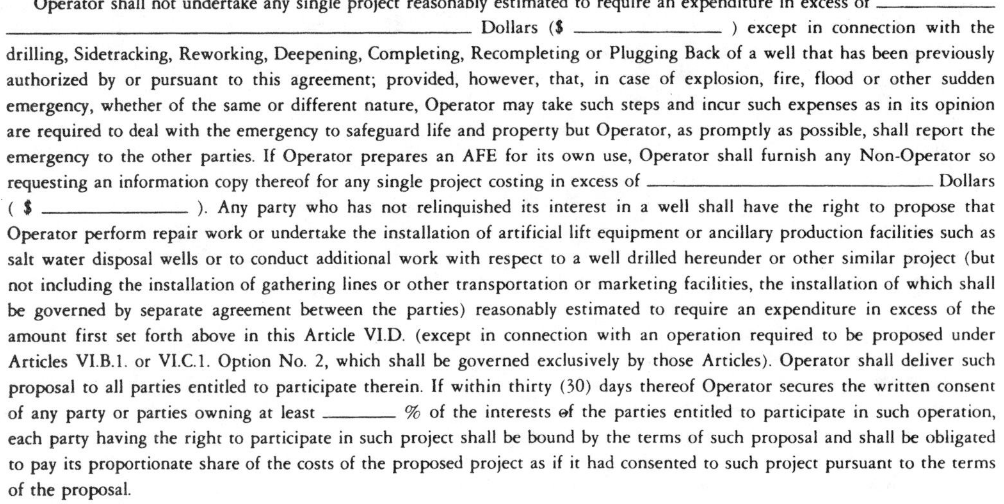

Operator shall not undertake any single project reasonably estimated to require an expenditure in excess of ____________ ____ Dollars ($ ______________) except in connection with the drilling, Sidetracking, Reworking, Deepening, Completing, Recompleting or Plugging Back of a well that has been previously authorized by or pursuant to this agreement; provided, however, that, in case of explosion, fire, flood or other sudden emergency, whether of the same or different nature, Operator may take such steps and incur such expenses as in its opinion are required to deal with the emergency to safeguard life and property but Operator, as promptly as possible, shall report the emergency to the other parties. If Operator prepares an AFE for its own use, Operator shall furnish any Non-Operator so requesting an information copy thereof for any single project costing in excess of ____________________________ Dollars ($ ______________). Any party who has not relinquished its interest in a well shall have the right to propose that Operator perform repair work or undertake the installation of artificial lift equipment or ancillary production facilities such as salt water disposal wells or to conduct additional work with respect to a well drilled hereunder or other similar project (but not including the installation of gathering lines or other transportation or marketing facilities, the installation of which shall be governed by separate agreement between the parties) reasonably estimated to require an expenditure in excess of the amount first set forth above in this Article VI.D. (except in connection with an operation required to be proposed under Articles VI.B.1. or VI.C.1. Option No. 2, which shall be governed exclusively by those Articles). Operator shall deliver such proposal to all parties entitled to participate therein. If within thirty (30) days thereof Operator secures the written consent of any party or parties owning at least ________ % of the interests of the parties entitled to participate in such operation, each party having the right to participate in such project shall be bound by the terms of such proposal and shall be obligated to pay its proportionate share of the costs of the proposed project as if it had consented to such project pursuant to the terms of the proposal.

E. Abandonment of Wells:

1. Abandonment of Dry Holes: Except for any well drilled or Deepened pursuant to Article VI.B.2., any well which has been drilled or Deepened under the terms of this agreement and is proposed to be completed as a dry hole shall not be plugged and abandoned without the consent of all parties. Should Operator, after diligent effort, be unable to contact any party, or should any party fail to reply within forty-eight (48) hours (exclusive of Saturday, Sunday and legal holidays) after delivery of notice of the proposal to plug and abandon such well, such party shall be deemed to have consented to the proposed abandonment. All such wells shall be plugged and abandoned in accordance with applicable regulations and at the cost, risk and expense of the parties who participated in the cost of drilling or Deepening such well. Any party who objects to plugging and abandoning such well by notice delivered to Operator within forty-eight (48) hours (exclusive of Saturday, Sunday and legal holidays) after delivery of notice of the proposed plugging shall take over the well as of the end of such forty-eight (48) hour notice period and conduct further operations in search of Oil and/or Gas subject to the provisions of Article VI.B.; failure of such party to provide proof reasonably satisfactory to Operator of its financial capability to conduct such operations or to take over the well within such period or thereafter to conduct operations on such well or plug and abandon such well shall entitle Operator to retain or take possession of the well and plug and abandon the well. The party taking over the well shall indemnify Operator (if Operator is an abandoning party) and the other abandoning parties against liability for any further operations conducted on such well except for the costs of plugging and abandoning the well and restoring the surface, for which the abandoning parties shall remain proportionately liable.

2. Abandonment of Wells That Have Produced: Except for any well in which a Non-Consent operation has been conducted hereunder for which the Consenting Parties have not been fully reimbursed as herein provided, any well which has been completed as a producer shall not be plugged and abandoned without the consent of all parties. If all parties consent to such abandonment, the well shall be plugged and abandoned in accordance with applicable regulations and at the cost, risk and expense of all the parties hereto. Failure of a party to reply within sixty (60) days of delivery of notice of proposed abandonment shall be deemed an election to consent to the proposal. If, within sixty (60) days after delivery of notice of the proposed abandonment of any well, all parties do not agree to the abandonment of such well, those wishing to continue its operation from the Zone then open to production shall be obligated to take over the well as of the expiration of the applicable notice period and shall indemnify Operator (if Operator is an abandoning party) and the other abandoning parties against liability for any further operations on the well conducted by such parties. Failure of such party or parties to provide

proof reasonably satisfactory to Operator of their financial capability to conduct such operations or to take over the well within the required period or thereafter to conduct operations on such well shall entitle Operator to retain or take possession of such well and plug and abandon the well.

Parties taking over a well as provided herein shall tender to each of the other parties its proportionate share of the value of the well's salvable material and equipment, determined in accordance with the provisions of Exhibit "C," less the estimated cost of salvaging and the estimated cost of plugging and abandoning and restoring the surface; provided, however, that in the event the estimated plugging and abandoning and surface restoration costs and the estimated cost of salvaging are higher than the value of the well's salvable material and equipment, each of the abandoning parties shall tender to the parties continuing operations their proportionate shares of the estimated excess cost. Each abandoning party shall assign to the non-abandoning parties, without warranty, express or implied, as to title or as to quantity, or fitness for use of the equipment and material, all of its interest in the wellbore of the well and related equipment, together with its interest in the Leasehold insofar and only insofar as such Leasehold covers the right to obtain production from that wellbore in the Zone then open to production. If the interest of the abandoning party is or includes an Oil and Gas Interest, such party shall execute and deliver to the non-abandoning party or parties an oil and gas lease, limited to the wellbore and the Zone then open to production, for a term of one (1) year and so long thereafter as Oil and/or Gas is produced from the Zone covered thereby, such lease to be on the form attached as Exhibit "B." The assignments or leases so limited shall encompass the Drilling Unit upon which the well is located. The payments by, and the assignments or leases to, the assignees shall be in a ratio based upon the relationship of their respective percentage of participation in the Contract Area to the aggregate of the percentages of participation in the Contract Area of all assignees. There shall be no readjustment of interests in the remaining portions of the Contract Area.

Thereafter, abandoning parties shall have no further responsibility, liability, or interest in the operation of or production from the well in the Zone then open other than the royalties retained in any lease made under the terms of this Article. Upon request, Operator shall continue to operate the assigned well for the account of the non-abandoning parties at the rates and charges contemplated by this agreement, plus any additional cost and charges which may arise as the result of the separate ownership of the assigned well. Upon proposed abandonment of the producing Zone assigned or leased, the assignor or lessor shall then have the option to repurchase its prior interest in the well (using the same valuation formula) and participate in further operations therein subject to the provisions hereof.

3. Abandonment of Non-Consent Operations: The provisions of Article VI.E.1. or VI.E.2. above shall be applicable as between Consenting Parties in the event of the proposed abandonment of any well excepted from said Articles; provided, however, no well shall be permanently plugged and abandoned unless and until all parties having the right to conduct further operations therein have been notified of the proposed abandonment and afforded the opportunity to elect to take over the well in accordance with the provisions of this Article VI.E.; and provided further, that Non-Consenting Parties who own an interest in a portion of the well shall pay their proportionate shares of abandonment and surface restoration costs for such well as provided in Article VI.B.2.(b).

F. Termination of Operations:

Upon the commencement of an operation for the drilling, Reworking, Sidetracking, Plugging Back, Deepening, testing, Completion or plugging of a well, including but not limited to the Initial Well, such operation shall not be terminated without consent of parties bearing ________ % of the costs of such operation; provided, however, that in the event granite or other practically impenetrable substance or condition in the hole is encountered which renders further operations impractical, Operator may discontinue operations and give notice of such condition in the manner provided in Article VI.B.1, and the provisions of Article VI.B. or VI.E. shall thereafter apply to such operation, as appropriate.

G. Taking Production in Kind:

☐ **Option No. 1: Gas Balancing Agreement Attached**

Each party shall take in kind or separately dispose of its proportionate share of all Oil and Gas produced from the Contract Area, exclusive of production which may be used in development and producing operations and in preparing and treating Oil and Gas for marketing purposes and production unavoidably lost. Any extra expenditure incurred in the taking in kind or separate disposition by any party of its proportionate share of the production shall be borne by such party. Any party taking its share of production in kind shall be required to pay for only its proportionate share of such part of Operator's surface facilities which it uses.

Each party shall execute such division orders and contracts as may be necessary for the sale of its interest in production from the Contract Area, and, except as provided in Article VII.B., shall be entitled to receive payment directly from the purchaser thereof for its share of all production.

If any party fails to make the arrangements necessary to take in kind or separately dispose of its proportionate share of the Oil produced from the Contract Area, Operator shall have the right, subject to the revocation at will by the party owning it, but not the obligation, to purchase such Oil or sell it to others at any time and from time to time, for the account of the non-taking party. Any such purchase or sale by Operator may be terminated by Operator upon at least ten (10) days written notice to the owner of said production and shall be subject always to the right of the owner of the production upon at least ten (10) days written notice to Operator to exercise at any time its right to take in kind, or separately dispose of, its share of all Oil not previously delivered to a purchaser. Any purchase or sale by Operator of any other party's share of Oil shall be only for such reasonable periods of time as are consistent with the minimum needs of the industry under the particular circumstances, but in no event for a period in excess of one (1) year.

Any such sale by Operator shall be in a manner commercially reasonable under the circumstances but Operator shall have no duty to share any existing market or to obtain a price equal to that received under any existing market. The sale or delivery by Operator of a non-taking party's share of Oil under the terms of any existing contract of Operator shall not give the non-taking party any interest in or make the non-taking party a party to said contract. No purchase shall be made by Operator without first giving the non-taking party at least ten (10) days written notice of such intended purchase and the price to be paid or the pricing basis to be used.

All parties shall give timely written notice to Operator of their Gas marketing arrangements for the following month, excluding price, and shall notify Operator immediately in the event of a change in such arrangements. Operator shall maintain records of all marketing arrangements, and of volumes actually sold or transported, which records shall be made available to Non-Operators upon reasonable request.

In the event one or more parties' separate disposition of its share of the Gas causes split-stream deliveries to separate pipelines and/or deliveries which on a day-to-day basis for any reason are not exactly equal to a party's respective proportionate share of total Gas sales to be allocated to it, the balancing or accounting between the parties shall be in accordance with any Gas balancing agreement between the parties hereto, whether such an agreement is attached as Exhibit "E" or is a separate agreement. Operator shall give notice to all parties of the first sales of Gas from any well under this agreement.

☐ **Option No. 2: No Gas Balancing Agreement:**

Each party shall take in kind or separately dispose of its proportionate share of all Oil and Gas produced from the Contract Area, exclusive of production which may be used in development and producing operations and in preparing and treating Oil and Gas for marketing purposes and production unavoidably lost. Any extra expenditure incurred in the taking in kind or separate disposition by any party of its proportionate share of the production shall be borne by such party. Any party taking its share of production in kind shall be required to pay for only its proportionate share of such part of Operator's surface facilities which it uses.

Each party shall execute such division orders and contracts as may be necessary for the sale of its interest in production from the Contract Area, and, except as provided in Article VII.B., shall be entitled to receive payment directly from the purchaser thereof for its share of all production.

If any party fails to make the arrangements necessary to take in kind or separately dispose of its proportionate share of the Oil and/or Gas produced from the Contract Area, Operator shall have the right, subject to the revocation at will by the party owning it, but not the obligation, to purchase such Oil and/or Gas or sell it to others at any time and from time to time, for the account of the non-taking party. Any such purchase or sale by Operator may be terminated by Operator upon at least ten (10) days written notice to the owner of said production and shall be subject always to the right of the owner of the production upon at least ten (10) days written notice to Operator to exercise its right to take in kind, or separately dispose of, its share of all Oil and/or Gas not previously delivered to a purchaser; provided, however, that the effective date of any such revocation may be deferred at Operator's election for a period not to exceed ninety (90) days if Operator has committed such production to a purchase contract having a term extending beyond such ten (10) -day period. Any purchase or sale by Operator of any other party's share of Oil and/or Gas shall be only for such reasonable periods of time as are consistent with the minimum needs of the industry under the particular circumstances, but in no event for a period in excess of one (1) year.

Any such sale by Operator shall be in a manner commercially reasonable under the circumstances, but Operator shall have no duty to share any existing market or transportation arrangement or to obtain a price or transportation fee equal to that received under any existing market or transportation arrangement. The sale or delivery by Operator of a non-taking party's share of production under the terms of any existing contract of Operator shall not give the non-taking party any interest in or make the non-taking party a party to said contract. No purchase of Oil and Gas and no sale of Gas shall be made by Operator without first giving the non-taking party ten days written notice of such intended purchase or sale and the price to be paid or the pricing basis to be used. Operator shall give notice to all parties of the first sale of Gas from any well under this Agreement.

All parties shall give timely written notice to Operator of their Gas marketing arrangements for the following month, excluding price, and shall notify Operator immediately in the event of a change in such arrangements. Operator shall maintain records of all marketing arrangements, and of volumes actually sold or transported, which records shall be made available to Non-Operators upon reasonable request.

ARTICLE VII.
EXPENDITURES AND LIABILITY OF PARTIES

A. Liability of Parties:

The liability of the parties shall be several, not joint or collective. Each party shall be responsible only for its obligations, and shall be liable only for its proportionate share of the costs of developing and operating the Contract Area. Accordingly, the liens granted among the parties in Article VII.B. are given to secure only the debts of each severally, and no party shall have any liability to third parties hereunder to satisfy the default of any other party in the payment of any expense or obligation hereunder. It is not the intention of the parties to create, nor shall this agreement be construed as creating, a mining or other partnership, joint venture, agency relationship or association, or to render the parties liable as partners, co-venturers, or principals. In their relations with each other under this agreement, the parties shall not be considered fiduciaries or to have established a confidential relationship but rather shall be free to act on an arm's-length basis in accordance with their own respective self-interest, subject, however, to the obligation of the parties to act in good faith in their dealings with each other with respect to activities hereunder.

B. Liens and Security Interests:

Each party grants to the other parties hereto a lien upon any interest it now owns or hereafter acquires in Oil and Gas Leases and Oil and Gas Interests in the Contract Area, and a security interest and/or purchase money security interest in any interest it now owns or hereafter acquires in the personal property and fixtures on or used or obtained for use in connection therewith, to secure performance of all of its obligations under this agreement including but not limited to payment of expense, interest and fees, the proper disbursement of all monies paid hereunder, the assignment or relinquishment of interest in Oil and Gas Leases as required hereunder, and the proper performance of operations hereunder. Such lien and security interest granted by each party hereto shall include such party's leasehold interests, working interests, operating rights, and royalty and overriding royalty interests in the Contract Area now owned or hereafter acquired and in lands pooled or unitized therewith or otherwise becoming subject to this agreement, the Oil and Gas when extracted therefrom and equipment situated thereon or used or obtained for use in connection therewith (including, without limitation, all wells, tools, and tubular goods), and accounts (including, without limitation, accounts arising from gas imbalances or from the sale of Oil and/or Gas at the wellhead), contract rights, inventory and general intangibles relating thereto or arising therefrom, and all proceeds and products of the foregoing.

To perfect the lien and security agreement provided herein, each party hereto shall execute and acknowledge the recording supplement and/or any financing statement prepared and submitted by any party hereto in conjunction herewith or at any time following execution hereof, and Operator is authorized to file this agreement or the recording supplement executed herewith as a lien or mortgage in the applicable real estate records and as a financing statement with the proper officer under the Uniform Commercial Code in the state in which the Contract Area is situated and such other states as Operator shall deem appropriate to perfect the security interest granted hereunder. Any party may file this agreement, the recording supplement executed herewith, or such other documents as it deems necessary as a lien or mortgage in the applicable real estate records and/or a financing statement with the proper officer under the Uniform Commercial Code.

Each party represents and warrants to the other parties hereto that the lien and security interest granted by such party to the other parties shall be a first and prior lien, and each party hereby agrees to maintain the priority of said lien and security interest against all persons acquiring an interest in Oil and Gas Leases and Interests covered by this agreement by, through or under such party. All parties acquiring an interest in Oil and Gas Leases and Oil and Gas Interests covered by this agreement, whether by assignment, merger, mortgage, operation of law, or otherwise, shall be deemed to have taken subject to the lien and security interest granted by this Article VII.B. as to all obligations attributable to such interest hereunder whether or not such obligations arise before or after such interest is acquired.

To the extent that parties have a security interest under the Uniform Commercial Code of the state in which the Contract Area is situated, they shall be entitled to exercise the rights and remedies of a secured party under the Code. The bringing of a suit and the obtaining of judgment by a party for the secured indebtedness shall not be deemed an election of remedies or otherwise affect the lien rights or security interest as security for the payment thereof. In addition, upon default by any party in the payment of its share of expenses, interests or fees, or upon the improper use of funds by the Operator, the other parties shall have the right, without prejudice to other rights or remedies, to collect from the purchaser the proceeds from the sale of such defaulting party's share of Oil and Gas until the amount owed by such party, plus interest as provided in "Exhibit C," has been received, and shall have the right to offset the amount owed against the proceeds from the sale of such defaulting party's share of Oil and Gas. All purchasers of production may rely on a notification of default from the non-defaulting party or parties stating the amount due as a result of the default, and all parties waive any recourse available against purchasers for releasing production proceeds as provided in this paragraph.

If any party fails to pay its share of cost within one hundred twenty (120) days after rendition of a statement therefor by Operator, the non-defaulting parties, including Operator, shall, upon request by Operator, pay the unpaid amount in the proportion that the interest of each such party bears to the interest of all such parties. The amount paid by each party so paying its share of the unpaid amount shall be secured by the liens and security rights described in Article VII.B., and each paying party may independently pursue any remedy available hereunder or otherwise.

If any party does not perform all of its obligations hereunder, and the failure to perform subjects such party to foreclosure or execution proceedings pursuant to the provisions of this agreement, to the extent allowed by governing law, the defaulting party waives any available right of redemption from and after the date of judgment, any required valuation or appraisement of the mortgaged or secured property prior to sale, any available right to stay execution or to require a marshalling of assets and any required bond in the event a receiver is appointed. In addition, to the extent permitted by applicable law, each party hereby grants to the other parties a power of sale as to any property that is subject to the lien and security rights granted hereunder, such power to be exercised in the manner provided by applicable law or otherwise in a commercially reasonable manner and upon reasonable notice.

Each party agrees that the other parties shall be entitled to utilize the provisions of Oil and Gas lien law or other lien law of any state in which the Contract Area is situated to enforce the obligations of each party hereunder. Without limiting the generality of the foregoing, to the extent permitted by applicable law, Non-Operators agree that Operator may invoke or utilize the mechanics' or materialmen's lien law of the state in which the Contract Area is situated in order to sercure the payment to Operator of any sum due hereunder for services performed or materials supplied by Operator.

C. Advances:

Operator, at its election, shall have the right from time to time to demand and receive from one or more of the other parties payment in advance of their respective shares of the estimated amount of the expense to be incurred in operations hereunder during the next succeeding month, which right may be exercised only by submission to each such party of an itemized statement of such estimated expense, together with an invoice for its share thereof. Each such statement and invoice for the payment in advance of estimated expense shall be submitted on or before the 20th day of the next preceding month. Each party shall pay to Operator its proportionate share of such estimate within fifteen (15) days after such estimate and invoice is received. If any party fails to pay its share of said estimate within said time, the amount due shall bear interest as provided in Exhibit "C" until paid. Proper adjustment shall be made monthly between advances and actual expense to the end that each party shall bear and pay its proportionate share of actual expenses incurred, and no more.

D. Defaults and Remedies:

If any party fails to discharge any financial obligation under this agreement, including without limitation the failure to make any advance under the preceding Article VII.C. or any other provision of this agreement, within the period required for such payment hereunder, then in addition to the remedies provided in Article VII.B. or elsewhere in this agreement, the remedies specified below shall be applicable. For purposes of this Article VII.D., all notices and elections shall be delivered only by Operator, except that Operator shall deliver any such notice and election requested by a non-defaulting Non-Operator, and when Operator is the party in default, the applicable notices and elections can be delivered by any Non-Operator. Election of any one or more of the following remedies shall not preclude the subsequent use of any other remedy specified below or otherwise available to a non-defaulting party.

1. Suspension of Rights: Any party may deliver to the party in default a Notice of Default, which shall specify the default, specify the action to be taken to cure the default, and specify that failure to take such action will result in the exercise of one or more of the remedies provided in this Article. If the default is not cured within thirty (30) days of the delivery of such Notice of Default, all of the rights of the defaulting party granted by this agreement may upon notice be suspended until the default is cured, without prejudice to the right of the non-defaulting party or parties to continue to enforce the obligations of the defaulting party previously accrued or thereafter accruing under this agreement. If Operator is the party in default, the Non-Operators shall have in addition the right, by vote of Non-Operators owning a majority in interest in the Contract Area after excluding the voting interest of Operator, to appoint a new Operator effective immediately. The rights of a defaulting party that may be suspended hereunder at the election of the non-defaulting parties shall include, without limitation, the right to receive information as to any operation conducted hereunder during the period of such default, the right to elect to participate in an operation proposed under Article VI.B. of this agreement, the right to participate in an operation being conducted under this agreement even if the party has previously elected to participate in such operation, and the right to receive proceeds of production from any well subject to this agreement.

2. Suit for Damages: Non-defaulting parties or Operator for the benefit of non-defaulting parties may sue (at joint account expense) to collect the amounts in default, plus interest accruing on the amounts recovered from the date of default until the date of collection at the rate specified in Exhibit "C" attached hereto. Nothing herein shall prevent any party from suing any defaulting party to collect consequential damages accruing to such party as a result of the default.

3. Deemed Non-Consent: The non-defaulting party may deliver a written Notice of Non-Consent Election to the defaulting party at any time after the expiration of the thirty-day cure period following delivery of the Notice of Default, in which event if the billing is for the drilling of a new well or the Plugging Back, Sidetracking, Reworking or Deepening of a well which is to be or has been plugged as a dry hole, or for the Completion or Recompletion of any well, the defaulting party will be conclusively deemed to have elected not to participate in the operation and to be a Non-Consenting Party with respect thereto under Article VI.B. or VI.C., as the case may be, to the extent of the costs unpaid by such party, notwithstanding any election to participate theretofore made. If election is made to proceed under this provision, then the non-defaulting parties may not elect to sue for the unpaid amount pursuant to Article VII.D.2.

Until the delivery of such Notice of Non-Consent Election to the defaulting party, such party shall have the right to cure its default by paying its unpaid share of costs plus interest at the rate set forth in Exhibit "C," provided, however, such payment shall not prejudice the rights of the non-defaulting parties to pursue remedies for damages incurred by the non-defaulting parties as a result of the default. Any interest relinquished pursuant to this Article VII.D.3. shall be offered to the non-defaulting parties in proportion to their interests, and the non-defaulting parties electing to participate in the ownership of such interest shall be required to contribute their shares of the defaulted amount upon their election to participate therein.

4. Advance Payment: If a default is not cured within thirty (30) days of the delivery of a Notice of Default, Operator, or Non-Operators if Operator is the defaulting party, may thereafter require advance payment from the defaulting party of such defaulting party's anticipated share of any item of expense for which Operator, or Non-Operators, as the case may be, would be entitled to reimbursement under any provision of this agreement, whether or not such expense was the subject of the previous default. Such right includes, but is not limited to, the right to require advance payment for the estimated costs of drilling a well or Completion of a well as to which an election to participate in drilling or Completion has been made. If the defaulting party fails to pay the required advance payment, the non-defaulting parties may pursue any of the remedies provided in this Article VII.D. or any other default remedy provided elsewhere in this agreement. Any excess of funds advanced remaining when the operation is completed and all costs have been paid shall be promptly returned to the advancing party.

5. Costs and Attorneys' Fees. In the event any party is required to bring legal proceedings to enforce any financial obligation of a party hereunder, the prevailing party in such action shall be entitled to recover all court costs, costs of collection, and a reasonable attorney's fee, which the lien provided for herein shall also secure.

E. Rentals, Shut-in Well Payments and Minimum Royalties:

Rentals, shut-in well payments and minimum royalties which may be required under the terms of any lease shall be paid by the party or parties who subjected such lease to this agreement at its or their expense. In the event two or more parties own and have contributed interests in the same lease to this agreement, such parties may designate one of such parties to make said payments for and on behalf of all such parties. Any party may request, and shall be entitled to receive, proper evidence of all such payments. In the event of failure to make proper payment of any rental, shut-in well payment or minimum royalty through mistake or oversight where such payment is required to continue the lease in force, any loss which results from such non-payment shall be borne in accordance with the provisions of Article IV.B.2.

Operator shall notify Non-Operators of the anticipated completion of a shut-in well, or the shutting in or return to production of a producing well, at least five (5) days (excluding Saturday, Sunday and legal holidays) prior to taking such action, or at the earliest opportunity permitted by circumstances, but assumes no liability for failure to do so. In the event of failure by Operator to so notify Non-Operators, the loss of any lease contributed hereto by Non-Operators for failure to make timely payments of any shut-in well payment shall be borne jointly by the parties hereto under the provisions of Article IV.B.3.

F. Taxes:

Beginning with the first calendar year after the effective date hereof, Operator shall render for ad valorem taxation all property subject to this agreement which by law should be rendered for such taxes, and it shall pay all such taxes assessed thereon before they become delinquent. Prior to the rendition date, each Non-Operator shall furnish Operator information as to burdens (to include, but not be limited to, royalties, overriding royalties and production payments) on Leases and Oil and Gas Interests contributed by such Non-Operator. If the assessed valuation of any Lease is reduced by reason of its being subject to outstanding excess royalties, overriding royalties or production payments, the reduction in ad valorem taxes resulting therefrom shall inure to the benefit of the owner or owners of such Lease, and Operator shall adjust the charge to such owner or owners so as to reflect the benefit of such reduction. If the ad valorem taxes are based in whole or in part upon separate valuations of each party's working interest, then notwithstanding anything to the contrary herein, charges to the joint account shall be made and paid by the parties hereto in accordance with the tax value generated by each party's working interest. Operator shall bill the other parties for their proportionate shares of all tax payments in the manner provided in Exhibit "C."

If Operator considers any tax assessment improper, Operator may, at its discretion, protest within the time and manner prescribed by law, and prosecute the protest to a final determination, unless all parties agree to abandon the protest prior to final determination. During the pendency of administrative or judicial proceedings, Operator may elect to pay, under protest, all such taxes and any interest and penalty. When any such protested assessment shall have been finally determined, Operator shall pay the tax for the joint account, together with any interest and penalty accrued, and the total cost shall then be assessed against the parties, and be paid by them, as provided in Exhibit "C."

Each party shall pay or cause to be paid all production, severance, excise, gathering and other taxes imposed upon or with respect to the production or handling of such party's share of Oil and Gas produced under the terms of this agreement.

ARTICLE VIII.
ACQUISITION, MAINTENANCE OR TRANSFER OF INTEREST

A. Surrender of Leases:

The Leases covered by this agreement, insofar as they embrace acreage in the Contract Area, shall not be surrendered in whole or in part unless all parties consent thereto.

However, should any party desire to surrender its interest in any Lease or in any portion thereof, such party shall give written notice of the proposed surrender to all parties, and the parties to whom such notice is delivered shall have thirty (30) days after delivery of the notice within which to notify the party proposing the surrender whether they elect to consent thereto. Failure of a party to whom such notice is delivered to reply within said 30-day period shall constitute a consent to the surrender of the Leases described in the notice. If all parties do not agree or consent thereto, the party desiring to surrender shall assign, without express or implied warranty of title, all of its interest in such Lease, or portion thereof, and any well, material and equipment which may be located thereon and any rights in production thereafter secured, to the parties not consenting to such surrender. If the interest of the assigning party is or includes an Oil and Gas Interest, the assigning party shall execute and deliver to the party or parties not consenting to such surrender an oil and gas lease covering such Oil and Gas Interest for a term of one (1) year and so long thereafter as Oil and/or Gas is produced from the land covered thereby, such lease to be on the form attached hereto as Exhibit "B." Upon such assignment or lease, the assigning party shall be relieved from all obligations thereafter accruing, but not theretofore accrued, with respect to the interest assigned or leased and the operation of any well attributable thereto, and the assigning party shall have no further interest in the assigned or leased premises and its equipment and production other than the royalties retained in any lease made under the terms of this Article. The party assignee or lessee shall pay to the party assignor or lessor the reasonable salvage value of the latter's interest in any well's salvable materials and equipment attributable to the assigned or leased acreage. The value of all salvable materials and equipment shall be determined in accordance with the provisions of Exhibit "C," less the estimated cost of salvaging and the estimated cost of plugging and abandoning and restoring the surface. If such value is less than such costs, then the party assignor or lessor shall pay to the party assignee or lessee the amount of such deficit. If the assignment or lease is in favor of more than one party, the interest shall be shared by such parties in the proportions that the interest of each bears to the total interest of all such parties. If the interest of the parties to whom the assignment is to be made varies according to depth, then the interest assigned shall similarly reflect such variances.

Any assignment, lease or surrender made under this provision shall not reduce or change the assignor's, lessor's or surrendering party's interest as it was immediately before the assignment, lease or surrender in the balance of the Contract Area; and the acreage assigned, leased or surrendered, and subsequent operations thereon, shall not thereafter be subject to the terms and provisions of this agreement but shall be deemed subject to an Operating Agreement in the form of this agreement.

B. Renewal or Extension of Leases:

If any party secures a renewal or replacement of an Oil and Gas Lease or Interest subject to this agreement, then all other parties shall be notified promptly upon such acquisition or, in the case of a replacement Lease taken before expiration of an existing Lease, promptly upon expiration of the existing Lease. The parties notified shall have the right for a period of thirty (30) days following delivery of such notice in which to elect to participate in the ownership of the renewal or replacement Lease, insofar as such Lease affects lands within the Contract Area, by paying to the party who acquired it their proportionate shares of the acquisition cost allocated to that part of such Lease within the Contract Area, which shall be in proportion to the interests held at that time by the parties in the Contract Area. Each party who participates in the purchase of a renewal or replacement Lease shall be given an assignment of its proportionate interest therein by the acquiring party.

If some, but less than all, of the parties elect to participate in the purchase of a renewal or replacement Lease, it shall be owned by the parties who elect to participate therein, in a ratio based upon the relationship of their respective percentage of participation in the Contract Area to the aggregate of the percentages of participation in the Contract Area of all parties participating in the purchase of such renewal or replacement Lease. The acquisition of a renewal or replacement Lease by any or all of the parties hereto shall not cause a readjustment of the interests of the parties stated in Exhibit "A," but any renewal or replacement Lease in which less than all parties elect to participate shall not be subject to this agreement but shall be deemed subject to a separate Operating Agreement in the form of this agreement.

If the interests of the parties in the Contract Area vary according to depth, then their right to participate proportionately in renewal or replacement Leases and their right to receive an assignment of interest shall also reflect such depth variances.

The provisions of this Article shall apply to renewal or replacement Leases whether they are for the entire interest covered by the expiring Lease or cover only a portion of its area or an interest therein. Any renewal or replacement Lease taken before the expiration of its predecessor Lease, or taken or contracted for or becoming effective within six (6) months after the expiration of the existing Lease, shall be subject to this provision so long as this agreement is in effect at the time of such acquisition or at the time the renewal or replacement Lease becomes effective; but any Lease taken or contracted for more than six (6) months after the expiration of an existing Lease shall not be deemed a renewal or replacement Lease and shall not be subject to the provisions of this agreement.

The provisions in this Article shall also be applicable to extensions of Oil and Gas Leases.

C. Acreage or Cash Contributions:

While this agreement is in force, if any party contracts for a contribution of cash towards the drilling of a well or any other operation on the Contract Area, such contribution shall be paid to the party who conducted the drilling or other operation and shall be applied by it against the cost of such drilling or other operation. If the contribution be in the form of acreage, the party to whom the contribution is made shall promptly tender an assignment of the acreage, without warranty of title, to the Drilling Parties in the proportions said Drilling Parties shared the cost of drilling the well. Such acreage shall become a separate Contract Area and, to the extent possible, be governed by provisions identical to this agreement. Each party shall promptly notify all other parties of any acreage or cash contributions it may obtain in support of any well or any other operation on the Contract Area. The above provisions shall also be applicable to optional rights to earn acreage outside the Contract Area which are in support of well drilled inside the Contract Area.

If any party contracts for any consideration relating to disposition of such party's share of substances produced hereunder, such consideration shall not be deemed a contribution as contemplated in this Article VIII.C.

D. Assignment; Maintenance of Uniform Interest:

For the purpose of maintaining uniformity of ownership in the Contract Area in the Oil and Gas Leases, Oil and Gas Interests, wells, equipment and production covered by this agreement no party shall sell, encumber, transfer or make other disposition of its interest in the Oil and Gas Leases and Oil and Gas Interests embraced within the Contract Area or in wells, equipment and production unless such disposition covers either:

1. the entire interest of the party in all Oil and Gas Leases, Oil and Gas Interests, wells, equipment and production; or

2. an equal undivided percent of the party's present interest in all Oil and Gas Leases, Oil and Gas Interests, wells, equipment and production in the Contract Area.

Every sale, encumbrance, transfer or other disposition made by any party shall be made expressly subject to this agreement and shall be made without prejudice to the right of the other parties, and any transferee of an ownership interest in any Oil and Gas Lease or Interest shall be deemed a party to this agreement as to the interest conveyed from and after the effective date of the transfer of ownership; provided, however, that the other parties shall not be required to recognize any such sale, encumbrance, transfer or other disposition for any purpose hereunder until thirty (30) days after they have received a copy of the instrument of transfer or other satisfactory evidence thereof in writing from the transferor or transferee. No assignment or other disposition of interest by a party shall relieve such party of obligations previously incurred by such party hereunder with respect to the interest transferred, including without limitation the obligation of a party to pay all costs attributable to an operation conducted hereunder in which such party has agreed to participate prior to making such assignment, and the lien and security interest granted by Article VII.B. shall continue to burden the interest transferred to secure payment of any such obligations.

If, at any time the interest of any party is divided among and owned by four or more co-owners, Operator, at its discretion, may require such co-owners to appoint a single trustee or agent with full authority to receive notices, approve expenditures, receive billings for and approve and pay such party's share of the joint expenses, and to deal generally with, and with power to bind, the co-owners of such party's interest within the scope of the operations embraced in this agreement; however, all such co-owners shall have the right to enter into and execute all contracts or agreements for the disposition of their respective shares of the Oil and Gas produced from the Contract Area and they shall have the right to receive, separately, payment of the sale proceeds thereof.

E. Waiver of Rights to Partition:

If permitted by the laws of the state or states in which the property covered hereby is located, each party hereto owning an undivided interest in the Contract Area waives any and all rights it may have to partition and have set aside to it in severalty its undivided interest therein.

F. Preferential Right to Purchase:

☐ (Optional; Check if applicable.)

Should any party desire to sell all or any part of its interests under this agreement, or its rights and interests in the Contract Area, it shall promptly give written notice to the other parties, with full information concerning its proposed disposition, which shall include the name and address of the prospective transferee (who must be ready, willing and able to purchase), the purchase price, a legal description sufficient to identify the property, and all other terms of the offer. The other parties shall then have an optional prior right, for a period of ten (10) days after the notice is delivered, to purchase for the stated consideration on the same terms and conditions the interest which the other party proposes to sell; and, if this optional right is exercised, the purchasing parties shall share the purchased interest in the proportions that the interest of each bears to the total interest of all purchasing parties. However, there shall be no preferential right to purchase in those cases where any party wishes to mortgage its interests, or to transfer title to its interests to its mortgagee in lieu of or pursuant to foreclosure of a mortgage of its interests, or to dispose of its interests by merger, reorganization, consolidation, or by sale of all or substantially all of its Oil and Gas assets to any party, or by transfer of its interests to a subsidiary or parent company or to a subsidiary of a parent company, or to any company in which such party owns a majority of the stock.

ARTICLE IX.
INTERNAL REVENUE CODE ELECTION

If, for federal income tax purposes, this agreement and the operations hereunder are regarded as a partnership, and if the parties have not otherwise agreed to form a tax partnership pursuant to Exhibit "G" or other agreement between them, each party thereby affected elects to be excluded from the application of all of the provisions of Subchapter "K," Chapter 1, Subtitle

"A," of the Internal Revenue Code of 1986, as amended ("Code"), as permitted and authorized by Section 761 of the Code and the regulations promulgated thereunder. Operator is authorized and directed to execute on behalf of each party hereby affected such evidence of this election as may be required by the Secretary of the Treasury of the United States or the Federal Internal Revenue Service, including specifically, but not by way of limitation, all of the returns, statements, and the data required by Treasury Regulations §1.761. Should there be any requirement that each party hereby affected give further evidence of this election, each such party shall execute such documents and furnish such other evidence as may be required by the Federal Internal Revenue Service or as may be necessary to evidence this election. No such party shall give any notices or take any other action inconsistent with the election made hereby. If any present or future income tax laws of the state or states in which the Contract Area is located or any future income tax laws of the United States contain provisions similar to those in Subchapter "K," Chapter 1, Subtitle "A," of the Code, under which an election similar to that provided by Section 761 of the Code is permitted, each party hereby affected shall make such election as may be permitted or required by such laws. In making the foregoing election, each such party states that the income derived by such party from operations hereunder can be adequately determined without the computation of partnership taxable income.

ARTICLE X.
CLAIMS AND LAWSUITS

Operator may settle any single uninsured third party damage claim or suit arising from operations hereunder if the expenditure does not exceed ______________________________ Dollars ($ ________________) and if the payment is in complete settlement of such claim or suit. If the amount required for settlement exceeds the above amount, the parties hereto shall assume and take over the further handling of the claim or suit, unless such authority is delegated to Operator. All costs and expenses of handling, settling, or otherwise discharging such claim or suit shall be at the joint expense of the parties participating in the operation from which the claim or suit arises. If a claim is made against any party or if any party is sued on account of any matter arising from operations hereunder over which such individual has no control because of the rights given Operator by this agreement, such party shall immediately notify all other parties, and the claim or suite shall be treated as any other claim or suit involving operations hereunder.

ARTICLE XI.
FORCE MAJEURE

If any party is rendered unable, wholly or in part, by force majeure to carry out its obligations under this agreement, other than the obligation to indemnify or make money payments or furnish security, that party shall give to all other parties prompt written notice of the force majeure with reasonably full particulars concerning it; thereupon, the obligations of the party giving the notice, so far as they are affected by the force majeure, shall be suspended during, but no longer than, the continuance of the force majeure. The term "force majeure," as here employed, shall mean an act of God, strike, lockout, or other industrial disturbance, act of the public enemy, war, blockade, public riot, lightning, fire, storm, flood or other act of nature, explosion, governmental action, governmental delay, restraint or inaction, unavailability of equipment, and any other cause, whether of the kind specifically enumerated above or otherwise, which is not reasonably within the control of the party claiming suspension.

The affected party shall use all reasonable diligence to remove the force majeure situation as quickly as practicable. The requirement that any force majeure shall be remedied with all reasonable dispatch shall not require the settlement of strikes, lockouts, or other labor difficulty by the party involved, contrary to its wishes; how all such difficulties shall be handled shall be entirely within the discretion of the party concerned.

ARTICLE XII.
NOTICES

All notices authorized or required between the parties by any of the provisions of this agreement, unless otherwise specifically provided, shall be in writing and delivered in person or by United States mail, courier service, telegram, telex, telecopier or any other form of facsimile, postage or charges prepaid, and addressed to such parties at the addresses listed on Exhibit "A." All telephone or oral notices permitted by this agreement shall be confirmed immediately thereafter by written notice. The originating notice given under any provision hereof shall be deemed delivered only when received by the party to whom such notice is directed, and the time for such party to deliver any notice in response thereto shall run from the date the originating notice is received. "Receipt" for purposes of this agreement with respect to written notice delivered hereunder shall be actual delivery of the notice to the address of the party to be notified specified in accordance with this agreement, or to the telecopy, facsimile or telex machine of such party. The second or any responsive notice shall be deemed delivered when deposited in the United States mail or at the office of the courier or telegraph service, or upon transmittal by telex, telecopy or facsimile, or when personally delivered to the party to be notified, provided, that when response is required within 24 or 48 hours, such response shall be given orally or by telephone, telex, telecopy or other facsimile within such period. Each party shall have the right to change its address at any time, and from time to time, by giving written notice thereof to all other parties. If a party is not available to receive notice orally or by telephone when a party attempts to deliver a notice required to be delivered within 24 or 48 hours, the notice may be delivered in writing by any other method specified herein and shall be deemed delivered in the same manner provided above for any responsive notice.

ARTICLE XIII.
TERM OF AGREEMENT

This agreement shall remain in full force and effect as to the Oil and Gas Leases and/or Oil and Gas Interests subject hereto for the period of time selected below; provided, however, no party hereto shall ever be construed as having any right, title or interest in or to any Lease or Oil and Gas Interest contributed by any other party beyond the term of this agreement.

☐ Option No. 1: So long as any of the Oil and Gas Leases subject to this agreement remain or are continued in force as to any part of the Contract Area, whether by production, extension, renewal or otherwise.

☐ Option No. 2: In the event the well described in Article VI.A., or any subsequent well drilled under any provision of this agreement, results in the Completion of a well as a well capable of production of Oil and/or Gas in paying quantities, this agreement shall continue in force so long as any such well is capable of production, and for an additional period of ______________ days thereafter; provided, however, if, prior to the expiration of such additional period, one or more of the parties hereto are engaged in drilling, Reworking, Deepening, Sidetracking, Plugging Back, testing or attempting to Complete or Re-complete a well or wells hereunder, this agreement shall continue in force until such operations have been completed and if production results therefrom, this agreement shall continue in force as provided herein. In the event the well described in Article VI.A., or any subsequent well drilled hereunder, results in a dry hole, and no other well is capable of producing Oil and/or Gas from the Contract Area, this agreement shall terminate unless drilling, Deepening, Sidetracking, Completing, Re-completing, Plugging Back or Reworking operations are commenced within ______________ days from the date of abandonment of said well. "Abandonment" for such purposes shall mean either (i) a decision by all parties not to conduct any further operations on the well or (ii) the elapse of 180 days from the conduct of any operations on the well, whichever first occurs.

The termination of this agreement shall not relieve any party hereto from any expense, liability or other obligation or any remedy therefor which has accrued or attached prior to the date of such termination.

Upon termination of this agreement and the satisfaction of all obligations hereunder, in the event a memorandum of this Operating Agreement has been filed of record, Operator is authorized to file of record in all necessary recording offices a notice of termination, and each party hereto agrees to execute such a notice of termination as to Operator's interest, upon request of Operator, if Operator has satisfied all its financial obligations.

ARTICLE XIV.
COMPLIANCE WITH LAWS AND REGULATIONS

A. Laws, Regulations and Orders:

This agreement shall be subject to the applicable laws of the state in which the Contract Area is located, to the valid rules, regulations, and orders of any duly constituted regulatory body of said state; and to all other applicable federal, state, and local laws, ordinances, rules, regulations and orders.

B. Governing Law:

This agreement and all matters pertaining hereto, including but not limited to matters of performance, non-performance, breach, remedies, procedures, rights, duties, and interpretation or construction, shall be governed and determined by the law of the state in which the Contract Area is located. If the Contract Area is in two or more states, the law of the state of ______________________________ shall govern.

C. Regulatory Agencies:

Nothing herein contained shall grant, or be construed to grant, Operator the right or authority to waive or release any rights, privileges, or obligations which Non-Operators may have under federal or state laws or under rules, regulations or orders promulgated under such laws in reference to oil, gas and mineral operations, including the location, operation, or production of wells, on tracts offsetting or adjacent to the Contract Area.

With respect to the operations hereunder, Non-Operators agree to release Operator from any and all losses, damages, injuries, claims and causes of action arising out of, incident to or resulting directly or indirectly from Operator's interpretation or application of rules, rulings, regulations or orders of the Department of Energy or Federal Energy Regulatory Commission or predecessor or successor agencies to the extent such interpretation or application was made in good faith and does not constitute gross negligence. Each Non-Operator further agrees to reimburse Operator for such Non-Operator's share of production or any refund, fine, levy or other governmental sanction that Operator may be required to pay as a result of such an incorrect interpretation or application, together with interest and penalties thereon owing by Operator as a result of such incorrect interpretation or application.

ARTICLE XV.
MISCELLANEOUS

A. Execution:

This agreement shall be binding upon each Non-Operator when this agreement or a counterpart thereof has been executed by such Non-Operator and Operator notwithstanding that this agreement is not then or thereafter executed by all of the parties to which it is tendered or which are listed on Exhibit "A" as owning an interest in the Contract Area or which own, in fact, an interest in the Contract Area. Operator may, however, by written notice to all Non-Operators who have become bound by this agreement as aforesaid, given at any time prior to the actual spud date of the Initial Well but in no event later than five days prior to the date specified in Article VI.A. for commencement of the Initial Well, terminate this agreement if Operator in its sole discretion determines that there is insufficient participation to justify commencement of drilling operations. In the event of such a termination by Operator, all further obligations of the parties hereunder shall cease as of such termination. In the event any Non-Operator has advanced or prepaid any share of drilling or other costs hereunder, all sums so advanced shall be returned to such Non-Operator without interest. In the event Operator proceeds with drilling operations for the Initial Well without the execution hereof by all persons listed on Exhibit "A" as having a current working interest in such well, Operator shall indemnify Non-Operators with respect to all costs incurred for the Initial Well which would have been charged to such person under this agreement if such person had executed the same and Operator shall receive all revenues which would have been received by such person under this agreement if such person had executed the same.

B. Successors and Assigns:

This agreement shall be binding upon and shall inure to the benefit of the parties hereto and their respective heirs, devisees, legal representatives, successors and assigns, and the terms hereof shall be deemed to run with the Leases or Interests included within the Contract Area.

C. Counterparts:

This instrument may be executed in any number of counterparts, each of which shall be considered an original for all purposes.

D. Severability:

For the purposes of assuming or rejecting this agreement as an executory contract pursuant to federal bankruptcy laws, this agreement shall not be severable, but rather must be assumed or rejected in its entirety, and the failure of any party to this agreement to comply with all of its financial obligations provided herein shall be a material default.

ARTICLE XVI.
OTHER PROVISIONS

IN WITNESS WHEREOF, this agreement shall be effective as of the ______ day of ______________,

19______.

ATTEST OR WITNESS:

____

____

OPERATOR

____

By ______________________________

____
Type or print name

Title ______________________________

Date ______________________________

Tax ID or S.S. No. ______________________________

NON-OPERATORS

____

____ By ____

____ ____

Type or print name

Title ____

Date ____

Tax ID or S.S. No. __________________________________

____

____ By ____

____ ____

Type or print name

Title ____

Date ____

Tax ID or S.S. No. __________________________________

____

____ By ____

____ ____

Type or print name

Title ____

Date ____

Tax ID or S.S. No. __________________________________

ACKNOWLEDGMENTS

Note: The following forms of acknowledgment are the short forms approved by the Uniform Law on Notarial Acts. The validity and effect of these forms in any state will depend upon the statutes of that state.

Individual acknowledgment:

State of ________________)

) ss.

County of ________________)

This instrument was acknowledged before me on

____ by ____.

(Seal, if any) ____

Title (and Rank) ______________________________

My commission expires: ________________________

Acknowledgment in representative capacity:

State of ________________)

) ss.

County of ________________)

This instrument was acknowledged before me on

____ by ____ as

________________ of ____.

(Seal, if any) ____

Title (and Rank) ______________________________

My commission expires: ________________________

The JOA delineates the duties and responsibilities of the operator and nonoperators. The JOA typically covers all phases (*i.e.*, exploration, development, and production) of operation of the joint property. The major subsections that may be included in a standard JOA are as follows:

1. **Definitions:** Defines basic terms used in the agreement. Examples include operator, nonoperator, contract area, AFE, oil and gas lease, drillsite, etc.
2. **Exhibits:** Contains a list of all of the exhibits that form the appendices of the agreement. For example, exhibit "A" contains the legal description of the properties, the parties to the agreement, and the percentage or fractional ownership interest of each owner. Exhibit "C" is the accounting procedure. Other exhibits that may be included are exhibits containing the gas balancing agreement, information about insurance, and information about the lease agreement.
3. **Interests of parties:** Defines how specific revenues, costs, and liabilities will be distributed according to the ownership interests specified in the *exhibits.*
4. **Titles:** Describes title examination requirements for drillsites, how the costs of the title process will be distributed, and how loss of title would be handled.
5. **Operator:** Designates the operator and the general rights and duties of the operator. Describes the process for the resignation or removal of the operator and the process for appointment of a successor. All records and reports—including the accounting records—to be maintained by the operator and provided to the nonoperators and governmental units are listed.
6. **Drilling and development:** Specifies procedures to be followed in drilling and development activities and termination operations. Includes the specific date and location for the drilling of the initial well. Participation in the initial well is typically obligatory for all parties. Describes the procedures that must be followed in undertaking subsequent drilling and operations, including a section providing for "operations by less than all parties." This section provides the process by which carried working interests are created and allows penalties in non-consent situations related to drilling, deepening, rework, and abandonment of wells. (Non-consent situations are discussed in more detail later in the chapter.)
7. **Expenditures and liabilities:** Specifies that any liabilities are to be several and not joint (*i.e.*, each party is individually responsible only for its own obligations and liabilities, and only its own proportionate share of development and operating costs). Gives the operator the right to demand and receive cash advances from the other working interest owners. Spells out the remedies that exist in the event that any owner fails to discharge its financial obligations related to the property. Provides for the payment of shut-in royalties, minimum royalties, delay rentals, and valorem taxes.
8. **Acquisition, maintenance, or transfer of interest:** Specifies procedures to follow when surrendering or renewing a lease or assigning interests. Also gives preferential rights of purchase to other working interest owners when one working interest owner wishes to sell part or all of its interest.

9. **Internal revenue code election:** States whether the joint venture is to be operated or be taxed as a partnership.
10. **Claims and lawsuits:** Defines the procedures to be followed in case of legal action relating to operations of the venture, and authorizes the operator to settle claims for uninsured damages up to a maximum amount.
11. **Force majeure:** Provides that if any party is unable to meet its nonmonetary obligations, such as the operator's obligation to proceed with drilling activities, because of circumstances beyond that party's control—act of God, war, fire, etc.—the party's obligation will be temporarily suspended.
12. **Notices:** States that all notices between parties will be in writing.
13. **Terms of agreement:** States that the agreement will remain in effect so long as the underlying lease is in effect whether through production, extension, renewal, or otherwise.
14. **Compliance with laws and regulations:** States that the agreement is subject to all applicable laws, and states that the operator cannot waive or release nonoperators from any rights, privileges, or obligations arising from governmental regulation.
15. **Miscellaneous:** Other provisions, including those concerning successors, severability, and the signature page.

In summary, the operator normally manages all developing, operating, recordkeeping, and administrative activities pertaining to the joint property. For major expenditures, such as drilling a well, the operator is generally required to obtain written authority from the nonoperators via an authorization for expenditure (AFE). This requirement provides a limit on expenditures without express written authority from the nonoperators. Periodic reports, which describe the activities performed by the operator, must generally be given to the nonoperators. The operator normally is required to pay all costs of developing and operating the joint property and then bill the nonoperators for their proportionate share. However, a provision for requiring advances from the nonoperators prior to cost incurrence or payment is generally included in the operating agreement.

THE ACCOUNTING PROCEDURE

The accounting procedure, which is normally attached to the JOA, is of extreme importance to the accountant because it details the procedures to be followed in charging costs to the joint operations. For this reason, a copy of an accounting procedure is shown in Figure 11-2. The accounting procedure illustrated applies to onshore joint operations (*COPAS Bulletin No. 22*).

The accounting procedure is an integral part of any JOA. The accounting procedure specifically addresses issues related to the maintenance of the **joint account**; specifically, the determination of appropriate charges and credits applicable to the joint operation.

Fig. 11-2 — *Accounting Procedure*

ACCOUNTING PROCEDURE
JOINT OPERATIONS

I. GENERAL PROVISIONS

1. Definitions

"Joint Property" shall mean the real and personal property subject to the agreement to which this Accounting Procedure is attached.

"Joint Operations" shall mean all operations necessary or proper for the development, operation, protection and maintenance of the Joint Property.

"Joint Account" shall mean the account showing the charges paid and credits received in the conduct of the Joint Operations and which are to be shared by the Parties.

"Operator" shall mean the party designated to conduct the Joint Operations.

"Non-Operators" shall mean the Parties to this agreement other than the Operator.

"Parties" shall mean Operator and Non-Operators.

"First Level Supervisors" shall mean those employees whose primary function in Joint Operations is the direct supervision of other employees and/or contract labor directly employed on the Joint Property in a field operating capacity.

"Technical Employees" shall mean those employees having special and specific engineering, geological or other professional skills, and whose primary function in Joint Operations is the handling of specific operating conditions and problems for the benefit of the Joint Property.

"Personal Expenses" shall mean travel and other reasonable reimbursable expenses of Operator's employees.

"Material" shall mean personal property, equipment or supplies acquired or held for use on the Joint Property.

"Controllable Material" shall mean Material which at the time is so classified in the Material Classification Manual as most recently recommended by the Council of Petroleum Accountants Societies.

2. Statement and Billings

Operator shall bill Non-Operators on or before the last day of each month for their proportionate share of the Joint Account for the preceding month. Such bills will be accompanied by statements which identify the authority for expenditure, lease or facility, and all charges and credits summarized by appropriate classifications of investment and expense except that items of Controllable Material and unusual charges and credits shall be separately identified and fully described in detail.

3. Advances and Payments by Non-Operators

A. Unless otherwise provided for in the agreement, the Operator may require the Non-Operators to advance their share of estimated cash outlay for the succeeding month's operation within fifteen (15) days after receipt of the billing or by the first day of the month for which the advance is required, whichever is later. Operator shall adjust each monthly billing to reflect advances received from the Non-Operators.

B. Each Non-Operator shall pay its proportion of all bills within fifteen (15) days after receipt. If payment is not made within such time, the unpaid balance shall bear interest monthly at the prime rate in effect at ____________ ____________ on the first day of the month in which delinquency occurs plus 1% or the maximum contract rate permitted by the applicable usury laws in the state in which the Joint Property is located, whichever is the lesser, plus attorney's fees, court costs, and other costs in connection with the collection of unpaid amounts.

4. Adjustments

Payment of any such bills shall not prejudice the right of any Non-Operator to protest or question the correctness thereof; provided, however, all bills and statements rendered to Non-Operators by Operator during any calendar year shall conclusively be presumed to be true and correct after twenty-four (24) months following the end of any such calendar year, unless within the said twenty-four (24) month period a Non-Operator takes written exception thereto and makes claim on Operator for adjustment. No adjustment favorable to Operator shall be made unless it is made within the same prescribed period. The provisions of this paragraph shall not prevent adjustments resulting from a physical inventory of Controllable Material as provided for in Section V.

5. **Audits**

A. A Non-Operator, upon notice in writing to Operator and all other Non-Operators, shall have the right to audit Operator's accounts and records relating to the Joint Account for any calendar year within the twenty-four (24) month period following the end of such calendar year; provided, however, the making of an audit shall not extend the time for the taking of written exception to and the adjustments of accounts as provided for in Paragraph 4 of this Section I. Where there are two or more Non-Operators, the Non-Operators shall make every reasonable effort to conduct a joint audit in a manner which will result in a minimum of inconvenience to the Operator. Operator shall bear no portion of the Non-Operators' audit cost incurred under this paragraph unless agreed to by the Operator. The audits shall not be conducted more than once each year without prior approval of Operator, except upon the resignation or removal of the Operator, and shall be made at the expense of those Non-Operators approving such audit.

B. The Operator shall reply in writing to an audit report within 180 days after receipt of such report.

6. **Approval By Non-Operators**

Where an approval or other agreement of the Parties or Non-Operators is expressly required under other sections of this Accounting Procedure and if the agreement to which this Accounting Procedure is attached contains no contrary provisions in regard thereto, Operator shall notify all Non-Operators of the Operator's proposal, and the agreement or approval of a majority in interest of the Non-Operators shall be controlling on all Non-Operators.

II. DIRECT CHARGES

Operator shall charge the Joint Account with the following items:

1. **Ecological and Environmental**

Costs incurred for the benefit of the Joint Property as a result of governmental or regulatory requirements to satisfy environmental considerations applicable to the Joint Operations. Such costs may include surveys of an ecological or archaeological nature and pollution control procedures as required by applicable laws and regulations.

2. **Rentals and Royalties**

Lease rentals and royalties paid by Operator for the Joint Operations.

3. **Labor**

A. (1) Salaries and wages of Operator's field employees directly employed on the Joint Property in the conduct of Joint Operations.

(2) Salaries of First Level Supervisors in the field.

(3) Salaries and wages of Technical Employees directly employed on the Joint Property if such charges are excluded from the overhead rates.

(4) Salaries and wages of Technical Employees either temporarily or permanently assigned to and directly employed in the operation of the Joint Property if such charges are excluded from the overhead rates.

B. Operator's cost of holiday, vacation, sickness and disability benefits and other customary allowances paid to employees whose salaries and wages are chargeable to the Joint Account under Paragraph 3A of this Section II. Such costs under this Paragraph 3B may be charged on a "when and as paid basis" or by "percentage assessment" on the amount of salaries and wages chargeable to the Joint Account under Paragraph 3A of this Section II. If percentage assessment is used, the rate shall be based on the Operator's cost experience.

C. Expenditures or contributions made pursuant to assessments imposed by governmental authority which are applicable to Operator's costs chargeable to the Joint Account under Paragraphs 3A and 3B of this Section II.

D. Personal Expenses of those employees whose salaries and wages are chargeable to the Joint Account under Paragraph 3A of this Section II.

4. **Employee Benefits**

Operator's current costs of established plans for employees' group life insurance, hospitalization, pension, retirement, stock purchase, thrift, bonus, and other benefit plans of a like nature, applicable to Operator's labor cost chargeable to the Joint Account under Paragraphs 3A and 3B of this Section II shall be Operator's actual cost not to exceed the percent most recently recommended by the Council of Petroleum Accountants Societies.

5. **Material**

Material purchased or furnished by Operator for use on the Joint Property as provided under Section IV. Only such Material shall be purchased for or transferred to the Joint Property as may be required for immediate use and is reasonably practical and consistent with efficient and economical operations. The accumulation of surplus stocks shall be avoided.

6. Transportation

Transportation of employees and Material necessary for the Joint Operations but subject to the following limitations:

A. If Material is moved to the Joint Property from the Operator's warehouse or other properties, no charge shall be made to the Joint Account for a distance greater than the distance from the nearest reliable supply store where like material is normally available or railway receiving point nearest the Joint Property unless agreed to by the Parties.

B. If surplus Material is moved to Operator's warehouse or other storage point, no charge shall be made to the Joint Account for a distance greater than the distance to the nearest reliable supply store where like material is normally available, or railway receiving point nearest the Joint Property unless agreed to by the Parties. No charge shall be made to the Joint Account for moving Material to other properties belonging to Operator, unless agreed to by the Parties.

C. In the application of subparagraphs A and B above, the option to equalize or charge actual trucking cost is available when the actual charge is $400 or less excluding accessorial charges. The $400 will be adjusted to the amount most recently recommended by the Council of Petroleum Accountants Societies.

7. Services

The cost of contract services, equipment and utilities provided by outside sources, except services excluded by Paragraph 10 of Section II and Paragraph i, ii, and iii, of Section III. The cost of professional consultant services and contract services of technical personnel directly engaged on the Joint Property if such charges are excluded from the overhead rates. The cost of professional consultant services or contract services of technical personnel not directly engaged on the Joint Property shall not be charged to the Joint Account unless previously agreed to by the Parties.

8. Equipment and Facilities Furnished By Operator

A. Operator shall charge the Joint Account for use of Operator owned equipment and facilities at rates commensurate with costs of ownership and operation. Such rates shall include costs of maintenance, repairs, other operating expense, insurance, taxes, depreciation, and interest on gross investment less accumulated depreciation not to exceed ______________ percent (______%) per annum. Such rates shall not exceed average commercial rates currently prevailing in the immediate area of the Joint Property.

B. In lieu of charges in paragraph 8A above, Operator may elect to use average commercial rates prevailing in the immediate area of the Joint Property less 20%. For automotive equipment, Operator may elect to use rates published by the Petroleum Motor Transport Association.

9. Damages and Losses to Joint Property

All costs or expenses necessary for the repair or replacement of Joint Property made necessary because of damages or losses incurred by fire, flood, storm, theft, accident, or other cause, except those resulting from Operator's gross negligence or willful misconduct. Operator shall furnish Non-Operator written notice of damages or losses incurred as soon as practicable after a report thereof has been received by Operator.

10. Legal Expense

Expense of handling, investigating and settling litigation or claims, discharging of liens, payment of judgements and amounts paid for settlement of claims incurred in or resulting from operations under the agreement or necessary to protect or recover the Joint Property, except that no charge for services of Operator's legal staff or fees or expense of outside attorneys shall be made unless previously agreed to by the Parties. All other legal expense is considered to be covered by the overhead provisions of Section III unless otherwise agreed to by the Parties, except as provided in Section I, Paragraph 3.

11. Taxes

All taxes of every kind and nature assessed or levied upon or in connection with the Joint Property, the operation thereof, or the production therefrom, and which taxes have been paid by the Operator for the benefit of the Parties. If the ad valorem taxes are based in whole or in part upon separate valuations of each party's working interest, then notwithstanding anything to the contrary herein, charges to the Joint Account shall be made and paid by the Parties hereto in accordance with the tax value generated by each party's working interest.

12. Insurance

Net premiums paid for insurance required to be carried for the Joint Operations for the protection of the Parties. In the event Joint Operations are conducted in a state in which Operator may act as self-insurer for Worker's Compensation and/or Employers Liability under the respective state's laws, Operator may, at its election, include the risk under its self-

13. Abandonment and Reclamation

Costs incurred for abandonment of the Joint Property, including costs required by governmental or other regulatory authority.

14. Communications

Cost of acquiring, leasing, installing, operating, repairing and maintaining communication systems, including radio and microwave facilities directly serving the Joint Property. In the event communication facilities/systems serving the Joint Property are Operator owned, charges to the Joint Account shall be made as provided in Paragraph 8 of this Section II.

15. Other Expenditures

Any other expenditure not covered or dealt with in the foregoing provisions of this Section II, or in Section III and which is of direct benefit to the Joint Property and is incurred by the Operator in the necessary and proper conduct of the Joint Operations.

III. OVERHEAD

1. Overhead - Drilling and Producing Operations

i. As compensation for administrative, supervision, office services and warehousing costs, Operator shall charge drilling and producing operations on either:

( ) Fixed Rate Basis, Paragraph 1A, or
( ) Percentage Basis, Paragraph 1B

Unless otherwise agreed to by the Parties, such charge shall be in lieu of costs and expenses of all offices and salaries or wages plus applicable burdens and expenses of all personnel, except those directly chargeable under Paragraph

3A, Section II. The cost and expense of services from outside sources in connection with matters of taxation, traffic, accounting or matters before or involving governmental agencies shall be considered as included in the overhead rates provided for in the above selected Paragraph of this Section III unless such cost and expense are agreed to by the Parties as a direct charge to the Joint Account.

ii. The salaries, wages and Personal Expenses of Technical Employees and/or the cost of professional consultant services and contract services of technical personnel directly employed on the Joint Property:

( ) shall be covered by the overhead rates, or
( ) shall not be covered by the overhead rates.

iii. The salaries, wages and Personal Expenses of Technical Employees and/or costs of professional consultant services and contract services of technical personnel either temporarily or permanently assigned to and directly employed in the operation of the Joint Property:

( ) shall be covered by the overhead rates, or
( ) shall not be covered by the overhead rates.

A. Overhead - Fixed Rate Basis

(1) Operator shall charge the Joint Account at the following rates per well per month:

Drilling Well Rate $ ____________________
(Prorated for less than a full month)

Producing Well Rate $ ____________________

(2) Application of Overhead - Fixed Rate Basis shall be as follows:

(a) Drilling Well Rate

(1) Charges for drilling wells shall begin on the date the well is spudded and terminate on the date the drilling rig, completion rig, or other units used in completion of the well is released, whichever is later, except that no charge shall be made during suspension of drilling or completion operations for fifteen (15) or more consecutive calendar days.

(2) Charges for wells undergoing any type of workover or recompletion for a period of five (5) consecutive work days or more shall be made at the drilling well rate. Such charges shall be applied for the period from date workover operations, with rig or other units used in workover, commence through date of rig or other unit release, except that no charge shall be made during suspension of operations for fifteen (15) or more consecutive calendar days.

(b) Producing Well Rates

(1) An active well either produced or injected into for any portion of the month shall be considered as a one-well charge for the entire month.

(2) Each active completion in a multi-completed well in which production is not commingled down hole shall be considered as a one-well charge providing each completion is considered a separate well by the governing regulatory authority.

(3) An inactive gas well shut in because of overproduction or failure of purchaser to take the production shall be considered as a one-well charge providing the gas well is directly connected to a permanent sales outlet.

(4) A one-well charge shall be made for the month in which plugging and abandonment operations are completed on any well. This one-well charge shall be made whether or not the well has produced except when drilling well rate applies.

(5) All other inactive wells (including but not limited to inactive wells covered by unit allowable, lease allowable, transferred allowable, etc.) shall not qualify for an overhead charge.

(3) The well rates shall be adjusted as of the first day of April each year following the effective date of the agreement to which this Accounting Procedure is attached. The adjustment shall be computed by multiplying the rate currently in use by the percentage increase or decrease in the average weekly earnings of Crude Petroleum and Gas Production Workers for the last calendar year compared to the calendar year preceding as shown by the index of average weekly earnings of Crude Petroleum and Gas Production Workers as published by the United States Department of Labor, Bureau of Labor Statistics, or the equivalent Canadian index as published by Statistics Canada, as applicable. The adjusted rates shall be the rates currently in use, plus or minus the computed adjustment.

B. Overhead - Percentage Basis

(1) Operator shall charge the Joint Account at the following rates:

(a) Development

____________ Percent (____ %) of the cost of development of the Joint Property exclusive of costs provided under Paragraph 10 of Section II and all salvage credits.

(b) Operating

____________ Percent (____ %) of the cost of operating the Joint Property exclusive of costs provided under Paragraphs 2 and 10 of Section II, all salvage credits, the value of injected substances purchased for secondary recovery and all taxes and assessments which are levied, assessed and paid upon the mineral interest in and to the Joint Property.

(2) Application of Overhead - Percentage Basis shall be as follows:

For the purpose of determining charges on a percentage basis under Paragraph 1B of this Section III, development shall include all costs in connection with drilling, redrilling, deepening, or any remedial operations on any or all wells involving the use of drilling rig and crew capable of drilling to the producing interval on the Joint Property; also, preliminary expenditures necessary in preparation for drilling and expenditures incurred in abandoning when the well is not completed as a producer, and original cost of construction or installation of fixed assets, the expansion of fixed assets and any other project clearly discernible as a fixed asset, except Major Construction as defined in Paragraph 2 of this Section III. All other costs shall be considered as operating.

2. **Overhead - Major Construction**

To compensate Operator for overhead costs incurred in the construction and installation of fixed assets, the expansion of fixed assets, and any other project clearly discernible as a fixed asset required for the development and operation of the Joint Property, Operator shall either negotiate a rate prior to the beginning of construction, or shall charge the Joint Account for overhead based on the following rates for any Major Construction project in excess of $ __________ :

A. __________ % of first $100,000 or total cost if less, plus

B. __________ % of costs in excess of $100,000 but less than $1,000,000, plus

C. __________ % of costs in excess of $1,000,000.

Total cost shall mean the gross cost of any one project. For the purpose of this paragraph, the component parts of a single project shall not be treated separately and the cost of drilling and workover wells and artificial lift equipment shall be excluded.

3. **Catastrophe Overhead**

To compensate Operator for overhead costs incurred in the event of expenditures resulting from a single occurrence due to oil spill, blowout, explosion, fire, storm, hurricane, or other catastrophes as agreed to by the Parties, which are necessary to restore the Joint Property to the equivalent condition that existed prior to the event causing the expenditures, Operator shall either negotiate a rate prior to charging the Joint Account or shall charge the Joint Account for overhead based on the following rates:

A. ________ % of total costs through $100,000; plus

B. ________ % of total costs in excess of $100,000 but less than $1,000,000; plus

C. ________ % of total costs in excess of $1,000,000.

Expenditures subject to the overheads above will not be reduced by insurance recoveries, and no other overhead provisions of this Section III shall apply.

4. **Amendment of Rates**

The overhead rates provided for in this Section III may be amended from time to time only by mutual agreement between the Parties hereto if, in practice, the rates are found to be insufficient or excessive.

IV. PRICING OF JOINT ACCOUNT MATERIAL PURCHASES, TRANSFERS AND DISPOSITIONS

Operator is responsible for Joint Account Material and shall make proper and timely charges and credits for all Material movements affecting the Joint Property. Operator shall provide all Material for use on the Joint Property; however, at Operator's option, such Material may be supplied by the Non-Operator. Operator shall make timely disposition of idle and/or surplus Material, such disposal being made either through sale to Operator or Non-Operator, division in kind, or sale to outsiders. Operator may purchase, but shall be under no obligation to purchase, interest of Non-Operators in surplus condition A or B Material. The disposal of surplus Controllable Material not purchased by the Operator shall be agreed to by the Parties.

1. **Purchases**

Material purchased shall be charged at the price paid by Operator after deduction of all discounts received. In case of Material found to be defective or returned to vendor for any other reasons, credit shall be passed to the Joint Account when adjustment has been received by the Operator.

2. **Transfers and Dispositions**

Material furnished to the Joint Property and Material transferred from the Joint Property or disposed of by the Operator, unless otherwise agreed to by the Parties, shall be priced on the following basis exclusive of cash discounts:

A. New Material (Condition A)

(1) Tubular Goods Other than Line Pipe

(a) Tubular goods, sized 2⅜ inches OD and larger, except line pipe, shall be priced at Eastern mill published carload base prices effective as of date of movement plus transportation cost using the 80,000 pound carload weight basis to the railway receiving point nearest the Joint Property for which published rail rates for tubular goods exist. If the 80,000 pound rail rate is not offered, the 70,000 pound or 90,000 pound rail rate may be used. Freight charges for tubing will be calculated from Lorain, Ohio and casing from Youngstown, Ohio.

(b) For grades which are special to one mill only, prices shall be computed at the mill base of that mill plus transportation cost from that mill to the railway receiving point nearest the Joint Property as provided above in Paragraph 2.A.(1)(a). For transportation cost from points other than Eastern mills, the 30,000 pound Oil Field Haulers Association interstate truck rate shall be used.

(c) Special end finish tubular goods shall be priced at the lowest published out-of-stock price, f.o.b. Houston, Texas, plus transportation cost, using Oil Field Haulers Association interstate 30,000 pound truck rate, to the railway receiving point nearest the Joint Property.

(d) Macaroni tubing (size less than 2⅜ inch OD) shall be priced at the lowest published out-of-stock prices f.o.b. the supplier plus transportation costs, using the Oil Field Haulers Association interstate truck rate per weight of tubing transferred, to the railway receiving point nearest the Joint Property.

(2) Line Pipe

(a) Line pipe movements (except size 24 inch OD and larger with walls ¾ inch and over) 30,000 pounds or more shall be priced under provisions of tubular goods pricing in Paragraph A.(1)(a) as provided above. Freight charges shall be calculated from Lorain, Ohio.

(b) Line pipe movements (except size 24 inch OD and larger with walls ¾ inch and over) less than 30,000 pounds shall be priced at Eastern mill published carload base prices effective as of date of shipment, plus 20 percent, plus transportation costs based on freight rates as set forth under provisions of tubular goods pricing in Paragraph A.(1)(a) as provided above. Freight charges shall be calculated from Lorain, Ohio.

(c) Line pipe 24 inch OD and over and ¾ inch wall and larger shall be priced f.o.b. the point of manufacture at current new published prices plus transportation cost to the railway receiving point nearest the Joint Property.

(d) Line pipe, including fabricated line pipe, drive pipe and conduit not listed on published price lists shall be priced at quoted prices plus freight to the railway receiving point nearest the Joint Property or at prices agreed to by the Parties.

(3) Other Material shall be priced at the current new price, in effect at date of movement, as listed by a reliable supply store nearest the Joint Property, or point of manufacture, plus transportation costs, if applicable, to the railway receiving point nearest the Joint Property.

(4) Unused new Material, except tubular goods, moved from the Joint Property shall be priced at the current new price, in effect on date of movement, as listed by a reliable supply store nearest the Joint Property, or point of manufacture, plus transportation costs, if applicable, to the railway receiving point nearest the Joint Property. Unused new tubulars will be priced as provided above in Paragraph 2 A (1) and (2).

B. Good Used Material (Condition B)

Material in sound and serviceable condition and suitable for reuse without reconditioning:

(1) Material moved to the Joint Property

At seventy-five percent (75%) of current new price, as determined by Paragraph A.

(2) Material used on and moved from the Joint Property

(a) At seventy-five percent (75%) of current new price, as determined by Paragraph A, if Material was originally charged to the Joint Account as new Material or

(b) At sixty-five percent (65%) of current new price, as determined by Paragraph A, if Material was originally charged to the Joint Account as used Material.

(3) Material not used on and moved from the Joint Property

At seventy-five percent (75%) of current new price as determined by Paragraph A.

The cost of reconditioning, if any, shall be absorbed by the transferring property.

C. Other Used Material

(1) Condition C

Material which is not in sound and serviceable condition and not suitable for its original function until after reconditioning shall be priced at fifty percent (50%) of current new price as determined by Paragraph A. The cost of reconditioning shall be charged to the receiving property, provided Condition C value plus cost of reconditioning does not exceed Condition B value.

(2) Condition D

Material, excluding junk, no longer suitable for its original purpose, but usable for some other purpose shall be priced on a basis commensurate with its use. Operator may dispose of Condition D Material under procedures normally used by Operator without prior approval of Non-Operators.

(a) Casing, tubing, or drill pipe used as line pipe shall be priced as Grade A and B seamless line pipe of comparable size and weight. Used casing, tubing or drill pipe utilized as line pipe shall be priced at used line pipe prices.

(b) Casing, tubing or drill pipe used as higher pressure service lines than standard line pipe, e.g. power oil lines, shall be priced under normal pricing procedures for casing, tubing, or drill pipe. Upset tubular goods shall be priced on a non upset basis.

(3) Condition E

Junk shall be priced at prevailing prices. Operator may dispose of Condition E Material under procedures normally utilized by Operator without prior approval of Non-Operators.

D. Obsolete Material

Material which is serviceable and usable for its original function but condition and/or value of such Material is not equivalent to that which would justify a price as provided above may be specially priced as agreed to by the Parties. Such price should result in the Joint Account being charged with the value of the service rendered by such Material.

E. Pricing Conditions

(1) Loading or unloading costs may be charged to the Joint Account at the rate of twenty-five cents (25¢) per hundred weight on all tubular goods movements, in lieu of actual loading or unloading costs sustained at the stocking point. The above rate shall be adjusted as of the first day of April each year following January 1, 1985 by the same percentage increase or decrease used to adjust overhead rates in Section III, Paragraph 1.A(3). Each year, the rate calculated shall be rounded to the nearest cent and shall be the rate in effect until the first day of April next year. Such rate shall be published each year by the Council of Petroleum Accountants Societies.

(2) Material involving erection costs shall be charged at applicable percentage of the current knocked-down price of new Material.

3. **Premium Prices**

Whenever Material is not readily obtainable at published or listed prices because of national emergencies, strikes or other unusual causes over which the Operator has no control, the Operator may charge the Joint Account for the required Material at the Operator's actual cost incurred in providing such Material, in making it suitable for use, and in moving it to the Joint Property; provided notice in writing is furnished to Non-Operators of the proposed charge prior to billing Non-Operators for such Material. Each Non-Operator shall have the right, by so electing and notifying Operator within ten days after receiving notice from Operator, to furnish in kind all or part of his share of such Material suitable for use and acceptable to Operator.

4. **Warranty of Material Furnished By Operator**

Operator does not warrant the Material furnished. In case of defective Material, credit shall not be passed to the Joint Account until adjustment has been received by Operator from the manufacturers or their agents.

V. INVENTORIES

The Operator shall maintain detailed records of Controllable Material.

1. **Periodic Inventories, Notice and Representation**

At reasonable intervals, inventories shall be taken by Operator of the Joint Account Controllable Material. Written notice of intention to take inventory shall be given by Operator at least thirty (30) days before any inventory is to begin so that Non-Operators may be represented when any inventory is taken. Failure of Non-Operators to be represented at an inventory shall bind Non-Operators to accept the inventory taken by Operator.

2. **Reconciliation and Adjustment of Inventories**

Adjustments to the Joint Account resulting from the reconciliation of a physical inventory shall be made within six months following the taking of the inventory. Inventory adjustments shall be made by Operator to the Joint Account for overages and shortages, but, Operator shall be held accountable only for shortages due to lack of reasonable diligence.

3. **Special Inventories**

Special inventories may be taken whenever there is any sale, change of interest, or change of Operator in the Joint Property. It shall be the duty of the party selling to notify all other Parties as quickly as possible after the transfer of interest takes place. In such cases, both the seller and the purchaser shall be governed by such inventory. In cases involving a change of Operator, all Parties shall be governed by such inventory.

4. **Expense of Conducting Inventories**

A. The expense of conducting periodic inventories shall not be charged to the Joint Account unless agreed to by the Parties.

B. The expense of conducting special inventories shall be charged to the Parties requesting such inventories, except inventories required due to change of Operator shall be charged to the Joint Account.

The main sections normally included in the accounting procedure are discussed below.

Section I. General Provisions

The General Provisions section deals with a variety of topics. Some of the typical terms are discussed below:

1. **Definitions:** This section lists terms used in the contract that are frequently subject to question or interpretation. Examples include first-level supervision, technical employee, and controllable material.
2. **Joint account records and currency exchange:** In international operations it is necessary to indicate which language and in which currency the joint account will be maintained. Commonly the accounts are maintained in both U.S. dollars and the local currency of the country where the operation is located. The operator must also maintain accurate records of production of crude oil and natural gas.
3. **Statements and billings:** The operator is to provide a monthly statement to all of the nonoperators. The statement should include a listing of all costs and expenditures incurred during the preceding month, the amount of advances received from the parties, each party's share of the costs and expenses, and the respective cash balances or deficits. The costs and expenditures incurred during the month should be identified as:
   a. costs and expenditures under the single expenditure limit and thus not requiring an AFE (these costs and expenditures should be categorized as either investment or expense)
   b. costs and expenditures relating to specific AFEs
   c. costs and expenditures relating to appropriations (authorized expenditures over the single expenditure limit but not related to a well)
4. **Payments and advances:** The operator is given the right to require the nonoperators to prepay or advance the next month's estimated cash outlays—referred to as a **cash call**. Cash calls are common in international oil and gas operations. In domestic operations, they are often encountered when there are a number of wells being drilled and cash outlays are large. In the event that the cash call for any given month exceeds the actual amount of cash required for that month, each party's share of the balance should be carried forward to reduce that party's cash call for the next month. If cash calls are not used, the operator will send a billing to the nonoperators itemizing costs for the month. The nonoperators must remit their proportionate share to the operator in a timely manner—typically within 10 days.
5. **Adjustments:** Payment of a billing or cash advance does not indicate that the nonoperator agrees with the correctness of a billing or statement. The nonoperator has 24 months from the end of the current year to raise exceptions to any charge. After that period, the statements are deemed to be true and correct.
6. **Audits:** All nonoperators have the right to audit the joint account and other records of the operator pertaining to the joint operation. The nonoperators have 24 months

from the end of the year in which the disputed charge occurred to raise an exception or make a claim to the operator.

7. **Allocations:** In the course of a joint operation, it frequently becomes necessary to allocate joint and/or common costs between the joint operation and other operations. For example, the operator may have solely-owned equipment that is used on jointly owned properties. The costs associated with the equipment may be allocated to all of the properties that it serves. Most contracts state such allocations are to be made on an equitable basis in accordance with accounting standards. Unfortunately, there are no formalized accounting standards relating to joint or common cost allocations in oil and gas operations. Therefore, it is necessary for the operator to be careful to use methods that are rational and equitable. The operator should be prepared to explain and justify such cost allocation methods if called upon to do so by the non-operators.

Section II. Direct Charges

All the working interest owners in a jointly owned property have an obligation to pay their proportionate share of the costs and expenses in return for a share of production. The term *joint account* refers to costs that have been identified with or allocated to a particular jointly owned property and therefore are the responsibility of that particular group of working interest owners. The term *joint account* does not necessarily refer to a separate account. Two specific types of costs are recognized and separated for the purpose of making charges to the joint account, **direct** and **indirect**.

Direct costs are costs that are specifically identified with the joint operation. Indirect costs are not individually identified with the joint operation *per se.* Rather, an operator is allowed to recover its indirect costs by charging the joint account some agreed upon amount, *e.g.*, a percentage of direct costs or a fixed amount per well drilled. A working knowledge of these two types of costs is necessary in order to understand the specific charges that can be made to the joint account.

Expenditures made for material and services on the joint property for the direct benefit of the joint property are the primary source of direct charges. The following are examples of general activities typically charged directly to the joint account: exploratory drilling, development drilling, installation of production equipment, operation, maintenance and repair of wells and equipment, and rentals.

Costs incurred off the property or at a general or administrative level are the primary sources of indirect or overhead charges. These costs benefit the joint property but in an indirect manner. Examples of general activities are: home office administration, data processing, office services, human resources, and legal support.

The following is a more detailed discussion of the costs that would be considered direct costs versus indirect costs.

1. **Licenses, permits, etc:** Generally all costs incurred by the operator in relation to the acquisition, maintenance, renewal or relinquishment of licenses, permits, or surface rights acquired by the joint operation are direct charges.
2. **Salaries, wages, and related costs:** The salaries, wages, and related costs of employees of the operator engaged in the joint operations whether temporarily or permanently assigned are direct charges.
   a. **First level of supervision:** A commonly encountered issue is what levels of the management organization are general administrative overhead and what levels are directly related to the property served. As management gets further and further removed from the physical operations the answers become less and less clear. Most operating agreements provide for the salaries and expenses of *first level supervision* in the field to be a direct charge to the joint account. The operations in the field include drilling wells, repairing wells, recompletion of wells, producing wells, constructing and operating facilities, etc. One frequently encountered problem within the industry is identifying the employees who qualify as first level supervision for these purposes. The *1974* and *1984 COPAS Accounting Procedures* defined first level supervision as those employees whose primary function in joint operations is the direct supervision of other employees and/or contract labor directly employed on the joint property in a field operating capacity. Generally, first level supervisors do not have engineering or administrative staff but rely upon the staff associated with the administrative or functional/technical office for these services.

      Defining first level supervisors is more complicated because each operator has their own organizational structure designed to optimize performance of the day-to-day functions necessary for drilling and producing operations. Differences exist in job classifications and work assignments. Further, one or more employees may perform first level supervision for multiple properties with different ownerships (*i.e.*, one foreman may provide direct supervision for multiple properties, each of which has different working interest owners). In addressing this matter of first level supervision, it is important to give consideration to *job functionality* or work being performed, and not just thejob classification.
   b. **Technical labor.** Charging the cost of the operator's technical employees to the joint account has been a long-standing issue in domestic joint venture accounting. Technical employees include such employees having special and specific engineering, geological, or other professional skills, and whose primary function in joint operations is the handling of specific operating conditions and problems. They usually have no supervisory authority except that required to resolve the particular problem to which they are assigned. Charging of the salary and wages of technical employees is expressly provided for in each COPAS accounting procedure beginning in 1962.
3. **Employee benefits:** Employee benefits are generally considered to be part of the employer's total labor cost. Most operating agreements allow the operator to charge

the joint account with the current cost of established employee benefit plans as long as they are made available to all employees on a regular basis. This charge is usually expressed as a percentage of the total labor chargeable to the joint account. Another method of charging employee benefits to the joint account is on a *when and as paid* basis. However, the percentage assessment is typically allowed. COPAS issued *Interpretation #11* to help clarify which costs are to be included in the percentage calculation and which are not. Costs that can be included in the percentage calculation include: bonus (*e.g.,* Christmas bonus), medical and dental insurance, business travel insurance, pensions, profit sharing plans, life insurance, tuition assistance, long-term disability insurance, and vision care plans. Among the costs not included are personal leaves, car pool subsidies, company car use, employee stock ownership plans, lay off benefits, and parking.

4. **Services:** The cost of services is generally broken down between those services provided by a third party and those provided by an affiliate of the operator. The cost of services performed by third parties for the benefit of the joint operation are direct charges, provided the transactions that resulted in the charges are derived pursuant to an arm's length transaction. The cost of professional administrative, scientific, or technical personnel services provided by an affiliate of the operator in lieu of services provided by the operator's own personnel are direct charges, if provided for the direct benefit of the joint operations. The rates charged must be equal to the actual cost of the services, must exclude any element of profit, and should not be higher than charges of third parties for comparable services performed under comparable conditions.
5. **Offices, camps, and miscellaneous facilities:** The costs of maintaining any offices, sub-offices, camps, warehouses, housing, shore-based facilities—or other facilities of the operator and/or affiliates of the operator that are directly serving the joint operations—are directly chargeable to the joint account. If any such facilities serve other operations in addition to the particular joint operation in question, or any business other than the petroleum operations, the net costs are to be allocated to the operations served on an equitable and consistent basis.
6. **Communications:** The costs of acquiring, leasing, installing, operating, repairing, or otherwise utilizing communication systems are direct charges if the equipment is necessary for the joint operations. Such equipment may include satellite, radio, and microwave facilities.
7. **Exclusively owned equipment and facilities of the operator:** The operator may use equipment and facilities it exclusively owns on a joint property. In that case, the operator is allowed to charge the joint account rental rates based on the actual cost incurred by the operator including factors relating to the cost of ownership. However, the rates charged may not exceed the average prevailing commercial rates of non-affiliated third parties for like equipment and facilities used in the same area.

8. **Ecological and environmental costs:** Ecological and environmental costs incurred in relation to a particular operation are generally considered to be a cost of operating jointly owned property and are directly chargeable to the joint account.
9. **Materials and supplies:** The costs of materials and supplies—net of any discounts—purchased or furnished by the operator for the joint operations are direct charges. The costs include, but are not limited to, export brokers' fees, transportation charges, loading and unloading fees, export and import duties, license fees, and in-transit losses not covered by insurance.
10. **Damages and losses:** Costs that can be associated with losses by casualty or theft are directly chargeable to the joint account. Any settlement received from an insurance carrier should be credited to the parties participating in any joint property insurance coverage.

Section III. Overhead

Various methods have been utilized in the industry in the past to recover the indirect costs or **overhead** associated with an oil or gas operation. The charging of overhead is the means by which the operator recovers its cost for such items as clerical, administrative, engineering, accounting, home office expenses, and other costs not allowed as direct charges. The accounting procedure typically provides for three types of overhead:

- overhead incurred in drilling and production operations
- overhead incurred in construction operations
- catastrophe overhead

The methods of computing overhead are either combined fixed rate or percentage basis. The **combined fixed rate** basis is the most commonly used method of computing overhead in domestic production and drilling operations. The rate is referred to as *combined* because it is meant to cover an operator's expenses at all levels (*i.e.*, home office, regional, district, etc.) and *fixed* because it does not vary in proportion to actual expenses. To determine the amount of overhead, the overhead rate for production operations is multiplied by the number of wells that are producing during all or any part of the month; the drilling overhead rate is multiplied times the number of wells being drilled, prorated for the number of days during the month that the wells are actually being drilled.

For example, assume the producing well rate is $100 per month and the drilling well rate is $300 per month. If there are 50 wells on the lease that produced at least one day during the month, then the total producing well overhead for the month would be $100 x 50 wells ($5,000). If there are 2 wells being drilled on the lease with one well having drilling in progress all month and the other having drilling operations for only 10 days, then the drilling well overhead is $400, *i.e.*, $300 + 10/30 x $300. The rates are determined through negotiation and are typically adjusted annually to reflect changes in prices and costs.

The less common alternate is to provide for production and drilling overhead to be computed as a percentage of direct costs. If the percentage basis is used, there is typically one rate for drilling and development overhead and another rate for production overhead, *e.g.*, 5% of all drilling and development costs plus 2% of all production expenditures.

The other types of overhead are construction overhead and catastrophe overhead. Construction overhead is provided to compensate the operator for indirect costs incurred while major construction projects are underway (*i.e.*, expansion of a gathering system or construction of a central tank battery). Catastrophe overhead compensates the operator for indirect costs incurred in the event of an unexpected event such as a hurricane, oil spill, explosion, fire, etc. Construction overhead and catastrophe overhead are typically calculated on a percentage sliding scale basis. For example:

5% of total costs up to and including \$100,000, plus
3% of costs in excess of \$100,000 but less than \$1,000,000, plus
2% of costs equal to or in excess of \$1,000,000

In international operations overhead rates are almost exclusively based on a sliding percentage of appropriate direct charges. Generally one set of rates applies to exploration operations, while a different set of rates apply to development operations and to production operations. The rates applicable to exploration operations are normally the highest while the rates applicable to production operations are the lowest. These rates are applied to the cumulative annual expenditures. The following is an example of international exploration operations overhead rates:

Annual Expenditures for Exploration Operations

Direct Charges Incurred and Charged to Joint Operations	Percentage Rate of Direct Charges Charged to Joint Operations as OH
\$0 to \$5,000,000	5%
\$5,000,001 to \$15,000,000	3%
\$15,000,001 to \$25,000,000	2%
more than \$25,000,000	1%

Section IV. Pricing of Joint Account Material Purchases, Transfers, and Dispositions

The operator is responsible for the provision and disposition of material to be used in the joint operations, and should make proper and timely charges and credits for all material movements affecting the joint property. In addition, the operator is obligated to make timely disposition of any idle and/or surplus material, with such disposal being made either through sale of the material to the operator or nonoperators, division in kind, or sales to outsiders. The operator may purchase, but is under no obligation to purchase, the interest of nonoperators in surplus material.

1. **Purchases:** Material that is purchased from a third party should be charged to the joint account at the price paid by the operator after the deduction of all discounts received. In some cases, by purchasing in bulk, the operator may be able to qualify for quantity discounts. The operator should pass a *pro rata* share of such discounts to the joint account.

2. **Material transfer pricing:** Material owned by one of the parties that is moved *to* the joint property, and material transferred *from* the joint property or disposed of by the operator, should be priced on the following basis exclusive of cash discounts (unless otherwise agreed to by the parties):

   - **New material (Condition A):** New material should be priced at the current new price in effect at the date of movement, as listed by a reliable supply store near the joint property or near the manufacturer. If applicable, transportation costs to the receiving point nearest the joint property are also charged to the lease.
   - **Good used material (Condition B):** Condition B material is material in sound and serviceable condition that is suitable for reuse without reconditioning. Material moved *to* the joint property is priced at 75% of the current new price. Material used on and moved *from* the joint property should be charged at 75% of the current new price. If the material was originally charged to the joint account as good used material (Condition B), the material should be transferred at 65% of the current new price.
   - **Used material (Condition C):** Condition C material is not in sound and serviceable condition, and is not suitable for its original function until after reconditioning. Condition C material transferred *to* a joint property should be priced at 50% of the current new price. The cost of reconditioning should be charged to the receiving property, provided condition C value, plus the cost of reconditioning, does not exceed condition B value.
   - **Condition D:** Material, excluding junk, no longer suitable for its original purpose, but usable for some other purpose should be priced on a basis commensurate with its use. In most cases, the operator may dispose of Condition D material without prior approval of nonoperators.
   - **Condition E:** Junk should be priced at prevailing junk prices.

   Transportation costs from the receiving point nearest the joint property (for condition A material) or from the sending property to the receiving property (for all other Conditions) should be charged to the joint account of the receiving property. The cost of the equipment itself and the transportation costs may be recorded in different sub-accounts to separately identify the costs for cost management purposes.

3. **Disposition of material:** The disposition of surplus material can occur by any of the three following methods:

a. **Material purchased by operator or nonoperator:** The operator or nonoperators may purchase material from the joint account on the basis of condition value.
b. **Division in kind:** The operator and nonoperators may divide surplus material from the joint property in proportion to each party's interest.
c. **Sales to outsiders:** Sales are accomplished on the basis of competitive bidding. Sales to outsiders normally occur after it has been determined that neither the operator nor nonoperators has a need for the material.

Section V. Inventories

The accounting procedure requires the operator to maintain detailed records of controllable material and to conduct regular physical inventories. The operator is responsible for maintaining an accurate record of controllable material. The listing of controllable material should be compared with a physical examination of existing assets at reasonable intervals, and appropriate action taken when discrepancies are identified.

Expense of conducting periodic inventories. Any expenses incurred by the operator in conducting periodic inventories should be charged to the joint account. If nonoperators elect to have a representative present they do so at their own cost and expense.

Special inventories. Special inventories are generally required whenever there is change in the operator. The expenses related to conducting special inventories resulting from a change of operator are normally charged to the joint account. Special inventories may be requested due to a sale or change of interest in the joint property. The expenses related to conducting other special inventories are generally borne by the parties requesting such an inventory. In either event, all parties must agree to the results of the inventory.

Joint Interest Accounting

Most joint ventures are accounted for using the proportionate consolidation method. Under the proportionate consolidation method, each owner accounts for its *pro rata* portion of the assets, liabilities, revenues, and expenses of the venture. The discussion and examples below illustrate this process

Booking charges to the joint account: accumulation of joint costs in operator's regular accounts

The most common method of booking joint costs is to initially record all costs in the operator's regular accounts (*i.e.*, lease operating expense, wells and related E&F, wells-in-progress, etc.). The costs are associated with specific properties via a system of property identification numbers. At the end of each month, the operator identifies all costs that have been charged to the jointly owned properties it operates. The operator then recognizes a receivable for the nonoperators' share of those costs and credits or *cuts back* its regular accounts for the nonoperators' portion of the costs. The operator's own portion of the costs incurred during the month is thus left in its regular accounts. This process is illustrated in the following example:

EXAMPLE

Accumulation of Joint Costs in Regular Accounts

Lucky Company owns 60%, South Company owns 10%, and North Company owns 30% of the joint working interest property 1002. Lucky Company is the operator. Lucky Company incurs the following costs during October 2008, in connection with Lease Number 1002.

Salaries and wages, field employees	$ 5,000
Contract service, reacidizing	2,500
Purchase and installation of compressor unit	900
Property taxes paid	500
Equipment from operator's inventory installed on lease	600
Allowed overhead charge (2 wells at $1,200 per well)	2,400

Entries during month by operator (Lucky Company)

Lease operating expense—joint lease	5,000	
Wages payable		5,000
Lease operating expense—joint lease	2,500	
A/P		2,500
Wells and related E&F—joint lease	900	
A/P		900
Lease operating expense—joint lease	500	
Property taxes payable		500
Wells and related E&F—joint lease	600	
Materials and supplies		600

Entries at end of month

Lucky Company		
Lease operating expense—joint lease	2,400	
Overhead expense—control account*		*2,400
A/R—South Company (10% x $11,900)	1,190	
A/R—North Company (30% x $11,900)	3,570	
Lease operating expense—joint lease (40% x $10,400)		4,160
Wells and related E&F—joint lease (40% x $1,500)		600
South Company		
Lease operating expense	1,040	
Wells and related E&F	150	
A/P—Lucky Company.		1,190
North Company		
Lease operating expense	3,120	
Wells and related E&F	450	
A/P—Lucky Company.		3,570

*The actual overhead costs were charged to the overhead expense—control account when incurred, with the allowed overhead charges billed to the lease at the end of the month.

Booking charges to the joint account: distribution of joint costs as incurred

Another method that might be used by the operator is to record the distribution of joint costs as incurred. Using this approach, the operator charges its regular (nonjoint interest) accounts and recognizes a receivable from the nonoperators for their portion of the costs as each transaction occurs. The operator, however, actually bills the nonoperators for their portion of the costs only once a month.

EXAMPLE

Distribution of Joint Costs as Incurred

Assume the same ownership and operator as in the previous example. Lucky Company incurs minor workover costs of $10,000.

Entry

Lease operating expense (60% x $10,000)	6,000	
Accounts receivable—South (10% x $10,000) . . .	1,000	
Accounts receivable—North (30% x $10,000). . .	3,000	
Cash		10,000

Non-consent operations

A situation that frequently occurs requiring considerable accounting effort is a **non-consent operation.** Non-consent operations arise when one or more of the working interest owners do not consent to the drilling, deepening, reworking, or abandonment of a well. Section VII.B. of the JOA provides the procedures the operator must follow when one or more of the parties decide to go non-consent. Since non-consent operations occur frequently in relation to drilling a well, this discussion focuses on drilling situations. Non-consent operations related to deepening, reworking, or abandonment are similar to drilling operations.

First, the party wishing to drill—usually the operator—must give written notice to all of the working interest owners of the proposed drilling operation. The parties have a period of time, typically 30 days, to reply. If one or more of the parties elects not to participate, the consenting parties are re-notified and given the election to pay only their proportionate share of the costs, or in addition to their proportionate share, to pay all or part of the non-consenting party's share. The party electing not to participate is referred to as a **carried working interest** or **carried party**. The working interest owners who agree to pay the carried party's share of the costs are referred to as the **carrying parties**. If none of the working interest owners agree to participate in drilling the well, the operator can either drill the well and carry all of the other owners himself or not drill the well.

When a working interest owner goes non-consent it does not relinquish its interest in the property. Rather its interest reverts temporarily to the carrying party(ies). The carrying parties are allowed to sell and keep the revenue from the carried party's share of oil and gas produced from the well until they have recovered the costs that they paid on behalf of the carried party plus a penalty. The penalty is provided in VI.B.2(I) of the JOA. A common penalty is 200% of costs, resulting in a total recovery of 300% of costs, *i.e.*, a recovery of costs (100%)

plus the penalty (200%). In practice, a 200% penalty is typically stated as a 300% penalty; thus, in practice the penalty is actually stated in terms of total cost recovery. When the carrying parties have recovered the cost they paid on behalf of the carried party plus the penalty, the carried party is said to have reached **payout**. From that point forward all of the working interest owners participate in costs and revenues at their pre-non-consent percentages.

EXAMPLE

Carried Working Interest

Lucky Company, South Company, and North Company each own 33.33% of the joint working interest property 1002. The royalty interest is 1/8. Lucky, the operator, notified North Company and South Company of its plans to drill Well No. 2 at an anticipated cost of $100,000. South Company elects to go non-consent. Lucky and North agree to carry their proportionate share of South Company's costs. If the well is successful, they will be allowed to recover from South Company's share of production, the costs they carried plus a penalty of 300%. Upon payout, South Company will resume participation at 33.33%.

Lucky Company and North Company determine their proportionate share of South Company's costs and revenues in the following manner:

	Interest of Consenting Parties	Proportion of South's Costs and Revenues
Lucky Company	33.3333%	33.3333/66.6666 = 50%
North Company	33.3333%	33.3333/66.6666 = 50%
Total	66.6666%	

Assume that the well is drilled at a cost of $100,000. Lucky Company and North Company each pay $50,000 of which $33,333 is their own share and $16,667 ($100,000 x 33.3333% x 50%) is their portion of South Company's share.

On July 25, 2004, Well No. 2 is completed. Production and operating costs for the first three months of production are: (Severance taxes have been ignored in this problem.)

Month	Sales	Sales Price	Total Operating Expenses
August	5,000 bbl	$20/bbl	$10,000
September	9,000 bbl	20/bbl	15,000
October	6,000 bbl	20/bbl	25,000

The payout calculation made by Lucky Company is as follows:

	Aug.	Sept.	Oct.
Sales Volume	5,000	9,000	6,000
Price/bbl	$ 20	$ 20	$ 20
Gross Sales ($)	100,000	180,000	120,000
Net of Royalty	7/8	7/8	7/8
Net Sales ($)	87,500	157,500	105,000
Operating Expense ($)	(10,000)	(15,000)	(25,000)
Net Revenue ($)	$ 77,500	$ 142,500	$ 80,000
Revenue to Lucky:			
Total to Lucky (50%)	$ 38,750	$ 71,250	$ 40,000
Lucky's portion (33.3333%)	25,833	47,500	26,667
South's portion (16.6667%)	$ 12,917	$ 23,750	$ 13,333

Payout:

South Company's share of well	$ 33,333
Lucky's proportion of South's cost	50%
Amount paid by Lucky Company	16,667
Penalty	x 300%
Recoverable by Lucky Company	$ 50,000

Amount to be recovered	$ 50,000
Recovered in August	(12,917)
Balance to be recovered	$ 37,083
Recovered in September	(23,750)
Balance to be recovered	$ 13,333
Recovered in October	(13,333)
Balance to be recovered	$ 0

In November, South Company would be treated as a 33.33% working interest owner, paying its proportionate share of operating costs and receiving its proportionate share of revenues.

Entries for Lucky Company for July and August

July:

Wells and related E&F (50% x $100,000)	50,000	
A/R—North Company (50% x $100,000)	50,000	
Accounts payable		100,000

August:

Lease operating expense (50% x $10,000)	5,000	
A/R—North Company (50% x $10,000)	5,000	
Cash		10,000
Cash	100,000	
Royalty payable ($100,000 x 1/8)		12,500
Accounts payable—North Company (50% x $100,000 x 7/8).		43,750
Oil revenue (50% x $100,000 x 7/8)		43,750

Note: Lucky Company does not set up a receivable from South Company for South's carried interest. Also, Lucky Company recognizes its share of revenue retained from South Company's interest as revenue—not as a reduction in its capitalized costs.

The operator provides the carried party with a payout statement on a monthly basis so both parties know when payout is near. Ideally, the operator would be able to determine the precise point during the month when payout occurs, and immediately begin to distribute operating costs and revenue to the carried party. Realistically, it is very difficult if not impossible to track payout that precisely. Accordingly, the carried party is usually notified at the end of the month and some type of adjustment is made to equalize the costs and revenues. For example, in the problem above, South Company could have received a cash payment from Lucky and North or Lucky and North could pay a portion of South's operating expenses during November.

Accounting for materials

One of the most difficult and challenging problems facing a joint interest operator is the pricing of material transferred *to* and *from* the property being operated. Material to be used on a joint property can be purchased directly for the specific property, moved to or from the operator's warehouse, or transferred from another property. Material purchased directly for a property is charged to that property at the cost of the material, less all discounts received. Any transportation charges are also charged to the property. The following example illustrates the purchase of new material for a joint property:

EXAMPLE

Purchase of New Material

Lucky Company purchased casing for a joint property at a net price of $100,000. Loading, hauling, and unloading costs from the railway railhead to the wellsite were $8,000. Lucky is the operator on the lease and has a 60% working interest.

Entry

A/R—nonoperators (40% x $108,000)	43,200	
W/P—L&WE (casing) (60% x $100,000)	60,000	
W/P—L&WE (freight) (60% x $8,000)	4,800	
Cash		108,000

The preceding example gives the *net result* at the end of the month after distribution of joint costs.

Transfer from warehouse—Condition A. When material is transferred from the operator's warehouse or a joint property to another joint property it must be priced according to the accounting procedure. The price used to record the transfer should approximate the current market price of the material or equipment. Several methodologies may be used to obtain market prices used to record transfers. These are listed in the *1995 COPAS Accounting Procedure* and include a COPAS database called Computerized Equipment Pricing System (CEPS) that contains a generic price calculated for each piece of material or equipment. The database is updated annually by a historical price multiplier. Other methods that can be used to approximate the current market price are vendor quotes, historical purchase prices, or manually applying the historical price multiplier to a published price. In addition, transfer prices may be mutually agreed upon by the parties.

Condition A is new material and the joint property would normally be charged with the current market price of the material plus transportation charges from the warehouse to the property.

The following example illustrates the transfer of a pump from the operator's warehouse to a joint property owned 60% by Lucky, the operator. Note that the asset being transferred is inventory, not wells and equipment.

EXAMPLE

Transfer from Warehouse—Condition A

Assume that Lucky Company has a submersible pump in its warehouse with a cost of $9,000, including shipping and handling to the warehouse. The pump is new and is to be installed in a newly drilled well. The current market value is $9,000. The pump is transferred from Lucky's warehouse to a jointly owned property in which Lucky has a 60% working interest. Transportation costs from the warehouse to the property total $200.

Entry to record transfer of pump

A/R—nonoperators ($9,000 x 40%)	3,600	
W/P—L&WE (pump) ($9,000 x 60%)	5,400	
Warehouse inventory		9,000

Entry to record transportation costs

A/R—nonoperators ($200 x 40%)	80	
W/P—L&WE (transportation)($200 x 60%) . . .	120	
A/P—freight company		200

Transfer from warehouse to joint property—Condition B. Condition B material is material in sound and serviceable condition that may be reused for its original purpose without reconditioning. This material should be charged from the sending property or warehouse *to* the receiving property at 75% of the current market price. The following example assumes the same facts as the previous example, except the material is in Condition B rather than A and the current market value is $10,000 rather than $9,000. Since the material is being transferred from a warehouse, the difference between the operator's share of the original cost of the material and the condition value of the material is recorded as a miscellaneous revenue or expense.

EXAMPLE

Transfer from Warehouse to Joint Property—Condition B

Assume the pump in the previous example that was carried on the books of Lucky's warehouse at $9,000 has a current market price of $10,000. Also assume that the pump is used, is in sound condition, and is being installed in a newly drilled well. Ignore transportation costs.

The Condition B value of the pump would be $10,000 x 75% = $7,500.

Entry

A/R—nonoperators (40% x $7,500)	3,000	
W/P—L&WE ($7,500 x 60%).	4,500	
Other revenue/expense (($10,000 – $7,500) x 60%)	1,500	
Warehouse inventory		9,000

Transfer from one property to another property—Condition B. When material is transferred *from* one property to another property, the percentage used in determining the credit to the sending property depends on the original condition of the material when it was charged to that property. If Condition B material was originally Condition A material when it was first charged to the property, the material will be removed at 75% of the current market price. However, if Condition B material was originally Condition B material when it was first charged to the property, the material will be removed at 65% of the current market price. The reduction represents an assumed reduction in service value.

When material is transferred between jointly owned properties with different working interests, the nonoperators of the sending property are credited with their share of the condition value of the equipment. The operator removes the equipment from its accounts at the equipment's original historical cost. Since neither the SE or FC rules allow gain or loss recognition in this case, the operator's share of the difference between the original cost of the equipment and the condition value is debited or credited to the accumulated DD&A account for the lease.

EXAMPLE

Transfer from Property to Another Property—Condition B

A pump originally costing \$32,000 is transferred from Alpha Lease, a joint property on which Lucky Company serves as operator and has a 60% interest, to Beta Lease, another joint property operated by Lucky in which Lucky has a 90% interest. At the time of transfer, the current market price of the pump is \$40,000. The pump is Condition B and was Condition A material when first charged to the property. Ignore transportation charges. Assume that the pump is being installed in a newly drilled well.

The condition value of the equipment is \$40,000 x 75% = \$30,000.

Entry

A/R—Beta Lease nonoperators (\$30,000 x 10%)	3,000	
W/P—L&WE—Beta Lease (\$30,000 x 90%) . . .	27,000	
Accumulated DD&A—Alpha Lease (\$32,000 – \$30,000) x 60%)	1,200	
A/R*—Alpha Lease nonoperators (\$30,000 x 40%).		12,000
L&WE—Alpha Lease (\$32,000 x 60%)		19,200

* When material is transferred off a lease, the customary accounting treatment is to credit (reduce) the account receivable from the nonoperators rather than to credit an account payable.

Transfer from property to another property—Condition C. As discussed earlier, Condition C material is not in sound and serviceable condition and is not suitable for its original use without reconditioning. Condition C material transferred *to* a property should be charged to the property at 50% of the current market price. The cost of any reconditioning is paid by the receiving property. Condition C material transferred off a property may be handled in two different ways, depending upon which property pays the reconditioning costs. Condition C material can be transferred at 50% of the current market price, with the receiving property paying for the recondition costs; or the material can be transferred at 75% of the current new price, with the sending property paying all of the recondition costs. (If the sending property reconditions the material, the material would then be in Condition B and would thus be transferred at 75% of the current market price.) Condition C material that is reconditioned should never have a total cost in excess of Condition B value.

EXAMPLE

Transfer from Joint Property to Another Joint Property—Condition C

Lucky Company is the operator on both Lease A and Lease B, and has a 40% working interest in Lease A and a 80% working interest in Lease B. A piece of equipment originally costing $30,000 is transferred from Lease A to Lease B. The current market price of the equipment is $36,000 and the equipment is transferred at Condition C. The working interest owners of Lease B will pay all costs of reconditioning.

The condition value of the equipment is $36,000 x 50% = $18,000.

Entry

A/R—nonoperators—Lease B (20% x $18,000) . .	3,600	
Wells and related E&F—L&WE—Lease B (80% x $18,000)	14,400	
Accumulated DD&A—Lease A (40% x [$30,000 – $18,000]).	4,800	
A/R—nonoperators—Lease A (60% x $18,000)		10,800
Wells and related E&F—L&WE—Lease A (40% x $30,000).		12,000

Assume that the cost of reconditioning the equipment is $5,000. This entire amount would be charged to Lease B.

A/R—nonoperators—Lease B (20% x $5,000) . .	1,000	
Wells and related E&F—L&WE—Lease B (80% x $5,000)	4,000	
A/P—Vendor.		5,000

In the preceding example, Lucky's accumulated DD&A account is debited for $4,800, which is Lucky's share of the difference between the original cost of the equipment and the condition value of the equipment: 40%(30,000 – (50% x 36,000)).

It is possible for Condition C equipment to be used in some manner other than its original purpose. For example, production tubing that is Condition C could not be re-used as production tubing without reconditioning. It could, however, be used as a flowline since a flowline would not have to withstand the pressure that tubing used in a well would. In that case, the equipment would be transferred as Condition C and there would be no reconditioning cost incurred by the receiving property.

In some cases, material transferred between properties may be priced in excess of the appropriate condition value of the current prices for the same new material. Premium pricing occurs when the operator's cost, either through reconditioning or actual price paid, exceeds the condition value of current prices for new material. The nonoperators must always be notified when premium pricing occurs.

Offshore Operations

Offshore operations are commonly joint working interest ventures because of the large dollar amounts and large amount of risk involved. Domestically, these operations are conducted in either federal- or state-owned waters. Bidding on leases from either the state or the federal government is normally required to obtain a lease. Bidding on offshore federal leases is done by sealed bids, with generally a separate bid for each tract.

Some different types of costs are incurred for offshore operations in comparison to onshore operations. These include the costs of using mobile rigs, fixed platforms, helicopter costs, and special safety equipment designed for offshore use. Many offshore costs require allocation between differing tracts, *i.e.*, fuel costs, boat costs, helicopter costs, etc. COPAS bulletins have addressed these problems.

Normal offshore joint operations are accounted for by using the offshore accounting procedure designed by COPAS (*COPAS Bulletin No. 15*).

Joint Interest Audits

Nonoperators in joint venture operations have the right to audit the accounts and records of the operator within 24 months following the close of a year. Companies involved in joint venture operations normally have full-time joint interest auditors on staff to audit the charges made by operating companies in joint venture operations. Audits are generally not called more than once each calendar year. The audit is normally initiated by the nonoperating partner with the largest interest in the venture. This nonoperator usually provides the lead auditor in conducting the audit. The other nonoperators may furnish auditors or share the expenses of the audit. The nonoperators must give written notice to the operator that an audit is requested, after which a date to begin the audit will be confirmed.

The joint interest auditor's role in an audit of costs is limited to examining the charges made to the joint accounts and the support for those charges. The procedures used in conducting joint interest audits will vary widely, but the following suggested procedures are discussed in *COPAS Bulletin No. 3* (1980)*:

* The current *COPAS Bulletin No. 3* does not discuss these procedures in as much detail as the 1980 Bulletin.

1. **Operating agreement and AFE:** Review and brief the agreement for compliance of activities, review the AFE and compare it to actual expenditures.
2. **Minutes of operator's meetings:** Review and note significant and unusual actions that often give rise to errors.
3. **Company labor:** Obtain labor rates, contracts, and other labor policies. Make a test of payroll time sheets giving special attention to labor rates, reasonableness of allocations to properties, and charges for holidays and vacation. On a selective basis, compare names, occupations, rates, hours, and distribution reflected on payroll time sheets to the daily time reports.
4. **Materials and services purchased externally:** Check that all invoices are properly authorized, all discounts taken, prices are reasonable, and supporting data exists for unusual charges.
5. **Material and supplies transferred to or from the operation:** Check all charges, paying particular attention to the condition of the material transferred and the prices assigned.
6. **Warehousing:** Determine whether periodic inventories are being taken, check all inventory adjustments, and make a periodic physical check of selected items of warehouse stock, tracing these items to the warehouse records.
7. **Overhead:** Review overhead charges and make sure they are in accordance with the joint operating agreement.
8. **Services and facilities:** Check the fairness of rates being charged for services and facilities.
9. **Capital and maintenance jobs:** Check that all work has been properly authorized, bids are being sought for jobs, costs are reasonable, and all expenditures in excess of the amount allowed under the joint operating agreement have been properly approved.
10. **Taxes and insurance:** Make sure taxes are assessed against the nonoperators fairly and that insurance is being carried in accordance with the joint operating agreement.
11. **Allocation of income and expenditures:** Insure that all charges and credits to the nonoperators are based on the ownership percentages specified in the joint operating agreement.
12. **Capital assets:** Insure that records are being properly maintained, that assets charged to the joint accounts are actually being used on the property(ies), and that these tangible assets are necessary on the projects.

Following the audit, the joint interest auditors prepare a report for the operator. There is no standard format for this report, but it will include information such as the property description, audit period, names of the auditors, and scope of the audit. Also included will be a list of audit exceptions and supporting materials for charges to which the auditors seek adjustment. The operator will then review the exceptions and make adjustments where appropriate. The operator has 180 days following receipt of the audit report to reply regarding adjustments.

The economic benefits from performing an audit often exceed the cost of the audit, especially in ventures where many drilling and development costs are being incurred. Consequently, joint interest audits by nonoperators are common, creating a separate field of expertise within oil and gas accounting.

While joint interest audits have traditionally concentrated on costs, audits of revenue are becoming more common. The need for revenue audits has increased in part because of the deregulation of natural gas pricing. Gas can now be sold through brokers or marketers, on the spot market, to transmission companies, or directly to end-users. Consequently, the complexity involved in tracking numerous gas sales can cause errors to be made.

As with audits of costs, nonoperators have the right to audit the accounts and records of the operator relating to revenues within 24 months following the close of a year. Revenue audits concentrate on production information and sales agreements, and require a thorough understanding of the measurement and flow of the product. Revenue auditing is a relatively new field but will continue to become more important because of the complexities of today's markets.

PROBLEMS

1. When two or more parties own a joint working interest, who will usually manage the property?

2. When do parties normally enter into an operating agreement?

3. What are the duties of the operator? Nonoperator?

4. What is a carried working interest? Payout?

5. What is the difference between a direct cost and overhead?

6. What is *combined fixed rate* overhead?

7. Aramus Company operates the Able Lease. The accounting procedure attached to the JOA allows Aramus to recoup its overhead by the use of a combined fixed rate-well basis of $1,000 per producing well and $10,000 per drilling well.

   REQUIRED:

   a. How much total overhead would Aramus bill the joint account if the Able Lease had four wells that produced every day the previous month?
   b. What if three wells produced every day and only one produced for five days?
   c. What if the only operation on the lease the previous month was the drilling of a well? Drilling operations commenced on the first day of the month, operations were sus-

pended for four days on the 20th, and operations commenced again on the 24th and continued through the end of the month. A month is considered to be 30 days.

8. Simms Company owns 70%, Duster Company owns 20%, and Ross Company owns 10% of the working interest property 1004. Assume Simms Company is the operator and incurs the following costs during the month of July 2003, in connection with the property:

Salaries and wages, field employees	$ 6,000
Salaries and wages, first-level field supervisors	2,000
Operator's cost of holiday, vacation, sickness, and disability benefits, 8% of above	640
Social Security tax, 7.5% of above	600
Employee benefits, group life insurance	1,920
Material installed on property from Simms' inventory . .	300
Transportation of material and employees 1,200 miles @ $0.25 per mile	300
Contract service, reacidizing (workover)	2,500
Purchase and installation of compressor unit	1,500
Repair of Christmas tree	500
Property taxes paid	500
Insurance premium paid	800
Overhead, 2 wells @ $700 per well	1,400

REQUIRED: Give the entries to record and distribute the costs, assuming regular accounts are used.

9. Tiger Oil Company, the operator of Lease A, purchased casing with a list price of $60,000 for a joint interest property in which it has a 40% WI. The vendor gives a discount of 10% off list price and also has credit terms of 2/10, n/30. Loading, hauling and unloading costs amounted to $4,000. Prepare the entry to record the purchase.

10. Aggie Oil Company transferred an item of equipment from its wholly owned warehouse to a jointly owned lease in which it has a 70% WI. The item of equipment is in Condition B, and the current market price for the equipment is $50,000. The item of equipment was carried on Aggie's books at $40,000. Give the entry to record the transfer, ignoring transportation charges.

11. Ames, Inc. owns 60%, Garza Company owns 30%, and Nance Company owns 10% of the working interest property number 2008. Ames, Inc. is the operator and bills Garza and Nance monthly for their portion of costs incurred. During May, 2006, Ames incurred costs as follows:

Salaries and wages, field employees	$ 8,000
Salaries and wages, first-level field supervisors	4,000
Social security taxes, 7.5% of above	900
New separator, purchased and installed	20,000
Repairs to Christmas tree	1,000
Property taxes paid	2,000

Other Costs:
Employee benefits—21% of salaries and wages
Transportation of material and employees,
  1,500 miles at $0.25 per mile
Overhead, 22% of all costs listed above

REQUIRED: Give the entries to record and distribute the costs, assuming regular accounts are used.

12. List and briefly discuss the main points of the COPAS accounting procedures.

13. Cain Oil Company owns a piece of equipment, originally costing $60,000 that is currently being used on Lease A. Cain Oil Company owns a 40% working interest in Lease A and serves as the operator of the lease. The company plans to use the equipment on a lease wholly owned by Cain Oil Company. The equipment is transferred to the company's warehouse.

    REQUIRED: Prepare the entry to record the transfer under each of the following independent situations assuming severance tax is ignored:

    a. The equipment is in Condition B and originally was Condition A when transferred to the property. The current market price is $80,000.
    b. The equipment is in Condition C and Cain Oil Company will pay for the reconditioning. The current market price is $80,000.

14. Hamilton Oil Company is the operator of Leases A and B and has a 60% WI in each lease. A piece of equipment, which originally cost $30,000, is transferred from Lease A to Lease B. The current market price of the equipment is $40,000. The equipment is in Condition C. The working interest owners of Lease B will pay for reconditioning.

    REQUIRED: Prepare an entry to record the transfer.

15. The Norwood Lease has the following working interest owners: Shamrock Company 50%, Diamond Company 25%, and Heart Company 25%. There is a 1/8 royalty on the lease. On April 1, 2001, Shamrock Company, the operator, receives notice that Diamond Company is going non-consent on the drilling of the Gusher No. 2. Shamrock Company

and Heart Company agree to carry Diamond's share proportionately. The non-consent penalty is 300%. On August 1, the Gusher No. 2, which was drilled and completed at a cost of $250,000, goes on production. The production and operating information for the next few months is as follows:

Month	Production	Operating Costs	Sales Price
August	8,000 bbl	$25,000	$20
September	12,000 bbl	$40,000	$19
October	15,000 bbl	$50,000	$18
November	20,814 bbl	$75,000	$20

REQUIRED: Assuming severance tax is ignored:

a. Determine Shamrock Company and Heart Company's proportionate share of drilling and equipping costs.
b. Prepare a table determining when Diamond Company will payout.
c. Prepare the journal entry that Shamrock Company will make during August to book its share of production revenue.

16. Ajax Oil and Gas owns a 33.3% working interest in a lease in Northern Oklahoma. Tye D. Boll, a local farmer, owns a 1/8 royalty interest in the lease. Ajax is the operator and its partners, Balbo, Inc. and Comet Oil each own 33.3% of the working interest. Ajax analyzed the prospects for the lease and proposed drilling a gas well. Comet agreed but Balbo decided to stand out. Ajax and Comet both agreed to proportionately carry Balbo's working interest. The joint operating agreement stipulates that a 150% drilling and completion cost penalty will be assessed on any partner choosing not to participate in drilling the well.

    On July 1, 2004, the Wildcat No. 2 was drilled and completed at a total cost of $100,000. The following information is available concerning production and sales. Assume each company contracts to sell its gas for $2.00/mcf.

Month	Total Sales Volume	Operating Expenses	Sold by Ajax	Sold by Comet
August	60,000 Mcf	$15,000	55,000 Mcf	5,000 Mcf
September	62,000 Mcf	15,000	50,000 Mcf	12,000 Mcf
October	68,000 Mcf	15,000	58,000 Mcf	10,000 Mcf
November	60,000 Mcf	15,000	40,000 Mcf	20,000 Mcf

REQUIRED: Ignoring severance tax:

a. Determine when Balbo will payout if payout is calculated based on the quantity actually sold. Hint: Ajax and Comet would have to compute payout separately.

b. Determine when Balbo will payout if payout is calculated using the amount that each partner is entitled to.

17. The Alpha lease is operated by APC Company. The foreman of the Alpha Lease must make a decision regarding the replacement of a pumping unit on the lease. The foreman has identified three possible alternatives:

Alternative 1:

Purchase a new pumping unit. The foreman has located a vendor willing to sell the unit needed for a delivered cost of $65,000 FOB destination (*i.e.*, the manufacturer pays freight to the destination point).

Alternative 2:

The Beta Lease, which is also operated by APC Company, has a surplus pumping unit that could be used. If the Beta Lease equipment is used, the Beta Lease would be charged with the cost of repairing the unit that would then be transferred to the Alpha Lease at Condition B.

Alternative 3:

The Gamma Lease, which is also operated by APC Company, has a surplus pumping unit that could be used. If the Gamma Lease unit is used, the equipment would be transferred unrepaired at Condition C.

Other information:

	APC's Working Interest	**Unit Repair Costs**	**Trucking Costs**
Alpha Lease	100%	n/a	$ 0
Beta Lease	60%	$10,000	$2,000
Gamma Lease	30%	$ 5,000	$3,000

The current market price for the unit is $70,000.

REQUIRED: Determine what the foreman on the Alpha Lease should do.

chapter twelve

Conveyances

In the oil and gas industry, companies often sell, trade, or exchange their interests in oil and gas properties to other parties. Collectively these transactions are referred to as *conveyances.* Mineral interests are generally conveyed by three basic methods: by leasing, sales or exchanges, or sharing arrangements. The basic oil and gas mineral lease whereby an oil and gas company contracts with the owner of the mineral rights in a property—creating a working interest and a royalty interest—was discussed in chapter 1. Accounting for leasing activities was discussed in chapter 4. This chapter discusses the accounting that is required when working interests are subsequently conveyed by sales, exchanges, and sharing arrangements under both the successful-efforts and full-cost methods of accounting.

Mineral Interests

The mineral interest in a property may be divided into operating (working) interests and nonoperating (nonworking) interests. The previous chapters dealt primarily with simple situations that included only a basic working interest and a royalty interest created through leasing. A working interest can also be purchased, sold, carried, pooled, or unitized. In addition, when a working interest is conveyed, a nonoperating interest is often created. These nonoperating interests, which are in addition to the RI, can be classified as overriding royalty interests, production payment interests, or net profits interests. These interests are normally created or conveyed through purchases, exchanges, or through sharing arrangements. The interests are typically created to spread risks, share costs, obtain financing, secure sufficient acreage to meet state spacing requirements, and to perform secondary or tertiary recovery operations.

The next section contains a discussion and definition of different types of interests including examples which illustrate how revenues and costs would be distributed among the various interest owners.

Types of interests

Basic working interest (WI). One WI owner and one lease make up the basic working interest.

Joint WI (JI). Two or more parties each own an undivided fraction of the working interest in a single lease. A joint working interest may result from one of the three methods mentioned above, *i.e.,* (1) leasing, (2) sales or exchanges, or (3) sharing arrangements. The joint interest format is very popular because it provides a means of sharing the high risk and high capital investment associated with oil and gas ventures.

One party—usually the party with the largest percentage of the working interest—has the responsibility of developing and operating the property. This party is known as the operator. In a joint interest situation, the WI parties enter into a joint operating agreement that specifies the rights and obligations of each party. Accounting for joint interest operations was discussed in chapter 11.

EXAMPLE

Joint WI

Lucky Oil Company and Duster Oil Company signed a lease agreement with Farmer Brown. Farmer Brown received a $^1/_8$ RI and Lucky Oil Company and Duster Oil Company each acquired an undivided 50% WI in a 3,000-acre lease in Texas. Subsequently, the two companies signed a joint operating agreement designating Lucky Oil Company as the operator of the lease. The two companies will each pay 50% of the cost to explore, develop, and operate the property. Farmer Brown will receive $^1/_8$ of the gross revenue from the property with Lucky Oil Company and Duster Oil Company sharing equally the remaining $^7/_8$ of the revenue from the property.

Basic Royalty Interest (RI). The basic type of interest retained by a mineral rights owner or fee interest owner when leasing the property to another party.

Overriding royalty interest (ORI). Frequently, working interest owners, through conveyance, create an ORI—a nonoperating interest created out of the WI. An ORI is similar to a basic royalty interest in that the owner is not responsible for the cost of exploring, developing, or producing the property. The fundamental difference between a royalty interest and an ORI is that the royalty interest was created from the original mineral rights and the ORI is created from the WI. The ORI's share of revenue is a stated percentage of the share of revenue belonging to the WI from which it was created. An ORI is created by either being retained or carved out. A retained ORI is created when the WI owner sells or conveys its working interest in a property and in the same transaction retains an ORI. A carved out ORI is created when the WI owner keeps the working interest but creates a ORI that is conveyed to another party.

EXAMPLE

ORI

Lucky Oil Company signed a lease agreement with Farmer Brown covering 640 acres in Texas. Farmer Brown received a $^1/_8$ RI and Lucky Oil Company received 100% of the WI. A few months later Lucky Oil Company decides that at the present time it is not financially able to develop the property.

a. **Retained ORI.** Lucky Oil Company decides to sell its working interest to Simms Oil Company for $5,000 and retains a $^1/_7$ overriding royalty. As a result of this transaction, Simms Oil Company will pay 100% of the cost to explore, develop, and produce the property. Farmer Brown will receive $^1/_8$ of the gross revenue from the property, Lucky Oil Company will receive $^7/_8$ x $^1/_7$ of the revenue from the property, and Simms Oil Company will receive $^7/_8$ x $^6/_7$ of the revenue. Since Lucky Oil Company sold its WI and retained an ORI, the ORI is referred to as a retained ORI.

b. **Carved out ORI.** Assume instead that Lucky Oil Company decides to go ahead and develop the property itself. Since Lucky Oil Company is short of capital, it approaches a local investor who agrees to pay Lucky $10,000, and in exchange, Lucky Oil Company conveys a $^1/_{16}$ ORI in the lease to the investor. The result of this transaction is that Lucky Oil Company will pay 100% of the cost to explore, develop, and produce the property. Farmer Brown will receive $^1/_8$ of the gross revenue from the property, the investor will receive $^7/_8$ x $^1/_{16}$, and Lucky Oil Company will receive $^7/_8$ x $^{15}/_{16}$. Since

Lucky Oil Company kept its WI and created an ORI, the ORI is referred to as a carved out ORI.

Production payment interest (PPI). A PPI is a nonoperating interest created out of the WI and is usually expressed in terms of a certain amount of money, a certain period of time, or a certain quantity of oil or gas. In other words, a PPI is limited to a specified amount of money, time, or quantity of oil or gas, after which the PPI ceases to exist. Therefore, unlike the other nonoperating interests, a PPI generally terminates before the reservoir is depleted. The owner of a PPI is not responsible for any of the cost of exploring, developing, or producing a property. If a PPI is payable with money, the payment is stated as a percentage of the working interest's share of revenue since the PPI was created from the WI. If a PPI is payable in product (*i.e.*, oil, gas, etc.), payment is typically stated as a percentage of the working interest's share of current production. Like ORIs, PPIs are created by being carved out or by being retained.

EXAMPLE

PPI

Lucky Oil Company signed a lease agreement with Farmer Brown covering 640 acres in Texas. Farmer Brown received a $^1/_8$ RI and Lucky Oil Company received 100% of the WI. A few months later Lucky Oil Company decides that at the present time it is not financially able to develop the property.

a. **Retained PPI.** Lucky Oil Company decides to sell its working interest to Smith Company for a consideration of $500,000 and a payment of 100,000 barrels of oil to be paid out of the first 20% of Smith Company's WI share of production. During the first year of production, the lease produces 40,000 barrels, of which 5,000 barrels (40,000 x $^1/_8$) belongs to Farmer Brown. Lucky Company will receive 7,000 (40,000 barrels x $^7/_8$ x 0.20), leaving the production payment balance owed to Lucky Oil Company of 93,000 barrels (100,000 barrels – 7,000 barrels). Smith Company's share of production is what is left after paying the RI and PPI: 40,000 bbl x $^7/_8$ x 80% = 28,000 bbl. Smith Company pays 100% of the costs to explore, develop, and produce the lease. Since Lucky Oil Company sold its WI and retained a PPI, the PPI is referred to as a retained PPI.

b. **Carved out PPI.** Assume instead that Lucky Oil Company decides to go ahead and develop the property itself. Since Lucky Oil Company is short of capital it approaches a local investor who on 1/1/2004 gives Lucky Oil Company $100,000. In exchange, Lucky Oil Company agrees to repay the investor the principle of $100,000 plus interest at a rate of 10% out of the first 20% of Lucky Oil's share of production. The first month's production is sold for $200,000. Farmer Brown will receive $25,000 ($200,000 x $^1/_8$) and Lucky Oil Company will pay the investor (the PPI owner) $35,000 ($200,000 x $^7/_8$ x .20). The remaining balance payable to the PPI owner is $166,667 ($200,000 plus interest of $1,667 less $35,000) payment: $200,000 + ($200,000 x 10% x $^1/_{12}$) – $35,000). Lucky Oil Company will pay 100% of the cost to explore, develop, and produce the property. Since Lucky Oil Company kept its WI and created a PPI, the PPI is referred to as a carved out PPI.

Net profits interest. A net profits interest is a nonoperating interest normally created out of the WI by either carve-out or retention, but more commonly by retention. A net profits interest may also be created when the mineral rights owner leases his interest. This type of interest is common offshore. A net profits interest is similar to an ORI except rather than being paid a percentage of the WI's share of production, the net profits interest owner receives a stated share of the WI owner's share of the net profits from the property. The holder of this type of interest is not responsible for paying his share of losses; however, such losses may be recovered by the working interest owner out of future net profit payments. The calculation of net profits, *i.e.,* the allowed deductions from gross revenue to compute net profit, should be clearly indicated in the contract.

EXAMPLE

Net Profits Interest

Lucky Company obtains an offshore lease, specifically a net-profit share lease, from the U.S. government. The lease contract specifies that the government will receive a lump sum bonus and a 10% share of the net profits produced from the lease. The contract provides for a detailed list of allowable charges from revenue in arriving at net profit. Lucky Oil Company owns the WI and, accordingly, must pay 100% of the cost to explore, develop, and produce the property.

CONVEYANCES

SFAS No. 19 provided the first authoritative guidance for accounting for conveyances. Although *SFAS No. 19* was suspended by *SFAS No. 25,* the SEC adopted all of the conveyance provisions in *SFAS No. 19.* Thus, these provisions apply to all publicly traded companies including those using the successful-efforts method and, with some minor modification, those using the full-cost method. These provisions are discussed in the following sections. The specific provisions for accounting for conveyances under FC accounting are discussed later in the chapter.

The main accounting question to be answered in recording a conveyance is whether a gain or loss should be recognized. The conveyance rules contained in *SFAS No. 19* can generally be summarized as follows:

1. Generally, no gain or loss should be recognized in transactions where:
    a. Oil and gas assets are exchanged for other oil and gas assets (*SFAS No. 19,* par. 44a).
    b. The transaction involves the pooling of assets in a joint undertaking (*SFAS No. 19,* par. 44b).
2. Generally, no gain should be recognized, but a loss may be recognized in transactions where:
    a. Part of an interest in an oil and gas asset is sold and substantial uncertainty exists about the recovery of costs applicable to the retained interest (*SFAS No. 19,* par. 45a).
    b. Part of an interest is sold and the seller has a substantial obligation for future performance without proportional reimbursement (*SFAS No. 19,* par. 45b).
3. For transactions other than those listed above, gain or loss recognition is required so long as it is appropriate under GAAP. Examples include the sale of an entire interest in a proved property, the sale of the entire interest in an unproved property, and certain production payments.
4. Some transactions are in effect borrowings repayable in cash or cash equivalents. These transactions should be accounted for as borrowings regardless of their legal form. Typically, the most common examples of these transactions include some type of production payment.

The above types of transactions and the paragraphs in *SFAS No. 19* that specify the appropriate accounting treatment for these transactions are discussed below. When attempting to determine the appropriate accounting treatment for a transaction, it is necessary to carefully study the characteristics of the transaction rather than the particular name that a conveyance may go by. Paragraphs 47a through 47m of *SFAS No. 19* should then be examined to determine which paragraph(s) actually addresses that particular transaction.

While the principles for accounting for conveyances generally apply to companies using either the successful–efforts or the full–cost accounting method, the examples provided in the discussion below illustrate the accounting entries that would be made to record various types of conveyances by companies using the successful-efforts method. The specific provisions that apply to full-cost companies are discussed later in this chapter.

Exchanges and poolings

In the oil and gas industry there are numerous types of transactions where companies exchange assets used in oil and gas producing operations for other assets used in oil and gas producing operations. These transactions range from the exchange of a single piece of equipment to the conveyance of the entire working interest in a property. These sharing arrangements are nonmonetary exchanges, since the assets are not bought or sold. Examples include farm-ins/farm-outs, free wells, carried working interests, poolings, and unitizations. The following discussion describes these different types of sharing arrangements and explains the accounting treatment prescribed by *SFAS No. 19*:

Farm-ins/farm-outs

A farm-in/farm-out is an arrangement in which the owner of a WI (the farmor) assigns all or part of the WI to another party (the farmee) in return for the exploration and development of the property. These arrangements may be structured in many different ways. For example, the farmor may convey its WI to the farmee and retain an ORI. The farmee in return agrees to pay all the costs to explore, develop, and produce the property. This arrangement is a farm-out from the perspective of the farmor who conveys the working interest and retains an ORI. It is a farm-in from the perspective of the farmee who earns the WI by drilling a well, etc.

Farm-in/farm-out agreements are nonmonetary transactions for which there is no gain or loss recognition. Par. 47b of *SFAS No. 19* describes the accounting for a farm-in/farm-out where the farmor retains an ORI.

> *An assignment of the operating interest in an unproved property with retention of a non-operating interest in return for drilling, development, and operation by the assignee is a pooling of assets in a joint undertaking for which the assignor shall not recognize gain or loss. The assignor's cost of the original interest shall become the cost of the interest retained. The assignee shall account for all costs incurred as specified by paragraphs 15-41 and shall allocate none of those costs to the mineral interest acquired. If oil or gas is discovered, each party shall report its share of reserves and production (paragraphs 50-56). (SFAS No. 19, par. 47b)*

In the example below, Lucky Oil Company owns 100% of the working interest in the unproved lease. Lucky exchanged its working interest in the unproved property for a non-working interest in the property and in doing so avoids the risk and cost associated with drilling the well and developing the property. Lucky Oil Company (the farmor) initially has the cost of the unproven property on its books. When the WI is conveyed to Frisco Oil Company (the farmee), an entry is made on the books of Lucky Oil Company moving the balance from an unproved property with a WI to an unproved property with an ORI-type account. If Frisco Oil Company is successful in finding oil and gas reserves, Lucky Oil Company will transfer the balance in the account to a proved ORI account. In contrast, Frisco Oil Company will have no balance in the unproved property account. Instead, Frisco Oil Company will debit wells-in-progress as the well is drilled. Later the costs in that account would be transferred to wells and related E&F or dry hole expense depending on the outcome of the drilling.

EXAMPLE

Farm-in/Farm-out

Lucky Oil Company signed a lease agreement with Sally Fielding covering 400 acres in Oklahoma. Ms. Fielding received a bonus of $20,000 and a $^1/_8$ RI and Lucky Oil Company received 100% of the WI.

A few months later Lucky Oil Company decides that at the present time it is not financially able to develop the property. Lucky Oil Company enters into an agreement with Frisco Oil Company wherein Lucky Oil Company assigns its WI to Frisco Oil Company and retains a $^1/_7$ ORI. Frisco agrees to drill a well on the property within the next six months and to pay all of the costs of drilling the well, as well as other costs to develop the property. As a result of this agreement Lucky Oil Company owns a $^1/_7$ ORI interest in the well but Frisco Oil Company must pay all of the costs and bear all of the risks of drilling. Frisco drills a successful exploratory well at a cost of $100,000.

Applicable Paragraph: Paragraph No. 47b

Entries

Lucky Oil Company:
On the date of the conveyance:

Unproved ORI	20,000	
Unproved property		20,000

When and if the property becomes proved:

Proved ORI	20,000	
Unproved ORI		20,000

Frisco Oil Company:

On the date of the conveyance:
No entry
As the well is drilled:

W/P	100,000	
Cash		100,000

When and if the well is successful:

Wells and related E&F	100,000	
W/P		100,000

Farm-ins/farm-outs with a reversionary working interest

A common variation of the basic farm-in/farm-out agreement provides for a share of the WI to revert back to the farmor at some point in time. For example, a WI owner may convey its WI to another party and retain an ORI. The other party in return agrees to pay all of the cost to explore, drill, and develop the property. When the farmee has generated enough income from the property to recover the costs it expended on exploring, drilling, and developing the property, then the farmor's ORI reverts back to a WI and the two parties share the working interest in the property according to pre-agreed upon percentages. This type of conveyance is referred to as a farm-in/farm-out with a reversionary working interest.

EXAMPLE

Farm-in/Farm-out with Reversionary WI

Lucky Oil Company signed a lease agreement with Sally Fielding covering 400 acres in Oklahoma. Ms. Fielding received a bonus of \$20,000 and a $^1/_8$ RI and Lucky Oil Company received 100% of the WI.

On 1/1/2002, Lucky Oil Company enters into an agreement with Frisco Oil Company wherein Lucky Oil Company assigns its WI and retains a 1/7 ORI.

Frisco agrees to drill a well on the property within the next six months and to pay all of the costs of drilling the well. If the well is successful, Frisco Oil Company will pay all the operating costs and retain the net profit (after payment of the royalty, ORI, and operating expenses) until it has recovered the cost of drilling and completing the well. At that point Lucky Oil Company's ORI will revert to a 50% WI.

During November 2002, Frisco Oil Company drills Well No. 1 at a cost of $500,000. The well is successful. Estimated gross proved reserves total 400,000 barrels and proved developed reserves are 200,000 barrels. During 2003 and 2004, 25,000 gross barrels per year are produced and sold for $20 per barrel. Operating costs are $5 per barrel. Severance tax is ignored.

Applicable Paragraph: Paragraph No. 47b.

Computation of Barrels for Payout:

$$(\tfrac{7}{8} \times \tfrac{6}{7} \times \$20X) - \$5X = \$500,000$$
$$X = \underline{\underline{50,000 \text{ bbl}}}$$

where X = number of gross barrels produced before payout

Number of WI barrels produced before payout = $50,000 \times \tfrac{7}{8} \times \tfrac{6}{7}$
= $\underline{\underline{37,500 \text{ bbl}}}$

Computation of Proved Reserves:

Total WI share of proved reserves remaining after payout:
$\tfrac{7}{8}(400,000 - 50,000)$ = 306,250 barrels

Lucky Company's portion of proved reserves:

Reserves to payout (ORI):	$50,000 \times \tfrac{7}{8} \times \tfrac{1}{7}$ =	6,250 bbl
Remaining after payout:	306,250 x 50% =	153,125 bbl
Lucky's total proved reserves:		159,375 bbl

Note: Lucky Company is the carried party and so has only property costs. The property costs will be depreciated over the total proved reserves.

Computation of Proved Developed Reserves:

Total WI share of proved developed reserves remaining after payout:
$\tfrac{7}{8}(200,000 - 50,000)$ = 131,250 barrels

Frisco Oil Company's portion of proved developed reserves:

Reserves to payout:	50,000 x $^7/_8$ x $^6/_7$ =	37,500 bbl
Remaining after payout:	131,250 x 50% =	65,625 bbl
Frisco's total proved developed reserves:		103,125 bbl

Note: Frisco Oil Company is the carrying party and so has only well costs and no property costs. Additionally, any WI barrels before payout belong to Frisco, as the carrying party.

Computation of Payout:

	2003	2004
Sales Volume	25,000	25,000
Price/bbl	$ 20	$ 20
Gross Sales ($)	$500,000	$500,000
Net of Royalty	7/8	7/8
	437,500	437,500
Net of ORI	6/7	6/7
Net Revenue to WI	375,000	375,000
Operating Expense	(125,000)	(125,000)
Net Profit	$250,000	$250,000

	Payout:
Amount to be recovered	$500,000
Recovered in Year 1	(250,000)
Balance to be recovered	$250,000
Recovered in Year 2	(250,000)
Balance to be recovered	$ 0

Lucky Oil Company:

1/1/2002:		
Unproved ORI	20,000	
Unproved property		20,000
(to transfer the WI to a non-WI)		

11/2002:		
Proved ORI	20,000	
Unproved ORI		20,000

(to reclassify the property)

During 2003:

Cash (25,000 x 7/8 x 1/7 x \$20)	62,500	
Production revenue		62,500
(to record revenue)		
DD&A expense	392	
Accumulated DD&A—proved ORI		392

$$\frac{\$20{,}000}{159{,}375} \times 3{,}125^{*} = \underline{\underline{392}}$$

* Lucky's share of production = 25,000 x $^7/_8$ x $^1/_7$ = 3,125 bbl

During 2004:

Cash (25,000 x 7/8 x 1/7 x \$20)	62,500	
Production revenue		62,500
(to record revenue)		
DD&A expense	392	
Accumulated DD&A—Proved ORI		392

$$\frac{\$20{,}000 - \$392}{159{,}375 - 3{,}125} \times 3{,}125 = \underline{\underline{392}}$$

Proved property	19,216	
Accumulated DD&A—proved ORI	784	
Proved ORI		20,000
(to transfer the non-WI to a WI)		

Frisco Oil Company:

On the date of the conveyance:

No entry

11/2002—As the well is drilled:

W/P	500,000	
Cash		500,000

11/2002:

Wells and related E&F	500,000	
W/P		500,000

During 2003:

Cash (25,000 x \$20)	500,000	
Royalty payable (25,000 x $^1/_8$ x \$20)		62,500
ORI payable (25,000 x $^7/_8$ x $^1/_7$ x \$20)		62,500
Production revenue (25,000 x $^7/_8$ x $^6/_7$ x \$20) . .		375,000
(to record revenue)		
Operating expense (25,000 x \$5)	125,000	
Cash		125,000
(to record operating expense)		
DD&A expense	90,909	
Accumulated DD&A—wells and related E&F . .		90,909

$$\frac{\$500{,}000}{103{,}125} \times 18{,}750^* = \underline{\underline{90{,}909}}$$

* Frisco's share of production = 25,000 x $^7/_8$ x $^6/_7$ = 18,750 bbl

During 2004:

Cash (25,000 x \$20)	500,000	
Royalty payable (25,000 x $^1/_8$ x \$20)		62,500
ORI payable (25,000 x $^7/_8$ x $^1/_7$ x \$20)		62,500
Production revenue (25,000 x $^7/_8$ x $^6/_7$ x \$20) . .		375,000
(to record revenue)		
Operating expense (25,000 x \$5)	125,000	
Cash		125,000
(to record operating expense)		
DD&A expense	90,909	
Accumulated DD&A—wells and related E&F . .		90,909

$$\frac{\$500{,}000 - \$90{,}909}{103{,}125 - 18{,}750} \times 18{,}750 = \underline{\underline{\$90{,}909}}$$

12/31/2004:

Frisco Oil Company would adjust its property database and records to indicate that 50% of the WI reverted back to Lucky Oil Company and the property is now a jointly owned property.

Free wells

Sharing arrangements that result in free wells can take different forms. For example, the owner of a WI transfers a portion of its WI to a second party in exchange for the second party agreeing to drill (and possibly equip) a well free of cost to the assignor. The result is that the assignor ultimately has an interest in a well for which it has incurred no costs. These agreements may include more than one well, or the free well may need to be a producing well.

Accounting for a free well arrangement is similar to accounting for a farm-in/farm-out described previously. Each party accounts for the costs that they actually expend and no gain or loss recognition is allowed.

> *An assignment of a part of an operating interest in an unproved property in exchange for a "free well" with provision for joint ownership and operation is a pooling of assets in a joint undertaking by the parties. The assignor shall record no cost for the obligatory well; the assignee shall record no cost for the mineral interest acquired. All drilling, development, and operating costs incurred by either party shall be accounted for as provided in paragraphs 15-41 of this Statement. If the conveyance agreement requires the assignee to incur geological or geophysical expenditures instead of, or in addition to, a drilling obligation, those costs shall likewise be accounted for by the assignee as provided in paragraphs 15-41 of this Statement. If reserves are discovered, each party shall report its share of reserves and production (paragraphs 50-56). (SFAS No. 19, par. 47c)*

EXAMPLE

Free Well

In 2006, Lucky Oil Company entered into a lease agreement with Lenny Franks paying Mr. Franks a $50,000 bonus and a 1/8 RI. In 2008, Lucky Oil Company assigned 25% of its WI to Daring Company in return for Daring Company agreeing to drill and equip a well on that lease. Daring Company drilled the well in 2009 at a cost of $600,000 for IDC and equipment. The well was successful. Gross proved reserves were initially estimated to be 400,000 barrels and gross proved developed reserves were estimated to be 300,000 barrels. Gross production during 2009 was 10,000 barrels, production costs were $5 per barrel, and the selling price was $20 per barrel. Assume Daring Company is the operator and that the purchaser of the oil assumes the responsibility of distributing royalty and 5% severance taxes.

Applicable Paragraph: Paragraph No. 47c

Entries:

Lucky Oil Company:

2008
No entry for conveyance

2009

a. Proved property	50,000	
Unproved property		50,000
(to reclassify property)		
b. A/R (10,000 x \$20 x $^7/_8$ x 0.95 x 0.75)	124,688	
Severance tax expense (10,000 x \$20 x $^7/_8$ x 0.05 x 0.75)	6,562	
Oil sales (10,000 x \$20 x $^7/_8$ x 0.75)		131,250
(to record revenue)		
c. Operating expense (10,000 x \$5 x 0.75)	37,500	
Payable to Daring Company		37,500
(to record lifting costs)		
d. DD&A expense	1,250	
Accumulated DD&A		1,250

$$\text{Acquisition costs only: } \$50{,}000 \times \frac{10{,}000 \times 0.75 \times 7/8}{400{,}000 \times 0.75 \times 7/8} = \underline{\underline{\$1{,}250}}$$

Daring Company (operator):

2008
No entry for conveyance

2009

a. Wells and related E&F	600,000	
Cash		600,000
b. A/R (10,000 x \$20 x $^7/_8$ x 0.95 x 0.25)	41,562	
Severance tax expense (10,000 x \$20 x $^7/_8$ x 0.05 x 0.25)	2,188	
Oil sales (10,000 x \$20 x $^7/_8$ x 0.25)		43,750
c. Operating expense (10,000 x \$5 x 0.25)	12,500	
Joint interest receivable	37,500	
Cash (10,000 x \$5)		50,000
d. DD&A expense	20,000	
Accumulated DD&A		20,000
(to record DD&A)		

$$\text{Drilling costs only: } \$600{,}000 \times \frac{10{,}000 \times 0.25 \times 7/8}{300{,}000 \times 0.25 \times 7/8} = \$20{,}000$$

Carried interests

The most common occurrences of **carried interests** arise in joint interest operations when one or more of the working interest owners elect not to participate in the drilling of a well. Any party electing not to participate is referred to as a **carried interest** or **carried party.** The working interest owners who agree to pay the carried party's share of the costs are referred to as the **carrying parties.** When a working interest is carried, the carried working interest owner does not permanently relinquish its interest in the property. Rather its interest is temporarily transferred to the carrying parties and will revert back when the carrying parties reach payout. In order to minimize the likelihood that a working interest owner will not consent to the drilling of a well in order to avoid the risk of drilling, most joint operating agreements include a penalty provision whereby the carrying parties are allowed to recover the carried party's share of drilling and completion costs plus a penalty (often 100 to 200% of the carried party's share of costs). Carried working interests are discussed in detail in chapter 11.

Another situation where a carried interest may arise is a farm-in/farm-out where the farmor assigns a portion of his working interest in exchange for the farmee agreeing to drill and equip a well. Since the farmor has a working interest but does not participate in the drilling of the well, his working interest is said to be carried by the farmee. In these arrangements the farmee recoups the farmor's share of drilling costs (which the farmee paid) from the farmor's share of net profits. In other words, payout is computed. In a carried interest arising from this type of arrangement, there is no penalty assessed on the farmor's share of costs.

These two types of arrangements are accounted for in the same manner. Moreover, the accounting is identical to a farm-in/farm-out with a reversionary interest except that there is no ORI created.

> *A part of an operating interest in an unproved property may be assigned to effect an arrangement called a "carried interest" whereby the assignee (the carrying party) agrees to defray all costs of drilling, developing, and operating the property and is entitled to all of the revenue from production from the property, excluding any third party interest, until all of the assignee's costs have been recovered, after which the assignor will share in both costs and production. Such an arrangement represents a pooling of assets in a joint undertaking by the assignor and assignee. The carried party shall make no accounting for any costs and revenue until after recoupment (payout) of the carried costs by the carrying party. Subsequent to payout the carried party shall account for its share of revenue, operating expenses, and (if the agreement provides for subsequent sharing of costs rather than a carried interest) subsequent development costs. During the payout period the carrying party*

> *shall record all costs, including those carried, as provided in paragraphs 15-41 and shall record all revenue from the property including that applicable to the recovery of costs carried. The carried party shall report as oil or gas reserves only its share of proved reserves estimated to remain after payout, and unit-of-production amortization of the carried party's property cost shall not commence prior to payout. Prior to payout the carrying party's reserve estimates and production data shall include the quantities applicable to recoupment of the carried costs (paragraphs 50-56). (SFAS No. 19, par. 47d)*

In accounting for a carried interest no gain or loss is to be recognized. The carrying party pays all of the cost to drill and equip a well, the full amount of which is capitalized to wells in progress (W/P). The carried party pays and records none of the drilling-related costs. If the well is dry the carrying party clears the W/P account to dry hole expense. If the well is successful the costs are cleared to the carrying party's wells and related E&F account. As production occurs the carrying party keeps all of the proceeds from selling the production and pays all of the royalty, taxes, and operating expenses. As was illustrated in chapter 11, the carrying party is entitled to keep the carried party's share of net profit until the carrying party has recouped the carried party's share of drilling and completion costs which the carrying party paid plus any applicable penalty. At the point that payout is reached, the carried party resumes its role as a regular working interest partner in the operation.

Another important and difficult feature of this type of arrangement is the computation of DD&A before payout. Before payout, the carried party does not recognize DD&A because it did not bear any of the cost of drilling or completing the well and because it is not receiving any revenue from production. In the carrying party's computation of DD&A, the estimate of proved developed reserves to be used is the carrying party's normal WI percentage of reserves after payout *plus* the amount of the carried party's share of reserves that the carrying party estimates it will be entitled to in order to recover the carried party's share of costs plus any penalty, *i.e.,* until payout.

EXAMPLE

Carried Interest Created by a Farm-In/Farm-Out

Lowfund Company owns 100% of the working interest in a lease that it acquired from Joseph Tyler. Lowfund paid a $100,000 bonus to Mr. Tyler who agreed to a $^1/_5$ RI. Lucky Oil Company agrees to drill a well on Lowfund Company's unproved lease in return for 50% of the WI in the lease and the right to recover all of its costs. In Year 1, Lucky Oil Company incurs $275,000 in IDC and equipment costs in drilling the well. The well is successful and finds gross proved reserves of 900,000 barrels and gross proved developed reserves of 400,000 barrels. During each of the first 5 years, 12,500 gross barrels of oil per year are produced and sold for $20 per barrel from the lease and lifting costs totaled $5 per barrel. Ignore severance taxes and assume reserve estimates do not change.

Applicable Paragraph: Paragraph No. 47d

Computation of Barrels for Payout:

$$(\tfrac{4}{5} \times \$20X) - \$5X = \$275,000$$
$$X = 25,000 \text{ bbl}$$

Where X = number of barrels produced before payout

Number of WI barrels produced before payout = 25,000 x $\tfrac{4}{5}$ = 20,000 barrels

Computation of Proved Reserves:

Total WI share of proved reserves remaining after payout:
$\tfrac{4}{5}(900,000 - 25,000) = 700,000$ barrels

Lowfund Company's portion of proved reserves remaining after payout:
700,000 x 50% = 350,000 barrels

Lowfund Company's total proved reserves = 350,000 bbl

Note: Lowfund Company is the carried party and so has only property costs and no well costs. Additionally, any WI barrels before payout belong to the carrying party.

Computation of Proved Developed Reserves:

Total WI share of proved developed reserves remaining after payout:
$\tfrac{4}{5}(400,000 - 25,000) = 300,000$ barrels

Lucky Oil Company's portion of proved developed reserves remaining after payout:
(300,000 x 50%) = 150,000 barrels

Lucky Oil Company's total proved developed reserves =
150,000 + (25,000 x $\tfrac{4}{5}$) = 170,000

Note: Lucky Company is the carrying party and so has only well costs and no property costs. Additionally, any WI barrels before payout belong to Lucky, as the carrying party.

Computation of Payout:

	Yr 1	Yr 2
Sales Volume	12,500	12,500
Price/bbl	$ 20	$ 20
Gross Sales ($)	250,000	250,000
Net of Royalty	4/5	4/5
Net Sales ($)	200,000	200,000
Operating Expense	(62,500)	(62,500)
Net Profit	$137,500	$137,500

Net Profit to Lucky:

Total to Lucky (100%)	$137,500	$137,500
Lucky's portion (50%)	68,750	68,750
Net Profit	$ 68,750	$ 68,750

	Payout:
Cost of well	$275,000
Lowfund Company's proportion	50%
Lowfund's share of cost	$137,500
Amount to be recovered	$137,500
Recovered in Year 1	(68,750)
Balance to be recovered	$ 68,750
Recovered in Year 2	(68,750)
Balance to be recovered	$ 0

Entries

No entries for Lowfund Company until after payout (except to reclassify property as proved).

<u>Lucky Oil Company, YEAR 1:</u>

Wells and related E&F	275,000	
Cash		275,000

A/R—purchaser (12,500 x $20)	250,000	
Royalty payable (12,500 x $^1/_5$ x $20)		50,000
Sales revenue (12,500 x $^4/_5$ x $20)		200,000
Operating expense (12,500 x $5)	62,500	
Cash		62,500
DD&A expense	16,176	
Accumulated DD&A		16,176

Drilling costs only: $\dfrac{\$275{,}000}{170{,}000\text{ bbl}} \times (12{,}500 \times 4/5)\text{bbl}^* = \underline{\underline{\$16{,}176}}$

* WI share of production = 12,500 x $^4/_5$ = 10,000 bbl

Lucky Company has no property costs and therefore calculates DD&A only on its drilling costs.

<u>Lucky Oil Company, YEAR 2:</u>

A/R—purchaser (12,500 x $20)	250,000	
Royalty payable (12,500 x $^1/_5$ x $20)		50,000
Sales revenue (12,500 x $^4/_5$ x $20)		200,000
Operating expense (12,500 x $5)	62,500	
Cash		62,500
DD&A expense	16,176	
Accumulated DD&A		16,176

Drilling costs only: $\dfrac{\$275{,}000 - \$16{,}176}{170{,}000 - 10{,}000\text{ bbl}} \times 10{,}000\text{ bbl} = \underline{\underline{\$16{,}176}}$

Note that Lucky's proved developed reserves in the denominator have been reduced by its portion of production in Year 1.

<u>Lucky Oil Company, YEAR 3—After Payout:</u>

A/R—purchaser (12,500 x $20)	250,000	
Royalty payable (12,500 x $^1/_5$ x $20)		50,000
A/P—Lowfund (12,500 x $^4/_5$ x $20 x 50%)		100,000
Sales revenue (12,500 x $^4/_5$ x $20 x 50%) . .		100,000
Operating expense (12,500 x $5 x 50%)	31,250	
A/R—Lowfund (12,500 x $5 x 50%)	31,250	
Cash		62,500

DD&A expense	8,088	
Accumulated DD&A		8,088

Drilling costs only: $\dfrac{\$275{,}000 - \$16{,}176 - \$16{,}176}{170{,}000 - 10{,}000 - 10{,}000 \text{ bbl}} \times 5{,}000 \text{ bbl}^{*} = \underline{\underline{\$8{,}088}}$

* Lucky's share of production = 12,500 bbl x $^{4}/_{5}$ x 50% = 5,000 bbl

Note that once payout is reached, Lowfund Company receives 50% of the WI's share of revenue and pays 50% of the costs incurred.

<u>Lowfund Company, YEAR 3:</u>

A/R—Lucky Oil Co. (12,500 x $^{4}/_{5}$ x \$20 x 50%)	100,000	
Sales revenue		100,000
Operating expense (12,500 x \$5 x 50%)	31,250	
A/P—Lucky Oil Co		31,250
DD&A expense	1,429	
Accumulated DD&A		1,429

Leasehold costs only: $\dfrac{\$100{,}000}{350{,}000 \text{ bbl}} \times 5{,}000 \text{ bbl} = \underline{\underline{\$1{,}429}}$

Lowfund Company has no well costs and therefore calculates DD&A only on its leasehold costs.

Joint venture operations

In chapter 11 accounting for joint venture operations was discussed in detail. It was noted that in a joint venture operation each company accounts for its proportionate share of costs and revenue. Par. 47e of *SFAS No. 19* describes a situation that is a joint venture situation, specifically where working interest owners swap their working interests in different properties. This type of transaction is a pooling of interest in a joint undertaking to find and produce oil and gas; thus, no gain or loss is to be recognized by any of the parties.

> *A part of an operating interest owned may be exchanged for a part of an operating interest owned by another party. The purpose of such an arrangement, commonly called a joint venture in the oil and gas industry, often is to avoid duplication of facilities, diversify risks, and achieve operating efficiencies. Such reciprocal conveyances represent exchanges of similar productive assets, and no gain or loss shall be recognized by either party at the time of the transaction. In some joint ventures, which may or may not involve an exchange of*

interests, the parties may share different elements of costs in different proportions. In such an arrangement a party may acquire an interest in a property or in wells and related equipment that is disproportionate to the share of costs borne by it. As in the case of a carried interest or a free well, each party shall account for its own cost under the provisions of this Statement. No gain shall be recognized for the acquisition of an interest in joint assets, the cost of which may have been paid in whole or in part by another party. (SFAS No. 19, par. 47e)

The following example illustrates a conveyance where a portion of the WI in one lease is traded for a portion of the WI in another lease. Each party merely transfers a portion of the investment in the property that the party is conveying to the property that the party is receiving.

EXAMPLE

Exchange of WIs

Lucky Oil Company acquired 100% of the working interest in Lease A for $80,000. Leveland Company acquired 100% of the working interest in Lease B for $200,000. Lucky and Leveland enter into an agreement whereby Lucky assigns 50% of its WI in Lease A to Leveland and in return Leveland conveys to Lucky Oil Company 30% of its working interest in Lease B.

Applicable Paragraph: Paragraph No. 47e

Entries to record conveyance:

Lucky Oil Company:		
Oil and gas properties—Lease B (50% x $80,000)	40,000	
Oil and gas properties—Lease A		40,000

Leveland Company:		
Oil and gas properties—Lease A (30% x $200,000)	60,000	
Oil and gas properties—Lease B		60,000

Poolings and unitizations

In a **pooling** or **unitization,** the working interests as well as the nonworking interests in two or more properties are typically combined, with each interest owner owning the same type of interest (but a smaller percentage) in the total combined property as they held previously in the separate property. The terms **pooling** and **unitization** are often used interchangeably; however, the most common usage of the term **pooling** is the combining of unproved properties, whereas the term **unitization** is commonly used to refer to a larger combination involving an entire producing field or reservoir for purposes of enhanced oil and gas recovery.

In a pooling two or more properties are combined to form a single operating unit. After the pooling, if the working interests are held by two or more parties, a joint operating agreement is entered into and one of the parties is designated as operator. Pooling may be voluntary or non-voluntary and usually brings together small tracts to obtain sufficient acreage so that a well may be drilled under a state's specific well spacing rules. As a rule, both working interests and nonworking interests are combined, and each party receives an interest stated as a percentage of ownership in the combined acreage of the same type as contributed. For example, a royalty interest owner receives a royalty interest in the combined acreage in exchange for the royalty interest in a particular property.

EXAMPLE

Pooling

Lucky Oil Company enters into an agreement to lease 80 acres from Rita Jones. Jones receives a \$15,000 bonus and a $^1/_8$ RI. Lucky owns 100% of the WI. Duster Oil Company enters into an agreement to lease 80 acres from Carolyn Pinkie. Pinkie receives a \$5,000 bonus and a $^1/_8$ RI. Duster owns 100% of the WI. The two leases are contiguous. In order to develop and operate the properties efficiently, Lucky and Duster agree to pool their leases and enter into a joint operating agreement. Ownership interest are recalculated based on acreage contributed.

Applicable Paragraph: Paragraph No. 47e

The pooled property consists of 160 acres. The parties' interest in the pooled property are:

Lucky Oil Company	50% WI
Duster Oil Company	50% WI
Rita Jones	$^1/_{16}$ RI
Carolyn Pinkie	$^1/_{16}$ RI

This transaction would result in no accounting entries for either party. The only action required would be to change the database or property records to reflect the change in property description and the change in interest.

Unitizations

A unitization is similar to a pooling, although unitization usually refers to the combination of leases that have been at least partially developed. In a unitization, the parties enter into a unitization agreement, which defines the areas to be unitized and specifies the rights and obligations of each party. One party, known as the unit operator, has the responsibility of operating the unit. The purpose of unitizations is more economical and efficient development and operation. In particular, a unitization may be necessary to conduct secondary or tertiary recovery operations.

Unitizations (or poolings) may be voluntary or mandatory according to state law. In some states, unitizations may be forced if some percentage, such as 65%, of the involved parties agrees to the unit. In other states, one party alone may instigate a unitization. Typically, both working interests and nonworking interests are combined with each party receiving an interest stated as a percentage of ownership in the unit of the same type as contributed.

Two significant issues must be resolved in a unitization. One problem is the determination of oil and gas reserves underlying each property contributed by each party. The relative amount of the reserves contributed by each party is typically used to determine the parties' **participation factors,** *i.e.,* percentage of working interests in the unitized property. The other problem in a unitization is the determination of the fair market value (FMV) to be assigned to each lease's wells and related equipment and facilities that are contributed to the unit. These contributions must be equalized because the participation factors do not account for the fact that the properties may be in varying stages of development. For example, if a working interest owner has a participation factor (percentage ownership interest) of 40%, then that party should have contributed to the unit 40% of the agreed-upon value of wells and equipment. If the party contributed less than 40%, then the party must pay cash; if the party contributed more than 40%, then the party will receive cash.

To equalize investment, *i.e.,* to equalize contributions, the participation factors are multiplied by the total agreed-upon FMV of the IDC and equipment contributed by all parties to arrive at the assigned value of each participant's interest in the unit-wide IDC and equipment. The participant's assigned value is matched against the FMV of the property transferred by the participant to determine the amount each participant will pay or receive in cash. Cash received is treated as a recovery of cost, and cash paid is treated as additional investment. No gain or loss is recognized.

In a unitization all the operating and nonoperating participants pool their assets in a producing area (normally a field) to form a single unit and in return receive an undivided interest (of the same type as previously held) in that unit. Unitizations generally are under-

taken to obtain operating efficiencies and to enhance recovery of reserves, often through improved recovery operations. Participation in the unit is generally proportionate to the oil and gas reserves contributed by each. Because the properties may be in different stages of development at the time of unitization, some participants may pay cash and others may receive cash to equalize contributions of wells and related equipment and facilities with the ownership interests in reserves. In those circumstances, cash paid by a participant shall be recorded as an additional investment in wells and related equipment and facilities, and cash received by a participant shall be recorded as a recovery of cost. The cost of the assets contributed plus or minus cash paid or received is the cost of the participant's undivided interest in the assets of the unit. Each participant shall include its interest in reporting reserve estimates and production data (paragraphs 50-56). (SFAS No. 19, par. 47f*)*

EXAMPLE

Unitization

Three companies own adjacent leases that share a common reservoir. The companies have the following costs on their books as of May 1, 2008:

Company	IDC	L&WE	Leasehold	Total
A	$ 50,000	$30,000	$10,000	$ 90,000
B	150,000	50,000	20,000	220,000
C	none	none	30,000	30,000
	$200,000	$80,000	$60,000	$340,000

On May 1, the three companies enter into a unitization agreement. After an engineering study and negotiations between the companies, the following participation factors and fair market values and are agreed upon:

Company	IDC	L&WE	Leasehold*	Total	Participation Factor
A	$ 60,000	$ 40,000	ignore	$100,000	40%
B	100,000	60,000	ignore	160,000	25%
C	none	none	ignore	0	35%
	$160,000	$100,000		$260,000	100%

Applicable Paragraph: Paragraph No. 47f
Equalization of Investment:

a	b	c	d	e	f
Company	Participation Factor (Given)	Agreed Value Contributed (Given)	Assigned Value: Factor x Total Agreed Value (b x $260,000)	Receive (c – d)	Pay (d – c)
A	40%	$100,000	$104,000		$ 4,000
B	25%	160,000	65,000	$95,000	
C	35%	0	91,000		91,000
	100%	$260,000	$260,000	$95,000	$95,000

Entries to record equalization of investments:

COMPANY A

Wells and related E&F	4,000	
Cash		4,000

COMPANY B

Cash	95,000	
Wells and related E&F.		95,000

COMPANY C

Wells and related E&F	91,000	
Cash		91,000

Entries to record additional development or production costs would be similar to joint interest examples.

* Most agreements do not provide for an equalization of leasehold investments.

Sales

A sale of an oil and gas property can involve either a proved property or an unproved property and can be of the entire interest or only of part of an interest. Further, the interest sold can be a working interest or a nonworking interest. How the sale is accounted for depends upon whether a proved or unproved property is involved and whether an entire or partial interest is sold. Figure 12–1 summarizes the accounting treatment of sales of oil and gas properties.

Fig. 12–1 — *Sale of Property*

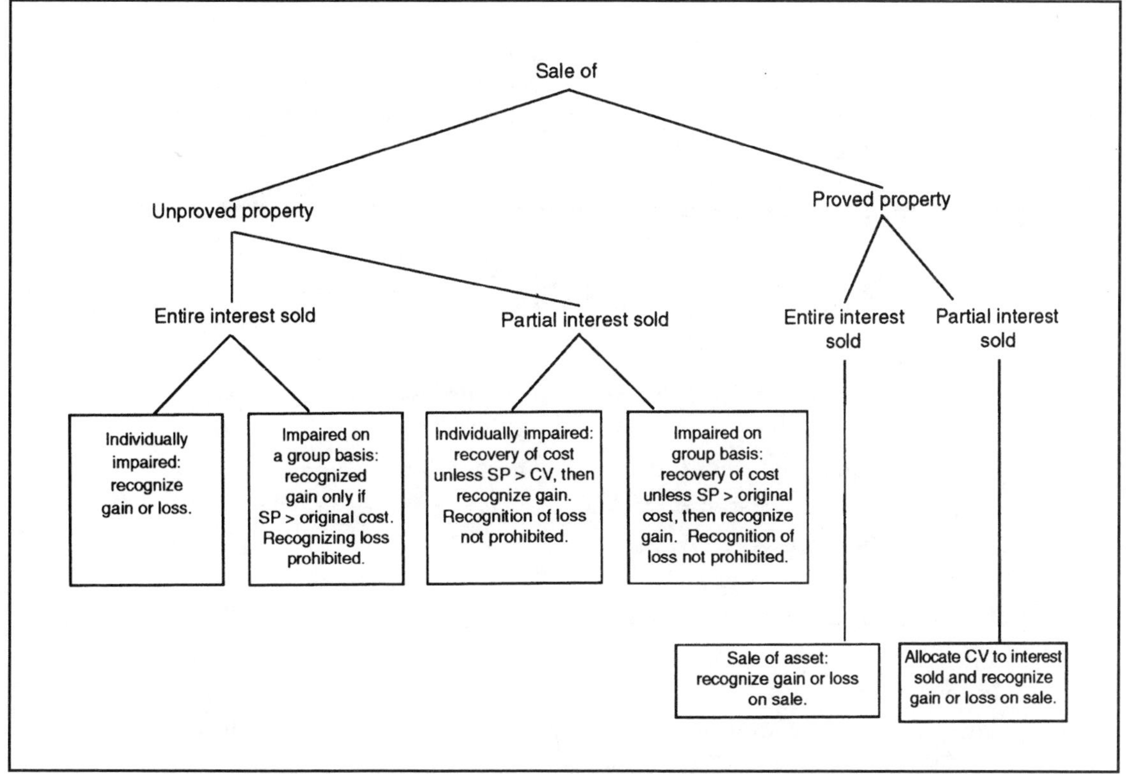

Unproved property sales

In order to understand the accounting treatment for sales of unproved properties it is necessary to divide the topic between sales of the entire interest in a property and sales involving only a partial interest in the unproved property.

Sales of entire interest in unproved property. The accounting treatment of a sale of the entire interest in an unproved property depends on whether impairment of the property has been assessed individually or on a group basis. If the entire interest in an unproved property that has been assessed on an individual basis is sold, a gain or loss should be recognized. If, however, the entire interest in an unproved property that has been assessed on a group basis is sold, a gain should be recognized only if the selling price exceeds the original cost. Recognizing a loss is prohibited in the case of a property assessed on a group basis because an individual carrying value is not known. To avoid recognizing a loss, the allowance for impairment account is debited.

If the entire interest in an unproved property is sold for cash or cash equivalent, recognition of gain or loss depends on whether, in applying paragraph 28 of this Statement, impairment had been assessed for that property individually or by amortizing that property as part of a group. If impairment was assessed individually, gain or loss shall be recognized. For a property amortized by providing a valuation allowance on a group basis, neither a gain nor loss shall be recognized when an unproved property is sold unless the sales price exceeds the original cost of the property, in which case gain shall be recognized in the amount of such excess. (SFAS No. 19, par. 47g)

EXAMPLE

Sale of Entire Unproved Property

Lucky Oil Company owns 100% of the working interest in an undeveloped lease in Texas for which it paid $80,000. In each case below, Lucky sells its entire interest in the property.

Applicable Paragraph: Paragraph No. 47g

Individually Impaired:
The property is individually significant and impairment of $30,000 had been recorded in a prior year. The following entry would be made if the property were sold for $60,000:

Cash	60,000	
Impairment allowance	30,000	
Unproved property		80,000
Gain		10,000

If, instead, the property were sold for $40,000 the following entry would be made:

Cash	40,000	
Impairment allowance	30,000	
Loss	10,000	
Unproved property		80,000

Property Impaired as a Group:
Now assume that the property was individually insignificant and had been amortized as part of a group of unproved properties. The following entry would be made if the property were sold for $60,000:

Cash	60,000	
Impairment allowance*	20,000	
Unproved property		80,000

If, instead the property were sold for $90,000 the following entry would be made:

Cash	90,000	
Unproved property		80,000
Gain		10,000

---

* The impairment allowance account is debited to avoid recognizing a loss.

Sales of a partial interest in unproved property. Sale of only part of an interest in an unproved property receives special accounting treatment because substantial uncertainty exists as to the future recoverability of the cost of the interest retained. In this type of conveyance, the selling price should be treated as a recovery of cost with no gain recognized. (Recognizing a loss is not prohibited.) However, if the property has been assessed on an individual basis and the selling price of the partial interest exceeds the net carrying value of the entire property, a gain should be recognized. If the property has been assessed on a group basis, a gain should be recognized if the selling price of the partial interest exceeds the original cost of the entire interest.

If a part of the interest in an unproved property is sold, even though for cash or cash equivalent, substantial uncertainty usually exists as to recovery of the cost applicable to the interest retained. Consequently, the amount received shall be treated as a recovery of cost. However, if the sales price exceeds the carrying amount of a property whose impairment has been assessed individually in accordance with paragraph 28 of this Statement, or exceeds the original cost of a property amortized by providing a valuation allowance on a group basis, gain shall be recognized in the amount of such excess. (SFAS No. 19, par. 47h)

EXAMPLE

Sale of Portion of Unproved Property—Individual Impairment

Lucky Oil Company has an unproved lease for which it paid $100,000. The property was individually significant and individual impairment of $20,000 had been assessed. Lucky Oil Company conveys 25% of the WI in return for $10,000 cash.

Applicable Paragraph: Paragraph No. 47h

Cash	10,000	
Unproved property		10,000

No gain or loss is recognized. Since the property is unproved there is substantial uncertainty regarding the costs applicable to the retained interest. However, since the property is individually significant, recognition of additional impairment may be called for since the 25% interest was sold for only $10,000.

Next, assume that the 25% interest is sold for $110,000. The entry to record the sale is:

Cash	110,000	
Allowance for impairment	20,000	
Unproved property		99,990*
Gain		30,010

* In order to maintain adequate control for the property in the accounts most companies leave some nominal amount (*e.g.*, $10) in the unproved property account.

In this case a gain is recognized because the selling price of the partial interest that was sold exceeds the net book value of the entire interest.

EXAMPLE

Sale of Portion of Unproved Property—Group Impairment

Lucky Oil Company has an unproved lease for which it paid $100,000. The property is not considered to be individually significant. Lucky Oil Company conveys 25% of the WI in return for $40,000 cash.

Applicable Paragraph: Paragraph No. 47h

Cash	40,000	
Unproved property		40,000

No gain or loss is recognized. Since the property is unproved there is substantial uncertainty regarding the costs applicable to the retained interest.

Next, assume that the 25% interest is sold for $110,000. The entry to record the sale is:

Cash	110,000	
Unproved property		99,990
Gain		10,010

In this case a gain is recognized because the selling price of the partial interest that was sold exceeds the original cost of the entire interest.

Par. 47h also applies to partial sales of an unproved property where the entire working interest is sold and a nonworking interest, such as an ORI, is retained. The recognition of gain or loss depends upon whether the interest in the working interest was individually impaired or amortized as part of a group.

EXAMPLE

Sale of WI Unproved Property with Retention of ORI

Lucky Oil Company owns 100% of the WI in an unproved lease that it acquired for $100,000. The property is individually significant and individual impairment of $20,000 had been assessed. Lucky Oil Company conveys the WI to another party and retains a $\frac{1}{7}$ ORI in return for $30,000 cash.

Applicable Paragraph: Paragraph No. 47h

Cash	30,000	
Allowance for impairment	20,000	
Unproved ORI	50,000	
Unproved property		100,000

No gain or loss is recognized. Since the property is unproved there is substantial uncertainty regarding the costs applicable to the ORI that was retained. The entire allowance balance is written off and the difference between the net book value and the cash received is attributed to the unproved ORI that was retained.

Now, assume that the working interest in the unproved property costing $100,000 is not considered to be individually significant. Lucky Oil Company sells the working interest and retains a $\frac{1}{7}$ ORI for a cash payment of $110,000.

Applicable Paragraph: Paragraph No. 47h

The entry to record the sale is:

Cash	110,000	
Unproved ORI	10*	
Unproved property		100,000
Gain		10,010

In this case a gain is recognized because the selling price of the working interest that was sold exceeds the original cost of the entire interest.

* In order to maintain adequate control for the property, $10 is placed in the unproved ORI account.

The rules regarding sales of partial and entire interests in unproved properties apply to partial and entire sales of nonworking interests as well as working interests. The accounting treatment for the sale of a partial or entire nonworking interest in an unproved property is the same as the accounting when a working interest is involved.

EXAMPLE

Sale of ORI in Unproved Property

Lucky Oil Company has an ORI in an unproved lease for which it paid $60,000. The ORI has not been impaired. In each case below, Lucky sells the entire ORI for cash.

Applicable Paragraph: Paragraph No. 47g

If Lucky Oil Company sells the entire ORI for a cash consideration of $50,000, a loss of $10,000 would be recognized.

Cash	50,000	
Loss	10,000	
Investment in ORI		60,000

However, if Lucky Oil Company sells the entire ORI for a cash consideration of $80,000, a gain of $20,000 would be recognized

Cash	80,000	
Investment in ORI		60,000
Gain		20,000

EXAMPLE

Sale of Portion of ORI in Unproved Property

Lucky Oil Company has an ORI in an unproved property for which it paid $80,000. The ORI has not been impaired. In each case below, Lucky sells a portion of the ORI for cash.

Applicable Paragraph: Paragraph No. 47h

If Lucky Oil Company sells 60% of the ORI for a cash consideration of $65,000, no gain or loss would be recognized.

Cash	65,000	
Investment in ORI		65,000

If Lucky Oil Company sells 60% of the ORI for a cash consideration of $105,000, a gain would be recognized and a nominal amount would be left in the ORI account for control purposes.

Cash	105,000	
Investment in ORI		79,990
Gain		25,010

Proved property sales

Sales of entire interest in proved property. If an entire proved property that has been accounted for as a separate amortization base is sold, a gain or loss should be recognized. This type of sale is treated the same as the sale of any depreciable asset.

The sale of an entire interest in a proved property that constitutes a separate amortization base is not one of the types of conveyances described in paragraphs 44 or 45. The difference between the amount of sales proceeds and the unamortized cost shall be recognized as a gain or loss. (SFAS No. 19, par 47i)

The following example illustrates the accounting when the entire interest in a proved property is sold:

EXAMPLE

Sale of Entire Interest in a Proved Property

Lucky Oil Company owns 100% of the WI in a proved property with the following costs:

Leasehold	$ 60,000
IDC	200,000
Equipment	75,000
Total accumulated DD&A	50,000

Lucky Oil Company sells this property for $300,000. The property constitutes a separate amortization base.

Applicable Paragraph: Paragraph No. 47i

Entry

Cash	300,000	
Accumulated DD&A – total	50,000	
Proved property		60,000
Wells and related E&F—IDC		200,000
Wells and related E&F—L&WE		75,000
Gain		15,000

The same rules apply to the sale of an entire nonworking interest in a proved property.

EXAMPLE

Sale of Entire ORI in Proved Property

Lucky Oil Company owns a $^{1}/_{7}$ ORI in a proved property with unrecovered costs of $40,000. The interest constitutes a separate amortization base. Lucky Oil Company sells the entire ORI for $30,000.

Applicable Paragraph: Paragraph No. 47i

The entry to record the sale is:

Cash	30,000	
Loss	10,000	
Investment in ORI, net		40,000

Sales of a partial interest in a proved property. Paragraph 47j describes the accounting when a partial interest in a proved property is sold. (Par. 47k describes the accounting for a more specific sale of a partial interest in a proved property where the WI is sold and a non-WI retained. This situation is discussed later.) When a partial interest in a proved property is sold, the unamortized cost of the property should be apportioned between the interest sold and the interest retained based on the relative fair market value of the two interests. Any difference between the selling price and the cost apportioned to the interest sold should be recognized as a gain or loss. If the interest sold is an undivided interest, then the fair market value of the interest sold and the FMV of the interest retained will be exactly the same proportion as the proportion of the working interest sold and retained. (See the following example.) The same apportionment and gain or loss recognition rules apply regardless of whether a working interest or a nonworking interest is retained.

The sale of a part of a proved property, or of an entire proved property constituting a part of an amortization base, shall be accounted for as the sale of an asset, and a gain or loss shall be recognized, since it is not one of the conveyances described in paragraphs 44 or 45. The unamortized cost of the property or group of properties a part of which was sold shall be apportioned to the interest sold and the interest retained on the basis of the fair values of those interests. However, the sale may be accounted for as a normal retirement under the provisions of paragraph 41 with no gain or loss recognized if doing so does not significantly affect the unit-of-production amortization rate. (SFAS No. 19, par. 47j)

EXAMPLE

Sale of Partial Interest in a Proved Property

Lucky Oil Company owns 100% of the WI in a proved property with the following costs:

Leasehold	$ 60,000
IDC	200,000
Equipment	75,000
Total accumulated DD&A	50,000

Lucky Oil Company conveys 75% of the working interest in this property for $300,000. The property constitutes a separate amortization base.

Applicable Paragraph: Paragraph No. 47j

In this case the costs would be allocated between the interest sold and the interest retained on the basis of relative fair market values. Since the interest is an undivided working interest, 75% of the costs will be allocated to the interest sold and 25% will be allocated to the interest retained.

Entry

Cash	300,000	
Accumulated DD&A—total (75% x $50,000)	37,500	
Proved property (75% x $60,000)		45,000
Wells and related E&F—IDC (75% x $200,000)		150,000
Wells and related E&F—L&WE (75% x $75,000)		56,250
Gain		86,250

In the less common situation where a divided interest is sold, the relative fair market values of the leasehold, equipment, and IDC must be determined and used to allocate the costs and determine any gain or loss.

EXAMPLE

Sale of Divided Interest in a Proved Property

Lucky Oil Company owns 100% of the WI in a 640 acre proved property with the following net, unamortized costs:

Leasehold	$ 55,000
IDC	190,000
Equipment	77,000

Lucky Oil Company sells 100% of the working interest including the wells and equipment in the western 320 acres of the property for $150,000. An appraisal is performed with the following results:

	FMV of Portion Sold	FMV of Portion Retained	Total FMV
Leasehold	$ 30,000	$ 90,000	$120,000
IDC	80,000	100,000	180,000
Equipment	40,000	30,000	70,000
Total	$150,000	$220,000	

Applicable Paragraph: Paragraph No. 47j

Allocation of cost to interest sold:

FMV_S = FMV of interest sold FMV_r = FMV of interest retained

$$\frac{FMV_S}{FMV_S + FMV_r} \times \text{CV of property} = \text{Cost assigned to interest sold}$$

Leasehold: $30,000/$120,000 x $55,000 = $13,750
IDC: $80,000/$180,000 x $190,000 = $84,444
Equipment: $40,000/$70,000 x $77,000 = $44,000

Entry

	Debit	Credit
Cash	150,000	
Proved property		13,750
Wells and related E&F—IDC		84,444
Wells and related E&F—L&WE		44,000
Gain		7,806

Sales of a WI in a proved property with retention of a non-WI. When the entire working interest in a proved property is sold and a nonworking interest is retained, a gain or loss is determined in the same manner as a partial sale of a proved property. The costs are allocated between interest sold and interest retained based on relative fair market values of the interests.

The sale of the operating interest in a proved property for cash with retention of a nonoperating interest is not one of the types of conveyances described in paragraphs 44 or 45.

Accordingly, it shall be accounted for as the sale of an asset, and any gain or loss shall be recognized. The seller shall allocate the cost of the proved property to the operating interest sold and the nonoperating interest retained on the basis of the fair values of those interests. (SFAS No. 19, par. 47k)

EXAMPLE

Sale of WI in a Proved Property with Retention of a Nonworking Interest

Lucky Oil Company sold its 100% WI in a proved property for $600,000 and retained an ORI. Lucky's net cost basis in the property was $555,000. The fair market value of the entire original WI was $900,000.

Lucky has the following net investment in its accounts:

Leasehold	$155,000
IDC	312,000
Equipment	88,000
Total	$555,000

Applicable Paragraph: Paragraph No. 47k

Since the fair market value of the entire property is $900,000 and the WI was sold for $600,000, the costs would be allocated as follows:

Allocation of costs to interest sold:
$600,000/$900,000 x $555,000 = $370,000

Allocation of costs to interest retained:
$300,000/$900,000 x $555,000 = $185,000

Entry

Cash	600,000	
Investment in ORI	185,000	
Proved property		155,000
Wells and related E&F—IDC		312,000
Wells and related E&F—L&WE		88,000
Gain ($600,000—$370,000)		230,000

The following example also illustrates the sale of a partial interest in a proved property; but in this example a non-WI is sold and a portion of the working interest is retained. In this situation, unlike the previous example, the accounts associated with the original working interest should remain on the books, but reduced by the amount of the non-WI sold.

EXAMPLE

Sale of ORI in Proved Property

Lucky Oil Company owns a 100% WI in the Alfalfa lease that has a $^1/_8$ RI. The property is proved and has the following capitalized costs:

Leasehold costs	$100,000
IDC	500,000
Equipment	125,000
Accumulated DD&A (on leasehold, IDC and equipment)	(225,000)
Total book value	$500,000

Lucky Company carved out an overriding royalty interest for a consideration of $300,000. The fair market value of the WI after the carve out was $1,200,000.

Applicable Paragraph: Paragraph No. 47j

Allocation

FMV_s = FMV of interest sold FMV_r = FMV of interest retained

$$\frac{FMV_s}{FMV_r + FMV_s} \times \text{CV of property} = \text{Cost assigned to interest sold}$$

$$\frac{\$300{,}000}{\$1{,}200{,}000 + \$300{,}000} \times \$500{,}000 = 0.2 \times \$500{,}000 = \$100{,}000$$

$$\frac{FMV_r}{FMV_r + FMV_s} \times \text{CV of property} = \text{Cost assigned to interest sold}$$

$$\frac{\$1{,}200{,}000}{\$1{,}500{,}000} \times \$500{,}000 = 0.8 \times \$500{,}000 = \$400{,}000$$

Entry

Cash	300,000	
Accumulated DD&A (0.2 x $225,000)	45,000	
Proved property (0.2 x $100,000)		20,000
Wells and related E&F—IDC (0.2 x $500,000)		100,000
Wells and related E&F—L&WE (0.2 x $125,000)		25,000
Gain ($300,000 – $100,000)		200,000

Fig. 12–2 — *Production Payment Interests*

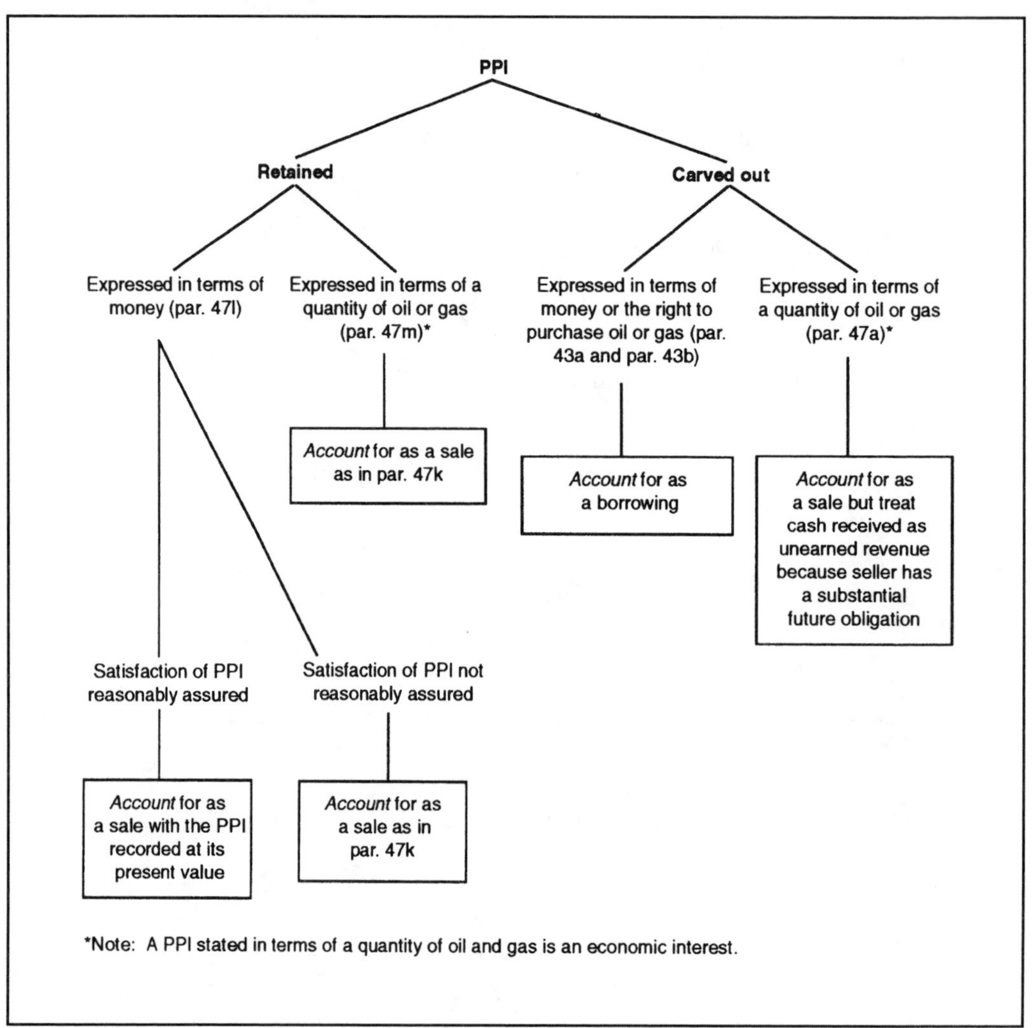

Production payments

Production payment interests may be created by being retained or carved out of the WI and may be satisfied by monetary payment or by delivery of a specified quantity of oil or gas. Accounting for a production payment depends upon whether the production payment is retained or carved out, how it will be paid, and whether the production payment is an economic interest. Generally only those production payment interests that are payable by delivery of a specified quantity of oil or gas are considered economic interests and are subject to DD&A. Production payment interests that are payable out of cash typically are not economic interests and are not subject to DDA.

Retained production payments. A retained production payment may be payable out of oil or gas or in money from a specified share of production from a proved property. The accounting treatment depends on whether the payment is stated in terms of money or in terms of a volume of production and whether the repayment of the production payment is reasonably assured or not reasonably assured. Each of these scenarios are discussed below:

> *The sale of a proved property subject to a retained production payment that is expressed as a fixed sum of money payable only from a specified share of production from that property, with the purchaser of the property obligated to incur the future costs of operating the property, shall be accounted for as follows:*
>
> *i. If satisfaction of the retained production payment is reasonably assured. The seller of the property, who retained the production payment, shall record the transaction as a sale, with recognition of any resulting gain or loss. The retained production payment shall be recorded as a receivable, with interest accounted for in accordance with the provisions of APB Opinion No. 21, "Interest on Receivables and Payables." The purchaser shall record as the cost of the assets acquired the cash consideration paid plus the present value (determined in accordance with APB Opinion No. 21) of the retained production payment, which shall be recorded as a payable. The oil and gas reserve estimates and production data, including those applicable to liquidation of the retained production payment, shall be reported by the purchaser of the property (paragraphs 50-56).*
>
> *ii. If satisfaction of the retained production payment is not reasonably assured. The transaction is in substance a sale with retention of an overriding royalty that shall be accounted for in accordance with paragraph 47(k). (SFAS No. 19, par. 47l)*

Retained production payments payable in money—reasonably assured. As seen above, when a retained production payment is payable out of a fixed sum of money, the accounting treatment depends upon whether the payment is reasonably assured or not. If the payment is reasonably assured, the seller records a receivable for the production payment in the amount of the present value of the production payment and recognizes a gain or loss on the sale of the

WI. The buyer records a payable equal to the present value of the production payment and capitalizes as the cost of the asset the present value of the production payment, plus any cash paid. Both parties compute the present value using their own cost of capital as the interest rate.

To compute the present value of the production payment, the future cash flows to the production payment owner must be estimated. To do this, a projection of future production over time must first be made. The production payment owner's share of revenue from that production is then discounted to arrive at the present value of the production payment. As the production payment is paid, the owner of the retained production payment records the payment received as interest revenue and a reduction to the receivable. The WI owner records the payment made as interest expense and a reduction to the liability. From the seller's standpoint, its production payment interest is not a mineral interest in the property and therefore is not subject to DD&A.

EXAMPLE

Retained Production Payment Payable in Money—Reasonably Assured

Lucky Oil Company has a 100% WI in a fully developed lease subject to a ⅛ RI. As of January 1, 2009, Lucky's net capitalized costs related to the property are $500,000 and total gross proved developed reserves are 250,000 barrels.

On January 1, 2009, Lucky Oil Company sold its entire WI to Roberts Oil Company for $1,000,000 and a retained production payment of $360,000. The production payment is payable to Lucky Oil Company out of 70% of the proceeds of the WI's share of production. It is estimated that the production payment will be paid off in three years. Assume that the present value of the production payment of $360,000 would be $300,000 using Lucky Company's cost of capital (10%) and $325,000 using Roberts Company's cost of capital (5%). During 2009, gross production totaled 11,200 barrels, operating costs totaled $48,980, and the selling price was $20 per barrel. Ignore severance taxes.

Applicable Paragraph: Paragraph No. 47li

Entries:

Lucky Oil Company:

Cash	1,000,000	
Production payment receivable (at PV using 10%)	300,000	
Net capitalized costs of property		500,000
Gain on sale ($1,300,000 – $500,000)		800,000
(to record conveyance)		

Roberts Company:

Proved property, etc. ($1,000,000 + $325,000) . . .	1,325,000	
Production payment payable (at PV using 5%)		325,000
Cash		1,000,000
(to record conveyance)		

A/R—purchaser (11,200 x $20)	224,000	
Royalty payable (11,200 x $^1/_8$ x $20)		28,000
Oil revenue (11,200 x $^7/_8$ x $20)		196,000
(to record revenue for the sale of oil)		

Operating expense	48,980	
Cash		48,980
(to record production costs)		

DD&A expense	59,360	
Accumulated DD&A		59,360

Allocation:

Roberts' share of proved developed reserves:
250,000 bbl x $^7/_8$ = 218,750 bbl

Roberts' share of production:
11,200 bbl x $^7/_8$ = 9,800 bbl

$$\frac{\$1,325,000}{218,750} \times 9,800 = \underline{\underline{\$59,360}}$$

Production payment payable ($137,200 – $16,250)	120,950	
Interest expense ($325,000 x 5%)	16,250	
Cash (70% x 9,800 x $20)		137,200
(to record partial payment of production payment payable)		

Lucky Oil Company:

Cash (70% x 9,800 x $20)	137,200	
Interest revenue ($300,000 x 10%)		30,000
Production payment receivable ($137,200 – $30,000)		107,200
(to record partial collection of production payment receivable)		

Retained production payments payable in money—not reasonably assured. If the retained production payment is not reasonably assured, the conveyance is, in substance, a sale of a proved property with a retained ORI. The production payment's life in this instance is likely to be coextensive with the life of the lease and, therefore, has the same characteristics as an ORI. Consequently, this type of production payment should be treated as an ORI rather than as a receivable or a payable, and even though the production payment is payable in cash, the production payment is considered to be an economic interest subject to DD&A. To the extent that the production payment's FMV can be determined, the unamortized cost of the WI should be allocated between the interest sold and the interest retained, and a gain or loss recognized on the interest sold.

EXAMPLE

Retained Production Payment Payable in Money—not Reasonably Assured

Lucky Oil Company owns a 100% WI in a proved property with net capitalized costs of $100,000. Lucky Oil Company sells the WI for $425,000 cash and a retained production payment of $200,000. The production payment is payable out of the first 70% of the proceeds of the WI's share of production. Satisfaction of the retained production payment is not reasonably assured, but its fair market value is estimated to be $75,000.

Applicable Paragraph: Paragraph No. 47l(ii)

Allocation:

Capitalized costs allocated to interest retained:

$$\frac{\$75{,}000}{\$425{,}000 + \$75{,}000} \times \$100{,}000 = \underline{\underline{\$15{,}000}}$$

Capitalized costs allocated to interest sold:

$$\frac{\$425{,}000}{\$425{,}000 + \$75{,}000} \times \$100{,}000 = \underline{\underline{\$85{,}000}}$$

Entry

Cash	425,000	
PPI	15,000	
Net capitalized costs		100,000
Gain ($425,000 – $85,000)		340,000

SFAS No. 19 does not specifically specify the treatment of a sale of an unproved property subject to a retained production payment expressed as a fixed sum of money. However, if the property is unproved, the production payment cannot be reasonably assured. It appears that the appropriate accounting treatment would be to treat the production payment as an ORI, with no gain or loss recognized on the sale unless the selling price exceeds the carrying value of the original interest if assessed individually, or exceeds the original cost if assessed on a group basis.

Retained production payments payable in production. A retained PPI payable by delivery of a specified quantity of petroleum products should be accounted for as a sale, with the unamortized cost of the property allocated between the interest sold and the interest retained, based on relative fair market values. A gain or loss is recognized on the interest sold. This type of production payment is an economic interest and is therefore subject to DD&A.

> *The sale of a proved property subject to a retained production payment that is expressed as a right to a specified quantity of oil or gas out of a specified share of future production shall be accounted for in accordance with paragraph 47(k). (SFAS No. 19, par. 47m)*

EXAMPLE

Retained Production Payment Payable in Product

Lucky Oil Company owns a WI in a proved property with the following capitalized costs:

Leasehold	$160,000
IDC	400,000
Equipment	140,000
Total accumulated DD&A	(100,000)
Total book value	$600,000

Lucky Oil Company sells the WI to Stephens Company for $800,000 and retains a production payment of 20,000 barrels of oil payable out of 40% of the WI's share of production. The fair market value of the entire original WI was $1,000,000.

Applicable Paragraph: Paragraph No. 47m

Allocation:

Cost allocated to interest sold:

$$\frac{\$800{,}000}{\$800{,}000 + \$200{,}000^{*}} \times \$600{,}000 = \underline{\underline{\$480{,}000}}$$

* $1,000,000 – $800,000

Cost allocated to interest retained:

$$\frac{\$200{,}000}{\$800{,}000 + \$200{,}000} \times \$600{,}000 = \underline{\underline{\$120{,}000}}$$

Entry:

Lucky Oil Company:

Cash	800,000	
PPI	120,000	
Accumulated DD&A	100,000	
Proved property, net		160,000
Wells and related E&F—IDC		400,000
Wells and related E&F—L&WE		140,000
Gain ($800,000 – $480,000)		320,000

Carved-out production payments

In a carved-out production payment the working interest owner keeps its working interest in the property and carves out or creates a production payment in exchange for cash. The buyer of the production payment interest agrees to receive either money or a stated amount of product in the future. The accounting treatment for carved-out production payments depends on whether the payment is stated in money or in product.

Carved-out production payments payable in money. Carved-out production payments payable in terms of money are generally designed as a means of financing. They may take many forms, including a production loan or an exploration advance.

A production loan may be negotiated between an oil and gas company and a lending institution. The lending institution loans the company money and the company agrees to repay the loan from the proceeds from the production from a particular property. However, if the production from the property is insufficient to liquidate the loan, then the company must repay the loan with money from other sources. This type of transaction should be accounted for as a payable from the perspective of the oil and gas company and a receivable from the perspective of the lending institution. There is no gain or loss recognition and no revenue effect recognized on the books of the oil and gas company.

Occasionally a pipeline company or utility company may advance funds to an oil or gas company in exchange for the right to purchase all or part of the oil or gas produced from a property. In this case, the receipt of cash is recorded as a payable on the books of the oil and gas company. As the product is produced and delivered to the production payment owner, revenue is recognized and the liability is liquidated.

> *Enterprises seeking supplies of oil or gas sometimes make cash advances to operators to finance exploration in return for the right to purchase oil or gas discovered. Funds advanced for exploration that are repayable by offset against purchases of oil or gas discovered, or in cash if insufficient oil or gas is produced by a specified date, shall be accounted for as a receivable by the lender and as a payable by the operator. (SFAS No. 19, par. 43a)*

The next example illustrates money advanced to the operator in return for the right to purchase oil or gas. This situation, although sometimes referred to as a carved-out production payment is, in fact, a borrowing.

EXAMPLE

Production Advance

Lucky Oil Company receives $100,000 from Myers Company in return for the right to purchase oil and gas in the future.

Applicable Paragraph: Paragraph No. 43a

Entries

Lucky Oil Company:

Cash	100,000	
Production payment payable		100,000

Myers Company:

Production payment receivable	100,000	
Cash		100,000

A third variation of a carved-out production payment payable in money occurs when an oil and gas company carves out and sells a production payment interest in a property in order to obtain financing from a lender. Unlike the first two cases, here the financing is repayable exclusively from the property from which it was carved out. In other words, the borrowing is without recourse to the company's other assets. The production payment is to be repaid if, when, and as production occurs. Clearly, the intent of the oil and gas company is to obtain financing—not to sell production. In addition, the lender is clearly making a loan rather than purchasing an interest in oil and gas reserves. Accordingly, the transaction is to be accounted for as a loan from both the perspective of the oil and gas company and the lender.

In all of these cases the oil and gas company includes the reserves necessary to liquidate the production loan or exploration advance among their reserves for purposes of disclosure and for computing amortization. In other words, the PPI is not accounted for as a mineral interest.

Funds advanced to an operator that are repayable in cash out of the proceeds from a specified share of future production of a producing property, until the amount advanced plus interest at a specified or determinable rate is paid in full, shall be accounted for as a borrowing. The advance is a payable for the recipient of the cash and a receivable for the party making the advance. Such transactions, as well as those described in paragraph 47(a) below, are commonly referred to as production payments. The two types differ in substance, however, as explained in paragraph 47(a). (SFAS No. 19, par. 43b)

EXAMPLE

Carved-Out Production Payment Payable in Money

Lucky Oil Company owns 100% of the WI in a fully developed lease that has a $^1/_8$ RI. The lease has the following capitalized costs and reserve data as of January 1, 2008:

Unrecovered costs	$400,000
Proved developed reserves, gross	800,000 bbl

On January 1, 2008, Lucky Oil Company carves out a $900,000 production payment to ABC Bank. The production payment is payable to ABC Bank out of 70% of the proceeds of Lucky's share of production with interest of 10% on the unpaid balance. During 2008, gross production totaled 16,000 barrels of oil, production costs were $5 per barrel, and the selling price was $20 per barrel. Ignore production taxes.

Applicable Paragraph: Paragraph No. 43b

Entries

Lucky Oil Company:

	Debit	Credit
Cash	900,000	
Production payment payable		900,000
(to record production payment)		

	Debit	Credit
Cash	320,000	
Royalty payable (16,000 x $^1/_8$ x \$20)		40,000
Oil revenue (16,000 x $^7/_8$ x \$20)		280,000
(to record revenue)		

	Debit	Credit
Operating expense (\$5 x 16,000)	80,000	
Cash		80,000
(to record production costs)		

	Debit	Credit
Production payment payable (\$196,000 – \$90,000)	106,000	
Interest expense (10% x \$900,000)	90,000	
Cash (\$280,000 x 0.70)		196,000
(to record partial payment of production payment)		

	Debit	Credit
DD&A expense	8,000	
Accumulated DD&A		8,000
(to record DD&A)		

Allocation:

Lucky's share of proved developed reserves:
800,000 x $^7/_8$ = 700,000

Lucky's share of production:
16,000 x $^7/_8$ = 14,000

$$\frac{\$400,000}{700,000} \times 14,000 \text{ bbl} = \underline{\underline{\$8,000}}$$

ABC Bank:

	Debit	Credit
Production payment receivable	900,000	
Cash		900,000
(to record production payment)		

	Debit	Credit
Cash	196,000	
Production payment receivable		106,000
Interest revenue		90,000
(to record partial collection of production payment)		

Carved-out production payments payable in product. A carved-out production payment payable by a specified quantity of oil or gas out of future production is, in essence, a sale of reserves in-place. However, the seller has a substantial obligation for future performance *i.e.,* to produce the oil or gas and transfer a portion of it to the buyer of the production payment. As a result, no gain or loss should be recognized at the time of the conveyance. In this situation, the seller records the funds received not as income, but as unearned revenue. As production occurs and the production payment is made, the seller recognizes a portion of the unearned revenue as earned revenue. The buyer or owner of the production payment capitalizes the cost of the production payment at the amount paid and amortizes the capitalized cost as reserves are produced and delivered to the buyer of the production payment.

> *Some production payments differ from those described in paragraph 43(b) in that the seller's obligation is not expressed in monetary terms but as an obligation to deliver, free and clear of all expenses associated with operation of the property, a specified quantity of oil or gas to the purchaser out of a specified share of future production. Such a transaction is a sale of a mineral interest for which gain shall not be recognized because the seller has a substantial obligation for future performance. The seller shall account for the funds received as unearned revenue to be recognized as the oil or gas is delivered. The purchaser of such a production payment has acquired an interest in a mineral property that shall be recorded at cost and amortized by the unit-of-production method as delivery takes place. The related reserve estimates and production data shall be reported as those of the purchaser of the production payment and not of the seller (paragraphs 50-56). (SFAS No. 19, par. 47a)*

Par. 47a has been the source of debate since *SFAS No. 19* was issued. The FASB clearly indicates that the transaction is to be recorded by the seller as unearned revenue. Revenue is to be recognized as the oil or gas is produced and the production payment delivered to the buyer. Related reserves are to be reported by the buyer of the production payment interest. Applied literally, this treatment results in a mismatching of expenses and revenues for the seller. As the production payment is delivered and the unearned revenue is recognized as revenue, there is no matching of DD&A expense against the production payment revenue. This results from the fact that the capitalized cost of the property from which the production payment is carved-out is not changed when the production payment is carved out. But as production occurs the capitalized cost is amortized over the proved developed reserves minus the production payment reserves. Thus, there is no DD&A expense related to the reserves which are carved out and then subsequently recognized as revenue as production occurs.

One possible solution, although not explicitly allowed for in *SFAS No. 19* is to divide the capitalized costs between the WI interest retained and the production payment carved out. (Note that the use of this approach does not change total capitalized costs, it simply separates the costs into different accounts.) Then, as the production payment is paid and the unearned revenue is recognized as revenue, the capitalized costs allocated to the carved-out production payment are amortized and matched with revenue. The other capitalized costs for the property are amortized over the non-production payment reserves. Remember, *SFAS No. 19* states that the reserves related to the production payment must be reported by the buyer of the payment and not by the seller.

The owner or buyer of the carved-out production payment capitalizes the cost when the interest is acquired. The production payment is an economic interest subject to DD&A and the company amortizes the production payment interest as the barrels are produced and delivered. This process is illustrated below:

EXAMPLE

Carved-Out Production Payment Payable in Product

Lucky Oil Company owns 100% of the WI in a fully developed lease with a $^1/_8$ RI. The lease has the following capitalized costs and reserve data as of January 1, 2008.

Unrecovered costs	$400,000
Proved developed reserves, gross	800,000 bbl

On January 1, 2008, Lucky Oil Company carves out a $540,000 production payment to Alpha Company. The production payment is payable to Alpha Company by delivery of 30,000 barrels out of the first 70% of Lucky's share of production. During 2008, gross production totaled 16,000 barrels of oil, production costs totaled $80,000, and the average selling price was $20 per barrel. Ignore production taxes.

Applicable Paragraph: Paragraph No. 47a

Allocations:
If the FMV of the 30,000 barrels carved-out production payment is $540,000 ($18/bbl), then the FMV of the retained interest in the proved developed reserves can be approximated at $12,060,000, *i.e.,* (800,000 x $^7/_8$ – 30,000) x $18.

The capitalized costs could be allocated to the interest carved out and the interest retained as follows:

$$\frac{\$540{,}000}{\$12{,}060{,}000 + \$540{,}000} \times \$400{,}000 = \underline{\underline{\$17{,}143}} \text{ to carved-out PPI}$$

$$\frac{\$12{,}060{,}000}{\$12{,}060{,}000 + \$540{,}000} \times \$400{,}000 = \underline{\underline{\$382{,}857}} \text{ to retained WI}$$

Entries

Lucky Oil Company:

Cash	540,000	
Unearned revenue		540,000
(to record production payment)		
Carved-out production payment	17,143	
Capitalized costs		17,143
(to record allocation of capitalized costs to carved-out production payment)		
Cash (16,000 x $^1/_8$ x $20)		
+ (16,000 x $^7/_8$ x .30 x $20)	124,000	
Royalty payable (16,000 x $^1/_8$ x $20)		40,000
Oil revenue (16,000 x $^7/_8$ x .30 x $20)		84,000
(to record sale of production)		
Unearned revenue (16,000 x $^7/_8$ x .70 x $18)	176,400	
Oil revenue		176,400
(to record delivery of the PPI's share of production at $18/bbl, *i.e.*, $540,000/30,000 bbl)		
Operating expense	80,000	
Cash		80,000
(to record production costs)		
DD&A expense	2,400	
Accumulated DD&A		2,400
(to record DD&A on capitalized costs retained)		

$$\frac{\$400{,}000 - \$17{,}143}{(800{,}000 \times 7/8) - 30{,}000} \times (16{,}000 \times 7/8 \times .30) = \underline{\underline{\$2{,}400}}$$

DD&A expense	5,600	
Carved-out production payment		5,600
(to record amortization of costs allocated to carved-out production payment)		

$$\frac{\$17{,}143}{30{,}000} \times (16{,}000 \times 7/8 \times .70) = \underline{\underline{\$5{,}600}}$$

Note the 30,000 barrels of proved developed reserves related to the production payment would be reported by Alpha Company and not by Lucky Oil Company.

Alpha Company:

Production payment interest	540,000	
Cash		540,000
(to record production payment)		
Cash	196,000	
Oil revenue (16,000 x 7/8 x .70 x $20)		196,000
(to record income for 9,800 barrels received and sold at $20)		
DD&A expense	176,400	
Production payment interest		176,400

$$\frac{\$540{,}000}{30{,}000} \times 9{,}800 = \underline{\underline{\$176{,}400}}$$

If Lucky Oil Company does not transfer $17,143 to a carved-out production payment account then that amount would remain with the other capitalized costs and would be amortized over Lucky's share of proved developed reserves minus the 30,000 barrels related to the production payment.

CONVEYANCES UNDER FULL COST

As indicated earlier, most conveyances are accounted for in the same way under full cost as under successful-efforts. The differences are discussed in the following paragraphs.

The full-cost accounting method provides for cost accumulation on a country-by-country basis. As discussed in an earlier chapter, acquisition, exploration, and development costs are capitalized in a full-cost cost pool (a country). Therefore, the individual costs theoretically lose their particular lease, field, or reservoir character. Sales and abandonments of individual properties should generally give rise only to adjustments in the cost pool. In general, no gains or losses should be recognized under FC accounting for sales of oil and gas properties unless such adjustments to capitalized costs arising from these transactions would materially distort the amortization rate. In *Reg. SX 4-10,* the SEC states a material distortion will generally not occur unless 25% or more of the reserves are sold.

(i) Sales and abandonments of oil and gas properties. Sales of oil and gas properties, whether or not being amortized currently, shall be accounted for as adjustments of capitalized costs, with no gain or loss recognized, unless such adjustments would significantly alter the relationship between capitalized costs and proved reserves of oil and gas attributable to a cost center. For instance, a significant alteration would not ordinarily be expected to occur for sales involving less than 25% of the reserve quantities of a given cost center. If gain or loss is recognized on such a sale, total capitalized costs within the cost center shall be allocated between the reserves sold and reserves retained on the same basis used to compute amortization, unless there are substantial economic differences between the properties sold and those retained, in which case capitalized costs shall be allocated on the basis of the relative fair values of the properties. Abandonments of oil and gas properties shall be accounted for as adjustments of capitalized costs, that is, the cost of abandoned properties shall be charged to the full cost center and amortized (subject to the limitation on capitalized costs in paragraph (b) of this section). (Reg. SX 4-10(i)(6)(i))

The following example illustrates the sale and abandonment of properties when the amortization rate is not materially distorted:

EXAMPLE

Sale and Abandonment—Unproved and Proved Properties

On December 31, 2005, Lucky Company had total capitalized costs of $6,000,000 in the U.S. cost pool and accumulated DD&A of $2,000,000. During January 2006, the following transactions occurred:

Unproved property abandoned:	
Original cost	$ 60,000
Proved property abandoned:	
Leasehold cost	30,000
Equipment	300,000
IDC	400,000
Salvage value of equipment	40,000

Proved property sold:

Selling price	$ 100,000	
Leasehold cost	20,000	
Equipment	150,000	
IDC	200,000	

Entry

Cash	100,000	
Inventory	40,000	
Accumulated DD&A (plug)	1,020,000	
Unproved properties		60,000
Proved properties, leasehold		50,000
Wells and related E&F—L&WE		450,000
Wells and related E&F—IDC		600,000

Theoretically, under full-cost accounting a write-off of capitalized amounts is not necessary. However, most companies would remove the costs that apply to properties sold or abandoned as shown in the preceding example. If the individual costs are not removed, then the proceeds received from the sale of the proved property and the items salvaged during abandonment would be debited to the cash account and credited to either the accumulated DD&A account or the cost pool account.

Sometimes a sale is great enough to cause a significant alteration in the relationship between capitalized costs and proved reserves within a cost center. When this happens, a gain or loss should be recognized as shown in the following example.

EXAMPLE

Sale of Significant Reserves: DD&A Rate Distortion

Lucky Company had total capitalized costs in the United States of $6,000,000 and accumulated DD&A of $1,800,000. Reserves in the United States amounted to 600,000 barrels. Two hundred thousand (200,000) barrels of reserves were sold for $3,000,000. The selling price per barrel was $15 ($3,000,000/200,000) vs. an amortization rate of $7 per barrel ([$6,000,000-$1,800,000]/600,000). Further, ⅓—more than 25% of the reserves—have been sold. Consequently, a significant change has occurred between capitalized costs and proved reserves, and a gain or loss should be recognized. Since ⅓ of the reserves have been sold, ⅓ of the capitalized costs should be allocated to the reserves sold.

Entry

Cash	3,000,000	
Accumulated DD&A (⅓ x $1,800,000)	600,000	
Proved properties (⅓ x $6,000,000)		2,000,000
Gain on sale of proved properties		1,600,000

The gain was $8/bbl ($15 – $7) on the 200,000 barrels sold for a total gain of $1,600,000.

Another problem in full-cost accounting for sales and abandonments concerns transactions involving unproved properties that were acquired for resale or promotion purposes. Prior to May 1984, gain or loss was recognized on these sales of inventory unproved property. In May 1984, the SEC revised *Reg. SX 4-10* (i)(6)(iii)(A) to read as follows:

> *Except as provided in subparagraph (i)(6)(i) of this section, all consideration received from sales or transfers of properties in connection with partnerships, joint venture operations, or various other forms of drilling arrangements involving oil and gas exploration and development activities (e.g., carried interest, turnkey wells, management fees, etc.) shall be credited to be the full cost account, except to the extent of amounts that represent reimbursement of organization, offering, general and administrative expenses, etc., that are identifiable with the transaction, if such amounts are currently incurred and charged to expense.*

Under this new rule, sales, abandonments, and drilling arrangements should, for almost all circumstances, not give rise to recognition of a gain or loss. Even if the unproved property is acquired for the purpose of selling or transferring to a drilling fund, no gain or loss should generally be recognized. The only entry is an adjustment to the full-cost pool. The new rules also severely limit the recognition of income from management fees or service activities where an interest is held in the property. (For more information see Appendix B.)

In comparison of the treatment of conveyances under SE versus FC, any conveyance that gives rise to a gain or loss under SE generally is treated as an adjustment to the cost pool under FC. All other conveyances are usually handled the same under FC as under SE accounting.

Problems

Assume all production and reserves are gross barrels or Mcf.

1. What is the major purpose of unitization?

2. Distinguish between a carried interest situation, a free-well transaction, and a farm-in/farm-out transaction.

3. Identify the types of interests that are created in the following situations. If the interest is an overriding royalty or a production payment interest, also state whether it is a retained or carved-out interest.

    a. Zeke Company owns the WI in a proved property with net capitalized costs of $100,000. Zeke sold the lease for $250,000 cash and a payment of $150,000 plus interest of 10% to be paid out of the first 60% of the oil produced.

    b. Wildcat Oil Company acquired an undeveloped lease for which it paid $30,000. Financially unable to develop the lease, Wildcat agreed to allow Friendly Company to earn a 30% working interest by paying 100% of the cost of drilling and completing a well.

    c. Young Oil Company owns a 100% WI in Lease A which has a ⅛ RI. On February 1, 2003, Young carved out a $400,000 payment, payable out of 60% of the net proceeds of the WI's share of production plus interest of 10% on the unpaid balance.

    d. Four companies own adjacent leases that share a common reservoir. The companies decide to operate the properties as one in order to obtain improved operating efficiency. Following negotiations by engineers, geologists, etc., the companies agree upon participation factors and market values of contributed IDC and equipment.

    e. Mabel Oil Company acquired an unproved property at a cost of $50,000. Mabel later sold the WI and kept a nonworking interest. As a result, Mabel will receive $^1/_{16}$th of the revenue of the working interest from which the interest was created.

    f. Company ABC assigned a 40% WI in an unproved property to Company XYZ in return for Company XYZ bearing all costs of drilling, developing, and operating the property. Company XYZ is entitled to all of the revenue from production (net of royalty) until Company XYZ has recovered all of its costs, at which time the property becomes a joint WI.

4. The following transactions occurred during 2008:

a. Joyner Oil Company and Brown Oil Company jointly purchased a 2,000 acre lease in Oklahoma for $60,000. Joyner has a 60% WI and Brown a 40% WI. Joyner will be the operator of the lease.

b. Rayburn Oil Company owns 100% WI in a lease (with a $\frac{1}{6}$ RI) with capitalized costs of $600,000. Rayburn assigns the WI to Fugate Oil for $800,000 and keeps a 15% overriding royalty interest.

c. Rayburn Oil Company owns 100% WI in a lease (with a $\frac{1}{6}$ RI) in Oklahoma and transfers a $\frac{1}{16}$ overriding royalty interest to the controller of the company.

d. Sells Oil Company owns the WI in a lease in Nieces County, Texas and assigns 40% of the WI to the Knight Oil Company in return for Knight drilling and equipping a well on the property. After the well is completed, Sells and Knight will share revenues and costs.

e. Knight Oil Company assigns the WI in Lease A to Sells Oil Company for $700,000 and, in return, Knight will receive 20,000 barrels of oil from the first 30% of Sells' production.

f. Sells Oil Company sells a PPI of 20,000 barrels of oil for $300,000, which is to be paid from the first 30% of production.

g. Cantu Oil Company owns a WI in Harris County, Texas and transfers 60% of the WI to Stephens Oil Company in exchange for Stephens bearing all costs of drilling, completing, and operating the property until payout. Stephens Oil Company will receive 100% of the WI's share of revenue until drilling and developing costs have been recovered, at which time the property will be operated as a joint property.

REQUIRED: Identify the following types of interests created by the 2008 transactions. If the interest is an overriding royalty or a production payment interest, also state whether it is a retained or carved-out interest.

5. Hays, Bush, and King signed a lease agreement with Big Pink, the owner of the mineral rights. Big Pink received a $\frac{1}{7}$ RI. The companies' WIs are 50%, 30%, and 20%, respectively. The companies signed an operating agreement designating Hays as the operator of the lease. Assuming revenues of $42,000 and costs of $14,000 the first year of operations, determine how much each party will receive in revenue and pay in costs the first year of operations.

6. Yale Oil Company owns the WI in a small lease in Louisiana that has a $\frac{1}{6}$ RI. Yale also owns the WI in numerous leases in Texas. Not having the facilities in Louisiana to develop the Louisiana lease, Yale assigns his entire WI in the lease to Smith for a consideration of $300,000 and retains a $\frac{1}{15}$ overriding royalty. Assuming revenues of $42,000 and costs

of $12,000 the first year of operations, determine how much each economic interest owner will receive in revenue and pay in costs the first year of operations.

7. Mabel Oil Company owns 100% of the WI in a lease that has a 1/7 RI. Needing additional funds to develop the property, Mabel sold Pitt Company 60,000 barrels of oil for a consideration of $800,000. The oil is to be paid out of the first 20% of the WI's share of production. During the first four years of production, 140,000 barrels are produced each year. How many barrels of oil does each interest receive in each of the first four years?

8. Wildcat Oil Company acquired an undeveloped lease for which it paid $30,000. The lease is burdened with a 1/8 royalty. Financially unable to develop the lease, Wildcat sold 60% of its WI to two parties for $200,000 ($100,000 each), agreeing to use the money to drill and equip a well. When the well is completed, each of the three companies will share future development and operating costs. The well cost $200,000 and was successful. Estimated proved reserves were 500,000 barrels, and proved developed reserves were 350,000 barrels (12/31). Wildcat is the operator, and 10,000 barrels were produced and sold in the first year of operations. The selling price was $20 per barrel, operating costs were $5 per barrel, and the severance tax rate was 5%. The purchaser assumed the responsibility of paying severance taxes and the RI owner.

   a. Determine how much revenue and operating costs each party should record for the first year of operations.
   b. Give all entries necessary for Wildcat Oil Company and the buyers.
   c. Give the entries assuming Wildcat Oil Company is an FC company.

9. Lomax Company assigned 40% of the WI in an unproved property with a 1/8 RI to Mabel Company in return for Mabel Company bearing all costs of drilling, developing, and operating the property (until payout). Mabel Company is entitled to all of the revenue from production until Mabel Company has recovered all of its costs, at which time the property becomes a joint working interest. Acquisition costs of the property incurred by Lomax Company were $80,000. Mabel Company incurs $450,000 in IDC and equipment costs in drilling the well. The well is successful. Production begins early in the second year, and 36,000 barrels are produced during each of the first four years of production. The selling price was $20 per barrel and operating costs were $5 per barrel. Determine how much revenue and operating costs each part should record for the first three years of operations.

10. Company Z (a successful-efforts company) owns 100% WI in a lease in which Dudley Smith owns a 1/8 RI. Company Z conveys to Company Q 30% of the WI in exchange for a cash consideration of $10,000. The undeveloped lease has capitalized costs of $40,000 and an impairment allowance of $15,000.

a. Give the entry for Company Z to record the conveyance.

b. Give the entry if the interest had sold for $30,000.

11. Company Z owns a 100% WI in a proved property with net capitalized costs of $100,000. Company Z sold the lease for $250,000 cash and a production payment of $150,000 plus interest of 10% to be paid out of the proceeds from the first 60% of the oil produced. Satisfaction of the retained production payment is not reasonably assured. The fair market value of the PPI is estimated to be $50,000. Give the entry to record the conveyance assuming that Company Z uses the successful-efforts method.

12. Philco Company owns a proved property with the following costs:

Leasehold	$ 80,000
IDC	500,000
Equipment	120,000
Total accumulated DD&A (separate amortization base)	250,000

Philco Company sells 100% of the WI in the property to Company Q for $600,000.

a. Give the entry for Philco Company to record the sale assuming that Philco uses the SE method.

b. Give the entry for Philco Company to record the sale if the property had sold for $400,000.

c. Give the entry for a. and b. assuming Philco Company is an FC company and the reserves sold constitute 15% of Philco's share of reserves in the cost center.

13. Wildcat Oil Company leased undeveloped acreage from David Jones for $30,000 with Jones receiving a $^1/_8$ RI. Financially unable to develop the lease, Wildcat enters into a farm-in/farm-out agreement with Jayhawk Company. Jayhawk agrees to drill and complete a well in return for 60% of the WI and the right to recover all of its costs. Jayhawk drills and completes the well for $100,000. Estimated proved reserves are 500,000 barrels, and proved developed reserves are 100,000 barrels. Jayhawk is the operator, and production totals 4,000 barrels per month for the first six months. Assume that the average selling price is $20 and lifting costs average $5 per barrel. Ignore severance taxes and assume reserve estimates do not change. Jayhawk assumes the responsibility of paying the RI owner.

a. Calculate payout.

b. Assuming that both companies are successful-efforts companies, give all of the entries including monthly DD&A expense for the first three months that would be made by Wildcat Oil Company and by Jayhawk Company.

14. Bingo Oil Company owns 100% of the WI in a fully developed lease on which there is a $^1/_8$ RI. The lease has the following capitalized costs and reserve data as of January 1, 2005.

Unrecovered costs	$600,000
Proved developed reserves	800,000 bbl

On January 1, 2005, Bingo Oil Company carves out a $500,000 production payment to Capital Bank. The production payment is payable to Capital Bank out of 60% of the proceeds of the Bingo's share of production with interest of 10% on the unpaid balance. During 2005, production totaled 20,000 barrels of oil, production costs were $5 per barrel, and the selling price was $20 per barrel. Ignore production taxes and assume Bingo pays the RI owner.

a. Give all the entries made by Bingo Oil Company (a SE company) relating to the above lease and to account for the carved-out production payment during 2005.

b. Give all the entries made by Capital Bank to account for the production payment during 2005.

15. Zink Company owns 100% of the WI in a fully developed lease on which there is a 1/8 RI. The lease has the following capitalized costs and reserve data as of January 1, 2007.

Unrecovered costs	$500,000
Proved developed reserves	500,000 bbl

On January 1, 2007, Zink Company carves out a $400,000 production payment to Delta Company. The production payment is payable to Delta Company by delivery of 25,000 barrels out of the first 50 percent of Zink's share of production. During 2007, production totaled 16,000 barrels of oil, production costs totaled $50,000, and the average selling price was $20 per barrel. Ignore production taxes and assume Zink pays the RI owner.

a. Give all the entries made by Zink Company (a SE company) relating to the above lease and to account for the carved-out production payment during 2007.

b. Give all the entries made by Delta Company (a SE company) to account for the production payment during 2007.

16. In 2006, Beta Company purchased the WI of an unproved lease for $50,000. In 2007, Beta Company recognized impairment of $20,000 on this lease. In 2008, Beta Company sold the entire WI to Company Q for:

a. $25,000

b. $55,000

Assume instead that the property was assessed on a group basis and sold for:

c. $25,000

d. $55,000

Give the entry to record the sale in each of the above situations assuming that Beta Company is a successful-efforts company.

17. Mair Company sold 50% of the WIs in a proved lease to Company Q for $450,000. Mair Company's net cost basis in this proved property was $200,000. Mair Company uses the SE method. Give the entry for Mair Company to record the sale.

18. Four companies own adjacent leases that share a common reservoir. The companies each have 100% WIs in their respective leases in which they have the following investment:

Company	IDC	L&WE	Leasehold	Total
A	$200,000	$ 80,000	$ 30,000	$ 310,000
B	100,000	60,000	20,000	180,000
C	300,000	150,000	40,000	490,000
D	0	0	50,000	50,000
	$600,000	$290,000	$140,000	$1,030,000

The companies decide to unitize in order to obtain improved operating efficiency. Following negotiations by engineers, geologists, etc., the following participation factors and market values were agreed upon:

Company	IDC	L&WE	Total	Participation Factor
A	$230,000	$100,000	$330,000	10%
B	140,000	50,000	190,000	40%
C	260,000	180,000	440,000	20%
D	0	0	0	30%
	$630,000	$330,000	$960,000	100%

a. Determine equalization of investment and prepare entries for all the parties assuming all of the companies are SE companies.

b. Give the entries, assuming Company D is a full-cost company.

19. Zepher Company acquired 100% of the WI in an unproved property at a cost of $50,000. Zepher later sold the WI, retaining an overriding royalty interest (ORI). Give the entry to record the conveyance of the WI, assuming Zepher is a SE company and received $30,000.

20. Brown Company, a SE company, has a ⅛ royalty interest in an unproved property. Assuming Brown Company's net capitalized cost in the property is $100,000, give the entry to record the sale of the entire royalty interest for:

    a. $80,000

    b. $110,000

21. Greene Oil Company, a SE company, owns 100% of the WI in a 320 acre proved property with the following net, unamortized costs:

Leasehold	$ 60,000
IDC	200,000
Equipment	88,000

    Greene Oil Company sells 100% of the working interest, including the wells and equipment, on the western 160 acres of the property for $200,000. An appraisal is performed with the following results:

	FMV of Portion Sold	FMV of Portion Retained	Total FMV
Leasehold	$ 30,000	$ 80,000	$110,000
IDC	130,000	90,000	220,000
Equipment	40,000	10,000	50,000
Total	$200,000	$180,000	

    Give the entry to record the sale.

22. Wolfforth Company sold its 100% WI in a proved property for $600,000 and retained an ORI. Wolfforth's net cost basis in the property was $500,000. The fair market value of the entire original WI was $700,000.

    Wolfforth had the following net investment in its accounts:

Leasehold	$100,000
IDC	300,000
Equipment	100,000
Total	$500,000

    Prepare Wolfforth Company's entry to record the sale assuming Wolfforth is a successful-efforts company.

23. Flower Company owns a 100% WI in an unproved property for which it paid $80,000. The property is burdened with a ⅛ royalty. Flower Company agrees to farm-out the WI

to Barrel Company and retain a $\frac{1}{7}$ ORI in return for Barrel Company agreeing to drill, develop, and operate the property. During 2005, Barrel Company incurs costs of $200,000 in drilling a well and $100,000 to equip the well. The total proved reserves are estimated to be 300,000 barrels and proved developed reserves are 100,000 barrels. Production during 2005 totaled 20,000 barrels, which was sold for $20 per barrel. Ignore severance tax and assume Barrel Company pays the RI and ORI owners.

a. Give the entries that would be made by Flower Company assuming it uses the SE method.

b. Give the entries that would be made by Barrel Company assuming it uses the SE method.

24. French Company signed a lease agreement with Rita Mack covering 900 acres in Oklahoma. Ms. Mack received a bonus of $50,000 and a $\frac{1}{5}$ RI and French Company received 100% of the WI.

    On 1/1/2008, French Company enters into an agreement with Donald Oil Company wherein French Company assigns its WI and retains a $\frac{1}{4}$ ORI. Donald agrees to pay all of the cost to drill a well on the property. If the well is successful, Donald Oil Company will pay all of the operating costs and retain the net profit (after payment of the royalty, ORI, and operating expenses) until it has recovered the cost of drilling and completing the well. At that point French Company's ORI will revert to a 45% WI.

    During November 2008, Donald Oil Company drills Well No. 1 at a cost of $210,000. The well is successful. Estimated proved reserves total 400,000 barrels and proved developed reserves are 200,000 barrels. During 2009 and 2010, 15,000 barrels per year are produced and sold for $20 per barrel. Operating costs are $5 per barrel. Ignore severance tax and assume reserve estimates do not change. Donald pays the RI and ORI owners.

    Compute payout. Prepare all of the entries that would be made by both French Company and Donald Oil Company during 2008, 2009, and 2010 assuming both companies are SE companies.

25. Fielder Oil Company, a SE company, has an unproved lease for which it paid $150,000. The property was individually significant and individual impairment of $50,000 had been assessed. Make the journal entries to record Fielder Oil Company's conveyance of 50% of the WI in return for:

    a. $10,000 cash

    b. $160,000 cash

26. Bammel Oil Company, a SE Company, has an unproved lease for which it paid $50,000.

The property is not considered to be individually significant. Make the journal entries to record Bammel Oil Company's conveyance of 20% of the WI in return for:

a. $10,000 cash

b. $70,000 cash

27. Gamble Company, a full-cost company, has an unproved lease for which it paid $100,000. Give the entry to record the sale of the property, assuming Gamble Company sold the property for:

a. $80,000

b. $130,000

Give the entry to record the sale, assuming Gamble Company sold only 25% of the WI in the property and received:

c. $60,000

d. $110,000

28. Tiger Oil Company, a successful-efforts company, owns an ORI in an unproved property that cost $20,000. The ORI has not been impaired. Assume Tiger sold its entire interest in the ORI for the following amounts:

a. $15,000

b. $25,000

Prepare journal entries to record the sales under the different assumptions.

29. Higgins Company receives $200,000 from Garza Company in payment for the right to purchase natural gas in the future.

Prepare journal entries for each company assuming that they both use the successful-efforts method.

30. Hein Oil Company, a SE company, owns 100% of the WI in a proved property that has the following capitalized costs:

Proved Property	$240,000
IDC	600,000
Equipment	400,000
Total Accumulated DD&A	200,000

A $^1/_8$ royalty on the property is owned by Sammy Jones. Hein sells the WI for $1,200,000 and retains a PPI of $300,000. The PPI is payable from the proceeds from 30% of the

WI's share of production. The fair market value of the entire original WI is $1,500,000.

Prepare the journal entry by Hein to record the above transaction, assuming that payment of the PPI is not reasonably assured.

31. Carpenter Oil Company owns a 100% WI in a proved property in Wise County, Texas. Carpenter sells the WI to Knight Oil Company for $400,000 plus a retained PPI of $300,000 payable in cash from 60% of the WI's portion of the revenue. The capitalized cost of the WI owner's proved property is $800,000 and accumulated DD&A is $200,000. The present value of the PPI is $250,000 and is reasonably assured of payout.

    a. Prepare the entry to record the sale of the proved property, assuming that Carpenter uses the SE method of accounting.

    b. Prepare the entry for Carpenter, assuming that the FC method of accounting is being used.

32. Sells Oil Company, a FC company, has total capitalized costs in Venezuela of $20,000,000 and accumulated DD&A of $4,000,000. Proved reserves in Venezuela are 2,000,000 barrels. Reserves of 500,000 barrels are sold to Oyona Oil Company for $8,000,000.

    Prepare the entry for Sells Oil Company.

chapter thirteen

Oil *and* Gas Disclosures

This chapter is based on "A Comprehensive Look at *FASB Statement 69*," by Gallun and Pearson.[1]

Statement of Financial Accounting Standards No. 69, "Disclosures about Oil and Gas Producing Activities," requires both successful-efforts and full-cost companies to prepare a comprehensive set of disclosures dealing with historical-based and future-based information. Companies required to present this information are publicly traded companies with significant oil and gas producing activities that meet one or more of the following criteria as specified in *SFAS No. 69*, amended by *SFAS No. 131*:

a. Revenues from oil and gas producing activities (including both sales to unaffiliated customers and sales or transfers to the enterprise's other operations) are 10% or more of the combined revenues (sales to unaffiliated customers and sales or transfers to the enterprise's other operations) of all the enterprise's industry segments. An industry segment is a component of an enterprise engaged in providing a product or service or a group of related products or services primarily to external customers for a profit.

b. Results of operations for oil and gas producing activities, excluding the effect of income taxes, are 10% or more of the greater of:

1. the combined operating profit of all industry segments that recognized a profit
2. the combined operating loss of all industry segments that recognized a loss

c. The identifiable assets relating to oil and gas producing activities (including an allocated portion of assets used jointly with other operations) are 10% or more of the assets of the enterprise, excluding assets used exclusively for general corporate purposes.

REQUIRED DISCLOSURES

SFAS No. 69 requires publicly traded companies with significant oil and gas producing activities to disclose supplementary information in their annual financial statements related to the following items:

Historical

1. Proved reserve quantity information
2. Capitalized costs relating to oil and gas producing activities
3. Costs incurred for property acquisition, exploration, and development activities
4. Results of operations for oil and gas producing activities

Future

5. A standardized measure of discounted future net cash flows relating to proved oil and gas reserve quantities
6. Changes in the standardized measure of discounted future net cash flows relating to proved oil and gas reserve quantities

All six disclosures must be presented in the aggregate. Disclosures 1, 3, 4, and 5 must also be presented for each geographical area where the company has significant operations.

Disclosures 1 through 4, the historical-based disclosures, are relatively uncomplicated and require little explanation in addition to the example presented in this chapter. Note, however, that disclosures 1 and 3 are the same for both FC and SE companies, while disclosures 2 and 4 differ for FC and SE companies.

Disclosures 5 and 6 (the future-based disclosures) are much more complicated and require significantly more computations than the historical-based disclosures. Although precision is precluded by the very nature of future-based information, exact computational approaches (along with some simplifying alternative approaches) are given for the future-based disclosures. Approaches that yield *exact* answers are given rather than simpler (and possibly more practical approaches) because the authors feel an exact approach facilitates a better understanding of the required disclosures.

Disclosure 5 presents the standardized measure, which is a present value amount related to estimated future *net* revenues from estimated future production of proved oil and gas reserves. The standardized measure is calculated as follows:

Future cash inflows (year-end prices x estimated future production)	$XX,XXX
Future development and production costs (estimated costs to be incurred in developing and producing proved reserves, based on year-end costs)	(XX,XXX)
Future pretax net cash flows	$XX,XXX
Future income tax (year-end statutory tax rate applied to future pretax net cash flows giving effect to tax deductions, tax credits, and allowances)	(XX,XXX)
Future net cash flows	X,XXX
Discount (future net cash flows discounted at 10%)	(X,XXX)
Standardized measure of discounted future net cash flows	$ X,XXX

Note that the discount factor above is 10%. *SFAS No. 69* requires all present value calculations to be at a rate of 10% a year.

The following brief example, which calculates the nondiscounted and discounted value of estimated future revenue only, illustrates the basic approach used in determining the standardized measure.

EXAMPLE

Future Revenue at 12/31/XA

	20XB	20XC	20XD	Total
Future production, bbl	30,000	40,000	50,000	120,000
x Year-end price/bbl	$ 22	$ 22	$ 22	$ 22
Future revenue	660,000	880,000	1,100,000	2,640,000
x PV* factors at 10%	0.9091	0.8264	0.7513	
PV* (future revenue)	$600,006	$727,232	$ 826,430	$2,153,668

* present value

Disclosure 6 presents the significant sources of change to the standardized measure. As discussed previously, the standardized measure is determined for a year using year-end estimates and year-end prices and costs. The standardized measure for the following year would be determined for that year using new year-end quantity estimates and new year-end prices and costs. Thus, any changes in estimates, prices, and costs cause the standardized measure to change. The following sources of change must be reported separately if individually significant:

a. Net change in sales and transfer prices and in production costs related to future production
b. Changes in estimated future development costs
c. Sales and transfers of oil and gas produced during the period
d. Net change due to extensions, discoveries, and improved recovery
e. Net change due to purchases and sales of minerals in place
f. Net change due to revisions in quantity estimates
g. Previously estimated development costs incurred during the period
h. Accretion of discount
i. Net change in income taxes
j. Other, unspecified

Some of these changes may be easily visualized and understood by referring to the above example. If the selling price of oil increased from $22 at the end of 20XA to $26 at the end of 20XB, there would be a change in the standardized measure due to a change in prices and costs of $4 per barrel. If total estimated production at the end of 20XB were 95,000 barrels instead of 90,000, there would be a change due to revision of quantity. If the total barrels to be produced stayed the same but the timing of production was changed—25,000 bbl in 20XC and 65,000 bbl in 20XD—there would be a change due to the change in production timing. This last change is typically included in the "Other" category.

An important source of change is accretion of discount. Accretion of discount is a change inherent in the concept of present value. If, for instance, the preceding example were redone at the end of 20XB, assuming no changes except that a year had passed, the cash flows would be one year closer. Therefore the 20XB values would be multiplied by 1, 20XC values by 0.9091, and 20XD values by 0.8264. Accretion of discount is calculated by multiplying the discount rate of 10% times the beginning-of-the-year standardized measure. A beginning standardized measure at 12/31/XA, plus or minus all changes to the standardized measure, should equal the ending standardized measure at 12/31/XB. The effect of adding accretion of discount to the beginning standardized measure, which is discounted to a present value as of 12/31/XA, is to restate the beginning measure to a present value as of 12/31/XB. The ending standardized measure is already at a present value as of 12/31/XB; therefore, to avoid adding and subtracting *mixed dollars*, all other changes should be discounted to 12/31/XB, as diagramed below:

Beginning standardized measure, PV at 12/31/XA	$XXXX
Accretion of discount	XXX
Beginning standardized measure, restated, PV at 12/31/XB	XXXX
Plus or minus all other changes, PV at 12/31/XB	XXX
Ending standardized measure, PV at 12/31/XB	$XXXX

Another source of change is designated as "Other" and is the plug amount necessary to account for the total change in the standardized measure. "Other" consists of changes not individually identified and errors that are a result of simplifications in assumptions or computations. Simplifications may be made in practice because the increased precision obtained without simplifications may not be felt in many cases to warrant the additional computations and time that may be required.

Case Study: Lucky Oil Company

Lucky Oil Company, a successful-efforts company, began operations early in 20XA with the acquisition of three unproved leases; Leases A, B, and C. Each of the leases is burdened with a $^1/_5$ royalty interest. During 20XA, Lucky incurred G&G costs and began drilling an exploratory well on each of the three leases. The well on Lease A was successfully completed, but no oil was produced and sold during 20XA. The well on Lease B was in progress at year-end, and the well on Lease C was a dry hole. No equipment was salvaged.

During 20XB, Lucky successfully completed the well on Lease B and produced and sold oil from both Lease A and Lease B. All production was sold in the year produced. All reserves on Leases A and B were fully developed at the end of 20XB. Lucky plans to further explore Lease C and paid a delay rental on Lease C. The facts representing Lucky Company's first two years of operations, which occurred solely in the United States, follow. All reserve quantities and production data given apply only to Lucky's interest.

Remember that the working interest's (WI) revenue is based on the WI's share of total production, in this case gross barrels x $^4/_5$. However, since the WI pays 100% of the costs, production costs are based on total gross barrels. Thus, if production costs are given as a dollar amount per barrel, total production costs would normally be gross barrels times the production costs. In this chapter, in order to simplify and make more understandable some of the calculations, production costs per WI barrel are given. Production costs per WI barrel is calculated by dividing total production costs by the WI's share of the barrels produced.

EXAMPLE

12/31/20XA

	Item	Lease A	Lease B	Lease C
a.	Acquisition costs	$ 30,000	$ 20,000	$ 40,000
b.	G&G costs	$ 50,000	$ 40,000	$ 20,000
c.	Drilling costs: IDC Tangible	 $200,000 90,000	 $100,000	 $120,000 10,000
d.	Drilling results	Proved reserves	Uncompleted	Dry
e.	Estimated production of estimated proved reserves, bbl Total proved reserves	20XB– 30,000 20XC– 40,000 20XD– 50,000 120,000		
f.	Reserve estimate, bbl 12/31/XA proved, at 12/31/ (decreases by estimated production) proved developed at 12/31/ (decreases by estimated production and increases as a result of estimated development)	 20XA–120,000 20XB– 90,000 20XC– 50,000 20XA– 55,000 20XB– 60,000 20XC– 50,000		
g.	Estimated tangible development costs	20XB–$45,000 20XC–$25,000		
h.	Estimated dismantlement costs	20XD–$15,000		
i.	Current market price of oil	$22/bbl		

Item	Lease A	Lease B	Lease C
j. Estimated total future production costs based on year-end costs	20XB–$150,000 20XC– 200,000 20XD– 250,000		
k. Estimated current production costs per WI barrel	$5/bbl		
	12/31/20XB		
a. Drilling costs Tangible		$50,000	
b. Drilling results		proved reserves	
c. Actual and estimated production, bbl	20XB– 32,000 (actual) 20XC– 43,000 (revised) 20XD– 48,000 (revised)	20XB– 35,000 (actual) 20XC– 50,000 estimated	
d. Actual and estimated tangible development costs	20XB– $46,000 (actual) 20XC– $0 (revised)	20XB–$50,000 (actual)	
e. Estimated dismantlement costs	20XD–$15,000		
f. Market price of oil sold	$25/bbl	$25/bbl	
g. Current market price of oil at end of year	$26/bbl	$26/bbl	
h. Total production cost of oil sold	$192,000	$210,000	

	Item	Lease A	Lease B	Lease C
i.	Production cost per WI barrel of oil sold	$6/bbl	$6/bbl	
j.	Estimated total future production costs at year-end	20XC–$258,000 20XD– 288,000	20XC–$300,000	
k.	Current production cost per WI barrel at year end	$6/bbl	$6/bbl	
l.	Delay rental			$ 5,000

Assume a tax rate of 40% and that Lucky does not qualify for percentage depletion because it is an integrated producer. For purposes of the required capitalization and amortization of 30% of IDC, assume nine months of amortization in 20XA. Because of the relatively short lives of the properties, also assume Lucky elects for tax purposes to use the unit-of-production method for calculating depreciation. Lucky uses proved reserves for calculating tax depletion and proved developed reserves for tax depreciation. In the case solution, the alternative minimum tax is ignored. Deferred income taxes are also ignored. Instead, it is assumed that income tax expense is based on pretax financial accounting income multiplied by the tax rate. Further, it is assumed that carryforwards are permitted when it is likely that the carryforward benefit will be realized.

CASE SOLUTION*

Disclosures for 20XA and 20XB are presented together for each disclosure given. The reader may find going through one year at a time more meaningful. Disclosures are presented assuming that Lucky is using:

- the SE method of accounting with a property as the cost center
- the FC method of accounting, with all possible costs included in DD&A

* Note to the instructor: If, because of time constraints, part of this chapter is deleted, the authors suggest that

the computation of future income tax be deleted. This deletion results in significantly fewer computations but still retains the basic concepts underlying the future based disclosures. Two homework problems are included in which the computation of future income tax is not required.

Historical-based disclosures

Disclosures 1 through 4 follow. Explanations are given in notes to the disclosures. Disclosure 1 reports only reserve *quantity* information. No costs, either capitalized or expensed, are reported. Consequently, this disclosure would be the same regardless of whether Lucky uses the successful-efforts method of accounting or the full-cost method of accounting.

DISCLOSURE 1

Reserve Quantity Information
for the Years Ended December 31, 20XB and 20XA

	20XB	20XA
Proved Reserves:		
Beginning of year	120,000	0
Revisions of previous estimates.	3,000[a]	0
Improved recovery	0	0
Extensions and discoveries	85,000	120,000
Production	(67,000)	0
Purchases of minerals in place	0	0
Sales of minerals in place	(0)	(0)
End of year	141,000	120,000
Proved developed reserves:		
Beginning of year	55,000	0
End of year	141,000	55,000
Quantities applicable to long-term supply agreements in which the company acts as producer or operator:		
Proved reserves—		
end of year	X	X
Received during the year	X	X
Equity in proved reserves of equity investees	X	X

Notes:

20XB

a. Estimate at 12/31/XB of previously discovered reserves in place 12/31/XB, Lease A	91,000 bbl
Plus: 20XB production of previously discovered reserves	32,000
Estimate of previously discovered reserves in place on 12/31/XA, made 12/31/XB	123,000
Less: Estimate of reserves, made 12/31/XA of reserves in place 12/31/XA	120,000
Revisions of previous estimates, increase	3,000 bbl

Note that since the estimate made in 20XA involved Lease A only, calculation of the revision to that previous estimate can only involve Lease A.

The components of the reserve quantity disclosure illustrated above are defined in *SFAS No. 69*, paragraph 11 as follows:

a. ***Revisions of previous estimates.*** *Revisions represent changes in previous estimates of proved reserves, either upward or downward, resulting from new information (except for an increase in proved acreage) normally obtained from development drilling and production history or resulting from a change in economic factors.*

b. ***Improved recovery.*** *Changes in reserve estimates resulting from application of improved recovery techniques shall be shown separately, if significant. If not significant, such changes shall be included in revisions of previous estimates.*

c. ***Purchases of minerals in place.***

d. ***Extensions and discoveries.*** *Additions to proved reserves that result from (1) extension of the proved acreage of previously discovered (old) reservoirs through additional drilling in periods subsequent to discovery and (2) discovery of new fields with proved reserves or of new reservoirs of proved reserves in old fields.*

If a company has both oil reserves and gas reserves and the quantities of each mineral are significant, the reserve quantity information should be reported separately. The reserve information should not be reported based on equivalent units.

Reserves included should be those related to properties in which the company has both working interests and nonworking interests, if the information is available. Reserve quantities for both working interests and nonworking interests owned should be included in the disclosure *net*. Reserves not owned by the company should never be included, *i.e.*, the reserves relating to interests of owners of other economic interests should not be included. For example, assume a company has the following interests:

- 70% of the working interest in a property with a $\frac{1}{8}$ royalty interest and 10,000 total gross barrels of oil reserves
- a $\frac{1}{5}$ royalty interest in a property with 15,000 total gross barrels of oil reserves

The net reserves would be computed as follows:

Property 1:	70% x $\frac{7}{8}$ x 10,000	=	6,125 barrels
Property 2:	$\frac{1}{5}$ x 15,000	=	3,000 barrels
Net reserves		=	9,125 barrels

Disclosure 2 presents capitalized costs at year-end. Unproved property costs, if significant, must be separately disclosed. Costs of support equipment and facilities may be disclosed either separately or included with the capitalized costs of proved and unproved properties. Disclosure 2a presents the capitalized costs, assuming Lucky is a successful-efforts company.

DISCLOSURE 2a

Capitalized Costs Relating to Oil and Gas Producing Activities at December 31, 20XB and 20XA

	20XB	20XA
Capitalized costs:		
Unproved oil and gas properties.	$ 40,000	$160,000[a]
Proved oil and gas properties	586,000[d]	320,000[b]
Total capitalized costs	626,000	480,000
Less: Accumulated DD&A	(189,710)[e]	(0)[c]
Net capitalized costs	$436,290	$480,000
Enterprise's share of equity method investees' net capitalized costs	X	X

Notes:

20XA	Lease B	Lease C	Total
a. Acquisition costs	$ 20,000	$ 40,000	$ 60,000
Wells in progress	100,000		100,000
Total	$120,000	$ 40,000	$160,000

	Lease A
b. Acquisition costs	$ 30,000
Wells and related E&F	290,000
Total	$320,000

c. No production during 20XA.

20XB	Lease B	Lease C	Total
d. Acquisition costs	$ 30,000	$ 20,000	$ 50,000
Wells and related E&F	290,000	150,000	440,000
Development costs	46,000	50,000	96,000
Total	$366,000	$220,000	$586,000

e. All reserves on both Lease A and Lease B are fully developed at 12/31/XB; all production was from Lease A and B.

Lease A: Costs to be amortized at 12/31/XB include acquisition costs, costs of drilling the well, development costs, and future restoration costs. Remember costs of dismantlement, restoration, and abandonment must be included in DD&A computations by both SE and FC companies. Proved reserves at 12/31/XB are 43,000 plus 48,000 barrels (see letter c of problem statement).

Lease B: Costs to be amortized include acquisition costs, cost of drilling the well, and development costs. Proved reserves at 12/31/XB are 50,000.

Lease A:

$$\frac{\$30,000 + \$290,000 + \$46,000 + \$15,000}{91,000 + 32,000} \times 32,000 \text{ bbl} = \$\ 99,122$$

Lease B:

$$\frac{\$20,000 + \$150,000 + \$50,000}{50,000 + 35,000} \times 35,000 \text{ bbl} = \underline{90,588}$$

$$\underline{\underline{\$189,710}}$$

Disclosure 2b presents the capitalized costs, assuming Lucky is a full-cost company and assuming Lucky amortizes all possible costs.

DISCLOSURE 2b

Capitalized Costs Relating to Oil and Gas Producing Activities at December 31, 20XB and 20XA

	20XB	20XA
Capitalized costs:		
Unproved oil and gas properties	$195,000[d]	$350,000[a]
Proved oil and gas properties	676,000[e]	370,000[b]
Total capitalized costs	871,000	720,000
Less: Accumulated DD&A	(285,394)[f]	(0)[c]
Net capitalized costs	$585,606	$720,000
Enterprise's share of equity method investees' net capitalized costs	X	X

Notes:

20XA	Lease B	Lease C	Total
a. Acquisition costs	$ 20,000	$ 40,000	$ 60,000
G&G costs	40,000	20,000	60,000
Drilling costs	100,000	130,000	230,000
Total	$160,000	$190,000	$350,000

	Lease A
b. Acquisition costs	$ 30,000
G&G costs	50,000
Wells and related E&F	290,000
Total	$370,000

c. No production during 20XA.

	Lease C
d. Acquisition costs	$ 40,000
G&G costs	20,000
Drilling costs	130,000
Delay rental	5,000
Total	$195,000

20XB	Lease A	Lease B	Total
e. Acquisition costs	$ 30,000	$ 20,000	$ 50,000
G&G costs	50,000	40,000	90,000
Wells and related E&F	290,000	150,000	440,000
Development costs	46,000	50,000	96,000
Total	$416,000	$260,000	$676,000

f. Reserves are fully developed on 12/31/XB (all capitalized costs relating to all three leases, plus dismantlement costs, are amortized).

$$\frac{\$871{,}000 + \$15{,}000}{141{,}000 + 67{,}000} \times 67{,}000 = \underline{\underline{\$285{,}394}}$$

Disclosure 3 below presents the acquisition, exploration, and development costs incurred during the year, regardless of whether the costs are capitalized or expensed. Consequently, the disclosure would be the same if Lucky were a successful-efforts company or a full-cost company. Disclosure of production costs is not reported in this disclosure; instead, it is reported in the disclosure of results of operations.

DISCLOSURE 3

Costs Incurred in Property Acquisition, Exploration and Development Activities for the Years Ended December 31, 20XB and 20XA

	20XB	20XA
Costs incurred:		
Acquisition of proved properties	$ 0	$ 0
Acquisition of unproved properties	0	90,000[a]
Exploration	55,000[c]	630,000[b]
Development	96,000[d]	0
Total costs incurred	$151,000	$720,000

Enterprise's share of equity method investees' costs of acquisition, exploration, and development	X	X

Notes:

20XA	**Lease A**	**Lease B**	**Lease C**	**Total**
a. Acquisition costs	$ 30,000	$ 20,000	$ 40,000	$ 90,000

	Lease A	**Lease B**	**Lease C**	**Total**
b. G&G costs	$ 50,000	$ 40,000	$ 20,000	$110,000
Exploratory drilling	290,000	100,000	130,000	520,000
Total	$340,000	$140,000	$150,000	$630,000

20XB	**Lease B**	**Lease C**	**Total**
c. Exploration drilling	$ 50,000		$ 50,000
Delay rental		$ 5,000	5,000
Total	$ 50,000	$ 5,000	$ 55,000

	Lease A	**Lease B**	**Total**
d. Development costs	$ 46,000	$ 50,000	$ 96,000

Note that this disclosure reports costs incurred only during the year and classifies the costs according to how they were accounted for when incurred. For example in 20XA, although Lease A is proved by year-end, at the time of acquisition Lease A was unproved. Therefore, the cost of acquisition is reported as an acquisition of unproved property cost. In 20XB, although Lucky now has both unproved and proved properties on its books, no costs of acquisition are reported because the properties were acquired in the previous year, not in 20XB.

Disclosure 4 presents results of operations for Lucky Company. Since some costs that are expensed under SE are capitalized under FC accounting, this disclosure would be different if Lucky were a successful-effort company versus a full-cost company. Disclosure 4a below presents the results of operations assuming that the accounting method being used is successful-efforts.

DISCLOSURE 4a

Results of Operations from Oil and Gas Producing Activities for the Years Ended December 31, 20XB and 20XA

	20XB	20XA
Revenues from oil and gas producing activities		
Sales to unaffiliated parties	$1,675,000[c]	$ 0
Transfers to affiliated entities	0	0
Revenues	1,675,000	(0)
Production (lifting) costs	(402,000)[d]	(0)
Exploration expenses	(5,000)[e]	(240,000)[a]
Depreciation, depletion, amortization, and valuation provisions	(189,710)[f]	(0)
Pretax income from producing activities	1,078,290	(240,000)
Income tax expenses/estimated loss carry forward benefit	(431,316)[g]	96,000[b]
Results of oil and gas producing activities (excluding corporate overhead and interest costs)	$ 646,974	$(144,000)
Enterprise's share of equity method investees' results of operations	X	X

Notes:

20XA	Lease A	Lease B	Lease C	Total
a. G&G costs	$50,000	$40,000	$ 20,000	$110,000
Dry exploratory hole			130,000	130,000
Total	$50,000	$40,000	$150,000	$240,000

b. Pretax loss	$(240,000)
Tax rate	40%
Estimated loss carryforward benefit	$ 96,000*

* The estimated loss carryforward benefit is recognized in the current period because realization is more likely than not. It is assumed that adequate reserves by Lucky Company provide evidence that this benefit will be realized. With respect to deferred taxes (which are being ignored in this chapter), the $240,000 is a timing difference that generates a future tax benefit. The $240,000 may be considered analogous to a future deductible amount. A deferred tax asset of $96,000 would be recognized, and since a loss situation exists, income tax expense would be credited by the same amount.

20XB

c. $1,675,000 = 32,000 bbl x $25 + 35,000 bbl x $25

d. $192,000 + $210,000 (production costs include severance taxes)

e. $5,000 = delay rental

f. See *note e* to Disclosure 2a

g.

Pretax income	$1,078,290
Tax rate	40%
Income tax expense	$ 431,316

Disclosure 4b presents the results of operations for Lucky Company, assuming the accounting method being used is full cost. Note that Lucky does not report any exploration expenses, because companies using full-cost accounting generally capitalize all exploration costs incurred.

DISCLOSURE 4b

Results of Operations from Oil and Gas Producing Activities for the Years Ended December 31, 20XB and 20XA

	20XB	20XA
Revenues from oil and gas producing activities		
Sales to unaffiliated parties	$1,675,000[b]	$ 0
Transfers to affiliated entities	0	0
Revenues	1,675,000	(0)
Production (lifting) costs	(402,000)[c]	(0)
Exploration expenses	(0)	(0)
Depreciation, depletion, amortization, and valuation provisions	(285,394)[d]	(0)
Pretax income from producing activities	987,606	(0)
Income tax expenses/estimated loss carry-forward benefit	(395,042)[e]	0[a]
Results of oil and gas producing activities (excluding corporate overhead and interest costs)	$ 592,564	$ 0

	20XB	20XA
Enterprise's share of equity method investees' results of operations	X	X

Notes:

20XA

a.	Pretax income	$ 0
	Tax rate	40%
	Current income tax	$ 0

20XB

b. $1,675,000 = 32,000 bbl x $25 + 35,000 bbl x $25

c. $192,000 + $210,000 (production costs include severance taxes)

d. See *note f* to Disclosure 2b

e.	Pretax income	$ 987,606
	Tax rate	40%
	Current income tax	$ 395,042

The results of operations for oil and gas producing activities are defined by *SFAS No. 69* as revenues less production costs, exploration expenses, DD&A expense, valuation provisions—such as impairment expense for SE companies or ceiling-test write-downs for FC companies—and income tax expenses. General corporate overhead and interest costs should not be deducted in computing the results of operations. However, some expenses incurred at a company's central administrative office may not be general corporate expenses, but rather may be operating expenses of oil and gas producing activities, and should be reported as such. The nature of an expense rather than the location where incurred should determine whether it is an operating expense. Only those expenses identified by their nature as operating expenses should be allocated as operating expenses in computing the results of operations for oil and gas producing activities.

Future value-based disclosures

Disclosures 5 and 6 and the supporting computations are given in the following sections.

Disclosure 5

Disclosure 5 presents the standardized measure for 20XA and 20XB. The standardized measure (SM) is the present value of estimated future net revenues from future production of proved reserves. This disclosure requires significant computations and explanations. Supporting computations for amounts shown for 20XA are presented in Tables 13–1 and 13–2. The supporting computations for the 20XB amounts in Disclosure 5 are presented in Tables 13–3 and 13–4. If the computations for future income tax were being ignored, the present value of future net inflows before income tax would be calculated instead of the standardized measure.

DISCLOSURE 5

**Standardized Measure of Discounted Future Net Cash Flows
Relating to Proved Oil and Gas Reserves
at December 31, 20XB and 20XA**

	20XB	20XA
Future cash inflows	$3,666,000	$2,640,000
Future costs		
Production	(846,000)	(600,000)
Transfers to affiliated entities	(0)	(70,000)
Dismantlement and abandonment	(15,000)	(15,000)
Future net inflows before income tax	2,805,000	1,955,000
*Future income taxes	(1,006,031)	(551,000)
*Future net cash flows	1,798,969	1,404,000
*10% discount factor	(213,801)	(248,797)
*Standardized measure of discounted net cash flows	$1,585,168	$1,155,203
Enterprise's share of equity method investees' standardized measure of discounted future net cash flows relating to proved oil and gas reserves	X	X
* Alternate: PV (Future cash inflows before income tax)	$1,562,774	$1,591,360

The components of Disclosure 5 are defined by *SFAS No. 69* as follows:

a. **Future cash inflows.** These are to be computed by applying year-end prices of oil and gas relating to the enterprise's proved reserves to the year-end quantities of those reserves. Future price changes should be considered only to the extent provided by contractual arrangements in existence at year-end.

b. **Future development and production costs.** These costs are to be computed by estimating the expenditures to be incurred in developing and producing the proved oil and gas reserves at the end of the year, based on year-end costs and assuming continuation of existing economic conditions. If estimated development expenditures are significant, they are to be presented separately from estimated production costs.

c. **Future income tax expense.*** These expenses are to be computed by applying the appropriate year-end statutory tax rates, with consideration of future tax rates already legislated, to the future pretax net cash flows relating to the enterprise's proved oil and gas reserves, less the tax basis of the properties involved. The future income tax expenses should give effect to tax deductions and tax credits and allowances relating to the enterprise's proved oil and gas reserves.

d. **Future net cash flows.** These amounts are the result of subtracting future development and production costs and future income tax expenses from future cash inflows.

e. **Discount.** This amount is to be derived from using a discount rate of 10% a year to reflect the timing of the future net cash flows relating to proved oil and gas reserves.

f. **Standardized measure of discounted future net cash flows.** This amount is the future net cash flows less the computed discount.

---

* This disclosure deals with future *cash flows.* Therefore, although the term *income tax expense* is used, what is effectively meant is *future income tax payable.*

Table 13–1 — *Schedule of Estimated Future Cash Flows 12/31/XA*

	20XB	20XC	20XD	Total
a. **Undiscounted Values**:				
Future production, bbl (given)	30,000	40,000	50,000	120,000
Future revenue (production x $22)	$660,000	$880,000	$1,100,000	$2,640,000
Future production costs (given)	(150,000)	(200,000)	(250,000)	(600,000)
Future development costs (given)	(45,000)	(25,000)	(0)	(70,000)
Future dismantlement, restoration, and abandonment costs (given)	(0)	(0)	(15,000)	(15,000)
Future cash inflows before income tax	465,000	655,000	835,000	1,955,000
*Future income tax (Table 13–2)	(29,400)	(233,689)	(287,911)	(551,000)
*Future net cash inflows	$435,600	$421,311	$ 547,089	$1,404,000

	20XB	20XC	20XD	Total
b. **Present Values:** (PV factor x above values)				
PV factors	0.9091	0.8264	0.7513	
PV (future revenue)	$600,006	$727,232	$ 826,430	$2,153,668
PV (future production costs)	(136,365)	(165,280)	(187,825)	(489,470)
PV (future development costs)	(40,909)	(20,660)	(0)	(61,569)
PV (future dismantlement, restoration, and abandonment costs)	(0)	(0)	(11,269)	(11,269)
PV (future cash inflows before income tax)	422,732	541,292	627,336	1,591,360
*PV (future income tax)	(26,728)	(193,121)	(216,308)	(436,157)
*PV (future net cash inflows)	$396,004	$348,171	$ 411,028	$1,155,203
*Discount (future net cash inflow – PV (future net cash inflows))				$ 248,797

* Do not calculate if ignoring future income tax calculations.

In Disclosure 5 and throughout this example, the cash flows were assumed to occur at the end of each year, as indicated by the PV factors. For example, at the end of 20XA the present value of the revenue expected in 20XB was calculated using a present value factor at a 10% interest rate and one year. It would be reasonable to assume that the cash flows occurred evenly throughout the year, and therefore, to use a mid-year PV factor.

Table 13–2 — *Computation of Future Income Taxes 12/31/XA***

	20XB	20XC	20XD	Total
Future revenue, net of production costs (Table 13–1a)	$510,000	$680,000	$850,000	$2,040,000
IDC expensed	(0)	(0)	(0)	(0)
Amortization of IDC (Table 13–2b)	(18,000)	(18,000)	(18,000)	(54,000)
Depreciation (Table 13–2c)	(45,000)	(51,111)	(63,889)	(160,000)
Depletion (Table 13–2d)	(20,000)	(26,667)	(33,333)	(80,000)
Abandonment costs write-off	(0)	(0)	(15,000)	(15,000)
Loss carryforward (Table 13–2a)	(353,500)	(0)	(0)	(353,500)
Taxable income	73,500	584,222	719,778	1,377,500
Tax rate	40%	40%	40%	40%
Gross tax	29,400	233,689	287,911	551,000
Credits	(0)	(0)	(0)	(0)
Net Tax	$ 29,400	$233,689	$287,911	$ 551,000

a. Actual Income Tax, 12/31/XA

Revenue	$ 0
Production	(0)
IDC and dry hole	(340,000)*
Amortization of IDC	(13,500)†
Depreciation expense	(0)
Depletion expense	(0)
Taxable income (loss)	$(353,500)
Tax rate	40%
Net tax/carryforward	$(141,400)

* ($300,000 x 70% + $120,000 + $10,000)

† ($300,000 x 30% x 9/60)

b. Computation of Future Amortization of IDC, 12/31/XA

$300,000 x 30% x $^{12}/_{60}$ = $18,000

c. Computation of Future Depreciation, 12/31/XA: Computed using the unit-of-production method, proved developed reserves, and costs incurred to date plus future estimated tangible costs.

** Do not calculate if ignoring future income tax calculations.

Year	Rate*		Estimated Production		Depreciation
20XB	1.50000	x	30,000	=	$45,000
20XC	1.27778	x	40,000	=	51,111
20XD	1.27778	x	50,000	=	63,889

* Depreciation Rates:

20XB $$\frac{\$90{,}000 + \$45{,}000}{60{,}000 + 30{,}000} = \frac{\$135{,}000}{90{,}000} = 1.50000$$

20XC $$\frac{\$135{,}000 + \$25{,}000 - \$45{,}000}{50{,}000 + 40{,}000} = \frac{\$115{,}000}{90{,}000} = 1.27778$$

20XD $$\frac{\$115{,}000 - \$51{,}111}{50{,}000} = \frac{\$63{,}889}{50{,}000} = 1.27778$$

d. Computation of Future Depletion, 12/31/XA: Based on proved reserves and assumed all future production is sold in year produced.

Year	Rate*		Estimated Production		Depletion
20XB	0.66667	x	30,000	=	$20,000
20XC	0.66667	x	40,000	=	26,667
20XD	0.66667	x	50,000	=	33,333

* Tax basis and depletion rate:

Tax basis of Lease A: acquisition costs	$30,000
G&G costs	50,000
	$80,000

$$\text{Depletion rate} = \frac{\$80{,}000}{120{,}000} = 0.66667$$

Note that since costs and reserve estimates do not change, if a new rate were compute for 20XC and 20XD, it would be the same as the rate shown above, *i.e.*, 0.66667.

Table 13–3 presents the schedule of total estimated future cash flows at 12/31/XB. The cash flows are broken down to the flows relating to reserves discovered in prior years (Lease A), reserves discovered in the current year (Lease B), and total reserves.

Table 13–3 ***Schedule of Estimated Future Cash Inflows 12/31/XB***

	Relating to Reserves Proved in Prior Year		
	20XC	**20XD**	**Total**
A. **Undiscounted Values:**			
Future production (barrels)(given)	43,000	48,000	91,000
Future revenue (production x $26)	$1,118,000	$1,248,000	$2,366,000
Future production costs (given)	(258,000)	(288,000)	(546,000)
Future development costs (given)	(0)	(0)	(0)
Future dismantlement, restoration, and abandonment costs (given)	(0)	(15,000)	(15,000)
Future cash inflows before income tax	860,000	945,000	1,805,000
*Future income tax (Table 13–4)	—	—	—
*Future net cash inflows	—	—	—
B. **Present Values:** (PV factor x above values)			
PV factors	0.9091	0.8264	
PV (future revenue)	$1,016,374	$1,031,347	$2,047,721
PV (future production costs)	(234,548)	(238,003)	(472,551)
PV (future development costs)	(0)	(0)	(0)
PV (dismantlement, restoration, and abandonment costs)	(0)	(12,396)	(12,396)
PV (future cash inflows before income tax)	781,826	780,948	1,562,774
*PV (future income tax)	—	—	—
*PV (future net cash inflows)	—	—	—
* Discount (future net cash inflows – PV (future net cash inflows))			

* Do not calculate if ignoring future income tax calculations.

Relating to Reserves Proved in Current Year	Relating to Reserves Proved in Prior and Current Year		
20XC	**20XC**	**20XD**	**Total**
50,000	93,000	48,000	141,000
$1,300,000	$2,418,000	$1,248,000	$3,666,000
(300,000)	(558,000)	(288,000)	(846,000)
(0)	(0)	(0)	(0)
(0)	(0)	(15,000)	(15,000)
1,000,000	1,860,000	945,000	2,805,000
—	(668,948)	(337,083)	(1,006,031)
—	$1,191,052	$ 607,917	$1,798,969
0.9091	0.9091	0.8264	
$1,181,830	$2,198,204	$1,031,347	$3,229,551
(272,730)	(507,278)	(238,003)	(745,281)
(0)	(0)	(0)	(0)
(0)	(0)	(12,396)	(12,396)
909,100	1,690,926	780,948	2,471,874
—	(608,141)	(278,565)	(886,706)
—	$1,082,785	$ 502,383	$1,585,168
			$ 213,801

* Do not calculate if ignoring future income tax calculations.

Table 13–4 ***Computation of Future Income Taxes 12/31/XB*****

	20XC	20XD	Total
Future revenue, net of production costs (Table 13–3)	$1,860,000	$960,000	$2,820,000
Amortization of IDC	(18,000)	(18,000)	(36,000)
Depreciation (Table 13–4b)	(106,369)	(53,073)	(159,442)
Depletion (Table 13–4c)	(63,261)	(31,220)	(94,481)
Abandonment costs write-off	(0)	(15,000)	(15,000)
Taxable income	1,672,370	842,707	2,515,077
Tax rate	40%	40%	40%
Gross tax	668,948	337,083	1,006,031
Credits	(0)	(0)	(0)
Net tax	$ 668,948	$337,083	$1,006,031

a. Actual Income Tax, 12/31/XB

	Total
Revenue	$1,675,000
Production	(402,000)
IDC	(0)
Amortization of IDC	(18,000)
Depreciation expense	(76,559)*
Depletion expense	(45,519)†
Delay rental	(5,000)
Loss carryforward	(353,500)
Taxable income	774,422
Tax rate	40%
Net tax	$ 309,769

* Depreciation expense:

Lease A $\frac{\$90,000 + \$46,000}{91,000 + 32,000} \times 32,000 = \$35,382$

Lease B $\frac{\$50,000 + \$50,000}{50,000 + 35,000} \times 35,000 = \underline{41,176}$

$\underline{\underline{\$76,558}}$

† Depletion expense:

Lease A	$80,000/(91,000 + 32,000)	x	32,000	=	$20,813
Lease B	$60,000/(50,000 + 35,000)	x	35,000	=	24,706
					$45,519

** Note: Do not calculate if ignoring future income tax calculations.

b. Computation of Future Depreciation, 12/31/XB

	Year	Rate*		Estimated Production		Depreciation
Lease A	20XC	1.10569	x	43,000	=	$47,545
	20XD	1.10569	x	48,000	=	53,073
Lease B	20XC	1.17648	x	50,000	=	58,824

* Tax basis and depreciation rates:

Undepreciated tax basis of Lease A equipment, 12/31/20XC:

	Lease A	Lease B	
Tangible successful drilling	$ 90,000	$ 50,000	
Development costs	46,000	50,000	
Total	$136,000	$100,000	
Depreciation, 20XB actual	(35,382)	(41,176)	
Net tax basis	$100,618	$ 58,824	$159,442

Depreciation rates:

Lease A	20XC:	$100,618/(48,000 + 43,000)	= 1.10569
	20XD:	($100,618 – $47,545)/48,000	= 1.10569
Lease B	20XC:	$58,824/50,000	= 1.17648

c. Computation of Future Depletion, 12/31/XB

	Year	Rate*		Estimated Production		Depletion
Lease A	20XC	0.65041	x	43,000	=	$27,967
	20XD	0.65041	x	48,000	=	31,220
Lease B	20XC	0.70588	x	50,000	=	35,294

* Tax basis and depletion rates:

Undepleted tax basis of Lease A and B, 12/31/20XC:

	Lease A	Lease B	
Acquisition costs	$ 30,000	$ 20,000	
G&G Costs	50,000	40,000	
Depletion, 20XB actual .	(20,813)	(24,706)	
Net tax basis	$ 59,187	$ 35,294	$94,481

Depletion rates:

Lease A	$59,187/(43,000 + 48,000)	= 0.65041
Lease B	$35,294/(50,000)	= 0.70588

Disclosure 6

Changes in the standardized measure of discounted future net cash flows are presented in Disclosure 6. The disclosure is a reconciliation of the beginning standardized measure (SM) balance to the ending SM balance. Supporting computations and explanations to the disclosure are given in a schedule format. In two cases, an alternate simplified approach is presented in addition to an exact approach, because of the widespread use in practice of the simplified approach and the significant savings in calculation time. An analysis or proof of individual changes or grouped changes to the standardized measure based on the *exact* solution is also provided to show all changes to an item have been accounted for.

DISCLOSURE 6

Changes in Standardized Measure of Discounted Future Net Cash Flows from Proved Reserve Quantities for the Years Ended December 31, 20XB and 20XA

	20XB Exact Solution	**20XA Exact Solution**
Standardized measure, beginning of year*	$1,155,203	$ 0
Sales and transfers, net of production costs (Schedule A)	(1,273,000)	(0)
Net change in sales and transfer prices, net of production costs (Schedule B)	293,052	0
Extensions, discoveries, and improved recovery, net of future production and development cost (Schedule C)	1,524,100	1,591,360
Changes in estimated future development cost (Schedule D)	21,728	0
Development costs incurred during the period that reduced future development costs (Schedule E)	96,000	0
Revisions of quantity estimates (Schedule F)	59,490	0
Accretion of discount (Schedule G)	159,136	0
Net change in income taxes† (Schedule H)	(450,549)	(436,157)
Purchase of reserves in place	0	0
Sale of reserves in place	(0)	(0)
Other	8	0
Standardized measure, end of period*	$1,585,168	$1,155,203

* If ignoring future income tax calculations, use the PV (future net inflows before income tax) instead of the standardized measure.
† Do not calculate if ignoring future income tax calculations.

SFAS No. 69 specifies that the sources of change illustrated in the above disclosure, if individually significant, should be presented separately. The amounts shown in Disclosure 6 are supported by schedules and discussions presented in the following sections. In reviewing these schedules, remember that accretion of discount, which is one source of change, is added for 20XB to the beginning SM balance at 1/1/XB. The beginning SM is thus restated to a PV at 12/31/XB. All changes to the SM for 20XB are also discounted to 12/31/XB to avoid adding and subtracting mixed dollars.

Analysis of reasons for change in value of standardized measure 12/31/XB

In the discussion below and later in the chapter, net selling price per barrel is calculated. Net selling price is revenue minus production costs. Revenue accruing to the working interest owner is based on WI barrels only, *i.e.*, the working interest's share of production. Thus, in order to determine the net selling price per barrel, the production costs per barrel must also be based on WI barrels.

Sales and transfers, net of production costs. The production and sale of reserves decrease the amount and value of reserves in the ground. If the reserves had been discovered in a prior year, the production and sale of the reserves would also decrease the standardized measure since the SM is a valuation of reserves in the ground.

On 12/31/XA, it was estimated that 30,000 WI barrels would be sold during 20XB. Those barrels were valued net at $17 (selling price minus production costs per WI barrel: $22-$5) and discounted using a PV factor of 0.9091, resulting in a carrying value per barrel of $15.45 in the 12/31/XA standardized measure presented in Disclosure 5. The accretion of discount change increases the carrying value of $15.45 at 12/31/XA to $17 ($15.45 x 110%). If the estimate made on 12/31/XA had been accurate, *i.e.*, 30,000 barrels sold at a net price of $17, the sale of the reserves would have exactly canceled out the carrying value of those reserves in the 12/31/XA standardized measure, adjusted by accretion of discount. However, previously discovered reserves of 32,000 barrels were sold at a net price of $19 ($25-$6) per barrel. The actual barrels sold and the actual net price are used in Schedule A, "Sales and transfers." The differences between the previous estimates and the actual barrels sold and the actual selling price, are included in other schedules. Specifically, the difference between estimated and actual barrels produced is a revision of quantity or production timing, and the difference between estimated and actual net selling price is a change in prices and costs, as discussed later.

In addition, of the reserves discovered on Lease B in 20XB, 35,000 barrels were produced and sold during 20XB. Although these barrels are not included in either the beginning or ending SM, in practice they are included for informational purposes as an increase to proved reserves in Schedule C, "Extensions, discoveries, and improved recovery." Therefore, the barrels must also be included in Schedule A, "Sales and transfers." The effect of including the barrels in Schedule C is to increase the SM while the effect of inclusion in Schedule A is to decrease the SM. Thus, the effects of including the barrels cancel out as must be the case since the barrels discovered and produced in the same year did not affect either the beginning or ending SM.

Schedule A-20XB: Sales and transfers, net of production costs

Sales ([32,000 + 35,000] bbl x $25)	$1,675,000
Production costs ($192,000 + $210,000).	(402,000)
	$1,273,000

Sales and transfers for 20XA would be zero since there was no production or sales during 20XA.

Change in prices and costs. The following example, which illustrates the points made in the next two paragraphs, is a simple example (not related to the current problem) of the calculation of the effects of a change in prices (revenue) and costs (production costs) and a revision of quantity.

EXAMPLE

	Barrels	Net Price	Net Revenue
Future production, 12/31/XA	30,000	$5	$150,000
Future production, 12/31/XB (assume no production during 20XB)	20,000	4	80,000

Analysis: Reconciliation of beginning balance to ending balance

Beginning future net revenue (30,000 x $5)	$150,000	
Change in prices (30,000 x $1)	(30,000)	
	120,000	($4/bbl)
Revision in quantity (10,000 x $4)	(40,000)	
Ending future net revenue (20,000 x $4)	$ 80,000	

In order to determine the effect of a change in one variable, all other variables must be held constant. Thus, to isolate that portion of total change in future net revenues due to a change in prices and costs, the quantity of production is held constant. Therefore, the net change in prices and costs per barrel should be applied to the beginning-of-the-year estimate of future production. If the end-of-the-year production estimate is used, a change due to revision in quantity that is a separately identified change will be introduced into the changes in prices and costs calculation. This results in mis-stating the effect of a change in prices and costs as well as partially double-counting the effect of a revision in quantity.

The order of calculation of changes in the standardized measure is significant for certain calculations. As seen in the preceding example, the change in prices and costs was calculated first. The effect of this change was to restate all beginning barrels of future production from the beginning net price of $5/bbl to the year-end net price of $4/bbl. Revision in quantity, calculated next, was thus stated at the year-end net price of $4/bbl. *SFAS No. 69* requires that the effect of changes in prices and costs be computed before the effect of revisions in quantities so that revisions in quantities will be valued in year-end prices and costs.

In the case study of Lucky Oil Company, the change in prices and costs for 20XA is zero because Lucky is a new company and there were no estimates prior to 20XA. The calculation of the effect of changes in prices and costs for 20XB is more complicated than the preceding illustration. Lucky obtained and sold production during the year, and the net price of production sold during 20XB, $19/bbl (selling price minus production costs per WI barrel: $25-$6), differs from the year-end net price of $20/bbl ($26-$6). To get a precise measure of the effect of changes in prices and costs, the estimated 20XB production must be valued based on $19, not $20. Remember the effect of changes in prices and costs must be calculated before the effect of changes in quantities. Consequently, production estimates made as of 12/31/XA must be used instead of estimates made as of 12/31/XB.

In Version 1, in order to get an exact answer, the effects of the change are discounted on a year-by-year basis. In Version 2, the discount calculation is simplified by discounting the effects of the change by applying the average discount at year-end. The year-end average discount is used instead of the beginning of the year average discount because all changes, as discussed earlier, should be discounted to year-end dollars.

Schedule B-20XB: Net change in sales and transfer prices, net of production costs

Version 1 (discounted year-by-year, actual net price is used)

	20XB	**20XC**	**20XD**	**Total**
Future production, barrels				
(estimated 12/31/20XA)	30,000	40,000	50,000	120,000
x Net change per barrel	$ 2.00*	$ 3.00**	$ 3.00	
Net change................	60,000	120,000	150,000	$330,000
PV factor	1.00	0.9091	0.8264	
PV at 12/31/20XB				
(net change)	$60,000	$109,092	$123,960	$293,052

* 12/31/XA net price of $17 ($22-$5) vs. 20XB actual net price of $19 ($25-$6).

** 12/31/XA net price of $17 vs. 12/31/XB net price of $20 ($26-$6).

Version 2 (Same as Version 1, except discount calculation is simplified)

a. Calculation of reserves to which changes apply:
Estimate of reserves made 12/31/XA of reserves
in place 12/31/XA 120,000/bbl

b. Calculation of net change in price and cost factors:

For 20XB estimated production:	
Change in price ($25 actual – $22 beginning)	$ 3.00/bbl
Change in cost ($6 actual – $5 beginning)	1.00/bbl
Increase	$ 2.00/bbl
For all other reserves except 20XB estimated production:	
Change in price ($26 ending – $22 beginning)	$ 4.00/bbl
Change in cost ($6 ending – $5 beginning)	1.00/bbl
Increase	$ 3.00/bbl

c. Undiscounted net change due to change in price and cost:

30,000 bbl x $2	$ 60,000
90,000 bbl* x $3	270,000
	$330,000

* (120,000 bbl – 30,000 bbl)

d. Discounted net change due to change in price and cost:
(apply average discount to net change)

$$\frac{\text{PV (future revenues - future production costs) (12/31/XB) (Table 13–3B))}}{\text{(future revenues - future production costs) (12/31/XB) (Table 13–3A))}} \times \text{net change}$$

$$= \frac{\$3{,}229{,}551 - \$745{,}281}{\$3{,}666{,}000 - \$846{,}000} \times \$330{,}000 = \$290{,}712$$

If a company chooses not to perform the extra calculations involved in valuing 20XB estimated production at the actual net price for 20XB versus the year-end net price for 20XB, then production may be either (1) eliminated from the calculation with the change calculated on reserves remaining after production or (2) valued at the year-end net price.

The choice that yields the smallest error depends upon the facts of the situation, in particular whether the actual net price is closer in value to the beginning or year-end net price. If a decision is made to compute the change only on reserves after production, either actual or estimated production may be deducted from the beginning estimate. Again, the choice that yields the smallest error depends upon the facts of the situation.

Changes from extensions, discoveries, and improved recovery. Extension, discoveries, and improved recovery of proved reserves increase the amount and value of reserves in the ground and thus increase the standardized measure. The reserves are included in the SM net of production, development, and dismantlement costs. Calculation of the increase to the SM is relatively simple; however, the computation may be complicated slightly when part of the reserves discovered in a year are developed and/or produced in that same year.

In calculating the standardized measure at the end of the year, any development costs and production costs incurred and sales revenue earned during the current year are not included in the computation—those amounts are included in the historical cost data. If the development and production costs incurred and the sales revenue earned during the current year are related to reserves discovered during the same year, these costs and revenues would not have been estimated previously. These costs and revenues, therefore, do not affect either the beginning or the end-of-the year standardized measure.

Since there would be no change in the standardized measure associated with these items, the authors feel that it is more consistent and a truer reflection of the actual changes to ignore these costs and revenues in the calculation of changes to the standardized measure. However, because these costs and revenues are, in practice, being included in extensions, discoveries, and improved recovery, the authors have included these costs and revenue to be consistent with actual practice. Specifically, if the current-year development costs, production costs, and revenues that are related to current discoveries are included in the calculations of changes to the SM, then:

- the related costs and revenues must be included in valuing the change from extensions, discoveries, and improved recovery (Schedule C)
- the related development costs must be included as an increase to the change from development costs incurred during the period that reduced future development costs (Schedule E)
- the related sales revenue, net of production costs must be included in the change from sales and transfers (Schedule A)

These items would be included in the above schedules as follows:

Schedule C: + revenue – production costs – development costs
Schedule E: + development costs
Schedule A: – revenue + production costs

The net effect of including these items must be zero because these items do not affect either the beginning or ending SM. From above, it can be seen that the revenue and all of the costs cancel out and thus the net effect of including these items is zero.

Most of the actual calculations needed for Schedule C were done previously and can be found in Table 13–1 or Table 13–3. For 20XA refer to Table 13–1; for 20XB refer to that portion of Table 13–3 relating to reserves proved in the current year. The numbers from Table 13–3 relating to reserves proved in the current year however are only for the reserves still in the ground at 12/31/XB. In the case study, production and development costs were incurred and sales revenue was earned during 20XB relating to reserves discovered during 20XB. Therefore, to include the reserves on Lease B that were discovered and produced in 20XB, the numbers taken from Table 13–3 must be adjusted for the 35,000 barrels discovered and produced in the current year as shown below.

	From Table 13–3		20XB Production		
PV (future revenue)	$1,181,830	+	$875,000	=	$2,056,830
PV (future production costs)	(272,730)	+	(210,000)	=	(482,730)
PV (future development costs)	(0)	+	(50,000)	=	(50,000)
PV (future dismantlement and abandonment costs)	(0)	+	(0)	=	(0)

Schedule C-20XB, 20XA: Extensions, discoveries, and improved recovery, net of future production and development costs

	20XB	20XA
PV (future revenue)	$2,056,830	$2,153,668
PV (future production costs)	(482,730)	(489,470)
PV (future development costs)	(50,000)	(61,569)
PV (future dismantlement and abandonment costs)	(0)	(11,269)
	$1,524,100	$1,591,360

Note to the instructor: The present value amounts shown in Schedule C were calculated in Tables 13–1 and 13–3 as intermediate steps to determining the PV of the future net cash inflows. If an alternate Schedule C (for instructional purposes only) as follows is computed, the PV of the future net cash inflows can be determined more directly, resulting in substantial computational savings on the part of students.

PV (future revenue)	PV($1,300,000)	+	$875,000	=	PV($2,175,000)
PV (future production costs)	PV(300,000)	+	210,000	=	PV(510,000)
PV (future development costs)	(0)	+	50,000	=	PV(50,000)
PV (future dismantlement and abandonment costs)	(0)	+	0	=	0
	PV($1,000,000)	+	$615,000	=	PV($1,615,000)

	20XB	20XA
PV (future revenue)	PV($2,175,000)	PV($2,640,000)
PV (future production costs)	PV(510,000)	PV(600,000)
PV (future development costs)	(50,000)	PV(70,000)
PV (future dismantlement and abandonment costs)	(0)	PV(15,000)
	PV($1,615,000)	PV($1,955,000)

PV($1,955,000) = $1,591,360 (Table 3–1)
PV($1,615,000) = $909,100 (Table 3–3) + $615,000 = $1,524,000

Changes in estimated future development costs. Changes in future development costs (including dismantlement, restoration, and abandonment costs) result from three sources:

a. Revisions to previously estimated development costs, including revisions to those estimated for the current year
b. Future development costs associated with reserves discovered in current year
c. Development costs incurred in the current year, which were previously estimated and which reduced future development costs

Revisions to previously estimated development costs (source *a* above) arise from a change in the estimated amount of development activities or in the estimated cost of development activities. Source *b* is not included in the calculation of changes in estimated future development costs because it is included in changes from extensions, discoveries, and improved recovery (Schedule C). Source *c* also is not included in the calculation because it is presented as a separate category in the analysis of changes in the standardized measure (Schedule E).

The following schedule analyzes changes to estimated future development costs. The change in estimated future development costs is the present value of column *d.*

Schedule D-20XB: Changes in estimated future development costs (including dismantlement, restoration, and abandonment costs), Source *a* only

a	b	c	d	e	f
Year	Estimated as of 12/31/XA	Estimated as of 12/31/XB	Change In Estimates	PV Factors	PV of Change In Estimate
20XB	$45,000	$46,000	$ 1,000	1.0000	$ 1,000
20XC	25,000	0	(25,000)	0.9091	(22,728)
20XD	15,000	15,000	0	0.8264	0
	$85,000	$61,000	$(24,000)		$(21,728)*

* Increase to the standardized measure.

The $21,728 is a decrease in estimated future development *cost* and so is an increase in standardized measure.

Development costs incurred during the period that reduce future development costs. Development costs that were estimated previously and are incurred during the current period are no longer future costs; thus, they act to reduce estimated future development costs and increase the standardized measure. (Estimated future development costs are a deduction in the computation of the standardized measure.)

In the case study at 12/31/XA, future undiscounted development costs of $45,000 were estimated as of 12/31/XA to be incurred during 20XB. The standardized measure at 12/31/XA was reduced by $40,909, the present value of the development costs. Accretion of discount increases this value to $45,000. If the estimate made on 12/31/XA had been correct, then the estimated future development costs for 20XA would have been offset exactly by the actual development costs incurred, and the standardized measure would have increased by $45,000 from 20XA to 20XB with respect to this item.

In the case study, even though Lucky's actual development costs incurred during 20XB were $46,000, the net effect on the standardized measure is still an increase of $45,000. This is because only $45,000 had been estimated previously and deducted in arriving at the standardized measure. Thus, no matter what actual costs had been incurred—assuming the development activity itself had been performed—the standardized measure is increased by only the estimated development costs, not the actual cost incurred. Schedule E, "Development costs incurred during the period that reduced future development costs," reports the actual development costs incurred. The difference between the estimated and actual development costs, $1,000 in the case study, is included in the previous schedule, "Changes in estimated future development costs." In this case, the difference would be an increase to future development costs and thus a reduction to the standardized measure. Therefore, after including development costs incurred during the period ($46,000) and changes in estimated future development costs ($1,000), the standardized measure has been increased by $46,000 and reduced by $1,000 for a net increase of $45,000. This is shown as follows:

	Increase (Decrease) In Estimated Future Costs	**Increase (Decrease) In Standardized Measure**
Actual 20XB cost in excess of estimated 20XB costs (Schedule D)	$ 1,000	$ (1,000)
Development costs incurred during 20XB (Schedule E)	(46,000)	46,000
	$(45,000)	$45,000

As discussed earlier, revenues and costs from reserves both discovered and produced in 20XB are being included in Disclosure 6. Specifically the related development costs are included in Schedule C with a resulting decrease to the SM. To cancel that effect, the related development costs must also be included in Schedule E even though these costs had not been estimated previously and thus do not reduce future development costs. Schedule E's amounts must be increased in the amount of $50,000 for the development costs not previously estimated, which were incurred on Lease B in 20XB and were related to the reserves proved in 20XB.

Schedule E-20XB: Development costs incurred during the period that reduced future development costs

Development costs incurred during the period relating to reserves discovered in prior years actual dollars ($46,000 + $50,000)	$96,000

When trying to understand and account for the change in the standardized measure due to a particular item, it may be helpful to reconcile the change in that particular item or group of items from the beginning of the year to the end of the year. This reconciliation can serve as a proof, when a precise approach is used, that all changes to that item or group of items have been accounted for and accounted for correctly. Even when a precise approach is not used, the reconciliation can give an indication of whether all changes have been accounted for. Equally important, the reconciliation increases understanding of the standardized measure and changes in the standardized measure.

The following reconciliation analyzes the changes in development costs:

Analysis of Changes in Development Costs
(Including Dismantlement, Restoration, and Abandonment Costs)

PV of beginning estimate of future development costs including dismantlement, restoration, and abandonment costs ($61,569 + $11,269)(Table 13–1)	$ 72,838
Accretion of discount (10% x 72,838)	7,284
PV of changes in estimated future development costs and dismantlement, restoration, and abandonment costs, Schedule D	(21,728)
Development costs incurred during the period, which reduced future development costs (Schedule E) . .	(46,000)*
Other (round off error)	2
PV of ending estimate of future development costs including dismantlement, restoration, and abandonment costs ($0 + $12,396) (Table 13–3).	$ 12,396

* Since the $50,000 of development costs related to reserves both discovered and produced in 20XB was not previously estimated and is not included in either the beginning or ending estimate of future development costs, the $50,000 was omitted from the analysis.

Revision of quantity. The standardized measure and changes to the standardized measure are all reported as present values. Therefore, to obtain an exact answer, both the quantity and the timing of the revisions must be considered. In other words, to get an exact answer, the change in the standardized measure due to the change in the timing of production must be computed at the same time as the change due to the revisions of quantity. The alternative approach presented following the exact version does not compute the effect of a change in timing; the effect of a change in timing, then, would be included in the "Other" category.

Schedule F-20XB: Revisions of quantity estimates and changes in timing of production

Version 1	20XB	20XC	20XD	Total
Future production, bbl (estimated 12/31/XB)	32,000[a]	43,000	48,000	123,000
Future production, bbl (estimated 12/31/XA)	30,000	40,000	50,000	120,000
Revision, increase (decrease)	2,000	3,000	(2,000)	3,000
Price, net of production costs, 12/31/XB	$ 19[b]	$ 20[c]	$ 20	
Undiscounted change from quantity revisions	$38,000	$60,000	$(40,000)	$ 58,000
PV factor	1.000	0.9091	0.8264	
PV (change from quantity revision and change in timing)	$38,000	$54,546	$(33,056)	$ 59,490

[a] actual barrels produced in 20XB
[b] actual net price for 20XB production
[c] net price at 12/31/20XB

Note that, in the preceding schedule, part of the change in production timing/revision of quantity involved the difference between estimated and actual production for 20XB. In this case, 2,000 barrels more were actually produced and sold than estimated. Because the actual net selling price per WI barrel of these barrels was not the same as the year-end net price, the

2,000 barrels were valued at the actual net selling price in order to get an exact answer. *SFAS No. 69,* however, requires that the effects of changes in prices and costs be computed before the effect of changes in quantities so changes in quantities will be valued at year-end prices and costs. Therefore, the precise approach of Version 1, while yielding an exact answer, does not conform to the statement's requirements in this regard. (In the case study, if the actual barrels produced and sold had been less than the barrels estimated, then the barrels that had been over-estimated would still be on hand at the end of the year and would be valued at year-end prices.)

An alternate approach, presented as Version 2, calculates the change due to the revision of quantity estimates without regard to timing. The effects of the change are discounted by applying the average discount at year-end. The actual selling price is used to value the difference between actual and estimated production, and the year-end price is used to value the remainder of the revision.

Version 2 (Calculation of discount is simplified)

1. **Revision in quantity**

Proved reserves in place 12/31/XA, estimated on 12/31/XA		120,000 bbl
Reserves in place 12/31/XB, proved in prior years, estimated on 12/31/XB	91,000 bbl	
Production during 20XB	32,000	
Revised estimate of reserves in place 12/31/XA, estimated on 12/31/XB		123,000
Revision in estimate, increase (decrease)		3,000 bbl

2. **Undiscounted effect of revision**

2,000 bbl x ($25 – $6) (actual net selling price)	$38,000
1,000 bbl x ($26 – $6) (end-of-year price and cost)	20,000
Total undiscounted increase from revision	$58,000

3. Present value of revision (apply average discount to revision)

$$\frac{\text{PV (future revenues – future production costs) (12/31/XB) (Table 13–3B)}}{\text{(future revenues – future production costs) (12/31/XB) (Table 13–3A)}} \times \text{revision}$$

$$\frac{\$3{,}229{,}551 - \$745{,}281}{\$3{,}666{,}000 - \$846{,}000} \times \$58{,}000 = \underline{\underline{\$51{,}095}}$$

After determining the effect of revisions in quantity, a reconciliation of beginning and ending estimated future revenues relating to reserves discovered in prior years may be prepared as follows to verify that every source of change has been explained.

Analysis of Changes in Price and Cost and Quantity Revisions and Sales and Transfers

	$ (PV)	bbl
PV of beginning future revenues, net of production costs estimated 12/31/XA ($2,153,668 – $489,470)(Table 13–1)	$1,664,198	120,000
Accretion of discount (10% x $1,664,198)	166,420	
Sales, net of production costs (Schedule A)	(608,000)*	(32,000)
PV of net change in prices and production costs, calculated without PV simplification (Schedule B, Version 1)	293,052	
Revision of quantity and timing, calculated without PV simplification (Schedule F, Version 1)	59,490	3,000
Other (round off)	10	
PV of ending future revenues, net of production costs relating to reserves discovered in prior years ($2,047,721 – $472,551)(Table 13–3B)	$1,575,170	91,000

* Since the sales net of production costs related to reserves both discovered and produced in 20XB were not previously estimated and are not included in either the beginning or ending estimate of future revenues net of production costs, that amount was omitted from the analysis.

Accretion of discount. As discussed earlier, accretion of discount is a change that is inherent in the concept of present value, and one that adjusts the beginning standardized measure from a present value as of December 31, 20XA, to a present value as of December 31, 20XB. Accretion of discount is calculated by applying the 10% discount rate to the value to be adjusted. *SFAS No. 69* requires all changes except income taxes to be reported on a pretax basis. Therefore, accretion of discount should be computed on a pretax basis. There is thus another portion of accretion of discount associated with taxes that would be included with the change in income taxes.

Schedule G-20XB: Accretion of discount (excluding accretion on income tax)

PV of beginning future net cash flows before income taxes (Table 13–1b)	$1,591,360
Discount Rate	10%
Accretion of discount (increase)	$ 159,136

Net change in income taxes. The change in estimated future income tax for 20XB results from (1) income tax actually incurred during the current period (increase to the SM) and (2) the difference between (a) the estimate of future income taxes made on December 31, 20XB, plus actual income taxes from 20XB and (b) the estimate made on December 31, 20XA. The estimate made on December 31, 20XB, plus actual income taxes from 20XB is equivalent to an estimate made of future taxes as of 12/31/XA based on 12/31/XB facts. Since actual income taxes from 20XB are included in (1) above as an increase to the SM and in (2) above as a reduction to the SM, the actual income taxes from 20XB in (1) and (2) would cancel out and thus may be ignored.

Schedule H-20XB: Net change in income taxes*, including accretion of discount

PV of future income taxes estimated 12/31/XB (Table 13–3b).	$886,706
Less: PV of future income taxes estimated 12/31/XA (Table 13–1b).	(436,157)
Net change in future income taxes, increase (decrease) to standardized measure.	$450,549

* Do not calculate if ignoring future income tax calculations.

A reconciliation of the change in income tax is not provided because a reconciliation is unnecessary, given the method by which the change in income tax is calculated.

The net change in income taxes for 20XA, the first year of operations for Lucky, is computed as follows:

Schedule H-20XA: Net change in income taxes*

PV of future income taxes estimated 12/31/XA (Table 13–1b) . . .	$436,157
PV of future income taxes, 1/1/XA	0
Net change in future income taxes, increase (decrease) to standardized measure	$436,157

* Do not calculate if ignoring future income tax calculations.

Conclusion

A primary purpose of supplemental disclosures is usefulness to users of the financial statements. Wright and Brock recently evaluated all of the academic research that has been published regarding the relevance and reliability of reserve quantity disclosures and reserve value disclosures.[2] The studies investigating relevance of reserve *quantity* disclosures indicate the disclosures are useful. Regarding reliability, studies indicate that the disclosures are not perceived as being accurate; however, there is no evidence that the figures are biased. Since the reserve quantities are not particularly accurate or reliable, it is not surprising that the studies examining the reliability of reserve *value* disclosures concluded that the disclosures are likewise not considered accurate or reliable.

Regarding relevance, the studies are mixed. Most studies failed to find evidence that the standardized measure disclosure provided information content to users of financial statements. Recent studies that attempted to break the standardized measure or the change in standardized measure into their components and test the reason for the change found some evidence that the information was being used by investors; however, the evidence is weak and not convincing.

In summary, there appears to be strong support for reserve quantity disclosures. Support is not as strong or consistent as to the usefulness of the standardized measure and changes to the standardized measure disclosures. Moreover, it is these latter disclosures that require significant preparation time. Summing up the overwhelming nature of these disclosures, one analyst commented in a survey, "I would rather have gasoline a few cents cheaper at the pump than to have these disclosures."[3]

PROBLEMS

1. What companies are required to present the disclosures specified by *SFAS No. 69*?

2. Which disclosures are different for a successful-efforts company compared to a full-cost company?

3. Which disclosures must be presented for each geographic area?

4. Casing Oil, a successful-efforts company, began operations on January 1, 20XA. Assume the following facts about Casing's first two years of operations. All reserve and production quantities apply only to Casing Oil's interest. Prepare the required disclosures under *SFAS No. 69.*

12/31/20XA

	Item		**Lease R**	**Lease S**
a.	Acquisition costs		$ 20,000	$ 25,000
b.	G&G costs		$ 40,000	$ 50,000
c.	Drilling costs:			
	IDC		$250,000	$ 50,000
	Tangible		50,000	
	Life of equipment		10 years	
d.	Drilling results		Proved reserves	Dry
e.	Estimated production of estimated	20XB –	20,000	
	proved reserves, bbl	20XC –	15,000	
		20XD –	18,000	
f.	Reserve estimate, 12/31/XA			
	proved, at 12/31/	20XA –	53,000	
	(decreases by estimated production)	20XB –	33,000	
		20XC –	18,000	
	proved developed at 12/31/			
	(decreases by estimated production	20XA –	30,000	
	and increases as a result of estimated	20XB –	33,000	
	development)	20XC –	18,000	
g.	Estimated tangible development costs, $	20XB –	40,000	

Item		Lease R	Lease S
h. Estimated restoration costs, $	20XD –	20,000	
i. Current market price of oil		$23/bbl	
j. Estimated total future production costs based on year-end costs	20XB – 20XC – 20XD –	$80,000 60,000 72,000	
k. Estimated current production costs per WI barrel		$4/bbl	

12/31/20XB

Item		Lease R	Lease S
a. Actual and estimated production, bbl	20XB – 20XC – 20XD –	12,000 actual 20,000 revised 22,000 revised	
b. Estimated and actual tangible development costs, $	20XB –	35,000 actual	
c. Estimated restoration costs, $	20XD –	20,000	
d. Market price of oil sold		$24/bbl	
e. Current market price of oil at end of year		$25/bbl	
f. Total production cost of oil sold		$72,000	
g. Production cost per WI barrel of oil sold		$6/bbl	
h. Estimated total future production costs at year-end	20XC – 20XD –	$100,000 110,000	
i. Current production cost per WI barrel at year-end		$5/bbl	
j. Delay rental			$2,000

Assume a tax rate of 40%, and that Casing Oil does not qualify for percentage depletion because it is an integrated producer. For purposes of the required capitalization and amortization of 30% of IDC, assume nine months of amortization in 20XA. Because of the short life of Lease A, also assume Casing elects to use the unit-of-production method for calculating depreciation. Use proved reserves for depletion and proved developed reserves for depreciation. Ignore the alternative minimum tax and deferred taxes. (What is the significance of no estimated future development costs on Lease A as of 12/31/XB?)

5. Buckley Oil Company, a successful-efforts company, began operations January 1, 20XA. Assuming the following facts about Buckley's first two years of operations, prepare the required disclosures under *SFAS No. 69*. All reserve and production quantities apply only to Buckley Oil's interest. Ignore the computations for future income tax.

12/31/20XA

Item		Lease Q	Lease T
a. Acquisition costs		$ 50,000	$ 40,000
b. G&G costs		$ 30,000	$ 35,000
c. Drilling costs:			
IDC		$150,000	$100,000
Tangible		$ 80,000	$ 10,000
Life of equipment		10 years	
d. Drilling results		Proved reserves	Uncompleted
e. Estimated production of estimated	20XB –	15,000	
proved reserves, bbl	20XC –	30,000	
	20XD –	20,000	
f. Reserve estimate, 12/31/XA			
proved, at 12/31/	20XA –	65,000	
(decreases by estimated production)	20XB –	50,000	
	20XC –	20,000	
proved developed at 12/31/			
(decreases by estimated production	20XA –	55,000	
and increases as a result of	20XB –	50,000	
estimated development)	20XC –	20,000	
g. Estimated tangible development costs, $	20XB –	25,000 (life, 10 years)	
h. Estimated restoration costs, $	20XD –	15,000	
i. Current market price of oil		$26/bbl	

Item		Lease Q	Lease T
j. Estimated total future production costs based on year-end costs	20XB – 20XC – 20XD –	75,000 150,000 100,000	
k. Estimated current production costs per WI barrel		$5/bbl	

12/31/20XB

a. Drilling costs: Tangible			$ 60,000
b. Drilling results			Proved reserves
c. Actual and estimated production, bbl	20XB – 20XC – 20XD –	8,000 actual 25,000 revised 22,000 revised	20XC – 35,000 estimated
d. Estimated and actual tangible development costs, $	20XB –	18,000 actual (life, 10 years)	20XB – 30,000 actual
e. Estimated restoration costs, $	20XD –	15,000	
f. Market price of oil sold		$27/bbl	
g. Current market price of oil at end of year		$29/bbl	$29/bbl
h. Total production cost of oil sold		$56,000	
i. Production cost per WI barrel of oil sold		$7/bbl	
j. Estimated total future production costs at year-end	20XC – 20XD –	$175,000 154,000	20XC–$245,000
k. Current production cost per WI barrel at year-end		$7/bbl	$7/bbl

Assume a tax rate of 40%, and that Buckley does not qualify for percentage depletion because it is an integrated producer. Ignore deferred taxes and the alternative minimum tax. (What is the significance of no estimated future development costs on Lease A and Lease B as of 12/31/XB?)

6. Wildcat Oil Company began operations on January 1, 20XA. The following facts relate to Wildcat's first two years of operations. All reserve and production quantities apply only to Wildcat Oil's interest.

12/31/20XA

	Item	Lease R	Lease S	Lease T
a.	Acquisition costs	$ 50,000	$ 40,000	$ 60,000
b.	G&G costs	$ 30,000	$ 35,000	$ 40,000
c.	Drilling costs: IDC Tangible Life of equipment	 $160,000 80,000 10 years	 $100,000	 $135,000 10,000
d.	Drilling results	Proved reserves	Uncompleted	Dry
e.	Estimated production of estimated proved reserves, bbl	20XB – 25,000 20XC – 30,000 20XD – 20,000		
f.	Reserve estimate, 12/31/XA proved, at 12/31/ (decreases by estimated production) proved developed at 12/31/ (decreases by estimated production and increases as a result of estimated development)	 20XA – 75,000 20XB – 50,000 20XC – 20,000 20XA – 30,000 20XB – 40,000 20XC – 20,000		
g.	Estimated tangible development costs, $	20XB – 25,000 20XC – 30,000		
h.	Estimated restoration costs, $	20XD – 15,000		
i.	Current market price of oil	$28/bbl		
j.	Estimated total future production costs based on year-end costs	20XB – $125,000 20XC – 150,000 20XD – 100,000		
k.	Estimated current production costs per WI barrel	$5/bbl		

12/31/20XB

	Item	Lease R	Lease S	Lease T
a.	Drilling costs: Tangible		$ 60,000	
b.	Drilling results		Proved reserves	
c.	Actual and estimated production, bbl	20XB – 18,000 actual 20XC –25,000 revised 20XD –16,000 revised	20XB – 25,000 20XC – 35,000 estimated	
d.	Estimated and actual tangible development costs, $	20XB – 20,000 actual 20XC – 0 revised	20XB – 30,000 actual	
e.	Estimated restoration costs, $	20XD – 15,000		
f.	Market price of oil sold	$24/bbl	$24/bbl	
g.	Current market price of oil at end of year	$26/bbl	$26/bbl	
h.	Total production cost of oil sold	$108,000	$150,000	
i.	Production cost per WI barrel of oil sold	$6/bbl	$6/bbl	
j.	Estimated total future production costs at year-end	20XC – $175,000 20XD – 112,000	20XC –$245,000	
k.	Current production cost per WI barrel at year-end	$7/bbl	$7/bbl	
l.	Delay rental			$ 9,000

Assume a tax rate of 40%, and that Wildcat does not qualify for percentage depletion because it is an integrated producer. For purposes of the required capitalization and amortization of 30% of IDC, assume nine months of amortization in 20XA. Because of the short lives of Leases A and B, also assume Wildcat elects to use the unit-of-production method for calculating depreciation. Use proved reserves for depletion and proved developed reserves for depreciation. Ignore the alternative minimum tax and deferred taxes. (What is the significance of no estimated future development costs on Lease A and Lease B as of 12/31/XB?)

a. Prepare the required disclosures under *SFAS No. 69*, assuming Wildcat is a successful–efforts company.

b. Assume instead that Wildcat is a full-cost company that amortizes all possible costs. Prepare only those disclosures that would differ under FC compared to SE.

7. Tiger Oil began operations on January 1, 20XA. Assume the following facts about Tiger's first two years of operations. All reserve and production quantities apply only to Tiger Oil's interest. Ignore the computations for future income tax.

12/31/20XA

	Item		Lease M	Lease N	Lease O
a.	Acquisition costs		$ 70,000	$ 50,000	$ 40,000
b.	G&G costs		$ 90,000	$ 65,000	$ 80,000
	Test-well contribution				15,000
c.	Drilling costs:				
	IDC		$300,000	$200,000	$180,000
	Tangible		130,000		20,000
	Life of equipment		10 years		
d.	Drilling results		Proved reserves	Uncompleted	Dry
e.	Estimated production of	20XB –	55,000		
	estimated proved reserves,	20XC –	40,000		
	bbl	20XD –	30,000		
f.	Reserve estimate, 12/31/XA				
	proved, at 12/31/	20XA –	125,000		
	(decreases by estimated	20XB –	70,000		
	production)	20XC –	30,000		
	Reserve estimate, 12/31/XA				
	proved developed at 12/31/	20XA –	50,000		
	(decreases by estimated	20XB –	70,000		
	production and increases as a result of	20XC –	30,000		
	estimated development)				
g.	Estimated tangible	20XB –	60,000		
	development costs, $	20XC –	45,000		
h.	Estimated restoration costs, $	20XD –	20,000		

	Item		Lease M	Lease N	Lease O
i.	Current market price of oil		$20/bbl		
j.	Estimated total future production costs based on year-end costs	20XB – 20XC – 20XD –	$330,000 240,000 180,000		
k.	Estimated current production costs per WI barrel		$6/bbl		

12/31/20XB

	Item		Lease M	Lease N	Lease O
a.	Drilling costs: Tangible			$ 70,000	
b.	Drilling results			Proved reserves	
c.	Actual and estimated production, bbl	20XB – 20XC – 20XD –	28,000 actual 47,000 revised 35,000 revised	20XB – 40,000 20XC – 60,000 estimated	
d.	Estimated and actual tangible development costs, $	20XB – 20XC –	63,000 actual 0 revised	20XB – 75,000 actual	
e.	Estimated restoration costs, $	20XD –	20,000		
f.	Market price of oil sold		$21/bbl	$21/bbl	
g.	Current market price of oil at end of year		$23/bbl	$23/bbl	
h.	Total production cost of oil sold		$168,000	$240,000	
i.	Production cost per WI barrel of oil sold		$6/bbl	$6/bbl	
j.	Estimated total future production costs at year-end	20XC – 20XD –	$282,000 210,000	20XC – $360,000	
k.	Current production cost per WI barrel at year-end		$6/bbl	$6/bbl	
l.	Delay rental				$8,000

Assume a tax rate of 40%, and that Tiger does not qualify for percentage depletion because it is an integrated producer. Ignore deferred taxes and the alternative minimum tax.

a. Prepare the required disclosures under *SFAS No. 69*, assuming Tiger is a SE company.

b. Assume instead that Tiger is a FC company that amortizes all possible costs. Prepare only those disclosures that would differ under FC compared to SE.

NOTES

1. Gallun, Rebecca A., and Pearson, Della, "A Comprehensive Look at FASB Statement 69," *Journal of Extractive Industries Accounting*, Spring, 1984, pp. 115–165.

2. Wright and Brock, "Relevance Versus Reliability of Oil and Gas Reserve Quantity and Value Disclosures: The Results of Two Decades of Research," *Journal of Petroleum Accounting and Financial Management.* Fall/Winter 1999 Vol. 18, No. 3, pp. 86–110.

3. Avard, Stephen, "Oil and Gas Disclosures-Analysts' Perceptions of Usefulness," *Journal of Extractive Industries Accounting*, Fall/Winter, 1983, pp. 97–103.

chapter fourteen

Accounting *for* International Petroleum Operations

A company electing to engage in operations outside the U.S. will likely encounter many issues and difficulties not encountered in domestic operations. For example, a company must consider that the costs of operating in a foreign country are likely to be higher than undertaking similar operations domestically. In some locations there may be risks associated with factors such as the stability of the government, stability of the currency, adequacy of the supply of materials and equipment, and availability of a qualified labor pool. In addition, local laws and customs may affect the way business is conducted in the country. For example, an oil and gas exploration and production company will encounter different laws and customs regarding the ownership of the minerals in the earth. The local government rather than an individual or company will, with limited exception, own the minerals. Consequently, an oil and gas company must enter into a contract with the local government for the right to explore for and produce oil and gas reserves. These contracts are complex and vary widely from country to country.

All of these issues impact on oil and gas companies engaging in exploration and production activities outside the U.S. This chapter provides an overview of some of the issues and difficulties encountered in international oil and gas operations. Of special interest are contracts between oil and gas companies and governments that dictate how costs, revenues, and reserves are to be shared.

PETROLEUM FISCAL SYSTEMS

Since mineral rights are typically owned by the government of the foreign country or "host" country, the government and the oil and gas company must come to an agreement as to what collective payments are to be received by the government in return for allowing the company to operate there. Collectively these payments are referred to as the **fiscal system** of the country. Examples of such payments include:

- Up-front bonuses paid to the host country
- Royalties paid to the host country
- Federal and provincial income taxes
- Various other taxes collected by the host country including duties and special petroleum taxes
- Production sharing wherein oil or gas is allocated between the parties for the purpose of recovery of capital and/or operating costs
- Infrastructure development for the host country

The exact nature of payments that the government receives is determined by the legal system in the country. In the United States and in limited situations in Canada and Trinidad, it is possible for individuals to own mineral rights. Outside the U.S., Canada, and Trinidad, mineral rights are owned exclusively by the host government. Some countries allow non-government ownership of oil and gas after it is produced while others do not. Countries that allow non-government ownership of oil and gas are referred to as having **concessionary systems.** (The U.S. is a unique example of a concessionary system in that individuals can own minerals-in-place and can convey their rights to oil and gas companies.) With the limited exceptions noted above, in all other concessionary systems the government owns the rights to all minerals but will transfer ownership of the minerals to oil and gas companies when the minerals are produced or sold. In these countries, the contract between the government that owns the minerals and the oil and gas company wishing to explore, develop, and produce the minerals is referred to as a **concessionary contract** or **agreement**.

In concessionary systems the most common forms of payment made to mineral owners are up-front bonuses and royalties. In addition, the government collects income taxes and perhaps other taxes such as duties, production taxes, value added taxes (VAT) and other special petroleum taxes.

Countries where the government owns all mineral rights and retains title to all minerals produced are referred to as having **contractual systems**. In a contractual system the oil and gas company must contract with the local government for the right to share in revenue from oil and gas production. The government retains title to the minerals throughout exploration, development, and production, with the company either receiving revenues from the sale of its share of production or being allowed to receive oil and gas in the form of repayment for cost recovery, a share of profits, and/or services rendered.

The two most common contracts in a contractual system are **production sharing contracts** (PSCs) and **service contracts**. The fundamental difference between these two types of contracts relates to whether the oil and gas company is paid in the form of oil and/or gas or in the form of money. Conceptually, in a PSC the oil and gas company, referred to as the **contractor,** is allowed to recover certain costs and receives a share of the profits. The company typically receives payment in-kind (in the form of oil or gas). In a true service contract, the contractor receives money representing a fee for conducting exploration, development, and production activities. In practice, most contracts have terms and conditions that make it difficult to classify them as being either PSCs or service contracts. Most are simply referred to as PSCs. Each of these types of arrangements is discussed in detail below. It should be noted that the term *contractor* is used collectively to refer to one company or several companies who are acting together in a joint operation.

Almost every contract has its own unique terms and characteristics. Often it is not only difficult to classify the contracts as being PSCs or service contracts but also as being concessionary or contractual in nature. Each country negotiates the terms it believes are most beneficial to the citizens of the country. Consequently, contracts and agreements generated by any given country may be contractual in nature but have some aspects that resemble a concessionary contract and vice versa. It is important to remember that any discussion of these contracts reflects the most usual case given. Contracts can be very different since they are the product of intense negotiation and can include whatever clauses and conditions the negotiators agree to.

CONCESSIONARY SYSTEMS

As mentioned above, in all countries outside the U.S., Canada, and Trinidad, an oil and gas company must contract with the host government for the right to explore, develop, and produce. The contractor conducts exploration, drilling, development, and production activities at its sole risk and cost. In other words, the company pays all of the cost associated with exploration, drilling, development, and production without any prospect for cost recovery if oil or gas is not discovered or without any prospect for repayment by the government. If oil or gas is discovered and produced, title to the oil or gas in a concessionary system will, at some point, pass to the contractor who, in turn, pays a royalty to the government as production occurs. A contract under a concessionary system contains basic terms that are similar to a U.S. lease agreement.

Another form of revenue to the government is taxes. Generally, all countries have some form of income taxes. If the government decides to increase the petroleum-related payments received by the state, it may enact other types of taxes and levies assessed on petroleum exploration and production activities. These include for example, production or severance taxes, petroleum revenue taxes, value added taxes, and resource rent taxes. In a concessionary system, the primary source of payments to the state is in the form of royalties and taxes.

Consequently, the countries having concessionary systems are sometimes referred to as tax/royalty countries. The following example illustrates a simple concession agreement:

EXAMPLE

Concession Agreement

Lucky Company enters into a concession agreement with the Canadian government. Lucky pays the government, in U.S. dollars, a $5,000,000 signing bonus and agrees to pay the government royalties of 10% of gross production and a 5% severance tax. Lucky bears all of the costs associated with exploration, development, and production.

Lucky spends $5,000,000 on exploration and drilling costs and in 2002 has gross revenue of $7,000,000 and production costs of $1,000,000. The income tax laws allow deduction of all production costs with exploration and drilling costs deductible over a 5-year period. The tax rate is 40%.

Gross revenue for 2002 would be shared by the parties as follows:

	Lucky Company	To Government
Gross revenue	$7,000,000	
Royalty 10%	(700,000)	$ 700,000
Severance tax 5%	(350,000)	350,000
Net revenue	$5,950,000	
Operating expenses	(1,000,000)	
1/5 of exploration and drilling costs	(1,000,000)	
Taxable income	$3,950,000	
Income taxes 40%	(1,580,000)	1,580,000
Net to Lucky/Government	$2,370,000	$2,630,000

Concessionary agreements with the government participation

One variation of a concessionary agreement involves the host government participating in the oil and gas operations as a working interest owner. This type of arrangement is generally referred to as **government participation**. The government typically sets up a (government-owned) oil company to participate in the operation as a working interest owner. This arrangement may also be referred to as a joint venture arrangement. As with other joint operations, a joint operating agreement is typically executed between the parties. In this particular type of arrangement, the contractor typically agrees to pay 100% of the exploration-type expenditures and "carries" the government-owned company through the exploration phase. In other words, the contractor pays all of the costs related to exploration, exploratory drilling, and any other costs specified in the contract.

If commercial reserves are found, the government retains the right to participate or **back in** to the development and production operations as a working interest owner at an interest of up to 51%. This means the government can elect to become a joint venture partner with the contracting company after the results of initial exploration and drilling are known. If the government elects to participate, it then becomes liable for its proportionate share of all *future* drilling, development, and production costs. The agreement may allow the company to recover all or a portion of its up-front exploration-related expenditures. If this is the case, there are two methods of recovery. One is direct payment by the government to the company. The other, more frequently used method, is to allow the contractor to recover some or all of its costs by the contractor keeping the government's share of production until the contractor has recouped the allowed costs. Afterwards, the government-owned company shares in costs and production just like any other working interest owner. Under this arrangement, the government still receives a royalty on gross production along with income taxes and other fiscal obligations required by the laws and regulations of the country. Customs duties on the importation of materials and supplies and export duties on production are generally exempted. Similar to a straight concession agreement, title to a *pro rata* share of oil and gas passes to the contractor at some point.

Contractual Systems

In some countries the legal system as it pertains to mineral ownership is based on the principle that natural resources are owned by all citizens of the country, and the government should act in such a manner as to maximize the value of the resources flowing back to the people. As a result, the government owns and retains title to all minerals. The contractor is entitled to minerals or money from the sale of minerals as a "fee" for the services it performs (*i.e.*, exploration, drilling, development, and production). In some countries, the contractor may be entitled to an ownership interest in the minerals at the point of sale. In other instances the contractor may not ever receive title to the minerals, but instead is entitled to a share of the revenue from the sale of the minerals. Some agreements require that all or part of any oil or gas produced within the country be sold to the host government.

Government involvement in operations

In a contractual system the government plays an active role in exploration, development, and production typically by acting through a government-owned oil company. The contractor (or one of the contracting companies) usually acts as the operator. Contracts frequently call for the formation of a joint management group to oversee operations and vote on all major operating decisions. Such a joint management group is typically made up of representatives from the contractor (or each of the companies that compose the contractor), the government-owned oil company, and the government (*i.e.*, representatives from the Ministry of Petroleum or some other government agency). The contractor is typically required to submit an annual work program and budget to the joint management group for review and approval. In addition to the annual budget, the joint management group generally makes all major decisions regarding the management of the project including approval of all major expenditures, evaluation of the results of exploration, planning and drilling of wells, and determination of the commerciality of drilling results.

The contractor is typically required to provide all technology and financing. In most contractual systems, any equipment or facilities brought into the country by the contractor become the property of the local government. (This does not apply to equipment and facilities that are owned by service companies, equipment brought into the country temporarily, or to leased equipment.) In some cases title to the equipment and facilities passes to the government at the time the goods are brought into the country or upon installation. In other cases, title passes to the government when the costs of the equipment and facilities have been recovered by the contractor.

The contractor is allowed to bring some employees into the country; however, the majority of employees must be hired locally. The contractor must have a plan for training the local employees. This training obligation generally continues throughout the life of the contract and constitutes a significant cost of operations.

As mentioned earlier, the two types of agreements prevalent in contractual systems are production sharing and risk service contracts, with the production sharing contracts being, by far, the most common. These contracts are discussed below.

Production Sharing Contracts

Production sharing contracts first emerged in the 1960s when governments began to evaluate alternative strategies to maximize the value of their resources. Under a concessionary agreement (where the government is not a working interest owner), the government has little or no involvement in the management and decision-making related to day-to-day drilling and operations. Governments generally were looking for ways to increase their total share of petroleum-related revenues and profits. Governments were also frustrated due to their inability to be involved in the management of petroleum operations. In addition to the legal constraints related to the ownership of minerals, these factors largely led to the trend toward production sharing contracts and away from concessionary agreements.

Signature and production bonuses

A common feature of both concessionary agreements and PSCs is that the contractor agrees to pay the government an up-front bonus for signing the agreement. This bonus is often referred to as a **signing** or **signature bonus**. Typically, signature bonuses are payable in a lump sum of money but sometimes may involve payment in the form of equipment.

In some instances a lump sum of money is paid at the signing of the contract and subsequent payments are made to the government when production reaches an agreed upon level. These later payments are referred to as **production bonuses**. For example, a contractor may agree to pay the government, in U.S. dollars, a $3 million bonus at signing, $2 million when production reaches 2,000 barrels of oil per day, and $1 million when production exceeds 4,000 barrels of oil per day. In other words, rather than paying $6 million at signing, the bonuses are phased in as certain levels of production are achieved. This arrangement theoretically allows the contractor to use its capital in exploration and drilling with production bonuses being paid as money is coming in from the production and sale of oil or gas. On the other hand, the government is taking on some degree of risk since it receives a lower signing bonus and may not receive any additional bonuses if production does not occur or if it does not reach the specified level. However, that risk is balanced with the expectation or hope that exploration and development will increase and will result in higher royalties, etc.

Royalties

An interesting feature of production sharing contracts is the inclusion of a royalty provision. Payment of a royalty is a logical concept in a concessionary system where title to oil and gas passes to the contractor at some point. In a contractual system title to the oil and gas in-place never passes to the contractor; nonetheless, many PSCs contain royalty provisions. In practice, royalty provisions range from as low as zero to 15% or higher.

Since a royalty represents a payment to the government off the top of gross receipts, many argue that royalties can actually discourage capital investment in new drilling and development and, in marginal situations, actually work against further development and production. For example, if the royalty is 10%, only 90% of gross revenue is available to cover capital costs and operating costs. This may discourage an oil and gas company from developing a marginal field or may result in the abandonment of a marginal field earlier than would otherwise be the case. To help offset this effect to some extent, some contracts contain **sliding scale royalties**. A sliding scale royalty provides for a lower royalty amount when production is lower and increases as production increases. Thus, in marginal situations where production is lower, the lower royalty may allow production that would otherwise not have been profitable. The following example illustrates a sliding scale royalty:

EXAMPLE

Sliding Scale Royalty

Average Daily Production	Royalty
Up to 7,000 barrels per day	5%
7,001 to 14,000 barrels per day	10%
Above 14,000 barrels per day	15%

In this example, when production is low the royalty payment is low and when production increases the royalty increases. By using a sliding scale, when production is low there is more cash available to the parties for additional exploration and development.

Government participation

As in concessionary systems, some PSCs allow the government to participate in oil and gas projects through a government-owned oil company. In most cases, the contractor must pay all costs and bear all risks during the exploration phase of a project. The government does not reimburse the contractor for its share of the exploration costs. Instead, the contractor must look to production and cost recovery to recoup those costs. If there is no production, the contractor bears all of the costs without any provision for reimbursement.

If oil and gas is discovered, the government—through the government-owned oil company—can elect to participate at whatever level of working interest it chooses up to a maximum of 51%. If the government backs into the project, their role is similar to any other working interest owner. In other words, the government is responsible for its proportionate share of development and operating costs.

Cost recovery

Cost recovery is a feature common to most PSCs. The contract should specify which costs are recoverable, the order of recoverability, any limits on recoverability, and whether costs not recovered in one period can be carried forward into the next period. Since title to the oil or gas in-place never passes to the contractor, cost recoverability is the mechanism whereby the contractor is able to recoup the costs that it has expended on a project. The oil (or gas) that goes to the parties to allow them to recover their costs is referred to as **cost oil**. In some PSCs the contractor must pay 100% of the cost incurred in the exploratory phase.

In China, for example, this type of PSC is typical. If the field is declared a commercial field, then the government, through the government-owned oil company, can elect to participate as a working interest owner at any level it chooses up to a maximum of 51%. If sufficient production occurs, the contractor is eventually able to recover 100% of the exploration expenditures, its proportionate share of development expenditures, and its proportionate share of production expenditures.

Typically, there is a ceiling or maximum amount of revenue available for cost recovery. In most contracts (though not all), recoverable costs that are not recovered in any given year can be carried forward to future years. Some contracts *amortize* or *depreciate* the amount of capital costs recoverable in any year (*e.g.*, only 1/5 of allowable capital costs are recoverable per year in any of the first five years of production) while other contracts simply employ an annual maximum to cap the amount of recoverable capital costs in any given year.

Some contracts allow **interest cost recovery** or recovery of interest incurred on capital expenditures. The typical Chinese PSC, for example, allows interest cost recovery on "deemed" or imputed interest on costs incurred during the development phase but not during the exploration phase. The recovery of interest costs may be a contentious issue since governments generally assume it is the contractor's responsibility to acquire sufficient funds to cover capital requirements.

Government ownership of all equipment and facilities raises interesting questions regarding who is responsible for future abandonment and reclamation costs. Some PSCs require the companies to put money into a sinking fund to be used in the future to pay for abandonment and reclamation. If these sinking fund payments are considered to be recoverable costs, as with other exploration or development costs, the government is actually paying for dismantlement and reclamation. In other words, assuming the companies fully recover all of their costs, the companies deposit money into a fund and, ultimately, get to cost recover those deposits from future production. Since the companies are allowed cost recovery on the deposits, the companies' net cost is zero and there is money in the fund available to pay for dismantlement and reclamation as the costs are actually incurred.

Most contracts indicate the order in which costs recovery is to occur. The order of cost recovery is important since it determines how quickly the contractor is able to recover certain costs. For example, assume a contractor paid 100% of exploration expenditures and shared development expenditures with the government-owned company. Obviously the contractor would prefer to be able to recover all exploration expenditures before development expenditures are recoverable. Typically, cost recovery occurs in an order similar to the following:

- Current year operating costs
- Unrecovered exploration expenditures
- Unrecovered development expenditures
- Capitalized interest (if allowed)
- Any investment credit or capital uplift (defined later)
- Future abandonment cost fund

In many contracts exploration and development expenditures are referred to as *capital costs* and are *amortized.* This means the costs are recoverable over a fixed number of years. In most cases, if the amortized capital costs are not recovered in any given year, the unrecovered portion can be carried forward and recovered in subsequent years. In some cases, however, carryforward of unrecovered amortized capital costs is not allowed and recoverability is permanently lost.

Profit oil

The gross revenue from production typically goes first to pay royalties, next certain production-related taxes are taken out (*e.g.*, production taxes and VAT) and, finally, to cost recovery. The gross revenue remaining after deducting royalties, taxes, and cost recovery is referred to as **profit oil**. The profit oil is shared between the parties based on the terms and conditions set forth in the contract. For example, assume that a contract provides for a 5% production tax, 10% royalty, and cost recovery of 50% of gross production. The profit oil is equal to 35% [100%-(5%+10%+50%)]. This 35% of annual gross production is shared, as spelled out in the contract, by the contractor, the government-owned oil company, and the government. In some contracts, a specified percentage of the profit oil goes directly to the government with the companies (the contractor and the government-owned oil company) sharing the remainder in proportion to their working interests.

EXAMPLE

PSC Cost Recovery

Assume that Lucky Company, a U.S. company, is involved in petroleum operations in China. Lucky has a 49% WI while the Local Oil Company has a 51% WI. Annual gross production is to be split in the following order:

1. Royalty is 15% of annual gross production and is to be paid in-kind.
2. VAT is equal to 5% of annual gross production and is to be paid in-kind.
3. Cost oil is limited to 60% of gross production with costs to be recovered in the following order:
   a. Operating expenses
   b. Exploration costs (paid entirely by Lucky Company)

c. Development costs (paid 49% by Lucky Company and 51% by Local Oil Company).

4. Any excess production remaining after cost recovery becomes profit oil:
   a. The government receives 12% of profit oil.
   b. The remainder is split between Lucky and Local Oil Company based on their working interests.

For 2005 assume that:

- Recoverable operating costs total $5,000,000.
- Exploration costs (unrecovered to date) total $100,000,000.
- Development costs (unrecovered to date) total $200,000,000.
- The annual gross production for the year is 3,000,000 barrels of oil.
- The agreed upon price is $25 per barrel.

Note that in order to convert the costs into barrels it is necessary to divide by a price per barrel. The contract will specify how the parties are to negotiate and agree upon the price per barrel to be used for this purpose.

The allocation of production to the parties would be as follows:

	Barrels	Chinese Government (in bbl)	Local Company 51% (in bbl)	Lucky Company 49% (in bbl)
Annual Gross Production	3,000,000			
Royalty (15%)	450,000	450,000		
VAT (5%)	150,000	150,000		
Cost Oil (60% x 3,000,000)	1,800,000			
Operating costs:				
$5,000,000/$25 =				
200,000 bbl	200,000		102,000	98,000
Exploration costs				
$100,000,000/$25 =				
4,000,000 bbl (only due				
1,600,000 bbl to limit)	1,600,000			1,600,000
Development costs				
(none recoverable this				
year due to limit)	0			
Remainder	0			
Profit oil: 3,000,000 x				
[1-(5%+15%+60%)/100%]	600,000			
To Government:				
600,000 x 12%	72,000	72,000		
Allocable Profit Oil:				
(600,000 x 88%)	528,000			
528,000 x 51%			269,280	
528,000 x 49%				258,720
TOTAL	3,000,000	672,000	371,280	1,956,720

In this example, the cost oil was 60% x 3,000,000 bbl or 1,800,000 barrels. After allocating 200,000 barrels to the parties for recovery of operating costs, only 1,600,000 barrels were available for recovery of exploration expenditure. All 1,600,000 barrels were allocated to Lucky Oil Company since all exploration expenditures were paid by Lucky. Lucky's unrecovered exploration and development expenditures are carried forward to be recovered in future years.

Also note that Lucky Oil Company will be liable for income taxes on all oil and gas operations determined in accordance with the local income tax laws.

Other terms and fiscal incentives

Governments provide incentives to companies in an effort to maximize the amount of money the companies will invest in exploration, drilling, and development. These incentives may appear in PSCs or result from other negotiations.

Capital uplifts. A capital uplift is an incentive offered by the government to encourage the contractor to maximize investment. A capital uplift, sometimes referred to as an **investment credit**, is an additional amount of cost recovery on capital expenditures over and above actual amounts spent. For example, if a company spends $1,000,000 in recoverable capital expenditures and there is a 10% capital uplift in the contract, the company will be allowed to recover 110% of actual spending or $1,100,000.

Ringfencing. Generally each contract area stands alone when computing cost recovery. That is, in determining cost recovery, only costs that are expended relative to a particular contract area are recoverable from production from that specific contract area. This restriction is referred to as **ringfencing.** In other words, there is an imaginary boundary around the contract area—neither costs nor production can be transferred outside the boundary. If production in that area is insufficient to allow for full recovery, costs cannot be transferred to another contract area where production is higher and recovered from that production. In these instances the contract areas are said to be ringfenced. An incentive that governments may provide is to **un-ringfence** or allow **cross fence recovery.** This incentive is most effective when the government is seeking to increase exploration in a particular area by allowing a company to immediately recover certain exploration expenditures in the new, frontier area against production from a different, currently producing area.

Ringfencing and cross-fence allowance may also be used as tax-related incentives in computing certain petroleum taxes. In some petroleum tax regimes, gross revenue, deductions, and taxable income are determined on a project-by-project basis. If the projects are ringfenced, then deductible costs cannot be transferred outside the project area where they were incurred. If cross-fence (or cross-field) deduction is allowed, then deductible costs from one project area—perhaps where there is no taxable income—can be transferred to another project area and used to reduce taxable income there. This provision is one means by which governments may encourage new exploration. The cross-fence allowance permits a company exploring in a new area (where it has no taxable income and thus an expenditure is of no current tax benefit) to receive immediate tax benefit by deducting the exploration costs against the taxable income generated in another project area.

Domestic market obligation. Some contracts specify that a certain percentage of the contractor's share of profit oil be sold to the local government, typically at a price that is less than the current market price. This requirement is referred to as the **domestic market obligation**

and is often included in situations where the demand for crude oil in the country is greater than the government's share of production. The contractor's domestic market obligation reduces the shortfall, resulting in a lessening of the government's need to rely on imported oil or oil from more expensive sources.

Royalty holidays and tax holidays. **Royalty holidays** and **tax holidays** are incentives governments may use to encourage contractors to maximize investment early in the life of production. The government may specify a period of time (i.e., the first four years of production) during which the royalty provision is waived resulting in the contractor paying no royalty on production during that period of time. This incentive provides the contractor with a break from the payment of royalties or a **royalty holiday.** It also leaves the contractor with more money to reinvest in additional drilling and development. Therefore, the holiday benefits both the government and the contractor. Likewise, the government may specify a period of time during which the contractor is exempt from certain taxes. This is referred to as a **tax holiday**.

These are just a few of the most popular terms that may appear in PSCs. A discussion of all of the possible terms and variations would be too lengthy to include in this chapter.

SERVICE CONTRACTS

The second type of agreement prevalent in a contractual system is a service agreement. Service agreements can be classified as being either **risk service** contracts or **nonrisk service** contracts. In nonrisk service agreements, the contractor provides services in the form of such activities as exploration, development, and production and is paid a fee by the government that covers all costs. In practice nonrisk service agreements are rare. Risk service contracts are much more common. In a risk service contract, the contractor bears all of the costs and risks related to exploration, development, and production activities. In return, if production is achieved, the contractor is allowed to recover its costs as production is sold. In addition, the government pays the contractor a fee for its "services." The fee is typically based on production. The terms and features of risk service contracts are similar to those appearing in PSCs. During the past few years risk service agreements have become popular in many countries in South America. As with production sharing contracts, a significant task facing the accountant is the proper determination of recoverable operating and capital costs.

EXAMPLE

Risk Service Agreement

Lucky Company enters into a risk service agreement with the Bolivian government. Lucky pays the government, in U.S. dollars, a $1,000,000 signing bonus and bears all of the costs associated with exploration, development, and production. The contract defines costs incurred in the exploration and development phase of each project area as being capital costs (CAPEX) and all costs incurred in the production phase as being operating costs (OPEX).

Each year in which production occurs the government agrees to pay Lucky Company a fee comprised of the following:

- All OPEX incurred in the current year
- 1/10th of all unrecovered CAPEX
- \$0.50 per barrel on production from 0 to 4,000 barrels per day, \$0.75 per barrel on production from 4,001 to 10,000 barrels per day, and \$1.00 per barrel on production above 10,000 barrels per day

The maximum total fee that will be paid in any year is \$1.35 per barrel times the total number of barrels produced. Any unrecovered OPEX or CAPEX (unrecovered due to the maximum fee) can be carried forward to future years. In 2004, production begins on Field No. 1. To date Lucky has spent \$10,000,000 on CAPEX and during 2004 spends \$2,000,000 in OPEX. Production during 2004 equals 4,000,000 barrels or 4,000,000/365 = 10,959 barrels per day.

The fee that Lucky would receive for 2004 would be determined as follows:

OPEX	\$2,000,000
CAPEX \$10,000,000/10	1,000,000
4,000 x 365 days x \$.50	730,000
6,000 x 365 days x \$.75	1,642,500
959 x 365 days x \$1	350,035
Total fee	\$5,722,535

The total fee per barrel is computed as follows:

\$5,722,535/4,000,000 barrels = \$1.4306 per barrel

The \$1.4306 fee per barrel is greater than the maximum of \$1.35; therefore, the actual fee paid to Lucky Oil Company is:

\$1.35 x 4,000,000 = \$5,400,000

The difference between the calculated fee of \$5,722,535 and the maximum fee of \$5,400,000 is \$322,535 that is considered to be unrecovered CAPEX and is carried forward to the next year.

Lucky Oil Company will also be liable for income taxes on all oil and gas operations determined in accordance with the local income tax laws.

Joint Operating Agreements

As with domestic operations, when two or more international parties are involved in a joint operation they must execute some type of joint operating agreement. In international operations the function of the joint operating agreement depends on the parties and the detail included in the underlying concession agreement, PSC, or risk service contract. *Model Form International Joint Operating Agreements* are published by the Association of International Petroleum Negotiators (AIPN).

The contract between the government and the contractor provides the basis for all subsequent contracts and agreements. Sometimes the PSC or concession agreement is thoroughly written and effectively also serves as the joint operating agreement. In other cases a separate joint operating agreement is executed. Further, if the contractor is comprised of more than one company, the PSC, concession agreement, etc. may serve as the basic agreement between the government and the contractor with a separate joint operating agreement being executed between the parties acting as the contractor.

Whether the PSC, concession agreement, or risk service contract serves as the joint operating agreement or a separate joint operating agreement is executed, an accounting procedure must exist. Accounting procedures, as they appear in most domestic joint operating agreements, were discussed in detail in chapter 11. As detailed in chapter 11, the accounting procedure defines, among other things, how costs are to be shared between the parties, determination of overhead, material transfers and pricing, and inventories. An international joint operating agreement should address these same major categories. In addition, the accounting procedure included in a PSC or risk service contract should define which costs are recoverable and which costs are not recoverable.

Recoverable versus non-recoverable costs

In joint interest accounting, one of the key tasks is to determine the proper amount of costs and revenues to be shared by each of the parties. In accounting for a PSC or risk service contract, evaluation of costs and allocation of revenues are likewise issues that involve considerable accounting attention. As a general rule of thumb, costs that are defined as direct costs in domestic joint operating agreements are also recoverable under the PSC or risk service contract. In a domestic joint operating agreement, recoupment of indirect costs by the operator is specified via application of overhead rates. Overhead rates (typically sliding scale rates) are also frequently used in PSCs and risk service contracts to determine the amounts of recoverable indirect costs.

Chapter 11 provides a thorough discussion of the types of costs that are typically classified as direct in domestic joint operating agreements. Many of these costs would likewise appear in international accounting procedures as either direct costs or recoverable costs. There are a number of potentially sizeable costs encountered in international operations. A few of these are listed below:

- Transportation costs, in general, are likely to be much larger in international operations.
- Rig mobilization and demobilization costs are likely to be significant since rigs may have to be moved long distances.
- Companies typically have a home office in their home country. In addition they may have large, administrative offices within the country of operations. The costs associated with the in-country offices are generally direct and recoverable. If there is more than one project and/or different projects with different owners then the cost of the in-country office will have to be allocated to the projects on some reasonable basis.
- Costs associated with expatriate employees (*i.e.*, employees working in a foreign country) who are relocated into the country include, travel, moving, living quarters, education for dependent children, etc.
- Relocation costs associated with expatriate employees who move from one foreign assignment to another are frequently an issue that must be spelled out in the contract. For example, if an employee is relocated from an assignment in Poland to an assignment in China, and from there to an assignment in Thailand, how much of the total relocation costs would be chargeable to the operations in China?
- In most domestic joint operating agreements, charges relating to technical employees are chargeable to the joint operation only if the employee is physically on location at the joint property a minimum amount of time, typically eight hours. In international operations, technical employees working in offices located outside of the host country may nonetheless still be chargeable to the joint operation so long as their work is directly benefiting the joint property.

Financial Accounting Issues

Financial accounting versus contract accounting

SFAS No. 19 and the SEC's *Reg. SX 4-10* were written from the perspective of concessionary contracts. However, over the years, accountants have recognized that, while the contract terms and conditions are quite different for PSCs and risk service contracts, the financial accounting should be fundamentally the same. Consequently, financial accounting for PSCs and risk service contracts is no different from financial accounting for concessionary contracts. Even though in a PSC or risk service contract the government owns the oil and gas reserves, the exploration, development, and production activities should be accounted for as oil and gas producing activities and not as contractor services. The production revenue that accrues to the contractor is to be accounted for as revenue and not as cost recovery.

In chapters 2 through 7 of this book definitions that appear in *SFAS No. 19* and *Reg. SX 4-10* were discussed in detail (*e.g.*, exploratory wells and development wells). It should be noted that the definitions utilized in a contract are not necessarily the same as the definitions appearing in *SFAS No. 19* and *Reg. SX 4-10.* For example, in a PSC, an exploratory well may be defined as any well drilled in the exploratory phase, and a development well as any well drilled in the development phase. Depending on the specific circumstances, it is entirely possible that a well drilled in the exploratory phase as defined in the PSC could be a development well per *SFAS No. 19* and *Reg. SX 4-10,* or a well drilled in the development phase as defined in the PSC could be an exploratory well per the FASB and SEC definition. Obviously, the contract definitions apply when accounting for the contract and the financial accounting definitions apply when involved in financial accounting.

Disclosures of reserves under SFAS No. 69

Estimation of reserves. Reserve estimation under a PSC or risk service contract is much more complex than reserve estimation under a concessionary contract. Basically, if estimating working interest reserves under a concessionary contract, gross recoverable reserves would be estimated, reserves attributable to royalty interests or other non-operating interest would be subtracted, and the remainder would be allocated to the working interest owners based on their relative working interests.

When estimating working interest reserves under a PSC or risk service contract, the net reserves remaining after deducting reserves related to royalties and other non-operating interests would be allocated between the parties based on the amounts to which they are entitled per the contract terms. To compute *entitlement reserves* one would have to resolve issues such as:

- the amount of recoverable costs that have been or will be incurred
- the amount of costs that are capital versus operating costs
- the cost recovery terms in the contract
- the profit oil sharing terms in the contract
- the price assumptions to be used in converting the recoverable costs into quantities

A slight change in the prices that are used to convert recoverable costs to quantities of reserves can cause a significant shift in the total estimate of reserves that the working interest owners are entitled to.

Disclosure of reserves. *SFAS No. 69* recognizes that ownership of reserves is an issue that should be considered in determining how reserves are to be reported. Paragraph 102 describes the nature of reserves under certain contracting arrangements and indicates that since there is not *direct ownership interest,* reserves associated with agreements such as PSCs and risk service contracts are different in nature from other proved reserves.

In some countries, oil and gas producing enterprises can obtain access to oil and gas reserves only through such agreements [long-term supply, purchase, or similar agreements with a government or governmental authority] and not through direct acquisition of mineral interests. If an oil and gas producing enterprise participates in the operation of a property subject to such an agreement or otherwise serves as 'producer' of the reserves from the property, disclosure of the reserve quantities identified with, and quantities of oil or gas received under, that type of agreement with those governments or authorities provides useful information. The fact that the reserves are available to the enterprise requires their inclusion to give a complete presentation of the enterprise's reserve position. However, because of the different nature of those agreements (that is, they do not represent direct ownership interests in reserves), those reserve quantities are to be reported separately from the enterprise's own proved reserves. (SFAS No. 69 par. 102)

Paragraph 13 requires that these reserves and changes therein be disclosed separately:

Net quantities disclosed in conformity with paragraphs 10-12 shall not include oil or gas subject to purchase under long-term supply, purchase, or similar agreements and contracts, including such agreements with governments or authorities. However, quantities of oil or gas subject to such agreements with governments or authorities as of the end of the year, and the net quantity of oil or gas received under the agreements during the year, shall be separately disclosed if the enterprise participates in the operation of the properties in which the oil or gas is located or otherwise serves as the 'producer' of those reserves, as opposed, for example, to being an independent purchaser, broker, dealer, or importer. (SFAS No. 69 par. 13)

Operations in some countries are subject to risks associated with instability in the government that could result in the government taking over operations or otherwise forcing an operator out of the country. Due to these concerns, *SFAS No. 69* paragraph 12 requires companies to report their net proved reserves separately for each significant foreign location:

If some or all of its reserves are located in foreign countries, the disclosure of net quantities of reserves of oil and of gas and changes in them required by paragraph 10 and 11 shall be separately disclosed for (a) the enterprise's home country (if significant reserves are located there) and (b) each foreign geographic area in which significant reserves are located. Foreign geographic areas are individual countries or groups of countries as appropriate for meaningful disclosure in the circumstances. (SFAS No. 69 par. 12)

Taken together, paragraphs 12, 13, and 102 appear to impose the requirement that companies report their net proved reserves by major geographic location and also by type of contract. Practice within the industry generally does not reflect this level of detailed disclosure. Companies do generally disclose their net share of proved reserves by major geographic area, but few provide information further segregating the reserves related to PSCs or risk service contracts. This practice may result, in part, from the fact that the disclosure of the *Standardized Measure of Discounted Future Net Cash Flow From the Production of Proved Oil and Gas Reserves* (also specified by *SFAS No. 69*) does not require reporting reserves based on contract type.

International Accounting Standards

Regardless of the country in which the parent company is located, most foreign governments require companies operating within their country to file financial statements prepared in accordance with local GAAP. For example, a French company operating in Australia might be required to account for its Australian operations using Australian GAAP and, for purposes of parent company reporting, to also account for the Australian operations using French GAAP. Many foreign countries have adopted accounting standards issued by the International Accounting Standards Committee (IASC). This may be particularly helpful for companies that operate in several foreign countries, since using IASC standards allows them to apply one set of accounting standards rather than standards relating to each of the countries in which they operate. In 1999, the IASC steering committee, which was appointed to draft a standard for the extractive industries (including oil and gas and mining), began working on a draft standard. The standard will address such issues as whether to allow both full-cost and successful-efforts methods, specific provisions related to many issues including capitalization of exploration, drilling, development, and production costs, calculation of DD&A, accounting for conveyances, and reserve disclosures. It is anticipated that this standard will be finalized by 2002.

Problems

1. Define the following:
   - Domestic market obligation
   - Ringfencing
   - Cost oil
   - Profit oil
   - Capital uplift
   - Production bonus
   - Sliding scale royalty
   - Tax holiday
   - Royalty holiday
   - Government participation

2. Distinguish between the following:
   Concessionary system
   Contractual system

3. Distinguish between the following:
   Nonrisk service contract
   Risk service contract

4. Describe the similarities and differences between PSCs and risk service contracts.

5. Discuss the various methods that governments utilize to generate revenues and other benefits from mineral resources.

6. Explain how estimation of a company's net proved reserves differs when operating under a concessionary contract vs. a PSC.

7. Assume that Zapco Company, a U.S. Company, is involved in petroleum operations in China. Zapco Company has a 49% WI while the Local Oil Company has a 51% WI. Annual gross production is to be split in the following order:

   a. VAT is equal to 5% of annual gross production and is to be paid in-kind.
   b. Royalty is 12.5% of annual gross production and is to be paid in-kind.
   c. Cost oil is limited to 62% of gross production with costs to be recovered in the following order:

      1) Operating expenses
      2) Exploration costs (paid entirely by Zapco Company)
      3) Development costs (paid 49% by Zapco Company and 51% by Local Oil Company)

   d. Any excess remaining after cost recovery becomes profit oil:

      1) 10% of profit oil goes to the government.
      2) The remainder is split between Zapco Oil Company and Local Oil Company based on their working interests.

   For 2003 assume that:

   - Recoverable operating costs total $10,000,000
   - Exploration costs (unrecovered to date) total $50,000,000
   - Development costs (unrecovered to date) total $100,000,000
   - The annual gross production for the year is 7,000,000 barrels of oil
   - The agreed upon price is $20 per barrel

REQUIRED: Allocate the production between the parties

8. Assume that Protex Company, a U.S. company, is involved in petroleum operations in Thailand. Protex Company has a 49% WI while the Local Oil Company has a 51% WI. Annual gross production is to be split in the following order:

   a. Royalty is 15% of annual gross production and is to be paid in-kind.
   b. VAT is equal to 5% of annual gross production and is to be paid in-kind.
   c. Cost oil is limited to 50% of gross production with costs to be recovered in the following order:

      1) Operating expenses paid 80% by Protex Company and 20% by Local Oil Company.
      2) Exploration costs (paid entirely by Protex Company).
      3) Development costs: after completion of exploration Local Oil Company opted to participate at 20%; therefore, development and operating costs were paid 80% by Protex Company and 20% by Local Oil Company.

   d. Any excess remaining after cost recovery become profit oil:

      1) 10% of profit oil goes to the government.
      2) The remainder is split between Protex and Local Oil Company based on their working interests.

   For 2004 assume that:

   - Recoverable operating costs total $1,000,000
   - Exploration costs unrecovered to date total $100,000,000
   - Development costs unrecovered to date total $500,000,000
   - The annual gross production for the year is 10,000,000 barrels of oil
   - The agreed upon price is $25/barrel

   REQUIRED: Allocate the production between the parties

9. Ibis Company enters into a concession agreement with the British government. Ibis pays the government a $10,000,000 U.S. signing bonus and agrees to pay the government royalties of 8% of gross production and 5% severance tax. Ibis bears all of the costs associated with exploration, development, and production.

   During 2003, Ibis spends $7,000,000 on exploration and drilling costs. Gross revenue was $5,000,000 and production costs were $2,000,000. The income tax laws allow deduction of all production costs with exploration and drilling costs deductible over a 7-year period. The tax rate is 40%.

REQUIRED: Show how the gross revenue for 2003 would be shared by the parties.

10. Fortune Company enters into a risk service agreement with the Chilean government.Fortune pays the government, in U.S. dollars, a $5,000,000 signing bonus and also agrees to pay all of the costs associated with exploration, development, and production. (The contract defines costs incurred in the exploration and development phase of each project area as being capital costs [CAPEX] and all costs incurred in the production phase as being operating costs [OPEX]).

    Each year in which production occurs the government agrees to pay Fortune a fee comprised of the following:

    a. All OPEX incurred in the current year.
    b. 1/10th of all unrecovered CAPEX.
    c. $0.40 per barrel on production from 0 to 4,000 bbl per day, $0.60 per barrel on production from 4,001 to 10,000 bbl per day, and $0.90 per barrel on production above 10,000 bbl per day.

    The maximum total fee that will be paid in any year is $1.20 per barrel times the total number of barrels produced. Any unrecovered OPEX or CAPEX (unrecovered due to the maximum fee) can be carried forward to future years.

    Assume that in 2002, production commences on the Llama Field. At that time Fortune had spent $10,000,000 on CAPEX and during 2002 spends $3,000,000 in OPEX. Production equals 5,000,000 barrels or 5,000,000/365 = 13,699 barrel per day.

    REQUIRED: Compute the fee that Fortune Company would receive for 2002.

chapter fifteen

Analysis of Oil & Gas Companies' Financial Statements

The financial statements of oil and gas producing companies contain a great deal of information that is not contained in the financial statements of companies in other industries. In chapter 2, the reporting requirements in the U.S. were discussed, specifically the use of the two alternative historical cost accounting methods that are supplemented by extensive disclosures. These reporting requirements resulted primarily from the SEC's conclusion that while neither the successful-efforts method nor the full-cost method effectively communicate information that is useful in projecting the future cash flows of oil and gas producing companies, the extensive disclosures do provide that information. But, by allowing companies to use a historical-cost method to account for exploration, development, and production activities, the SEC preserved the historical-cost information for potential use by financial statement users in comparing oil and gas producing companies to companies in other industries.

In this chapter various ratios and techniques employed by financial statement users in evaluating the financial performance of oil and gas producing companies are discussed. The chapter focuses on the ratios that are unique to oil and gas producing companies.

Comparing Financial Statements

Financial statements of oil and gas companies are evaluated for several different purposes. Oil and gas companies compete in capital markets for new investment dollars. Investors and lenders must determine in which companies to invest and to which companies to lend money. In determining which companies to invest in, investors and lenders use various criteria, including the analysis of certain key ratios or performance measures. Investors and lenders then compare the ratios to identify the firms that have the greatest potential for successful performance in the future.

In addition, companies are continuously seeking to improve their own performance. One technique a company can use to evaluate its performance is to compare itself to other similar companies. This comparison is commonly referred to as **benchmarking**. The process involves identifying a pool of companies who are "peer" companies, *i.e.,* companies that are similar to the company in size, operations, and various other factors. Performance measures or key ratios are computed for all of those companies, then analyzed to determine how the company compares to its peer companies. In this way, comparative strengths and weaknesses can be identified. A plan can then be established and implemented to build on the strengths and correct the weaknesses. Benchmarking is a critical step in improving the quality of business processes.

Whether the financial statement analysis is performed by outside investors or for internal benchmarking, it is important to keep in mind that an analysis which involves trends over several years is typically much more informative than the calculation of ratios for a single year. Recall from chapter 2 that one of the unique characteristics of the oil and gas industry is the long time span between when exploration expenditures are made and when the results are known, *i.e.*, whether oil or gas has been found. The use of ratios computed over several years will not resolve the timing problem of expenditures made in one year and results known in another, but will somewhat mitigate the effects and is superior to single year analysis.

The ratios commonly used in the oil and gas industry for financial statement analysis and benchmarking are discussed in the remainder of this chapter. The ratios discussed are reserve ratios, reserve cost ratios, reserve value ratios, and financial ratios.

The source of the data necessary to compute most of the ratios unique to oil and gas companies is typically the *SFAS No. 69* disclosures. As discussed in chapter 13, *SFAS No. 69* disclosures must be presented in the annual reports of both successful-efforts and full-cost companies. Specifically, the following disclosures are required:

Historical

1. Proved reserve quantity information
2. Capitalized costs relating to oil and gas producing activities
3. Costs incurred for property acquisition, exploration, and development activities
4. Results of operations for oil and gas producing activities

Future

5. A standardized measure of discounted future net cash flows relating to proved oil and gas reserve quantities
6. Changes in standardized measure of discounted future net cash flows relating to proved oil and gas reserve quantities

Reserve Ratios

Reserve replacement ratio

One key performance measure is the reserve replacement ratio. The reserve replacement ratio measures a company's ability to continue to operate in the future by measuring the extent to which a company is replacing the reserves it is producing. A firm that is not replacing the reserves it produces will ultimately deplete its pool of available reserves and be forced to either purchase reserves in-place or simply cease to do business. Since purchasing reserves in-place is typically more expensive than acquisition of reserves through exploration and drilling, it is desirable for a company to consistently add to its reserves at a rate equal to or higher than its rate of production. Accordingly, a company replacing the reserves it produces should have a reserve replacement ratio of at least one.

One problem with the reserve replacement ratio is determining which categories of reserves to include in the calculation. Specifically, reserve additions resulting from revisions in previous estimates and purchases of reserves in-place are often treated in different ways. Consequently, there are three different ways to calculate the reserve replacement ratio. Each method is discussed below. The most basic formula is:

Reserve Replacement Ratio =

$$\frac{\text{Extensions and discoveries} + \text{Improved recovery}}{\text{Production}}$$

Computed in this manner, the reserve replacement ratio considers only reserve additions resulting from current discoveries, extensions, and improved recovery. Some analysts argue that economic conditions may result in revisions in previously discovered reserves, and that those revisions should not be overlooked in examining a company's ability to replace its reserves. On the other hand, many of the conditions that result in revisions in estimates are beyond the control of a company's management. Thus, inclusion of those reserves does not accurately reflect management's role in reserve replacement. An alternative reserve replacement ratio that includes any current year revisions in reserve estimates in the numerator of the formula is as follows:

Reserve Replacement Ratio =

$$\frac{\text{Extensions and discoveries} + \text{Improved recovery} + \text{Revisions in previous estimates}}{\text{Production}}$$

Still another calculation includes reserves that are purchased in-place (rather than acquired through discovery). Some analysts argue that, in certain situations, purchasing reserves is an alternative that may be in the best interest of the company. If reserves acquired by purchase are included in the numerator of the reserve replacement ratio, then any reserves sold in-place (as opposed to through production) should be added to the denominator as follows:

Reserve replacement ratio =

$$\frac{\text{Extensions and discoveries} + \text{Improved Recovery} + \text{Revisions in previous estimates} + \text{Purchases of reserves in-place}}{\text{Production} + \text{Sales of reserves in-place}}$$

Presumably, a firm could not consistently replace its reserves by purchasing reserves in-place; therefore, the above formula might only be useful during those periods when in-place purchases are material.

EXAMPLE

Reserve Replacement Ratio

Assume that the following information appears in a *SFAS No. 69* disclosure on Lucky Company's 2002 annual report. This basic data is used throughout the chapter to calculate various ratios.

Estimated Quantities of Net Proved Crude Oil and Natural Gas Liquids (World-Wide Totals) *in Thousands of Barrels*

Year ended Dec. 31	2000	2001	2002
Beginning of year PR	2,553	2,654	2,729
Revisions of previous estimates	60	130	221
Improved recovery	210	132	115
Purchases of reserves in-place	50	15	10
Sales of reserves in-place	(23)	(32)	(37)
Extensions and discoveries	53	73	65
Production	(249)	(243)	(275)
End of year PR	2,654	2,729	2,828

The reserve replacement ratios calculated using the three alternative methods for 2000, 2001, and 2002 are:

2000:

$$\text{Reserve replacement ratio} = \frac{53 + 210}{249} = 1.056$$

$$\text{With revisions} = \frac{53 + 210 + 60}{249} = 1.297$$

$$\text{With revisions and in-place sales} = \frac{53 + 210 + 60 + 50}{249 + 23} = 1.371$$

2001:

$$\text{Reserve replacement ratio} = \frac{73 + 132}{243} = 0.844$$

$$\text{With revisions} = \frac{73 + 132 + 130}{243} = 1.379$$

With revisions and in-

$$\text{place purchases and sales} = \frac{73 + 132 + 130 + 15}{243 + 32} = 1.273$$

2002:

$$\text{Reserve replacement ratio} = \frac{65 + 115}{275} = 0.655$$

$$\text{With revisions} = \frac{65 + 115 + 221}{275} = 1.458$$

With revisions and in-

$$\text{place purchases and sales} = \frac{65 + 115 + 221 + 10}{275 + 37} = 1.317$$

If a company has both oil and gas reserves, the reserve replacement ratio and the reserve life ratio (discussed below) typically are calculated separately for each mineral. The ratios may be calculated using equivalent units if called for by the particular analysis being done.

Reserve life ratio

The reserve life ratio is used to approximate or measure the number of years that production could continue at the current rate if no new reserves were added. The higher the reserve life ratio, the longer a firm could continue to generate enough cash flow to cover its financial obligations even if it curtailed exploration and discovery activities—assuming current production generates enough cash flow to cover financial obligations. The reserve life ratio is calculated as follows:

$$\text{Reserve Life Ratio} = \frac{\text{Total proved reserves at the beginning of the year}}{\text{Production}}$$

The ratio can be computed separately for oil reserves and for gas reserves. The average reserve life ratio for large companies is approximately eight to nine years for oil and longer than eight to nine years for gas.

EXAMPLE

Reserve Life Ratio

Using the basic data from the previous example for Lucky Company, the reserve life ratio for each year is:

2000	**2001**	**2002**
$\frac{2,553}{249} = 10.25$ years	$\frac{2,654}{243} = 10.92$ years	$\frac{2,729}{275} = 9.92$ years

Net wells to gross wells ratio

Companies typically disclose their total gross wells and net wells in their SEC filings. Gross wells is the number of wells in which the company has any working interest, regardless of how large or small the interest may be. Net wells is the net interest in the wells, *i.e.*, the total of all working interests in all wells. It is computed by multiplying each well in which a company has a working interest by the relevant working interest percentage and summing the result. For example, if a company had a 100% WI in 5 wells, a 75% working interest in 3 wells and a 50% WI in 2 wells, it would have 10 gross wells (*i.e.*, 5 + 3 + 2) and 8.25 net wells (*i.e.*, 5 x 1.0 + 3 x .75 + 2 x .50).

The net wells to gross wells ratio is calculated as follows:

$$\text{Net Wells to Gross Wells} = \frac{\text{Net Wells}}{\text{Gross Wells}}$$

The ratio of net wells to gross wells is used as a gauge of future profitability. The rationale is if a company owns a large interest in each well, the company is likely to *be* the operator, *benefit from* being the operator, and, as the operator, have a greater say in operations. In addition, the company is likely to be more profitable as a result of consolidated interests (*i.e.*, having relatively larger interests in fewer properties rather than having to spread its resources over more properties with smaller interests). A high ratio indicates that a company owns relatively large working interests in wells while a low ratio indicates that a company owns many small working interests.

EXAMPLE

Net Wells to Gross Wells Ratio

The table below shows Lucky's interests in various wells and the calculation of gross wells and net wells:

	Gross Wells	Working Interest	Net Wells
	15	100%	15
	10	75%	7.5
	15	60%	9
	20	50%	10
	5	40%	2
	2	20%	.4
	6	15%	.9
	2	10%	.2
Total	75		45

Lucky Company has an interest in 75 wells so its gross wells is 75. Lucky's net wells is 45 and its ratio of net wells to gross wells is:

$$\text{Net wells to gross wells} = \frac{45}{75} = 0.60$$

Average reserves per well ratio

The average reserves per well ratio is another measure used to evaluate a company's future profitability. The measure calculates the average reserves produced per well. A high average reserves per well ratio indicates that a given quantity of reserves can be produced with fewer wells and thus be produced more efficiently and more profitably. Consequently, the higher the average reserves per well ratio, the greater the company's future profit potential. The formula is:

Average reserves per well ratio =

$$\frac{\text{Total proved reserves at beginning of year}}{\text{Net wells}}$$

EXAMPLE

Average Reserves Per Well Ratio

Using the proved reserves presented earlier for Lucky Company and the net well calculation presented earlier, the average reserves per well for each of the three years are (assuming the net wells do not change over the years):

2000:

$$\frac{2{,}553{,}000}{45} = 56{,}733.333 \text{ bbl/well}$$

2001:

$$\frac{2{,}654{,}000}{45} = 58{,}977.778 \text{ bbl/well}$$

2002:

$$\frac{2{,}729{,}000}{45} = 60{,}644.444 \text{ bbl/well}$$

When a company has both oil and gas reserves, the average reserves per well ratio and the average daily production per well ratio (discussed below) are normally computed by converting the reserves to a common unit of measure based on energy content, specifically, barrels of energy (BOE). Most frequently the British thermal unit conversion (Btu) (approximately 6 to 1) is used to compute BOE (chapters 6 and 7). For example, if Lucky Company also had proved gas reserves at the beginning of 2000 of 3,000,000 Mcf, the average reserves per well ratio for 2000, assuming the same net wells would be:

$$\frac{2{,}553{,}000 + 3{,}000{,}000/6}{45} = 67{,}844.44 \text{ BOE/well}$$

Average daily production per well

The average daily production rate per well is yet another measure of a company's future profitability. The higher the average daily production per well, the more efficiently and profitably the reserves can be produced. The average daily production per well is computed as follows:

$$\text{Average daily production per well} = \frac{\text{Annual production}/365}{\text{Net wells}}$$

EXAMPLE

Average Daily Production Per Well

Using the reserve and production information presented thus far for Lucky Company, the average daily production per well for each of the three years is:

2000:

$$\frac{249{,}000/365}{45} = 15.1598 \text{ bbl/day/well}$$

2001:

$$\frac{243{,}000/365}{45} = 14.7945 \text{ bbl/day/well}$$

2002:

$$\frac{275{,}000/365}{45} = 16.7428 \text{ bbl/day/well}$$

Reserve Cost Ratios

The reserve cost ratios are based on costs and on reserves. When a company has both oil and gas reserves, the reserve cost ratios are computed by converting the reserves to a common unit of measure based on energy content (BOE).

Finding costs ratios

Finding costs per BOE is one of the most common, but difficult to define, performance measures used in evaluating oil and gas operations. Finding costs per BOE is the most frequently cited ratio utilized in evaluating the efficiency of a company in adding new reserves. The basic ratio consists of the *finding costs of adding new reserves* divided by the *new reserves added.* The difficulty with calculating and using the finding costs per BOE ratio results from several factors.

First, there is no consensus regarding which costs should be included as finding costs. Second, companies use different methods of accounting for oil and gas exploration and development operations, *i.e.*, full cost and successful–efforts. Consequently, even if there were a specific definition of finding costs, the amounts would likely still differ since the various accounting methods treat the costs differently in terms of expense and capitalization. Third, there is typically a timing difference between the period(s) when the finding costs were actually expended and when the new reserves are actually reported in the financial statements. This timing difference poses a difficulty in interpreting the finding costs per BOE ratio. Finally, there is some debate as to which reserve estimates should actually be used in the calculation. These issues are discussed below.

Two methods of accounting for exploration and development costs are currently accepted in practice, the successful-efforts method and the full-cost method. In general, under the successful-efforts method, geological and geophysical (G&G) exploration costs are written off as incurred. The costs of dry exploratory wells are written off when the determination is made that the well is dry. The costs of successful exploratory wells and successful and dry development wells are capitalized and amortized over production. Under the full-cost method, all costs incurred in exploration, drilling, and development are capitalized and amortized. Clearly, any attempt to calculate finding costs for a sample of both full-cost and successful-efforts companies would require a detailed evaluation of all of the reported costs to ensure that equivalent costs, whether expensed or capitalized, are used.

In computing finding costs per BOE, there should be correspondence or matching between the costs in the numerator and the reserves in the denominator. The difficulty is determining which reserves should be used to correspond to the costs in the numerator. Reserves categories include reserves added through discoveries and extensions, purchases of reserves in-place, revisions in reserve estimates, or enhanced recovery. If reserves added through extensions and discoveries are to be used then finding costs per BOE should be calculated as:

Formula 1 without revisions:

Finding costs per BOE =

$$\frac{\text{G\&G exploration costs + All exploratory drilling costs}}{\text{Reserve extensions and discoveries (excluding revisions)}}$$

or

Formula 1 with revisions:

Finding costs per BOE =

$$\frac{\text{G\&G explorations costs + All exploratory drilling costs}}{\text{Reserve extensions and discoveries (including revisions)}}$$

Some analysts include current reserve revisions while others exclude them. There is no single consistently used method. One possible solution that is frequently used is re-computing the previous year's finding costs per BOE by adjusting the reserves for that year by revisions reported in the current year. The logic is that, since the revisions relate to prior years rather than the current year, the revisions should be reflected in the ratio for the prior years. Obviously this solution assumes that finding costs per BOE would be analyzed in the context of several years' data rather than for a single year.

The next issue is whether to include reserves purchased in-place. If these reserves are to be included in the calculation, proved properties must be included in the numerator so that costs and reserves correspond. The formula would be:

Formula 2 without revisions:

Finding costs per BOE =

$$\frac{\text{G\&G exploration costs + Exploratory drilling costs + Proved properties}}{\text{Reserve extensions and discoveries + Purchases of reserves in-place}}$$

Formula 2 with revisions:

Finding costs per BOE =

$$\frac{\text{G\&G exploration costs + Exploratory drilling costs + Proved properties}}{\text{Reserve extensions and discoveries + Purchases of reserves in-place (including revisions)}}$$

Finally, sometimes finding costs per BOE is computed by attempting to include all costs necessary to replace reserves. This ratio includes all of the costs in the above ratio plus the cost of unproved properties and development drilling expenditures:

Formula 3 with revisions:

Finding costs per BOE =

$$\frac{\text{G\&G exploration costs + All exploratory and development drilling costs + Proved and unproved property acquisition costs}}{\text{All reserve additions (including revisions)}}$$

The purpose of a development well is not to find reserves but to produce previously discovered proved reserves. Consequently, when development drilling costs are included in the ratio, this ratio is sometimes referred to as a finding and development cost ratio or a reserve replacement cost ratio.

These different formulae can result in significantly different finding costs per BOE. Gaddis, Brock, and Boynton used the financial statements of several oil and gas producers to demonstrate the vastly different results in finding costs per BOE.[1]

- Method A: G&G costs and exploratory drilling costs only divided by reserve additions excluding revisions.
- Method B: G&G costs and exploratory drilling costs only divided by reserve additions including revisions.
- Method C: G&G costs, exploratory drilling costs and development drilling costs divided by all reserve additions including revisions.

The following results were computed using a five year period (1986-1990):

Company	Method A	Method B	Method C
Amoco	$6.59	$3.20	$6.15
ARCO	$4.19	$2.93	$4.46
Chevron	$8.75	$2.90	$5.23
Conoco	$5.58	$3.27	$6.69
Exxon	$4.53	$3.18	$9.27
Marathon	$2.68	$2.38	$5.68
Mobil	$5.97	$2.34	$3.54
Phillips	$3.53	$1.61	$2.74
Texaco	$4.71	$1.81	$4.04
Unocal	$2.04	$1.93	$4.43
Average	$4.86	$2.56	$5.22

Given the lack of correspondence between the timing of the expenditure and the addition of the reserves, Gaddis, Brock, and Boynton recommend computing finding costs using moving averages across several years.

EXAMPLE

Finding Costs Per BOE

Using the reserve disclosure data presented earlier for Lucky Company, finding costs per BOE are computed below using the various formulae. Assume the following costs (in thousands):

	2000	2001	2002
Unproved property acquisition	$ 25	$ 45	$ 80
Proved property acquisition	50	50	50
G&G	300	350	325
Exploratory drilling (including dry hole)	321	400	450
Development drilling	150	90	200

Formula 1 without revisions:

2000:

$$\frac{\$300 + \$321}{53} = \$11.717\ /\text{BOE}$$

2001:

$$\frac{\$350 + \$400}{73} = \$10.274\ /\text{BOE}$$

2002:

$$\frac{\$325 + \$450}{65} = \$11.923\ /\text{BOE}$$

Formula 1 with revisions:

2000:

$$\frac{\$300 + \$321}{53 + 60} = \$5.496\ /\text{BOE}$$

2001:

$$\frac{\$350 + \$400}{73 + 130} = \$3.695\ /\text{BOE}$$

2002:

$$\frac{\$325 + \$450}{65 + 221} = \$2.710 \text{ /BOE}$$

Formula 2 without revisions:

2000:

$$\frac{\$300 + \$321 + \$50}{53 + 50} = \$6.515 \text{ /BOE}$$

2001:

$$\frac{\$350 + \$400 + \$50}{73 + 15} = \$9.091 \text{ /BOE}$$

2002:

$$\frac{\$325 + \$450 + \$50}{65 + 10} = \$11.000 \text{ /BOE}$$

Formula 2 with revisions:

2000:

$$\frac{\$300 + \$321 + \$50}{53 + 50 + 60} = \$4.117 \text{ /BOE}$$

2001:

$$\frac{\$350 + \$400 + \$50}{73 + 15 + 130} = \$3.670 \text{ /BOE}$$

2002:

$$\frac{\$325 + \$450 + \$50}{65 + 10 + 221} = \$2.787 \text{ /BOE}$$

Formula 3 with revisions:

2000:

$$\frac{\$300 + \$321 + \$150 + \$50 + \$25}{53 + 50 + 60 + 210} = \$2.268 \text{ /BOE}$$

2001:

$$\frac{\$350 + \$400 + \$90 + \$50 + \$45}{73 + 15 + 130 + 132} = \$2.671 \text{ /BOE}$$

2002:

$$\frac{\$325 + \$450 + \$200 + \$50 + \$80}{65 + 10 + 221 + 115} = \$2.689 \text{ /BOE}$$

In summary, since there is no consensus regarding the most appropriate formula, it is important to consistently use the formula or formulas selected. Also, in any analysis, it is important to look at trends and averages over several years.

Lifting costs per BOE

Lifting costs per BOE is a very popular performance indicator. Lifting costs per BOE is a measure that may be used to evaluate the extent to which a company is controlling its operating costs and/or how efficiently the company is getting oil and gas out of the ground.

The basic formula is:

$$\text{Lifting costs per BOE} = \frac{\text{Total annual lifting costs}}{\text{Annual production in BOE}}$$

In order to use this ratio appropriately, it is important to understand the costs that may or may not be included and how the results should be interpreted.

Frequently the term lifting costs and production costs are used interchangeably. The FASB defines production costs as:

> *those costs incurred to operate and maintain an enterprise's wells and related equipment and facilities, including depreciation and applicable operating costs of support equipment and facilities (paragraph 26) and other costs of operating and maintaining those facilities. (SFAS No. 19, par. 24)*

Using this definition, examples of production costs would include labor, repairs and maintenance, supplies, materials, fuel, property taxes, and severance taxes.

If lifting cost per BOE is used to measure how effectively companies are controlling their costs, then it is important that the ratio include only those costs actually controllable at the field level. For example, depreciation, property taxes, and severance taxes are rarely within the control of field and operating personnel. Accordingly, if all production costs are included, then the ratio may actually measure overall operating efficiency rather than the control of costs at the field level.

There are several factors that should be considered when using lifting cost per BOE in benchmarking, or when comparing several companies. For example, the ratio is computed across oil and gas production; however, it typically costs more to produce oil than to produce gas. Therefore, if one company has primarily oil production while another company primarily gas production, the first company may appear to be much less efficient compared to the second company when, in fact, that may not be the case. Careful selection of peer companies will help mitigate this issue.

Lifting costs per BOE should not be used without giving adequate consideration to revenues and net income. A company having higher lifting cost per BOE may also have higher revenues and higher net income per BOE. For example, assume that Company A's lifting cost (not including depreciation) per BOE is \$4.50 and Company B's lifting cost per BOE is \$4.20. It would appear that Company B is more efficient and, therefore, the more profitable operator. Assume, however, that Company A's costs are largely driven by the fact that it is operating in an offshore location where the crude oil is of much higher quality. As a result Company A is able to sell its oil for \$21 per barrel while Company B's average revenue is \$19. Company A's gross profit per barrel is \$16.50 (*i.e.,* \$21 – \$4.50) while Company B's gross profit per barrel is \$14.80 (*i.e.,* \$19 – \$4.20). This example illustrates the need to utilize lifting cost per BOE in the context of other financial ratios as well as the need to choose peer companies with similar operations, *e.g.*, mostly offshore versus onshore, etc.

When used appropriately, lifting cost per BOE is an informative performance measure. When computing lifting cost per BOE by using financial statement information, *Statement No. 69: Disclosure of Results of Operations from Oil and Gas Producing Activities* is commonly used as a source of lifting or production costs and another disclosure— the *Reserve Quantity Information Disclosure*— is the source of current period production information.

EXAMPLE

Lifting Costs per BOE

Assume that the production costs that appear on Lucky Company's annual report are (in thousands):

	Lifting (Production) Costs	Depreciation & Depletion
2000	\$ 500	\$ 800
2001	525	775
2002	510	750

Using the reserve production information presented previously, lifting cost per BOE would be computed as follows:

2000:

$\frac{\$500}{249}$ = \$2.008 per BOE

2001:

$\frac{\$525}{243} = \$2.160$ per BOE

2002:

$\frac{\$510}{275} = \$1.855$ per BOE

DD&A per BOE

One means of dealing with depreciation, depletion, and amortization (DD&A) is to omit DD&A on all wells, equipment, and facilities from the lifting cost per BOE calculation and separately compute a ratio of DD&A per BOE. DD&A reflects the historical cost of finding and developing reserves and, as such, relates to prior periods, while lifting costs relate to the current period. DD&A per BOE is not helpful in assessing current period efficiencies; however, since DD&A appears on the income statement, it does affect current profitability. The formula is:

$$\text{DD\&A per BOE} = \frac{\text{Total annual depreciation, depletion, and amortization}}{\text{Annual production in BOE}}$$

EXAMPLE

DD&A per BOE

Using the information presented previously for Lucky Company, DD&A per BOE is:

2000:

$\frac{\$800}{249} = \$3.213$ per BOE

2001:

$\frac{\$775}{243} = \$3.189$ per BOE

2002:

$\frac{\$750}{275} = \$2.727$ per BOE

Reserve Value Ratios

When a company has both oil and gas reserves, the reserve value ratios are computed by converting the reserves to a common unit of measure based on energy content (BOE).

Value of proved reserve additions per BOE

One criticism often leveled at ratio analyses using reserve quantities is that the analyses do not convey information about the "quality" of the reserves added. Some reserves are of high quality, *i.e.*, they require relatively less in production, transportation, or refining costs while the production of other reserves requires extensive expenditures for lifting, transportation, or refining. While the disclosures mandated by *Statement No. 69* are controversial, the use of certain value-related amounts found in the disclosures in ratio analyses has become a popular means of incorporating reserve value. One such ratio is the value of proved reserve additions per BOE. This ratio is computed using certain elements of the changes in the standardized measure disclosure and the reserve quantity information disclosure. For example:

Formula 1 without revisions:

Value of proved reserve additions per BOE =

$$\frac{\text{Changes due to extensions, discoveries, and improved recovery}}{\text{Reserve extensions and discoveries + Improved recovery (excluding revisions)}}$$

or

Formula 1 with revisions:

Value of proved reserve additions per BOE =

$$\frac{\text{Changes due to extensions, discoveries, and improved recovery + Changes due to revisions in estimates}}{\text{Reserve extensions and discoveries + Improved recovery (including revisions)}}$$

or

Formula 2 without revisions:

Value of proved reserve additions per BOE =

$$\frac{\text{Changes due to extensions, discoveries, and improved recovery + Changes due to purchases of reserves in-place}}{\text{Reserve extensions and discoveries + Improved recovery + Purchases of reserves in-place (excluding revisions)}}$$

Formula 2 with revisions:

Value of proved reserve additions per BOE =

$$\frac{\text{Changes due to extensions, discoveries, and improved recovery} + \text{Changes due to purchases of reserves in-place} + \text{Changes due to revisions in estimates}}{\text{Reserve extensions and discoveries} + \text{Improved recovery} + \text{Purchases of reserves in-place (including revisions)}}$$

As is the case with finding costs per BOE, the calculation of value of proved reserve additions per BOE requires consistency between the value figures included in the numerator and the reserve categories included in the denominator. Again, for the best results, multiple year analyses should be used.

EXAMPLE

Value of Proved Reserve Additions per BOE

The following information is taken from *SFAS No. 69* disclosures in Lucky Company's annual report:

Changes in Standardized Measure of Discounted Future Net Cash Flows *in Thousands*

Year ended Dec. 31	2000	2001	2002
Beginning of year	20,355	20,212	19,792
Changes due to prices and costs	801	500	(902)
Changes due to purchases in-place	530	220	139
Changes due to extensions, discoveries, and improved recovery	599	890	712
Changes due to revisions in estimates	699	1,100	989
Changes due to sales in-place	(225)	(350)	(400)
Changes due to production	(2,700)	(2,980)	(3,000)
Accretion of discount	153	200	250
Balance end of year	20,212	19,792	17,580

Formula 1 without revisions:

2000:

$$\frac{\$599}{53 + 210} = \$2.278/\text{BOE}$$

2001:

$$\frac{\$890}{73 + 132} = \$4.341/\text{BOE}$$

2002:

$$\frac{\$712}{65 + 115} = \$3.956/\text{BOE}$$

Formula 1 with revisions:

2000:

$$\frac{\$599 + \$699}{53 + 210 + 60} = \$4.019/\text{BOE}$$

2001:

$$\frac{\$890 + \$1{,}100}{73 + 132 + 130} = \$5.940/\text{BOE}$$

2002:

$$\frac{\$712 + \$989}{65 + 115 + 221} = \$4.242/\text{BOE}$$

Formula 2 without revisions:

2000:

$$\frac{\$599 + \$530}{53 + 210 + 50} = \$3.607/\text{BOE}$$

2001:

$$\frac{\$890 + \$220}{73 + 132 + 15} = \$5.045/\text{BOE}$$

2002:

$$\frac{\$712 + \$139}{65 + 115 + 10} = \$4.479/\text{BOE}$$

Formula 2 with revisions:

2000:

$$\frac{\$599 + \$530 + \$699}{53 + 210 + 50 + 60} = \$4.901/\text{BOE}$$

2001:

$$\frac{\$890 + \$220 + \$1{,}100}{73 + 132 + 15 + 130} = \$6.314/\text{BOE}$$

2002:

$$\frac{\$712 + \$139 + \$989}{65 + 115 + 10 + 221} = \$4.477/\text{BOE}$$

Value added ratio

Another popular reserve value ratio is the value added ratio. First, the value of proved reserve additions per BOE ratio as described above is computed. This ratio is then compared to finding costs per BOE. Specifically, the value added ratio is computed by dividing the value of proved reserve additions per BOE ratio by the finding costs per BOE ratio. The objective of this analysis is to compare the cost of finding reserves with the value added by those reserves. Obviously such a comparison requires consistent use of costs and reserve categories in each formula. For example, assume:

Finding costs per BOE =

$$\frac{\text{G\&G exploration costs + All exploratory and development drilling costs + Proved and unproved property acquisition costs}}{\text{All reserve additions (including revisions)}}$$

The comparison of finding costs and value added would require that the value of proved reserve additions ratio per BOE include similar categories of values and reserves. Accordingly,

Value of proved reserve additions per BOE =

$$\frac{\text{Changes due to extensions, discoveries, and improved recovery + Changes due to purchases of reserves in-place + Changes due to revisions in estimates}}{\text{Reserve extensions and discoveries + Improved recovery + Purchases of reserves in-place (including revisions)}}$$

A value added ratio may then be computed:

$$\text{Value added ratio} = \frac{\text{Value of proved reserve additions per BOE}}{\text{Finding costs per BOE}}$$

Companies should strive to maximize the value added ratio. Companies having the highest value added ratio are adding maximum reserve value at minimum finding costs.

EXAMPLE

Value Added Ratio

The value added ratios for 2000, 2001, and 2002 for Lucky Company are computed below. (The finding costs per BOE and value of proved reserve additions per BOE ratios were computed previously.)

Finding costs per BOE: Formula 3 with revisions =

$$\frac{\text{G\&G exploration costs + All drilling costs + Proved and unproved property acquisition costs}}{\text{All reserve additions (including revisions)}}$$

Divided by

Value of proved reserve additions per BOE: Formula 2 with revisions =

$$\frac{\text{Changes due to extensions, discoveries and improved recovery + Changes due to purchases of reserves in-place + Changes in revisions in estimates}}{\text{Reserve extensions and discoveries + Improved recovery + Purchases of reserves in-place (including revisions)}}$$

Value added ratio:

2000:	**2001:**	**2002:**
$\frac{\$4.901}{\$2.268} = 2.161$	$\frac{\$6.314}{\$2.671} = 2.364$	$\frac{\$4.477}{\$2.689} = 1.665$

Financial Ratios

There are many other ratios that are frequently used in financial statement analysis. These ratios are used to evaluate companies in all industries, including oil and gas companies. These ratios include:

Liquidity ratios

$$\text{Current ratio} = \frac{\text{Current assets}}{\text{Current liabilities}}$$

$$\text{Quick ratio} = \frac{\text{Liquid current assets}}{\text{Current liabilities}}$$

Working capital = Current assets – Current liabilities

Financial strength

$$\text{Debt to stockholder's equity} = \frac{\text{Total long and short-term debt}}{\text{Total stockholder's equity}}$$

$$\text{Debt to assets} = \frac{\text{Total long and short-term debt}}{\text{Total assets}}$$

$$\text{Times interest earned} = \frac{\text{Net income}}{\text{Total interest expense}}$$

Profitability ratios

$$\text{Net income to sales} = \frac{\text{Net income}}{\text{Sales}}$$

$$\text{Return on stockholder's equity} = \frac{\text{Net income}}{\text{Total stockholder's equity}}$$

$$\text{Return on assets} = \frac{\text{Net income}}{\text{Total assets}}$$

$$\text{Cash flow from operations to sales} = \frac{\text{Cash flow generated in the current year}}{\text{Current year sales}}$$

$$\text{Price/Earnings ratio} = \frac{\text{Market value of common per share}}{\text{Adjusted earning per share}}$$

$$\text{Price/Cash flow ratio} = \frac{\text{Market value of common per share}}{\text{Cash flow per share}}$$

Some analysts incorporate *Statement No. 69* reserve values into some of these formulae. For example, in computing ratios that include total assets or stockholder's equity (*i.e.*, return on assets or return on stockholder's equity), the standardized measure of discounted future net cash flows may be substituted for proved properties as reported in the balance sheet. Additionally, a comparison between the book value of oil and gas assets and the standardized measure of discounted future net cash flows is sometimes used as a measure of the financial strength of underlying oil and gas assets.

PROBLEMS

1. What is benchmarking?

2. What is the reserve replacement ratio? What is the reserve replacement ratio attempting to measure? How would you interpret it?

3. What is the reserve life ratio? What is the reserve life ratio attempting to measure? How would you interpret it?

4. What does a high net wells to gross wells ratio indicate?

5. What does a low average reserves per well ratio indicate?

6. How do you interpret the average daily production per well ratio?

7. Discuss difficulties with computing and applying finding costs ratios.

8. How do you interpret finding cost ratios? Why are they so popular in financial statement analysis?

9. What does a high (low) lifting cost per BOE indicate? When does lifting cost per BOE indicate that costs are effectively being controlled at the field level?

10. How do you interpret DD&A per BOE?

11. What is the value of proved reserve additions ratio attempting to measure? How do you interpret the value of proved reserve additions ratio?

12. What is the value added ratio? What does a high (low) value added ratio indicate?

13. The following reserve table appeared in the financial statements of Beard Company:

Estimated Quantities of Net Proved Crude Oil and Natural Gas (World-Wide Totals only) *in Thousands of Barrels and Millions of cubic feet*

Year ended Dec. 31	2005		2006		2007	
	Oil	Gas	Oil	Gas	Oil	Gas
Beginning of year	171	779	234	783	335	724
Revisions of previous estimates	10	12	15	31	(11)	22
Improved recovery	21	30	25	23	15	50
Purchases of reserves in-place	0	0	12	12	0	24
Sales of reserves in-place	(12)	(12)	(20)	(99)	(70)	(24)
Extensions and discoveries	69	78	90	42	6	150
Production	(25)	(104)	(21)	(68)	(24)	(76)
End of year	234	783	335	724	251	870

YR	Net Wells	Gross Wells
2005	750	2000
2006	840	1910
2007	900	2050

REQUIRED: Compute the following ratios for all three years.

a. The reserve replacement ratio computed for all three methods and for oil and gas separately.
b. The reserve life ratio computed for oil and gas separately.
c. The net wells to gross wells ratio.
d. The average reserves per well ratio computed using BOE, *i.e.*, combining reserves based on relative energy content.
e. The average daily production per well computed using BOE.

14. Beard Company reported the following costs on its financial statements (in thousands):

	2005	2006	2007
Unproved property acquisition	$405	$200	$800
Proved property acquisition	0	500	350
G&G	500	650	425
Exploratory drilling (including dry hole)	221	700	650
Development drilling	50	90	300

REQUIRED: Using the reserve disclosure for Beard Company in Problem 13 and the data presented in this problem, compute finding costs per BOE using the various formulae.

15. Beard Company reported the following expenses in its financial statements (in thousands):

	Lifting Costs	DD&A
2005	$211	$500
2006	226	450
2007	183	525

REQUIRED: Using the reserve disclosure for Beard Company given in Problem 13 and the data presented in this problem:

a. Compute lifting costs per BOE.
b. Compute DD&A per BOE.

16. Beard Company's Statement No. 69 disclosures included the following information:

Changes in Standardized Measure of Discounted Future Net Cash Flows *in Thousands*

Year ended Dec. 31	2005	2006	2007
Beginning of year	$1,131	$ 643	$1,656
Changes due to prices and costs	(830)	550	200
Changes due to purchases in-place	0	156	139
Changes due to extensions, discoveries, and improved recovery	400	180	712
Changes due to revisions in estimates	99	310	39
Changes due to sales in-place	(22)	(35)	(40)
Changes due to production	(270)	(398)	(500)
Accretion of discount	135	250	199
Balance end of year	$ 643	$1,656	$2,405

REQUIRED: Using the information for Beard Company in Problems 13, 14, and 15 and in this problem:

a. Compute the value of proved reserve additions per BOE using the various formulae.
b. Compute the value added ratio for each year, utilizing formula 3 for finding costs per BOE, and formula 2 with revisions for value of reserves added per BOE.

17. Discuss the ratios computed for Beard Company in problems 13, 14, 15, and 16. What is your assessment of the performance and future potential of Beard Company?

NOTES

1. Gaddis, Dwight, Brock, Horace, and Boynton, Charles, "Pros, Cons of Techniques Used to Calculate Oil, Gas Finding Costs," *Oil and Gas Journal*, June, 1992, pp. 93-95.

chapter sixteen

Overview *of* Gas Pipeline Accounting

The authors gratefully acknowledge Sabra L. Harrington, CPA, Duke Energy Corporation for her material and significant contributions in preparing this chapter.

In today's market, pipeline companies primarily provide a transportation service. Natural gas belonging to a pipeline's customer is put into the pipeline (receipt point) and is delivered to a physical point designated by the customer (delivery point). The delivery point may be the customer's facilities, a natural gas storage field, or an interconnect with another pipeline. Because the actual measured flow of the gas received and the gas delivered will not be the same as the expected gas flow, imbalances between pipelines and their customers will be created. Pipeline companies and their customers also store gas in storage facilities in low usage months to meet demand in peak usage months. This chapter discusses the regulations and accounting treatment related to these issues.

General Regulations

Companies involved in the transportation of natural gas are regulated by governmental authorities. For companies that transport natural gas between states, *i.e.,* interstate transportation, the Federal Energy Regulatory Commission (FERC) is the governing body. (FERC is a division of the U.S. Department of Energy.) Companies that transport natural gas within states, *i.e.,* intrastate transportation, are typically governed by state commissions.

FERC, or other specific governing authorities:

- approve construction of new or expanded pipeline facilities
- approve the Tariff rates that are charged to customers
- administer various accounting and reporting requirements

FERC initially approves all interstate pipeline projects by reviewing proposed costs associated with construction and operation, market evidence that exhibits the need for additional pipeline capacity, and the environmental and operating plans of the projects. The rates that pipelines charge customers are recorded as revenues and are based on construction and operating costs and a return on equity that have been approved by the regulator. These rates are specified in a document called a **Tariff.** Tariff rates may vary for different areas of transportation, for instance market area zones and supply area zones, and are established for each type of transportation service provided.

The FERC's accounting and reporting requirements are published in the Code of Federal Regulations. The code is revised at least once each calendar year and includes a chart of accounts and an accounting methodology that must be used by all gas pipelines subject to the jurisdiction of FERC. This standardization provides FERC with financial data that is comparable among the pipelines, making it easier for FERC to ensure compliance and accuracy. A summary of the account numbers used in the FERC's chart of accounts (Table 16-1) is as follows:

Type of Account	Number
Balance Sheet (except Plant detail)	100–299
Assets	100–199
Liabilities	200–299
Plant Detail	300–399
Income	400–432, 434–435
Retained Earnings	433, 436–439
Operating Revenues	480–499
O&M Expenses	700–949

Table 16-1 — ***Summary of Account Numbers***

While interstate pipeline companies are required to report to the FERC, public pipeline companies must report to the SEC as well. Reports to the SEC and other financial statement reports provided to bankers or other external parties must comply with GAAP; reports to the FERC must use the chart of accounts summarized above. Consequently, many pipelines maintain a separate Chart of Accounts for GAAP and management reporting that can be cross-referenced or converted to the related FERC accounts. Thus, ultimately, every transaction is recorded to a specific FERC account. (While specific FERC versus GAAP differences have dwindled in the past 5 years, there are a few remaining differences in fairly complex areas —primarily taxes, purchase accounting, and various classification issues. This chapter focuses predominantly on FERC versus GAAP accounting.)

FERC's Code of Federal Regulations (CFR) contains detailed account definitions and specific instructions for recording various accounting transactions. The FERC rules are precise and rigid, allowing almost no flexibility in application. For example, for entries to Account 808.1 *Gas Withdrawn from Storage—Debit,* CFR states that "contra credits for entries to this account shall be made to Accounts 117.1 through 117.4, or Account 164.2 Liquefied Natural Gas Stored, as appropriate."

There are also specific industry GAAP rules for accounting in a regulated environment per *SFAS No. 71,* "Accounting for the Effects of Certain Types of Regulation." One of the primary elements of *Statement 71* is to allow regulated companies to capitalize certain costs that would otherwise be charged to expense. This treatment is allowed due to the reasonable assurance that these costs will be approved by the regulator for recovery from customers. This rule is why many natural gas pipeline companies have significant deferred assets, whereas most non-regulated entities are not able to carry large deferred assets on their balance sheets.

Table 16-2 identifies the primary FERC accounts that are used in the gas accounting for pipelines:

PIPELINE COMPANY—FERC ACCOUNTS

Income Statement

Revenue Accounts	**Account Number**
Transportation revenue—gathering	489.1
Transportation revenue—transmission	489.2
Transportation revenue—distribution	489.3
Storage revenue	489.4
Sale of extracted products	490
Gas processed by others	491
Other gas revenues	495
Expense Accounts	**Account Number**
Regulatory debits	407.3
Regulatory credits	407.4
Gas purchases—wellhead	800
Gas purchases—field line	801
Exchange gas	806
Gas withdrawn from storage	808.1
Gas delivered to storage	808.2
Gas used for compressor station fuel—credit	810
Gas used for products extraction—credit	811
Gas used for other utility operations—credit	812

Table 16-2 — *FERC Accounts*

Expense Accounts (continued)	**Account Number**
Other gas supply expenses	813
Compressor station labor and expenses	853
Gas used for compressor station fuel	854
Mains expenses	856
Transmission and compression of gas by others	858
Gas processed by others	777

Balance Sheet

Asset Accounts	**Account Number**
Customer accounts receivable	142
Miscellaneous current & accrued assets	174
Gas stored—current	164.1
Gas stored—base gas	117.1
System balancing gas	117.2
Gas stored—noncurrent	117.3
Gas owed to system gas	117.4
Other regulatory assets	182.3

Liability Accounts	**Account Number**
Accounts payable	232
Miscellaneous current & accrued liabilities	242
Other deferred credits	253

Transportation Services

In the past, pipeline companies purchased gas from a producer and transported this company-owned gas to the end user. In recent years, the number of industry players has increased, with gas producers selling directly to **end users** (the parties who actually use the natural gas). Additionally, marketers and brokers are acquiring gas for sales to each other, **local distribution companies (LDC),** industrial facilities, and to other end users. As a result, pipeline companies typically serve only as the transporter of the gas versus *both* the purchaser and the transporter of the gas. The customers are now responsible for the direct purchase of natural gas from producers or marketer/brokers to ensure that their gas supply needs are met.

Pipelines provide transportation services to their customers under contracts that are in accordance with the pipeline's FERC Tariff. These contracts have historically been long-term in nature—typically more than 10 years. As these long-term contracts expire, replacement contracts are being negotiated for shorter average periods, but the contracts can range anywhere from 30 days to 10 years. The transportation contract specifies the type of service to be provided, the points of receipt and delivery, the transportation fees, and other contractual terms such as maximum daily quantities, quality of gas, etc.

There are four types of agreements associated with transportation:

1. Transportation *for* Others – TFO
2. Operational Balancing Agreements – OBA
3. Transportation *by* Others – TBO
4. Exchange Agreements

A **Transportation for Others Agreement (TFO)** is the primary customer contract for transportation services and represents an agreement for the pipeline to transport a customer's gas on a pipeline. The *customer owns the gas* and can be a producer, marketer, broker, LDC, or end user. The pipeline company agrees to receive the customer's gas and deliver the gas at designated points along the pipeline system.

Since the amount of natural gas received and delivered cannot be precisely controlled, physical gas activity measured through a gas meter does not always match the expected gas flow. This mismatch creates imbalances between the pipeline and the customers. **Operational Balancing Agreements (OBAs)** are used to balance the difference between measured and expected gas flow at a particular meter. The natural gas delivered to a meter is allocated among the various shippers that had designated deliveries at that point based on the amount of gas that they requested, or nominated at that meter. The remaining difference of actual measured gas flow at the meter is then allocated to the OBA contract.

A **Transportation by Others Agreement (TBO)** is an arrangement between pipelines, *i.e.*, one pipeline becomes the shipper on another pipeline system. These agreements are necessary when one pipeline needs to lease space on another pipeline in order to provide customers with additional market access. TBOs are also referred to as "858" contracts because the expense recorded by the leasing pipeline is recorded in FERC Account 858.

Under **Exchange Agreements,** gas does not physically flow from one point to another. Instead, volumes received at one point will be "exchanged" at another point for a like volume of gas. Thus, the volume of gas does not need to be physically transported. Exchange Agreements can occur between two pipelines or between a pipeline and a customer. They are used to balance the system and increase operational efficiencies. Generally, no fees are charged for this service because the parties receive a mutual economic benefit.

A pipeline provides two basic types, or classes, of transportation services. These same classes of service apply to storage services that may also be provided.

1. *Firm Transportation.* Under the firm transportation service class, the customer *reserves* capacity on the pipeline for its use. The pipeline company in turn is committed to deliver the specified quantity of gas. Revenues received by the pipeline include:

- Demand Charges – revenues based on the amount of space reserved
- Commodity Charges – revenues based on the actual gas transported

2. *Interruptible Transportation.* An interruptible transportation service class is a lower priority service that can be interrupted if there is not enough capacity available. Usage by Firm customers is the first priority. Commodity charges are the only revenues received by the pipeline.

The pipeline's FERC Tariff defines the rates that the pipeline may charge customers relating to transportation and storage services. The rates include the following:

- *Transportation and storage charges*
  - Demand or reservation charges for firm capacity or storage space
  - Commodity charges per MMBtu transported
  - Fuel reimbursement (to run compressors, etc.)
  - Various surcharges to recover approved incremental costs

- *Balancing charges*
  - *Cash-out charges or credits.* Certain imbalances are settled each month by either paying or receiving cash at rates specified in the Tariff. These rates are usually based on the current natural gas market prices multiplied by a factor that is dependent on the percentage out of balance. For example, if a customer is less than 5% out of balance, it may be cashed out 1.00 times the market rate. However, if the imbalance is greater than 5% (but less than 10%), the customer may be cashed out at 1.10 times the market rate.
  - *Balancing penalties.* If imbalances exceed certain thresholds that have been identified as potentially harmful to the operations of the system, penalties may be assessed. Penalties are billed based on the imbalance volumes, *e.g.,* $25 per MMBtu of imbalance.

REVENUE ACCOUNTING

All charges billed to pipeline customers for transportation services are recorded to the following FERC revenue accounts:

- Account 489.1 Transportation revenue—gathering
- Account 489.2 Transportation revenue—transmission
- Account 489.3 Transportation revenue—distribution

The selection of the appropriate account depends on the nature of the facilities that transport the gas - gathering, transmission, or distribution. The majority of revenues earned by interstate pipelines are attributable to their transmission facilities, and are therefore recorded in FERC Account 489.2 *Transportation revenue—transmission.*

Because the demand for gas is highly seasonal, gas is stored in natural gas storage fields by the customer or the pipeline company during lower-usage periods in order to have enough gas available for peak demand periods. If this storage capacity were not available, pipeline companies would need to increase their pipeline capacity significantly to meet peak demand. For storage of the gas, pipelines usually own or lease space in depleted gas fields that are used for storage. Storage service charges billed to customers for storing their gas are recorded in FERC Account 489.4 *Storage revenue.*

Charges and credits for balancing transactions are usually recorded in FERC Account 495 *Other Gas Revenues,* although there are also numerous special instructions for recording balancing transactions.

EXAMPLE

Transportation Services

Lucky Pipeline Company provided various transportation and storage services during April. Charges for these services per the pipeline invoice are summarized below:

Description	Invoice Amount
Winter Storage	$ 54,829.38
Enhanced Firm Transportation	15,632.62
Peaking Storage	46,516.09
Enhanced Firm Transportation	11,447.20
In/Out Storage	30,569.40
Enhanced Firm Transportation	191,393.23
Enhanced Firm Transportation	76,851.04
	$427,238.96

Entry to record transportation revenues

Account 142 Customer accounts receivable	295,324	
Account 489.2 Transportation revenue—transmission		295,324

Entry to record storage revenues

Account 142 Customer accounts receivable	131,915	
Account 489.4 Storage revenue		131,915

Note that for each type of revenue, a receivable is debited and a revenue account is credited.

For FERC reporting purposes, the detail for transportation revenues must also be maintained by zone of receipt (gathering) and delivery (transmission) using the zone definitions in the pipeline's Tariff, and by rate schedule. For management reporting purposes, it may be beneficial to break these entries down further into demand and commodity charge components since the profit margins vary for each of these revenue components.

Gas Balance: Operating and Accounting Issues

In order to maintain appropriate operational, customer, and accounting controls, pipelines must be able to identify all sources and uses of gas flow on their systems. This involves physical measurement of gas received and delivered at various meters along the pipeline, as well as the allocation of the physical gas measured and delivered to the various customers that use each meter. Estimates of certain activity may also be made since precise measurement of all movement is not possible.

The tool, *i.e.,* document, used by pipelines to gather and track this data is called a **Gas Balance.** In a Gas Balance, gas is stated in physical units based on the heating value of natural gas. A Gas Balance is similar to a financial balance sheet in that the inputs and outputs should balance to zero.

The Accounting department is responsible for gathering the information from the various sources within the company to prepare the Gas Balance. Accounting groups will use the Gas Balance to ensure that all gas flow activity has been properly identified and accounted for in the financial statements of the pipeline. In addition, the Accounting department coordinates the review of the data among the various departments that utilize the Gas Balance.

For example, the Gas Balance is utilized by the operating departments to review the efficiency of the pipeline system. For instance, how does the fuel usage summarized in the Gas Balance compare to historical usage and industry standards? How much pipeline or storage field capacity remains available for customer or company use?

The marketing department can review Gas Balance information to assess customer utilization of the pipeline receipt points and delivery points. Which points are utilized most and are therefore more valuable to customers?

The primary components of a Gas Balance for a natural gas pipeline are as follows:

Sources/Receipts of Gas

- Storage Gas Withdrawals
- Gas Received from Customers for Delivery

- Gas Received from Customers for Fuel Reimbursement
- Gas Purchases
- Line Pack Decreases

Uses/Deliveries of Gas
- Storage Gas Injections
- Gas Delivered for Customers
- Gas Used in Operations
- Line Pack Increases
- Unaccounted-for Gas

The following sections review each of these operational areas and the related accounting treatment.

IMBALANCES
(Gas Received from Customers less Gas Delivered for Customers)

Most of the gas activity on a pipeline is associated with gas that is owned by the customers. The largest receipts and deliveries in a gas balance are the amount of gas received from customers for delivery and the gas delivered for customers. The net activity for each customer's account is called an **imbalance,** which represents the amount of natural gas owed to or by the pipeline. The FERC Code of Federal Regulations specifies that these imbalances are to be recorded as either a miscellaneous current asset or liability (Account 174 or 242) based on spot prices. The offsetting account is Account 806 *Exchange gas,* an operating expense account. Note that an entry is made to record only the imbalance. No entry is made to record the gross receipt or gross delivery of the gas because the pipeline does not own the gas; it is merely transporting the gas.

Specifically, an imbalance is the difference between the volume of gas received by the pipeline from the customer and the volume delivered by the pipeline to, or for, the customer. Most pipelines require that customers also provide enough gas to reimburse the pipeline for the amount of compression gas used to move customers' gas. Therefore, the calculation of an imbalance is as follows:

Receipt – Fuel Reimbursement – Delivery = Imbalance Liability/(Asset)

EXAMPLE

Imbalance

Ignoring the fuel component, if a customer nominates to have 20,000 MMBtu delivered, but only provides 18,000 MMBtu at the receipt point, the pipeline will deliver 20,000 MMBtu and record a net asset equal to the value of 2,000 MMBtu. In other words, the customer owes the pipeline 2,000 MMBtu in order to keep the pipeline whole.

Many pipelines record adjusting entries to the imbalance accounts each month to adjust the cumulative gas owed to the current market price of natural gas.

Storage Gas

Natural gas storage field capacity may be used by either a pipeline or its customers. Pipelines usually own and operate storage fields for pipeline operating needs, as well as for customers' use, but pipelines may also lease storage space from other pipelines' storage fields.

Storage gas owned by a pipeline can be separated into two broad categories:

1. *Base Gas.* Initial line pack (*i.e.,* the volume of gas necessary to fill up a new pipeline) is recorded as an asset in the plant accounts.
2. *System Gas.* System gas is the volume of gas that may be stored by the pipeline for system balancing purposes. This gas is capitalized in Account 117.2 *System balancing gas* when it is injected into the storage facility. Account 808.2, *Gas delivered to storage,* an expense account, is credited. (When gas is purchased, it is debited to an expense account. Thus, the injection entry offsets the expense impact of the purchase entry and creates an asset.)

Certain pipeline companies no longer maintain system gas inventory. Instead, they require customers to manage their purchase, storage, and transport activity such that there is always enough gas available for monthly requirements. For instance, if a customer requires more gas in February than it is able to obtain from producers in February, the customer would need to have stored enough gas to meet the excess demands.

For those pipelines that do maintain system gas inventory, there are two alternative methods of accounting that are approved by FERC for the investment in system gas.

- Inventory Method
- Fixed Asset Method

Accounting for System Gas

There are four different asset accounts used for System Gas – 117.1 through 117.4. The following definitions are based on definitions contained within FERC Order No. 581, issued September 28, 1995.

- Account 117.1 *Gas stored—current* includes the cost of gas volumes that may be recoverable from the storage field but are currently necessary to maintain the pressure in the field for deliverability requirements for the storage facility. (A storage field must be pressurized in order to be able to deliver gas out of the field.) This gas is also known as storage field operating line pack.
- Account 117.2 *System balancing gas* is used to record a pipeline's investment in any additional system gas volumes that may be needed for balancing, transportation, and other operational purposes.
- Account 117.3 *Gas stored—noncurrent* is used to record the cost of noncurrent company-owned stored gas not includable in Accounts 117.1 or 117.2.
- Account 117.4 *Gas owed to system gas* is primarily used by pipelines that account for system gas using the Fixed Asset model (see below). Account 117.4 reflects encroachments upon, *i.e.,* reductions of system gas inventory balances that result from transportation imbalances or other operational needs. It may also be used to reflect encroachments of base gas recorded in Account 117.1 for pipelines using the inventory method (see below).

The two methods of accounting used for system gas inventory are discussed in the following paragraphs.

Inventory Method

Under the inventory method, withdrawals of system gas from storage are to be credited to Account 117.2 *System balancing gas* (a current asset) at the inventory cost of gas with a corresponding debit to Account 808.1 *Gas Withdrawn From Storage* (an expense). Acceptable costing models for use in the inventory method include LIFO, FIFO, and weighted average cost methods. Although this withdrawal entry results in an expense being recorded, the remaining gas balance activity will usually include entries that offset the net income impact of the storage activity. The reason there is usually no net income impact from this operational activity is that FERC often approves recovery of any costs associated with this activity from the pipeline's customers. The offsetting entries are illustrated in the remaining discussion of gas balance activity.

EXAMPLE

Storage Gas

Lucky Pipeline Company operates a storage field in southeast Texas where it owns and stores gas to facilitate balancing of its pipeline system. The gas in storage is valued on a weighted average cost of gas basis (WACOG). Lucky's average cost of gas activity for the months is $1.30 per MMBtu. Lucky uses the inventory method for recording system gas activity. During the month of June 2006, the company injected 250,000 MMBtu and withdrew 20,000 MMBtu.

Entry to record injection

Account 117.2 System balancing gas (asset: 250,000 x $1.30)	325,000	
Account 808.2 Gas delivered to storage (operating expense)		325,000

Entry to record the withdrawal

Account 808.1 Gas withdrawn from storage (expense: 20,000 x $1.30)	26,000	
Account 117.2 System balance gas (current asset)		26,000

Fixed Asset Method

Under the fixed asset method, encroachments upon system gas are to be credited to Account 117.4 *Gas owed to system gas* (an asset) at the current market price of gas, with a corresponding charge to Account 808.1 *Gas withdrawn from storage* (an expense).

The FERC originally intended Account 117.4 to reflect the obligation of the pipeline to replace gas volumes that had been used to meet system needs. FERC sees this obligation as a constructive requirement if a pipeline intends to remain in the business as a transporter. However, since no GAAP liability actually exists, FERC acknowledged that this account is really more of a valuation account that attempts to reflect the estimated cost to be incurred to replace system gas.

Since encroachments are recorded at the current price of gas under the fixed asset method of accounting for system gas, a pipeline must maintain various supporting records detailing monthly encroachment volumes and unit prices unless the pipeline revalues its total encroachment balance monthly. If a pipeline revalues the balance in Account 117.4, it should debit or credit Account 813 *Other gas supply expenses* with the amount of the revaluation. To the extent that there are corresponding changes in the value of imbalance receivables or payables, the pipeline should make an appropriate adjustment to Account 174 *Miscellaneous current and accrued assets* or Account 242 *Miscellaneous current and accrued liabilities* with the offsetting entry to Account 813 *Other gas supply expense.*

Gas Purchases

Since pipelines primarily function as transporters, gas is purchased by a pipeline company only to the extent that there is an operational need for gas on that pipeline's system, *e.g.*, to be used as fuel to operate compressors or to meet customer imbalance needs. Gas purchases are typically made on the spot market, usually by working with gas marketers to obtain the best daily prices available for the locations requested. Gas purchases, which are immediately put into the pipeline, are recorded as expense using specific FERC expense accounts based on the point of receipt into the pipeline.

Many times, the cost of purchasing gas for operational needs is allowed to be recovered from customers. In other words, in addition to paying transportation Tariff rates, FERC may approve additional surcharges that require the customer to pay for specific current operating costs.

EXAMPLE

Purchases

Lucky Pipeline Company works with a gas marketer to identify gas available at the ABC receipt point on its pipeline. The spot price quoted by the marketer is \$1.50 per MMBtu. For the month of June 2006, Lucky received at the ABC receipt point 2,000,000 MMBtu and delivered this gas to its storage field. There was no other pipeline or storage activity during the month.

Entry

	Debit	Credit
Account 803 Natural gas transmission line purchases (expense)	3,000,000	
Account 232 Accounts payable—marketer		3,000,000

Entry

	Debit	Credit
Account 117.2 System balancing gas (current asset)	3,000,000	
Account 808.2 Gas delivered to storage (expense)		3,000,000

Note that gas purchases are recorded as an expense, but an offsetting entry is made to reflect the storage inventory activity, resulting in no net income impact.

Gas Used in Company Operations

Once gas is produced, it must be gathered and transported to market. During these stages, gas is used for or lost in various activities. Moving the gas is accomplished by compressing the gas at compressor stations located at intervals along the pipeline system. Gas flows by expanding in the pipe from the discharge/high-pressure side of a compressor station toward the suction/low-pressure side of the next compressor station. Natural gas from the pipeline is normally the source of fuel for operating the compressors. This gas is commonly called **compressor station fuel.**

Gas produced from some natural gas production fields may need to be processed in a gas-processing plant before entry into the pipeline. A gas plant, in processing the gas, removes liquid hydrocarbons from the natural gas stream, reducing or shrinking the total volume of gas. In addition to gas volume shrinkage because of processing, some gas is also used to operate the gas plants.

Still another usage or loss of gas is from pipeline drips. Pipeline drips occur as the gas cools in the pipeline, causing liquids to accumulate. These liquids are collected at collection points along the pipeline. Other uses of gas by a pipeline company include gas used as dehydrator fuel and gas lost because of leakage in the pipeline.

FERC requires that gas used in operations be recorded as a memo entry that debits and credits two different operating expense accounts based on spot prices. The specific entry to record gas used in company operations for fuel to run gas compressors is as follows:

Entry

Account 854 Gas used for compressor station fuel	XX,XXX	
Account 810 Gas used for compressor station fuel—credit		XX,XXX

These accounts are both expense accounts, resulting in no net income impact from this entry. FERC requires this entry in order to isolate and report the fuel activity. The net system gas inventory or imbalance impact of this activity is recorded separately.

EXAMPLE

Usage

Lucky Pipeline Company incurred the following usage on its system during June 2006:

Transmission compressor fuel—12,200 MMBtu
Shrinkage through gas plant—1,240 MMBtu

The monthly posted spot price was $1.30 per MMBtu.

Entry

	Debit	Credit
Account 854 Gas used for compressor station fuel (expense: 12,200 MMBtu x $1.30)	15,860	
Account 777 Gas processed by others (expense: 1,240 MMBtu x $1.30)	1,612	
Account 810 Gas used for compressor station fuel—credit		15,860
Account 811 Gas used for products extraction—credit		1,612

If, in the same period, Lucky retained 2,000 MMBtu from gas tendered by third parties for transportation, Lucky would reduce its gas compressor fuel expense and the offsetting fuel credit account by $2,600 ($1.30 x 2,000) to reflect the use of other parties' gas. (See next section.)

Account 854, debited in the entry above, is sometimes used in a generic sense to apply to several different types of fuel usage.

Reimbursed Fuel

Pipelines are reimbursed for compressor fuel used in the operation of the pipeline. Typically, this reimbursement is made in-kind. Fuel reimbursement rates are determined by historical fuel usage for each pipeline.

EXAMPLE

Reimbursed Fuel

A customer needs 10,000 MMBtu of gas delivered. The pipeline fuel rate is 3% of MMBtu received from the customer.

Fuel Volume	=	Delivery Volume / (1 – Fuel rate %) – Delivery Volume
	=	10,000 MMBtu / (1 – .03) – 10,000 MMBtu
	=	10,309 MMBtu – 10,000 MMBtu
	=	309 MMBtu
Required Receipt	=	10,309 MMBtu

The additional 309 MMBtu received by the pipeline compensates the pipeline for the fuel used to provide the transportation service to the customer.

The entry to record reimbursed gas is as follows:

Account 812 Gas used for utility operations—credit (expense)	XX,XXX	
Account 854 Gas used for compressor station fuel (expense)		XX,XXX

Again two expense accounts are debited and credited based on spot prices, resulting in no net income impact from the entry. In other words, this entry, which is essentially a memo entry required by FERC, offsets the fuel usage previously recorded.

LINE PACK

Line pack is the volume of gas required to initially fill a new pipeline up to the system's operating pressure. The volume of gas required to fill the pipeline will depend on the size and length of the pipe to be filled and the pressure at which the pipeline is operated. The volume required is computed by engineers and measured as it is put into the pipeline, *i.e.,* as it is *metered.* This gas is purchased by and is owned by the pipeline company. Therefore, the value of the gas volume in the pipeline is capitalized as part of the cost of the pipeline. The value of the line pack is included in the rate base and affects the pipeline's cost of service. The capitalized cost of the line pack is depreciated at the rate determined by FERC in its approval of the pipeline's Tariff rates.

If cold weather is forecast in the near future, the gas dispatchers may choose to increase the pressure of the pipeline as much as possible in order to have more pipeline space available to meet the expected demand. The change in line pack would be computed by the gas control dispatchers.

- Increase in line pressure, decreases line pack volume, and increases the amount of system capacity in the pipeline
- Decrease in line pressure, increases line pack volume, and decreases the amount of system capacity in the pipeline

The accounting entry to record line pack activity each month adjusts Account 117.1 *Gas stored-base gas,* with an offset to Account 808.1 or 808.2 expense accounts to reflect the withdrawal or injection. The adjustment is valued at inventory cost.

Unaccounted-for Gas

Unaccounted-for gas represents the difference between the sum of all the inputs and the sum of all the outputs on the pipeline system. Although there are meters, allocation procedures, and other controls designed to ensure that all MMBtu's are accounted for, natural gas cannot be precisely controlled. Pipelines usually recover from their customers the value of unaccounted-for gas losses as part of their fuel reimbursement mechanism. It is recognized in the industry that pipelines will experience some loss, but if the loss becomes significant, it could represent a competitive handicap to the pipeline by making fuel rates too high.

A company must be aware of the causes of unaccounted-for gas in order to take corrective action before the loss becomes too large. Operating issues that can affect unaccounted-for gas are:

- Measurement results—accuracy of measurement can be affected by gas flow conditions, gas quality, and improper maintenance of meters.
- Allocations—the allocation of all measured receipts and deliveries to transportation contracts is critical to accurate identification of unaccounted-for gas.

Since unaccounted-for gas represents another form of operational gas usage, the accounting entries for unaccounted-for gas are the same memo entries that are used for gas used in operations and are based on spot prices.

Regulatory Assets

Certain costs incurred by the pipeline are determined by FERC to be recoverable, or **tracked costs.** These costs may include the costs of net fuel used on the system (fuel used in operations minus reimbursed fuel), unaccounted-for gas, and gas purchases. Although initially recorded as expenses, these costs are recoverable costs and are deferred on the balance sheet pursuant to both FERC and *SFAS No. 71*. Thus, they will never impact the net income of a regulated company. Regulatory assets are normally recorded at inventory cost by debiting Account 182.3 *Other regulatory assets* and crediting Account 407.4 *Regulatory credits* (an expense account).

Typically, the tracking mechanism approved by FERC is reconciled annually by comparing actual costs and reimbursements, and adjustments are made to future fuel percentages, one-time billings, or surcharges under the pipeline's Tariff. Each pipeline's tracking mechanisms are the result of specific operating and Tariff conditions for that pipeline and, therefore, can be very different from one another. For instance, one pipeline may have a compressor fuel cost (only) tracker and another pipeline may have a tracker that provides for recovery (or crediting) of the value of *all* gas balance activity.

EXAMPLE

Fuel Usage, Cost Deferral, and Cost Recovery

Lucky Pipeline Company maintains a certain level of gas inventory for system operating purposes. During the 12 months ending June 2006, Lucky used 1,000,000 MMBtu to operate the compressor stations on its system. In addition, Lucky received 800,000 MMBtu from its customers for reimbursement of fuel over the same twelve-month period. Lucky filed a report with FERC in September 2006 that reported the excess fuel costs of 200,000 MMBtu and FERC approved the recovery of these costs from Lucky's customers via a one-time billing in December 2006. Spot prices averaged $2.00 during the twelve months of operations and the average cost of inventory gas was $2.25. Assume that this fuel activity was the only activity on the pipeline during this period and that Lucky uses the inventory method to account for system gas.

Entry to record fuel usage (1,000,000 x $2.00)

Account 854 Gas used for compressor station fuel (expense)	2,000,000	
Account 810 Gas used for compressor station fuel—credit (expense)		2,000,000

Entry to record fuel receipts from customers (800,000 x \$2.00)
Account 812 Gas used for utility
operations—credit (expense) 1,600,000
Account 854 Gas used for compressor
station fuel (expense) 1,600,000

Entry to record system gas net withdrawal (200,000 x \$2.25)
Account 808.1 Gas withdrawn from
storage (expense) 450,000
Account 117.2 System gas inventory (asset) . . 450,000

Net Income Impact = debit \$450,000

Entry to record deferral of costs
Account 182.3 Other regulatory assets 450,000
Account 407.4 Regulatory credits (expense) . 450,000

Net Income Impact = credit \$450,000
Cumulative Net Income Impact = \$0

Entry to record recovery of costs
Account 131.0 Cash 450,000
Account 182.3, Other regulatory assets 450,000

EXAMPLE

Gas Balance

Summarized descriptions of operational and financial activity for one month related to the various Gas Balancing events of a natural gas pipeline are given below. The Gas Balance Form shows the gas volumes corresponding to these descriptions. By summarizing all of the receipt and delivery, or source and use activity for the month, the unaccounted-for gas lost on the system can be computed. The associated accounting entries are given after the Gas Balance.

a. Early in the month, Pipeline anticipates that it will need additional system balancing gas and purchases 500,000 MMBtu on the spot market at \$2.00.

Later in the month, Pipeline realizes it will need additional quantities of gas and purchases an additional 800,000 MMBtu on the spot market at $2.50.

b. Pipeline receives 110,000,000 MMBtu from transport customers, retains 5,500,000 MMBtu for fuel usage, and delivers 106,000,000 to the customer's delivery points.
c. Pipeline uses 5,100,000 MMBtu for compressor station fuel to operate the pipeline.
d. Pipeline withdraws 5,000,000 MMBtu of system gas from storage at various times during the month, and injects 4,700,000 MMBtu of system gas into storage.
e. Unaccounted-for gas (UAF) lost on the system during the month is computed based on the above transactions to be 500,000 MMBtu, or less than 1% of customer receipts for delivery.

Sources/Receipts of Gas	**MMBtu**
Storage Gas Withdrawals	5,000,000
Gas Received from Customers for Delivery (110,000,000—5,500,000)	104,500,000
Gas Received from Customers for Fuel Reimbursement	5,500,000
Gas Purchases (500,000 + 800,000)	1,300,000
Total Receipts	116,300,000

Uses/Deliveries of Gas	
Storage Gas Injections	4,700,000
Gas Delivered for Customers	106,000,000
Gas Used in Operations	5,100,000
Unaccounted-for Gas	500,000
Total Deliveries	116,300,000

The following entries are created using the above data from Gas Balance. Assume the pipeline is using the inventory method for recording System Gas, and is using the weighted average cost of gas (WACOG) as its cost method. The WACOG rate is $2.10 and the average spot price for the month is $2.00. In addition, FERC has approved a *fuel* tracker mechanism for this pipeline company. The following entries are structured so that the net System Gas (storage) activity is all recorded in the last entry.

Purchase System Gas (500,000 MMBtu x $2.00 + 800,000 MMBtu x $2.50)

Account 803 Natural gas transmission line purchases (expense)	3,000,000	
Account 232 Accounts payable		3,000,000

Transport Customer Imbalances

(Imbalance: [110,000,000 – 5,500,000 – 106,000,000] x $2.00*)

Account 174 Miscellaneous current & accrued assets (imbalance receivable)	3,000,000	
Account 806 Exchange gas (expense)		3,000,000

* $2.00 is the average spot price for the month.

Net Fuel Receipts & Usage

([5,500,000—5,100,000] x $2.00; WACOG rate of $2.10 is used for Regulatory Asset)

(Note that this entry is reverse of the usual circumstances since there is a net receipt from customers)

Account 812 Gas used for utility operations—credit (expense)	800,000	
Account 854 Gas used for compressor station fuel (expense)		800,000
Account 407.3 Regulatory debits (expense)	840,000	
Account 182.3 Other regulatory assets (fuel tracker)		840,000

Unaccounted-for Gas

(500,000 x $2.00; WACOG rate of $2.10 is used for Regulatory Asset)

Account 854 Gas used for compressor station fuel (expense)	1,000,000	
Account 812 Gas used for utility operations—credit (expense)		1,000,000
Account 182.3 Other regulatory assets	1,050,000	
Account 407.4 Regulatory credits (expense) . .		1,050,000

Net System Gas Activity (The net system gas inventory entry is usually made last due to the pricing impacts of all of the other components.)

One method for determining the rate at which to record the withdrawal and injection activity is to summarize the net system gas activity for the month, determining an average rate, as follows:

	MMBtu	Cost
Purchases	1,300,000	$ 3,000,000
Imbalances	(1,500,000)	(3,000,000)
Fuel	400,000	840,000
UAF	(500,000)	(1,050,000)
Net System Gas Activity	(300,000)	(210,000)
Average System Gas Rate		$ 0.70

Account 117.2 System balancing gas—purchases	3,000,000	
Account 117.2 System balancing gas—fuel	840,000	
Account 117.2 System balancing gas—imbalances		3,000,000
Account 117.2 System balancing gas—UAF		1,050,000
Account 808.1 Gas withdrawn from storage (5,000,000 x $0.70)	3,500,000	
Account 808.2 Gas delivered to storage (4,700,000 x $0.70)		3,290,000

Account 808 entries are priced at the Average System Gas Rate.

By summarizing the expense impact of the above entries, it can be seen that the net impact is zero. The zero net income impact results because the pipeline has a tracking mechanism for this activity.

In addition to the above entries, re-pricing entries to the *Miscellaneous current and accrued assets* Account (*i.e.,* the imbalance account) would be needed to reflect the change in the spot rate.

(If the fixed asset method was being used, the primary differences would be that the System gas activity would be recorded to Account 117.4 *Gas owed to system gas,* instead of Account 117.2.)

FERC Reporting—Forms 2 and 11

The Natural Gas Act requires that periodic reports for interstate natural gas pipelines be provided to FERC. Generally, these forms are intended to provide a standard reporting mechanism for all pipelines so that FERC and other interested parties are able to compare and monitor the financial and operating structures of the various pipelines.

Two of the forms that must be prepared and submitted electronically are as follows:

1. **FERC Form No. 11, Natural Gas Pipeline Company Quarterly Statement of Monthly Data.** This quarterly report includes monthly data for revenues and throughput. As discussed earlier, FERC requires that revenue records be maintained in order to report revenues by rate schedule. In Texas Eastern's Form 11, for instance, January 1998 transportation revenues were classified into 28 different rate schedules.

Each rate schedule must then report the quantity of gas transported and the revenues must be split between transportation services and other specified surcharges for incremental costs.

2. **FERC Form No. 2, Annual Report of Major Natural Gas Companies.** This annual report includes schedules that contain general corporate information, financial statements and supporting schedules, a common section, and gas plant statistical data. There are 68 total schedules listed in the index for Form 2.

Since there is a wide spectrum of data provided in Form 2, many departments within a pipeline company are involved in compiling various pages within the Form. In addition to the other specific departments discussed below, the Regulatory and Legal Departments may also be keenly interested in reviewing and approving the Form 2 compilation to ensure that reporting and other regulatory requirements are being met.

The FERC Form 2 Financial Statements and Notes to Financial Statements must correspond or be reconcilable to the GAAP audited financial statements of the company. As discussed earlier, a pipeline may have two sets of accounts—one for GAAP and one for FERC. These differing account structures cause GAAP and FERC financial statements to sometimes give very different results, raising different sets of disclosure and reporting issues. The company is also required to include an opinion from the external auditors on the financial statements included in the Form 2.

The financial supporting schedules (balance sheet and income statement), as well as the common section, contain detail account activity for various sections of the financial statements. Most of the data for these supporting schedules would be obtained from the Accounting Department of a pipeline company. These schedules are very detailed and contain many cross-references between the schedules. Many companies keep up with these reporting requirements on a monthly basis in order to reduce the amount of year-end preparation time required.

The Gas Plant Statistical Data may be obtained from several different sources. The underlying source for this data is usually the Transmission, Engineering or Operating Departments, which have source records for the transmission lines and for all gas flowing through the system. The financial aspects of the gas flow, or gas balance, will be compiled by the Controller's department. Therefore, this section typically requires coordination between the groups to make sure that both sets of records are in agreement and that the Form 2 reporting pages accurately reflect the operational data.

In summary, Form 2 reporting requires a tremendous amount of time and coordination between various departments of the company. Once completed and filed, however, it can be a valuable reference or research document for FERC and other parties interested in understanding the financial and operating structure of a natural gas pipeline.

PROBLEMS

1. Grant Pipeline Company provided various transportation and storage services during July 2007. Charges for these services per the pipeline invoice are summarized below:

Description	Invoice Amount
Winter Storage	$ 50,000
Enhanced Firm Transportation	25,000
Enhanced Firm Transportation	200,000
Peaking Storage	40,000
Enhanced Firm Transportation	5,000
In/Out Storage	30,000
Enhanced Firm Transportation	60,000
	$410,000

REQUIRED:
Prepare the journal entries necessary to record the above services provided.

2. Cooper Power Company nominates to have 80,000 MMBtu delivered in May, but only provides 70,000 MMBtu at the receipt point. Grant Pipeline delivers 80,000 MMBtu.

REQUIRED:
If the fuel component is ignored, what net asset or liability value should Grant Pipeline record?

3. Garza Pipeline Company operates a storage field in far west Texas where it owns and stores gas to facilitate balancing of its pipeline system. Garza's current average cost of gas activity for the month is $1.60 per MMBtu. Garza uses the inventory method for recording system gas activity. During August 2009, Garza injected 300,000 MMBtu and withdrew 40,000 MMBtu.

REQUIRED:
Give the entries to record the injection and withdrawal of gas.

4. Aggie Pipeline Company works with a gas marketer to identify gas available at a receipt point on its pipeline. The spot price quoted by the marketer was $1.90 per MMBtu. During July 2009, Lucky received at the receipt point 4,000,000 MMBtu and delivered this gas to its storage field. There was no other pipeline or storage activity during the month.

REQUIRED:
Prepare the journal entries to record the above activities.

5. Cooper Power Company needs 40,000 MMBtu of gas delivered in July. Grant Pipeline's fuel rate is 2% of MMBtu received from a customer.

   REQUIRED:
   Determine the total MMBtu to be received by Grant in order to compensate Grant for the fuel used to provide the transportation service.

6. Basic Pipeline Company incurred the following usage on its system during August 2010:

   Shrinkage through gas plant—2,500 MMBtu
   Transmission compressor fuel—14,000 MMBtu

   The posted spot price for August was $1.60 per MMBtu.

   REQUIRED:
   Prepare the journal entries to record the above activities.

7. Smith Pipeline Company maintains a certain level of gas inventory for system operating purposes. During the twelve months ended May 2008, Smith used 900,000 MMBtu to operate the compressor stations on its system. Smith received 800,000 MMBtu from its customers for reimbursement of fuel over the same twelve-month period. Lucky filed a report with FERC in September 2008 that reported the excess fuel costs of 100,000 MMBtu and FERC approved the recovery of these costs from Smith's customers via a one-time billing in December 2008. Spot prices averaged $2.40 during the twelve months of operations and the average cost of inventory gas was $2.60. Assume that this fuel activity was the only activity on the pipeline during this period and that Smith uses the inventory method to account for system gas.

   REQUIRED:
   Prepare the journal entries necessary to record the above activities.

8. Summarized descriptions of operational and financial activity for August 2010 related to the various Gas Balancing events of Lomax Pipeline Company are given below.

   a. During the first days of August, Lomax purchased 600,000 MMBtu on the spot market at $2.80 in anticipation of additional system balancing gas needs.
   b. Pipeline receives 50,000,000 MMBtu from transport customers, retains 2,200,000 MMBtu for fuel usage, and delivers 46,800,000 to the customer's delivery points.
   c. Pipeline uses 2,600,000 MMBtu for compressor station fuel to operate the pipeline.
   d. Pipeline withdraws 2,500,000 MMBtu of system gas from storage at various times during the month, and injects 2,300,000 MMBtu of system gas into storage.

e. Unaccounted-for gas (UAF) lost on the system during the month is computed based on the above transactions to be ________ MMBtu, or less than ________% of customer receipts for delivery.

REQUIRED:

a. Prepare a Gas Balance Form showing the gas volumes that correspond to the descriptions.

b. Prepare the associated accounting entries. Assume that Lomax Pipeline uses the inventory method for recording System Gas, and is using the weighted average cost of gas (WACOG) as its cost method. The WACOG rate is \$2.50 and the average spot price for the month is \$2.70. In addition, FERC has approved a *fuel* tracker mechanism for this pipeline company. Structure the entries so that the net System Gas (storage) activity is all recorded in the last entry.

Appendices

appendix a

Authorization *for* Expenditure

When joint working interests are involved, the working interest owner designated as operator obtains approval for estimated expenditures by the use of an Authorization for Expenditure (AFE). The AFE generally includes cost estimates of the projected work in enough detail for the nonoperators to determine the reasonableness of the dollar amounts.

The following AFE may be used for drilling and completing a well. As shown, a completed AFE delineates the cost of the activities that are necessary in drilling and completing a well. The AFE breaks out these costs by IDC and equipment. The nonoperators are interested, for tax purposes, in the amounts to be incurred for IDC and thus potentially subject to an immediate write-off.

The AFE is a cost estimate; however, the authorization extends to the actual costs incurred. The authorization may apply to only dry hole costs (i.e., drilling costs only) and not to completion costs. In most instances, the nonoperators authorizing the expenditures would authorize both dry-hole and completion costs.

AUTHORITY FOR EXPENDITURE

WELL/PROSPECT: J. Stevenson #1 **FIELD** Austin Chalk

LOCATION: 467' FSL & 4625' FSWL of the T. Roberts A–1122

COUNTY: Nueces County **STATE** Texas

AFE DESCRIPTION Drill & Complete 8900' Test

INTANGIBLE COST ESTIMATE:	DRY HOLE	PRODUCER
01 Roads, Location, Survey	30,000	30,000
02 Legal, Damages, Cleanup	1,500	1,500
03 Drilling Contractor, MI, RU, RD		
04 Drilling Contractor, Footage, Daywork	126,000	145,000
05 Completion, Workover, Swab Unit		10,000
06 Mud Chemicals	25,000	25,000
07 Cement, Cementing Services	10,000	18,000
08 Float Equipment	2,000	5,000
09 Open Hole Logging & Evaluation	20,000	20,000
10 Drillstem Testing		
11 Mud Logging		
12 Directional Drilling		
13 Fishing Tools and Services		
14 Water, Fuel	9,000	9,000
15 Bits	2,500	2,500
16 Rentals	25,000	25,000
17 Trucking/Transportation	10,000	15,000
18 Boats/Dockage		
19 Csg. Crews, Tongs, Handling Tools, LD/PU Machine	3,000	6,500
20 BHP, GOR, Potential Tests		
21 Cased Hole Logging and Perforating		5,000
22 Stimulation		
23 Inspection, Testing (BOP's, Csg.,Tbg)		1,500
24 Misc. Labor and Materials	3,000	5,000
25 Supervision and Overhead	6,500	8,000
Contingencies: 3%	8,000	9,500
SUB–TOTAL	**$281,500**	**$341,500**

TANGIBLE COST ESTIMATE:					
26 Conductor OD:	14"	Footage:	40'	3,500	3,500
27 Surface OD:	9 5/8"	Footage:	2000'	26,000	26,000
28 Intermediate OD:	___	Footage:	___		
29 Liner/Hanger OD:	___	Footage:	___		
30 Production OD:	7"	Footage:	8900'		86,000
31 Tubing OD:	2 7/8"	Footage:	8500'		28,000
32 Subsurface Equip.					6,000
33 Wellhead				2,500	10,000
34 Surface Fac. and Installation Incl. Pumping Unit					25,000
Contingencies:					
SUB-TOTAL				**$32,000**	**$184,500**
TOTAL COST				**$313,500**	**$526,000**

APPROVE: ______________________

DISAPPROVAL: ______________________

COMPANY/INDIVIDUAL: ______________________

DATE: ______________________

SIGNATURE: ______________________ **PREPARED BY:** ______________________

NAME/TITLES: ______________________ **DATE:** ______________________

appendix b

Regulation SX 4-10

Reg. § 210.4-10. Financial Accounting and Reporting for Oil and Gas Producing Activities Pursuant to the Federal Securities Laws and the Energy Policy and Conservation Act of 1975

This section prescribes financial accounting and reporting standards for registrants with the Commission engaged in oil and gas producing activities in filings under the federal securities laws and for the preparation of accounts by persons engaged, in whole or in part, in the production of crude oil or natural gas in the United States, pursuant to Section 503 of the Energy Policy and Conservation Act of 1975 [42 U.S.C. 6383] ("EPCA") and section 11(c) of the Energy Supply and Environmental Coordination Act of 1974 [IS U.S.C. 796] ("ESECA"), as amended by section 505 of EPCA. The application of this section to those oil and gas producing operations of companies regulated for rate-making purposes on an individual-company-cost-of-service basis may, however, give appropriate recognition to differences arising because of the effect of the rate-making process.

Exemption. Any person exempted by the Department of Energy from any recordkeeping or reporting requirements pursuant to Section 11(c) of ESECA, as amended, is similarly exempted from the related provisions of this section in the preparation of accounts pursuant to EPCA. This exemption does not affect the applicability of this section to filings pursuant to the federal securities laws.

DEFINITIONS

(a) *Definitions.* The following definitions apply to the terms listed below as they are used in this section:

(1) *Oil and gas producing activities.*

(i) Such activities include:

(A) The search for crude oil, including condensate and natural gas liquids, or natural gas ("oil and gas") in their natural states and original locations.

(B) The acquisition of property rights or properties for the purpose of further exploration and/or for the purpose of removing the oil or gas from existing reservoirs on those properties.

(C) The construction, drilling and production activities necessary to retrieve oil and gas from its natural reservoirs, and the acquisition, construction, installation, and maintenance of field gathering and storage systems — including lifting the oil and gas to the surface and gathering, treating, field processing (as in the case of processing gas to extract liquid hydrocarbons) and field storage. For purposes of this section, the oil and gas production function shall normally be regarded as terminating at the outlet valve on the lease or field storage tank; if unusual physical or operational circumstances exist, it may be appropriate to regard the production functions as terminating at the first point at which oil, gas, or gas liquids are delivered to a main pipeline, a common carrier, a refinery, or a marine terminal.

(ii) Oil and gas producing activities do not include:

(A) The transporting, refining and marketing of oil and gas.

(B) Activities relating to the production of natural resources other than oil and gas.

(C) The production of geothermal steam or the extraction of hydrocarbons as a by-product of the production of geothermal steam or associated geothermal resources as defined in the Geothermal Steam Act of 1970.

(D) The extraction of hydrocarbons from shale, tar sands, or coal.

(2) *Proved oil and gas reserves.* Proved oil and gas reserves are the estimated quantities of crude oil, natural gas, and natural gas liquids which geological and engineering data demonstrate with reasonable certainty to be recoverable in future years from known reservoirs under existing economic and operating conditions, i.e., prices and costs as of the date the estimate is made. Prices include consideration of changes in existing prices provided only by contractual arrangements, but not on escalations based upon future conditions.

(i) Reservoirs are considered proved if economic producibility is supported by either actual production or conclusive formation test. The area of a reservoir considered proved includes (A) that portion delineated by drilling and defined by gas-oil and/or oil-water contacts, if any; and (B) the immediately adjoining portions not yet drilled, but which can be reasonably judged as economically productive on the basis of available geological and engineering data. In the absence of information on fluid contacts, the lowest known structural occurrence of hydrocarbons controls the lower proved limit of the reservoir.

(ii) Reserves which can be produced economically through application of improved recovery techniques (such as fluid injection) are included in the "proved" classification when successful testing by a pilot project, or the operation of an installed program in the reservoir, provides support for the engineering analysis on which the project or program was based.

(iii) Estimates of proved reserves do not include the following:

(A) oil that may become available from known reservoirs but is classified separately as "indicated additional reserves";

(B) crude oil, natural gas, and natural gas liquids, the recovery of which is subject to reasonable doubt because of uncertainty as to geology, reservoir characteristics, or economic factors;

(C) crude oil, natural gas, and natural gas liquids, that may occur in undrilled prospects; and

(D) crude oil, natural gas, and natural gas liquids, that may be recovered from oil shales, coal, gilsonite and other such sources.

(3) *Proved developed oil and gas reserves.* Proved developed oil and gas reserves are reserves that can be expected to be recovered through existing wells with existing equipment and operating methods. Additional oil and gas expected to be obtained through the application of fluid injection or other improved recovery techniques

for supplementing the natural forces and mechanisms of primary recovery should be included as "proved developed reserves" only after testing by a pilot project or after the operation of an installed program has confirmed through production response that increased recovery will be achieved.

(4) *Proved undeveloped reserves.* Proved undeveloped oil and gas reserves are reserves that are expected to be recovered from new wells on undrilled acreage, or from existing wells where a relatively major expenditure is required for recompletion. Reserves on undrilled acreage shall be limited to those drilling units offsetting productive units that are reasonably certain of production when drilled. Proved reserves for other undrilled units can be claimed only where it can be demonstrated with certainty that there is continuity of production from the existing productive formation. Under no circumstances should estimates, for proved undeveloped reserves, be attributable to any acreage for which an application of fluid injection or other improved recovery technique is contemplated, unless such techniques have been proved effective by actual tests in the area and in the same reservoir.

(5) *Proved properties.* Properties with proved reserves.

(6) *Unproved properties.* Properties with no proved reserves.

(7) *Proved area.* The part of a property to which proved reserves have been specifically attributed.

(8) *Field.* An area consisting of a single reservoir or multiple reservoirs all grouped on or related to the same individual geological structural feature and/or stratigraphic condition. There may be two or more reservoirs in a field which are separated vertically by intervening impervious strata, or laterally by local geologic barriers, or by both. Reservoirs that are associated by being in overlapping or adjacent fields may be treated as a single or common operational field. The geological terms "structural feature" and "stratigraphic condition" are intended to identify localized geological features as opposed to the broader terms of basins, trends, provinces, plays, areas-of-interest, etc.

(9) *Reservoir.* A porous and permeable underground formation containing a natural accumulation of producible oil and/or gas that is confined by impermeable rock or water barriers and is individual and separate from other reservoirs.

(10) *Exploratory well.* A well drilled to find and produce oil or gas in an unproved area, to find a new reservoir in a field previously found to be productive of oil or gas in another reservoir, or to extend a known reservoir. Generally, an exploratory well is any well that is not a development well, a service well, or a stratigraphic test well as

(E) Severance taxes.

(ii) Some support equipment or facilities may serve two or more oil and gas producing activities and may also serve transportation, refining, and marketing activities. To the extent that the support equipment and facilities are used in oil and gas producing activities, their depreciation and applicable operating costs become exploration, development or production costs, as appropriate. Depreciation, depletion, and amortization of capitalized acquisition, exploration, and development costs are not production costs but also become part of the cost of oil and gas produced along with production (lifting) costs identified above.

Successful Efforts Method

(b) A reporting entity that follows the successful efforts method shall comply with the accounting and financial reporting disclosure requirements of Statement of Financial Accounting Standards No. 19, as amended.

Full Cost Method

(c) *Application of the full cost method of accounting.* A reporting entity that follows the full cost method shall apply that method to all of its operations and to the operations of its subsidiaries, as follows:

(1) *Determination of cost centers.* Cost centers shall be established on a country-by-country basis.

(2) *Costs to be capitalized.* All costs associated with property acquisition, exploration, and development activities (as defined in paragraph (a) of this section) shall be capitalized within the appropriate cost center. Any internal costs that are capitalized shall be limited to those costs that can be directly identified with acquisition, exploration, and development activities undertaken by the reporting entity for its own account, and shall not include any costs related to production, general corporate overhead, or similar activities.

(3) *Amortization of capitalized costs.* Capitalized costs within a cost center shall be amortized on the unit-of-production basis using proved oil and gas reserves, as follows:

(i) Costs to be amortized shall include (A) all capitalized costs, less accumulated amortization, other than the cost of properties described in paragraph (ii) below;

(B) the estimated future expenditures (based on current costs) to be incurred in developing proved reserves; and (C) estimated dismantlement and abandonment costs, net of estimated salvage values.

(ii) The cost of investments in unproved properties and major development projects may be excluded from capitalized costs to be amortized, subject to the following:

(A) All costs directly associated with the acquisition and evaluation of unproved properties may be excluded from the amortization computation until it is determined whether or not proved reserves can be assigned to the properties, subject to the following conditions: (1) Until such a determination is made, the properties shall be assessed at least annually to ascertain whether impairment has occurred. Unevaluated properties whose costs are individually significant shall be assessed individually. Where it is not practicable to individually assess the amount of impairment of properties for which costs are not individually significant, such properties may be grouped for purposes of assessing impairment. Impairment may be estimated by applying factors based on historical experience and other data such as primary Lease terms of the properties, average holding periods of unproved properties, and geographic and geologic data to groupings of individually insignificant properties and projects. The amount of impairment assessed under either of these methods shall be added to the costs to be amortized. (2) The costs of drilling exploratory dry holes shall be included in the amortization base immediately upon determination that the well is dry. (3) If geological and geophysical costs cannot be directly associated with specific unevaluated properties, they shall be included in the amortization base as incurred. Upon complete evaluation of a property, the total remaining excluded cost (net of any impairment) shall be included in the full cost amortization base.

(B) Certain costs may be excluded from amortization when incurred in connection with major development projects expected to entail significant costs to ascertain the quantities of proved reserves attributable to the properties under development (e.g., the installation of an offshore drilling platform from which development wells are to be drilled, the installation of improved recovery programs, and similar major projects undertaken in the expectation of Significant additions to proved reserves). The amounts which may be excluded are applicable portions of (1) the costs that relate to the major development project and have not previously been included in the amortization base, and (2) the estimated future expenditures associated with the development project. The excluded portion of any common costs associated with the development project should be based, as is most appropriate in the circumstances, on a comparison of either (i) existing proved reserves to total proved reserves expected to be established upon completion of the project, or (ii) the number of wells to which proved reserves have been assigned and total number of wells expected to be drilled. Such costs may be excluded from costs to be amortized until the earlier determination of whether additional reserves are proved or impairment occurs.

(C) Excluded costs and the proved reserves related to such costs shall be transferred into the amortization base on an ongoing (well-by-well or property-by-property) basis as the project is evaluated and proved reserves established or impairment determined. Once proved reserves are established, there is no further justification for continued exclusion from the full cost amortization base even if other factors prevent immediate production or marketing.

(iii) Amortization shall be computed on the basis of physical units, with oil and gas converted to a common unit of measure on the basis of their approximate relative energy content, unless economic circumstances (related to the effects of regulated prices) indicate that use of units of revenue is a more appropriate basis of computing amortization. In the latter case, amortization shall be computed on the basis of current gross revenues (excluding royalty payments and net profits disbursements) from production in relation to future cross revenues, based on current prices (including consideration of changes in existing prices provided only by contractual arrangements), from estimated production of proved oil and gas reserves. The effect of a significant price increase during the year on estimated future gross revenues shall be reflected in the amortization provision only for the period after the price increase occurs.

(iv) In some cases it may be more appropriate to depreciate natural gas cycling and processing plants by a method other than the unit-of-production method.

(v) Amortization computations shall be made on a consolidated basis, including investees accounted for on a proportionate consolidation basis. Investees accounted for on the equity method shall be treated separately.

(4) *Limitation on capitalized costs*:

(i)For each cost center, capitalized costs, less accumulated amortization and related deferred income taxes, shall not exceed an amount (the cost center ceiling) equal to the sum of:

(A) the present value of estimated future net revenues computed by applying current prices of oil and gas reserves (with consideration of price changes only to the extent provided by contractual arrangements) to estimated future production of proved oil and gas reserves as of the date of the latest balance sheet presented, less estimated future expenditures (based on current costs) to be incurred in developing and producing the proved reserves computed using a discount factor of ten percent and assuming continuation of existing economic conditions; plus

(B) the cost of properties not being amortized pursuant to paragraph (i)(3)(ii) of this section; plus

(C) the lower of cost or estimated fair value of unproven properties included in the costs being amortized; less

(D) income tax effects related to differences between the book and tax basis of the properties referred to in paragraphs (i)(4)(i)(B) and (C) of this section.

(ii) If unamortized costs capitalized within a cost center, less related deferred income taxes, exceed the cost center ceiling, the excess shall be charged to expense and separately disclosed during the period in which the excess occurs. Amounts thus required to be written off shall not be reinstated for any subsequent increase in the cost center ceiling.

(5) *Production costs.* All costs relating to production activities, including workover costs incurred solely to maintain or increase levels of production from an existing completion interval, shall be charged to expense as incurred.

(6) *Other transactions.* The provisions of paragraph (h) of this section, "Mineral property conveyances and related transactions if the successful efforts method of accounting is followed," shall apply also to those reporting entities following the full cost method except as follows:

(i) *Sales and abandonments of oil and gas properties.* Sales of oil and gas properties, whether or not being amortized currently, shall be accounted for as adjustments of capitalized costs, with no gain or loss recognized, unless such adjustments would significantly alter the relationship between capitalized costs and proved reserves of oil and gas attributable to a cost center. For instance, a significant alteration would not ordinarily be expected to occur for sales involving less than 25 percent of the reserve quantities of a given cost center. If gain or loss is recognized on such a sale, total capitalization costs within the cost center shall be allocated between the reserves sold and reserves retained on the same basis used to compute amortization, unless there are substantial economic differences between the properties sold and those retained, in which case capitalized costs shall be allocated on the basis of the relative fair values of the properties. Abandonments of oil and gas properties shall be accounted for as adjustments of capitalized costs; that is, the cost of abandoned properties shall be charged to the full cost center and amortized (subject to the limitation on capitalized costs in paragraph (b) of this section).

(ii) *Purchases of reserves.* Purchases of oil and gas reserves in place ordinarily shall be accounted for as additional capitalized costs within the applicable cost center; however, significant purchases of production payments or properties with lives substantially shorter than the composite productive life of the cost center shall be accounted for separately.

(iii) *Partnerships, joint ventures and drilling arrangements.*

(A) Except as provided in subparagraph (i)(6)(i) of this section, all consideration received from sales or transfers of properties in connection with partnerships, joint venture operations, or various other forms of drilling arrangements involving oil and gas exploration and development activities (e.g., carried interest, turnkey wells, management fees, etc.) shall be credited to the full cost account, except to the extent of amounts that represent reimbursement of organization, offering, general and administrative expenses, etc., that are identifiable with the transaction, if such amounts are currently incurred and charged to expense.

(B) Where a registrant organizes and manages a limited partnership involved only in the purchase of proved developed properties and subsequent distribution of income from such properties, management fee income may be recognized provided the properties involved do not require aggregate development expenditures in connection with production of existing proved reserves in excess of 10% of the partnership's recorded cost of such properties. Any income not recognized as a result of this limitation would be credited to the full cost account and recognized through a lower amortization provision as reserves are produced.

(iv) *Other services.* No income shall be recognized in connection with contractual services performed (e.g. drilling, well service, or equipment supply services, etc.) in connection with properties in which the registrant or an affiliate (as defined in § 210.1-02(b)) holds an ownership or other economic interest, except as follows:

(A) Where the registrant acquires an interest in the properties in connection with the service contract, income may be recognized to the extent the cash consideration received exceeds the related contract costs plus the registrant's share of costs incurred and estimated to be incurred in connection with the properties. Ownership interests acquired within one year of the date of such a contract are considered to be acquired in connection with the service for purposes of applying this rule. The amount of any guarantees or similar arrangements undertaken as part of this contract should be considered as part of the costs related to the properties for purposes of applying this rule.

(B) Where the registrant acquired an interest in the properties at least one year before the date of the service contract through transactions unrelated to the service contract, and that interest is unaffected by the service contract, income from such contract may be recognized subject to the general provisions for elimination of intercompany profit under generally accepted accounting principles.

(C) Notwithstanding the provisions of (A) and (B) above, no income may be recognized for contractual services performed on behalf of investors in oil and gas producing activities managed by the registrant or an affiliate. Furthermore, no income may be recognized for contractual services to the extent that the consideration received for such services represents an interest in the underlying property.

(D) Any income not recognized as a result of these rules would be credited to the full cost account and recognized through a lower amortization provision as reserves are produced.

(7) *Disclosures.* Reporting entities that follow the full cost method of accounting shall disclose all of the information required by paragraph (k) of this section, with each cost center considered as a separate geographic area, except that reasonable groupings may be made of cost centers that are not significant in the aggregate. In addition:

(i) For each cost center for each year that an income statement is required, disclose the total amount of amortization expense (per equivalent physical unit of production if amortization is computed on the basis of physical units or per dollar of gross revenue from production if amortization is computed on the basis of gross revenue).

(ii) State separately on the face of the balance sheet the aggregate of the capitalized costs of unproved properties and major development projects that are excluded, in accordance with paragraph (i) (3) of this section, from the capitalized costs being amortized. Provide a description in the notes to the financial statements of the current status of the significant properties or projects involved, including the anticipated timing of the inclusion of the costs in the amortization computation. Present a table that shows, by category of cost, (A) the total costs excluded as of the most recent fiscal year; and (B) the amounts of such excluded costs, incurred (1) in each of the three most recent fiscal years and (2) in the aggregate for any earlier fiscal years in which the costs were incurred. Categories of cost to be disclosed include acquisition costs, exploration costs, development costs in the case of significant development projects and capitalized interest.

INCOME TAXES

(d) *Income taxes.* Comprehensive interperiod income tax allocation by a method which complies with generally accepted accounting principles shall be followed for intangible drilling and development costs and other costs incurred that enter into the determination of taxable income and pretax accounting income in different periods.

Excerpt *from* Regulation S-K

REGULATION C.F.R. 229.302(b)

(b) *Information about oil and gas producing activities.* Registrants engaged in oil and gas producing activities shall present the information about oil and gas producing activities (as those activities are defined in Regulation S-X, §210.4-10(a)) specified in paragraphs 9-34 of Statement of Financial Accounting Standards ("SFAS") No. 69, "Disclosures about Oil and Gas Producing Activities," if such oil and gas producing activities are regarded as significant under one or more of the tests set forth in paragraph 8 of SFAS No. 69.

Instructions to Paragraph (b).

1. (a) SFAS No. 69 disclosures that relate to annual periods shall be presented for each annual period for which an income statement is required,

   (b) SFAS No. 69 disclosures required as of the end of an annual period shall be presented as of the date of each audited balance sheet required, and (c) SFAS No. 69 disclosures required as of the beginning of an annual period shall be presented as of the beginning of each annual period for which an income statement is required.

2. This paragraph, together with §210.4-10 of Regulation S-X, prescribes financial reporting standards for the preparation of accounts by persons engaged, in whole or in part, in the production of crude oil or natural gas in the United States, pursuant to Section 503 of the Energy Policy and Conservation Act of 1975 [42 U.S.C. 6383] ("EPCA") and Section 11(c) of the Energy Supply and Environmental Coordination Act of 1974 [15 U.S.C. 796] ("ESECA") as amended by Section 506 of EPCA. The application of this paragraph to those oil and gas producing operations of companies regulated for ratemaking purposes on an individual-company-cost-of-service basis may, however, give appropriate recognition to differences arising because of the effect of the ratemaking process.

3. Any person exempted by the Department of Energy from any recordkeeping or reporting requirements pursuant to Section 11(c) of ESECA, as amended, is similarly exempted from the related provisions of this paragraph in the preparation of accounts pursuant to EPCA. This exemption does not affect the applicability of this paragraph to filings pursuant to the federal securities laws.

appendix d

Excerpt *from* Statement *of* Financial Accounting Standards No. 19

Excerpt from Statement of Financial Accounting Standards No. 19, "Financial Accounting and Reporting by Oil and Gas Producing Companies," December 1977. *SFAS 19* is copyrighted by the Financial Accounting Standards Board, 401 Merritt 7, P.O. Box 5116, Norwalk, Connecticut 06856-5116. This excerpt is reprinted with permission. Complete copies of this document are available from the FASB.

Basic Concepts

11. An enterprise's oil and gas producing activities involve certain special types of assets. Costs of those assets shall be capitalized when incurred. Those types of assets broadly defined are:

a. *Mineral interest in properties* (hereinafter referred to as *properties*), which include fee ownership or a lease, concession, or other interest representing the right to extract oil or gas subject to such terms as may be imposed by the conveyance of that interest. Properties also include royalty interest, production payments payable in oil or gas, and other nonoperating interests in properties operated by others. Properties include those agreements with foreign governments or authorities under which an enterprise participates in the operation of the related properties or otherwise serves as "producer" of the underlying reserves (see paragraph 53); but properties do not include other supply agreements or contracts that represent the right to *purchase* (as opposed to *extract*) oil and gas. Properties shall be classified as proved or unproved as follows:

i. *Unproved properties* - properties with no proved reserves.

ii. *Proved properties* - properties with proved reserves.

b. *Wells and related equipment and facilities*[1], the cost of which include those incurred to:

i. Drill and equip those exploratory wells and exploratory-type stratigraphic test wells that have found proved reserves.

ii. Obtain access to proved reserves and provide facilities for extracting, treating, gathering, and storing the oil and gas, including the drilling and equipping of development wells and development-type stratigraphic test wells (whether those wells are successful or unsuccessful) and service well.

c. *Support equipment and facilities used in oil and gas producing activities,* such as seismic equipment, drilling equipment, construction and grading equipment, vehicles, repair shops, warehouses, supply points, camps, and division, district, or field offices.

d. *Uncompleted wells, equipment, and facilities,* the costs of which include those incurred to:

i. Drill and equip wells that are not yet completed.

ii. Acquire or construct equipment and facilities that are not yet completed and installed.

12. The costs of an enterprise's wells and related equipment and facilities and the cost of the related proved properties shall be amortized as the related oil and gas reserves are produced. That amortization plus production (lifting) costs become part of the cost of oil and gas produced. Unproved properties shall be assessed periodically, and a loss recognized if those properties are impaired.

13. Some cost incurred in an enterprise's oil and gas producing activities do not result in acquisition of an asset and, therefore, shall be charged to expense. Examples include geological and geophysical costs, the costs of carrying and retaining undeveloped properties, and the costs of drilling those exploratory wells and exploratory-type stratigraphic test wells that do not find proved reserves.

14. The basic concepts in paragraphs 11-13 are elaborated on in paragraphs 15-41.

Accounting at the Time Costs Are Incurred

Acquisition of Properties

15. Costs incurred to purchase, lease, or otherwise acquire a property (whether unproved or proved) shall be capitalized when incurred. They include the costs of lease bonuses and options to purchase or lease properties, the portion of costs applicable to minerals when land including mineral rights is purchased in fee, broker's fees, recording fees, legal costs, and other costs incurred in acquiring properties.

Exploration

16. Exploration involves (a) identifying areas that may warrant examination and (b) examining specific areas that are considered to have prospects of containing oil and gas reserves, including drilling exploratory wells and exploratory-type stratigraphic test wells. Exploration cost may be incurred both before acquiring the related property (sometimes referred to in part as prospecting costs) and after acquiring the property.

17. Principal types of exploration cost, which include depreciation and applicable operating costs of support equipment and facilities (paragraph 26) and other costs of exploration activities, are:

a. Costs of topographical, geological, and geophysical studies, rights of access to properties to conduct those studies, and salaries and other expenses of geologists, geophysical crews, and others conducting those studies. Collectively, these are sometimes referred to as geological and geophysical (G&G) costs.
b. Costs of carrying and retaining undeveloped properties such as delay rentals, *ad valorem* taxes on the properties, legal costs for title defense, and maintenance of land and lease records.
c. Dry hole contributions and bottom hole contributions.
d. Costs of drilling and equipping exploratory wells.
e. Costs of drilling exploratory-type stratigraphic test wells.[2]

18. Geological and geophysical costs, costs of carrying and retaining undeveloped properties, and dry hole and bottom hole contributions shall be charged to expense when incurred.

19. The cost of drilling exploratory wells and the costs of drilling exploratory-type stratigraphic test wells shall be capitalized as part of the enterprise's uncompleted wells, equipment, and facilities, pending determination of whether the well has found proved reserves. If the well has found proved reserves (paragraphs 31-34), the capitalized costs of drilling the well shall become part of the enterprise's wells and related equipment and facilities (even though the well may not be completed as a producing well); if, however, the well has not found proved reserves, the capitalized costs of drilling the well, net of any salvage value, shall be charged to expense.

20. An enterprise sometimes conducts G&G studies and other exploration activities on a property owned by another party, in exchange for which the enterprise is contractually entitled to receive an interest in the property if proved reserves are found or to be reimbursed by the owner for the G&G and other costs incurred if proved reserves are not found. In that case, the enterprise conducting the G&G studies and other exploration activities shall account for those costs as a receivable when incurred and, if proved reserves are found, they shall become the cost of the proved property acquired.

Development

21. Development costs are incurred to obtain access to proved reserves and to provide facilities for extracting, treating, gathering, and storing the oil and gas. More specifically, development cost, including depreciation and applicable operating costs of support equipment and facilities (paragraph 26) and other costs of development activities, are costs incurred to:

a. Gain access to and prepare well locations for drilling, including surveying well locations for the purpose of determining specific development drilling sites, clearing ground, draining, road building, and relocating public roads, gas lines, and power line, to the extent necessary in developing the proved reserves.
b. Drill and equip development wells, development-type stratigraphic test wells, and service wells, including the costs of platforms and of well equipment such as casing, tubing, pumping equipment, and the wellhead assembly.
c. Acquire, construct, and install production facilities such as lease flow lines, separators, treaters, heaters, manifolds, measuring devices, and production storage tanks, natural gas cycling and processing plants, and utility and waste disposal systems.
d. Provide improved recovery systems.

22. Development costs shall be capitalized as part of the cost of an enterprise's wells and related equipment and facilities. Thus, all costs incurred to drill and equip development wells, development-type stratigraphic test wells, and service wells are development costs and

shall be capitalized, whether the well is successful or unsuccessful. Costs of drilling those wells and costs of constructing equipment and facilities shall be included in the enterprise's uncompleted wells, equipment, and facilities until drilling or construction is completed.

Production

23. Production involves lifting the oil and gas to the surface and gathering, treating, field processing (as in the case of processing gas to extract liquid hydrocarbons), and field storage. For purposes of the Statement, the production function shall normally be regarded as terminating at the outlet valve on the lease or field production storage tank; if unusual physical or operational circumstances exist, it may be more appropriate to regard the production function as terminating at the first point at which oil, gas, or gas liquids are delivered to a main pipeline, a common carrier, a refinery, or a marine terminal.

24. Production costs are those costs incurred to operate and maintain an enterprise's wells and related equipment and facilities, including depreciation and applicable operating costs of support equipment and facilities (par. 26) and other costs of operating and maintaining those wells and related equipment and facilities. They become part of the cost of oil and gas produced. Examples of production costs (sometimes called lifting costs) are:

a. Costs of labor to operate the wells and related equipment and facilities.
b. Repairs and maintenance.
c. Materials, supplies, and fuel consumed and services utilized in operating the wells and related equipment and facilities.
d. Property taxes and insurance applicable to proved properties and wells and related equipment and facilities.
e. Severance taxes.

25. Depreciation, depletion, and amortization of capitalized acquisition, exploration, and development costs also become part of the cost of oil and gas produced along with production (lifting) costs identified in paragraph 24.

Support Equipment and Facilities

26. The cost of acquiring or constructing support equipment and facilities used in oil and gas producing activities shall be capitalized. Examples of support equipment and facilities include seismic equipment, drilling equipment, construction and grading equipment, vehi-

cles, repair shops, warehouses, supply points, camps, and division, district, or field offices. Some support equipment or facilities are acquired or constructed for use exclusively in a single activity–exploration, development, or production. Other support equipment or facilities may serve two or more of those activities and may also serve the enterprise's transportation, refining, and marketing activities. To the extent that the support equipment and facilities are used in oil and gas producing activities, their depreciation and applicable operating costs become an exploration, development, or production cost, as appropriate.

Disposition of Capitalized Costs

27. The effect of paragraphs 15-26, which deal with accounting at the time costs are incurred, is to recognize as assets: (a) unproved properties; (b) proved properties; (c) wells and related equipment and facilities (which consist of all development costs plus the costs of drilling those exploratory wells and exploratory-type stratigraphic test wells that find proved reserves); (d) support equipment and facilities used in oil and gas producing activities; and (e) uncompleted wells, equipment, and facilities. Paragraphs 28-41 which follow deal with disposition of the costs of those assets after capitalization. Among other things, those paragraphs provide that the acquisition costs of proved properties and the costs of wells and related equipment and facilities be amortized to become part of the cost of oil and gas produced; that impairment of unproved properties be recognized; and that the costs of an exploratory well or exploratory-type stratigraphic test well be charged to expense if the well is determined not to have found proved reserves.

Assessment of Unproved Properties

28. Unproved properties shall be assessed periodically to determine whether they have been impaired. A property would likely be impaired, for example, if a dry hole has been drilled on it and the enterprise has no firm plans to continue drilling. Also, the likelihood of partial or total impairment of a property increases as the expiration of the lease term approaches if drilling activity has not commenced on the property or on nearby properties. If the results of the assessment indicate impairment, a loss shall be recognized by providing a valuation allowance. Impairment of individual unproved properties whose acquisition costs are relatively significant shall be assessed on a property-by-property basis, and an indicated loss shall be recognized by providing a valuation allowance. When an enterprise has a relatively large number of unproved properties whose acquisition costs are not individually significant, it may not be practical to assess impairment on a property-by-property basis, in which case the amount of loss to be recognized and the amount of the valuation allowance needed to provide for impairment of those properties shall be determined by amortizing those properties, either in the aggregate or by groups, on the

basis of the experience of the enterprise in similar situations and other information about such factors as the primary lease terms of those properties, the average holding period of unproved properties, and the relative proportion of such properties on which proved reserves have been found in the past.

Reclassification of an Unproved Property

29. A Property shall be reclassified from unproved properties to proved properties when proved reserves are discovered on or otherwise attributed to the property; occasionally, a single property, such as a foreign lease or concession covers so vast an area that only the portion of the property to which the proved reserves relate–determined on the basis of geological structural features or stratigraphic conditions–should be reclassified from unproved to proved. For a property whose impairment has been assessed individually in accordance with paragraph 28, the *net* carrying amount (acquisition cost minus valuation allowance) shall be reclassified to proved properties; for properties amortized by providing a valuation allowance on a group basis, the gross acquisition cost shall be reclassified.

Amortization (Depletion) of Acquisition Costs of Proved Properties

30. Capitalized acquisition costs of proved properties shall be amortized (depleted) by the unit-of-production method so that each unit produced is assigned a pro rata portion of the unamortized acquisition cost. Under the unit-of-production method, amortization (depletion) may be computed either on a property-by-property basis or on the basis of some reasonable aggregation of properties with a common geological structural feature or stratigraphic condition, such as a reservoir or field. When an enterprise has a relatively large number of royalty interests whose acquisition costs are not individually significant, they may be aggregated, for the purpose of computing amortization, without regard to commonality of geological structural features or stratigraphic conditions; if information is not available to estimate reserve quantities applicable to royalty interest owned (paragraph 50), a method other than the unit-of-production method may be used to amortize their acquisition costs. The unit cost shall be computed on the basis of the total estimated units of proved oil and gas reserves. (Joint production of both oil and gas is discussed in paragraph 38.) Unit-of-production amortization rates shall be revised whenever there is an indication of the need for revision but at least once a year; those revisions shall be accounted for prospectively as changes in accounting estimates–see paragraphs 31-33 of APB Opinion No. 20, "Accounting Changes."

Accounting When Drilling of an Exploratory Well is Completed

31. As specified in paragraph 19, the costs of drilling an exploratory well are capitalized as part of the enterprise's uncompleted wells, equipment, and facilities pending determination of whether the well has found proved reserves. That determination is usually made on or shortly after completion of drilling the well, and the capitalized costs shall either be charged to expense or be reclassified as part of the cost of the enterprise's wells and related equipment and facilities at that time. Occasionally, however, an exploratory well may be determined to have found oil and gas reserves, but classification of those reserves as proved cannot be made when drilling is completed. In those cases, one or the other of the following subparagraphs shall apply depending on whether the well is drilled in an area requiring a major capital expenditure, such as a trunk pipeline, before production from that well could begin:

a. *Exploratory wells that find oil and gas reserves in an area requiring a major capital expenditure, such as a trunk pipeline, before production could begin.* On completion of drilling, an exploratory well may be determined to have found oil and gas reserves, but classification of those reserves as proved depends on whether a major capital expenditure can be justified which, in turn, depends on whether additional exploratory wells find a sufficient quantity of additional reserves. That situation arises principally with exploratory wells drilled in a remote area for which production would require constructing a trunk pipeline. In that case, the cost of drilling the exploratory well shall continue to be carried as an asset pending determination of whether proved reserves have been found only as long as both of the following conditions are met:
   i. The well has found a sufficient quantity of reserves to justify its completion as a producing well if the required capital expenditure is made.
   ii. Drilling of the additional exploratory wells is under way or firmly planned for the near future.

   Thus if drilling in the area is not under way or firmly planned, or if the well has not found a commercially producible quantity of reserves, the exploratory well shall be assumed to be impaired, and its costs shall be charged to expense.

b. *All other exploratory wells that find oil and gas reserves.* In the absence of a determination as to whether the reserves that have been found can be classified as proved, the costs of drilling such an exploratory well shall not be carried as an asset for more than one year following completion of drilling. If, after that year has passed, a determination that proved reserves have been found cannot be made, the well shall be assumed to be impaired, and its costs shall be charged to expense.

32. Paragraph 31 is intended to prohibit, in all cases, the deferral of the costs of exploratory wells that find some oil and gas reserves merely on the chance that some event totally beyond the control of the enterprise will occur, for example, on the chance that the selling prices of oil and gas will increase sufficiently to result in classification of reserves as proved that are not commercially recoverable at current prices.

Accounting When Drilling of an Exploratory-Type Stratigraphic Test Well is Completed

33. As specified in paragraph 19, the costs of drilling an exploratory-type stratigraphic test well are capitalized as part of the enterprise's uncompleted wells, equipment, and facilities pending determination of whether the well has found proved reserves. When that determination is made, the capitalized costs shall be charged to expense if proved reserves are not found or shall be reclassified as part of the costs of the enterprise's wells and related equipment and facilities if proved reserves are found.

34. Exploratory-type stratigraphic test wells are normally drilled on unproved offshore properties. Frequently, on completion of drilling, such a well may be determined to have found oil and gas reserves, but classification of those reserves as proved depends on whether a major capital expenditure–usually a production platform–can be justified which, in turn, depends on whether additional exploratory-type stratigraphic test wells find a sufficient quantity of additional reserves. In that case, the cost of drilling the exploratory-type stratigraphic test well shall continue to be carried as an asset pending determination of whether proved reserves have been found only as long as both of the following conditions are met:

i. The well has found a quantity of reserves that would justify its completion for production had it not been simply a stratigraphic test well.
ii. Drilling of the additional exploratory-type stratigraphic test wells is under way or firmly planned for the near future.

Thus if associated stratigraphic test drilling is not under way or firmly planned, or if the well has not found a commercially producible quantity of reserves, the exploratory-type stratigraphic test well shall be assumed to be impaired, and its cost shall be charged to expense.

Amortization and Depreciation of Capitalized Exploratory Drilling and Development Costs

35. Capitalized cost of exploratory wells and exploratory-type stratigraphic test wells that have found proved reserves and capitalized development costs shall be amortized (depreciated) by the unit-of-production method so that each unit produced in assigned a pro rata portion of the unamortized costs. It may be more appropriate, in some cases, to depreciate natural gas cycling and processing plants by a method other than the unit-of-production method. Under the unit-of-production method, amortization (depreciation) may be computed either on a property-by-property basis or on the basis of some reasonable aggregation of properties with a common geological structural feature or stratigraphic condition, such as a reservoir or field. The unit cost shall be computed on the basis of the total estimated units of proved *developed* reserves, rather than on the basis of all proved reserves, which is the basis for amortizing acquisition costs of proved properties. If significant development costs (such as the cost of an offshore production platform) are incurred in connection with a planned group of development wells before all of the planned wells have been drilled, it will be necessary to exclude a portion of those development costs in determining the unit-of-production amortization rate until the additional development wells are drilled. Similarly it will be necessary to exclude, in computing the amortization rate, those proved developed reserves that will be produced only after significant additional development costs are incurred, such as for improved recovery systems. However, in no case should future development costs be anticipated in computing the amortization rate. (Joint production of both oil and gas is discussed in paragraph 38.) Unit-of-production amortization rates shall be revised whenever there is an indication of the need for revision but at least once a year; those revisions shall be accounted for prospectively as changes in accounting estimates–see paragraphs 31-33 of *APB Opinion No. 20.*

Depreciation of Support Equipment and Facilities

36. Depreciation of support equipment and facilities used in oil and gas producing activities shall be accounted for as exploration cost, development cost, or production cost, as appropriate (paragraph 26).

Dismantlement Costs and Salvage Values

37. Estimated dismantlement, restoration, and abandonment costs and estimated residual salvage values shall be taken into account in determining amortization and depreciation rates.

Amortization of Costs Relating to Oil and Gas Reserves Produced Jointly

38. The unit-of-production method of amortization requires that the total number of units of oil or gas reserves in a property or group of properties be estimated and that the number of units produced in the current period be determined. Many properties contain both oil and gas reserves. In those cases, the oil and gas reserves and the oil and gas produced shall be converted to a common unit of measure on the basis of their approximate relative energy content (without considering their relative sales values).However, if the relative proportion of gas and oil extracted in the current period is expected to continue throughout the remaining productive life of the property, unit-of-production amortization may be computed on the basis of one of the two minerals only; similarly, if either oil or gas clearly dominates both the reserves and the current production (with dominance determined on the basis of relative energy content), unit-of-production amortization may be computed on the basis of the dominant mineral only.

Information Available after the Balance Sheet Date

39. Information that becomes available after the end of the period covered by the financial statements but before those financial statements are issued shall be taken into account in evaluating conditions that existed at the balance sheet date, for example, in assessing unproved properties (paragraph 28) and in determining whether an exploratory well or exploratory-type stratigraphic test well had found proved reserves (paragraph 31-34).

Surrender of abandonment of Properties

40. When an unproved property is surrendered, abandoned, or otherwise deemed worthless, capitalized acquisition costs relating thereto shall be charged against the related allowance for impairment to the extent an allowance has been provided; if the allowance previously provided is inadequate, a loss shall be recognized.

41. Normally, no gain or loss shall be recognized if only an individual well or individual item of equipment is abandoned or retired or if only a single lease or other part of a group of proved properties constituting the amortization base is abandoned or retired as long as the remainder of the property or group of properties continues to produce oil or gas. Instead, the asset being abandoned or retired shall be deemed to be fully amortized, and its cost shall be charged to accumulated depreciation, depletion, or amortization. When the *last* well on an individual property (if that is the amortization base) or group of properties (if amortization is determined on the basis of an aggregation of properties with a common geological struc-

ture) ceases to produce and the entire property or property group is abandoned, gain or loss shall be recognized. Occasionally, the partial abandonment or retirement of a proved property or group of proved properties or the abandonment or retirement of wells or related equipment or facilities may result from a catastrophic event or other major abnormality. In those cases, a loss shall be recognized at the time of abandonment or retirement.

Mineral Property Conveyances and Related Transactions

42. Mineral interest in properties are frequently conveyed to others for a variety of reasons, including the desire to spread risks, to obtain financing, to improve operating efficiency, and to achieve tax benefits. Conveyances of those interests may involve the transfer of all or a part of the rights and responsibilities of operating a property (operating interest). The transferor may or may not retain an interest in the oil and gas produced that is free of the responsibilities and costs of operating the property (a nonoperating interest). A transaction may, on the other hand, involve the transfer of a nonoperating interest to another party and retention of the operating interest.

43. Certain transactions, sometimes referred to as conveyances, are in substance borrowings repayable in cash or its equivalent and shall be accounted for as borrowings. The following are examples of such transactions:

a. Enterprises seeking supplies of oil or gas sometimes make cash advances to operators to finance exploration in return for the right to purchase oil or gas discovered. Funds advanced for exploration that are repayable by offset against purchases of oil or gas discovered, or in cash if insufficient oil or gas is produced by a specified date, shall be accounted for as a receivable by the lender and as a payable by the operator.
b. Funds advanced to an operator that are repayable in cash out of the proceeds from a specified share of future production of a producing property, until the amount advanced plus interest at a specified or determinable rate is paid in full, shall be accounted for as a borrowing. The advance is a payable for the recipient of the cash and a receivable for the party making the advance. Such transactions, as well as those described in paragraph 47(a) below, are commonly referred to as production payments. The two types differ in substance, however, as explained in paragraph 47(a).

44. In the following types of conveyances, gain or loss shall not be recognized at the time of the conveyance:

a. A transfer of assets used in oil and gas producing activities (including both proved and unproved properties) in exchange for other assets also used in oil and gas producing activities.

b. A pooling of assets in a joint undertaking intended to find, develop, or produce oil or gas from a particular property or group of properties.

45. In the following types of conveyances, gain shall not be recognized at the time of the conveyance:

a. A part of an interest owned is sold and substantial uncertainty exists about recovery of the costs applicable to the retained interest.
b. A part of an interest owned is sold and the seller has a substantial obligation for future performance, such as an obligation to drill a well or to operate the property without proportional reimbursement for that portion of the drilling or operating costs applicable to the interest sold.

46. If a conveyance is not one of the types described in paragraphs 44 and 45, gain or loss shall be recognized unless there are other aspects of the transaction that would prohibit such recognition under accounting principles applicable to enterprises in general.

47. In accordance with paragraphs 44-46, the following types of transactions shall be accounted for as indicated in each example.[3] No attempt has been made to include the many variations of those arrangements that occur, but paragraphs 44-46 shall, where applicable, determine the accounting for those other arrangements as well.

a. Some production payments differ from those described in paragraph 43(b) in that the seller's obligation is not expressed in monetary terms but as an obligation to deliver, free and clear of all expenses associated with operation of the property, a specified quantity of oil or gas to the purchaser out of a specified share of future production. Such a transaction is a sale of a mineral interest for which gain shall not be recognized because the seller has a substantial obligation for future performance. The seller shall account for the funds received as unearned revenue to be recognized as the oil or gas is delivered. The purchaser of such a production payment has acquired an interest in a mineral property that shall be recorded at cost and amortized by the unit-of-production method as delivery takes place. The related reserve estimates and production data shall be reported as those of the purchaser of the production payment and not of the seller (paragraphs 50-56).
b. An assignment of the operating interest in an unproved property with retention of a nonoperating interest in return for drilling, development, and operation by the assignee is a pooling of assets in a joint undertaking for which the assignor shall not recognize gain or loss. The assignor's cost of the original interest shall become the cost of the interest retained. The assignee shall account for all costs incurred as specified by paragraphs 15-41 and shall allocate none of those costs to the mineral interest acquired. If oil or gas is discovered, each party shall report its share of reserves and production (paragraphs 50-56).
c. An assignment of a part of an operating interest in an unproved property in exchange for a "free well" with provision for joint ownership and operation is a pooling of assets in a

joint undertaking by the parties. The assignor shall record no cost for the obligatory well; the assignee shall record no cost for the mineral interest acquired. All drilling, development, and operating costs incurred by either party shall be accounted for as provided in paragraphs 15-41 of this Statement. If the conveyance agreement requires the assignee to incur geological or geophysical expenditures instead of, or in addition to, a drilling obligation, those costs shall likewise be accounted for by the assignee as provided in paragraphs 15-41 of this Statement. If reserves are discovered, each party shall report its share of reserves and production (paragraphs 50-56).

d. A part of an operating interest in an unproved property may be assigned to effect an arrangement called a "carried interest" whereby the assignee (the carrying party) agrees to defray all costs of drilling, developing, and operating the property and is entitled to all of the revenue from production from the property, excluding any third party interest, until all of the assignee's costs have been recovered, after which the assignor will share in both costs and production. Such an arrangement represents a pooling of assets in a joint undertaking by the assignor and assignee. The carried party shall make no accounting for any costs and revenue until after recoupment (payout) of the carried costs by the carrying party. Subsequent to payout the carried party shall account for its share of revenue, operating expenses, and (if the agreement provides for subsequent sharing of costs rather than a carried interest) subsequent development costs. During the payout period the carrying party shall record all costs, including those carried, as provided in paragraphs 15-41 and shall record all revenue from the property including that applicable to the recovery of costs carried. The carried party shall report as oil or gas reserves only its share of proved reserves estimated to remain after payout, and unit-of-production amortization of the carried party's property cost shall not commence prior to payout. Prior to payout the carrying party's reserve estimates and production data shall include the quantities applicable to recoupment of the carried costs (paragraphs 50-56).

e. A part of an operating interest owned may be exchanged for a part of an operating interest owned by another party. The purpose of such an arrangement, commonly called a joint venture in the oil and gas industry, often is to avoid duplication of facilities, diversify risks, and achieve operating efficiencies. Such reciprocal conveyances represent exchanges of similar productive assets, and no gain or loss shall be recognized by either party at the time of the transaction. In some joint ventures, which may or may not involve an exchange of interests, the parties may share different elements of costs in different proportions. In such an arrangement a party may acquire an interest in a property or in wells and related equipment that is disproportionate to the share of costs borne by it. As in the case of a carried interest or a free well, each party shall account for its own cost under the provisions of this Statement. No gain shall be recognized for the acquisition of an interest in joint assets, the cost of which may have been paid in whole or in part by another party.

f. In a unitization all the operating and nonoperating participants pool their assets in a producing area (normally a field) to form a single unit and in return receive an undivided interest (of the same type as previously held) in that unit. Unitizations generally are undertaken to obtain operating efficiencies and to enhance recovery of reserves, often through improved recovery operations. Participation in the unit is generally proportionate to the oil and gas reserves contributed by each. Because the properties may be in different stages of development at the time of unitization, some participants may pay cash and others may receive cash to equalize contributions of wells and related equipment and facilities with the ownership interests in reserves. In those circumstances, cash paid by a participant shall be recorded as an additional investment in wells and related equipment and facilities, and cash received by a participant shall be recorded as a recovery of cost. The cost of the assets contributed plus or minus cash paid or received is the cost of the participant's undivided interest in the assets of the unit. Each participant shall include its interest in reporting reserve estimates and production data (paragraphs 50-56).

g. If the entire interest in an unproved property is sold for cash or cash equivalent, recognition of gain or loss depends on whether, in applying paragraph 28 of this Statement, impairment had been assessed for that property individually or by amortizing that property as part of a group. If impairment was assessed individually, gain or loss shall be recognized. For a property amortized by providing a valuation allowance on a group basis, neither a gain nor loss shall be recognized when an unproved property is sold unless the sales price exceeds the original cost of the property, in which case gain shall be recognized in the amount of such excess.

h. If a part of the interest in an unproved property is sold, even though for cash or cash equivalent, substantial uncertainty usually exists as to recovery of the cost applicable to the interest retained. Consequently, the amount received shall be treated as a recovery of cost. However, if the sales price exceeds the carrying amount of a property whose impairment has been assessed individually in accordance with paragraph 28 of this Statement, or exceeds the original cost of a property amortized by providing a valuation allowance on a group basis, gain shall be recognized in the amount of such excess.

i. The sale of an entire interest in a proved property that constitutes a separate amortization base is not one of the types of conveyances described in paragraphs 44 or 45. The difference between the amount of sales proceeds and the unamortized cost shall be recognized as a gain or loss.

j. The sale of a part of a proved property, or of an entire proved property constituting a part of an amortization base, shall be accounted for as the sale of an asset, and a gain or loss shall be recognized, since it is not one of the conveyances described in paragraphs 44 or 45. The unamortized cost of the property or group of properties a part of which was sold shall be apportioned to the interest sold and the interest retained on the basis of the fair values of those interests. However, the sale may be accounted for as a normal retirement under the provisions of paragraph 41 with no gain or loss recognized if doing so does not significantly affect the unit-of-production amortization rate.

k. The sale of the operating interest in a proved property for cash with retention of a non-operating interest is not one of the types of conveyances described in paragraphs 44 or

45. Accordingly, it shall be accounted for as the sale of an asset, and any gain or loss shall be recognized. The seller shall allocate the cost of the proved property to the operating interest sold and the nonoperating interest retained on the basis of the fair values of those interests.

l. The sale of a proved property subject to a retained production payment that is expressed as a fixed sum of money payable only from a specified share of production from that property, with the purchaser of the property obligated to incur the future costs of operating the property, shall be accounted for as follows:

   i. *If satisfaction of the retained production payment is reasonably assured.* The seller of the property, who retained the production payment, shall record the transaction as a sale, with recognition of any resulting gain or loss. The retained production payment shall be recorded as a receivable, with interest accounted for in accordance with the provisions of APB Opinion No. 21, "Interest on Receivables and Payables." The purchaser shall record as the cost of the assets acquired the cash consideration paid plus the present value (determined in accordance with APB Opinion No. 21) of the retained production payment, which shall be recorded as a payable. The oil and gas reserve estimates and production data, including those applicable to liquidation of the retained production payment, shall be reported by the purchaser of the property (paragraphs 50-56).
   ii. *If satisfaction of the retained production payment is not reasonably assured.* The transaction is in substance a sale with retention of an overriding royalty that shall be accounted for in accordance with paragraph 47(k).

m. The sale of a proved property subject to a retained production payment that is expressed as a right to a specified quantity of oil or gas out of a specified share of future production shall be accounted for in accordance with paragraph 47(k).

NOTES

1. Often referred to in the oil and gas industry as "lease and well equipment" even though, technically, the property may have been acquired other than by a lease.

2. While the costs of drilling stratigraphic test wells are sometimes considered to be geological and geophysical costs, they are accounted for separately under this Statement for reasons explained in paragraphs 200-202.

3. Costs of unproved properties are always subject to an assessment for impairment as required by paragraph 28.

a p p e n d i x e

Kerr-McGee Corporation *1999* Annual Report

Contents

FINANCIAL REVIEW

Management's Discussion and Analysis

Kerr-McGee/Oryx Merger

On February 26, 1999, the merger between Kerr-McGee and Oryx was completed. Oryx was a worldwide independent oil and gas exploration and production company. Its operations have been merged into and reported with Kerr-McGee's exploration and production segment. All references to the "company" refer to the merged entity.

Under the merger agreement, each outstanding share of Oryx common stock was effectively converted into the right to receive 0.369 shares of newly issued Kerr-McGee common stock. The merger qualified as a tax-free exchange to Oryx's shareholders and has been accounted for as a pooling of interests. Accordingly, results of operations, financial position and cash flows for all prior periods have been restated to reflect the combined company as though it had always been in existence.

The merger with Oryx was the largest transaction in Kerr-McGee's history. The company has successfully incorporated the assets, staffs and operations of the two companies and met the projected annualized level of $100 million of pretax synergy savings.

Operating Environment and Outlook

Based on proved reserves at December 31, 1999, the company is one of the largest independent, nonintegrated oil and gas exploration and production companies based in the United States.

In the first two months of 2000, oil prices are in the $27 to $30 per-barrel range, the highest level in a decade. Oil prices have risen steadily from the spring of 1999 when OPEC took steps to reduce supplies. OPEC's actions, combined with higher consumption due to a robust economy, have resulted in historically low levels of crude oil inventories. OPEC is scheduled to meet in late March 2000, and many experts forecast that these producers will boost production in order to prevent product shortages which will result in lower prices. Management recognizes these risks to commodity pricing and believes prices will average between $22 and $25 per barrel in 2000.

In early 2000, natural gas prices are in the $2.50 to $2.80 per million BTU range, and those levels continue in the near-term futures markets. Natural gas consumption in the U.S. has continued to grow, currently representing approximately 25% of the nation's fuel needs. In 1998 and 1999, the U.S. experienced a downturn in drilling in the shallow Gulf of Mexico and onshore United States. The company and others have made significant discoveries in the deepwater Gulf of Mexico; however, these projects require long startup times. The increasing need for this environmentally friendly fuel for power generation, coupled with slow supply development, should contribute to strengthening prices. Management believes prices are stable at their current levels and may increase when the market recognizes more of the risks associated with increasing natural gas supply.

On February 14, 2000, the company reached agreements (subject to customary conditions and governmental approvals) with Kemira Oyj of Finland to purchase its pigment operations in Savannah, Georgia, and Botlek, the Netherlands. The two plants have combined capacity of 201,000 tonnes per year, which will increase Kerr-McGee's total equity pigment capacity by 60% to 535,000 tonnes per year. After the transaction is completed, the company will rank as the world's third-largest producer and marketer of pigment with about 16% of the world market. During 1998, the company introduced a new universal grade of pigment, which currently represents nearly 30% of the company's U.S. pigment production. A new grade of pigment for the industrial-coatings market will go into commercial production in the first half of 2000. Management believes that pigment consumption will increase by 2.5% to 3% annually during the next five years. The company is undertaking cost-reduction programs at its non-U.S. pigment facilities similar to a program that was implemented in 1999 at the U.S. plant where significant cost reductions have been achieved.

14 Kerr-McGee Corporation 1999 Annual Report

Results of Consolidated Operations

Net income (loss) and per-share amounts for each of the three years in the period ended December 31, 1999, were as follows:

(Millions of dollars, except per-share amounts)	**1999**	1998	1997
Net income (loss)	**$142**	$ (68)	$382
Income (loss) from continuing operations excluding special items	**296**	(24)	343
Net income (loss) per share –			
Net income (loss) –			
Basic	**1.64**	(.78)	4.40
Diluted	**1.64**	(.78)	4.38
Income (loss) from continuing operations excluding special items –			
Basic	**3.42**	(.28)	3.95
Diluted	**3.42**	(.28)	3.93

Net income (loss) was impacted by a number of special items in each of the years. In 1999, special items were both operating and nonoperating and were associated principally with the Oryx merger and transition and with pending and settled litigation matters. The 1998 special items related primarily to impairment write-downs reflecting the then current market value of certain of the company's oil and gas producing fields and certain chemical facilities. Other 1998 special items were principally nonoperating and reduced net income by $22 million. In 1997, special items were principally nonoperating and increased net income by $8 million. These special items affect comparability between the periods and are shown on an after-tax basis in the following table, which reconciles income (loss) from continuing operations excluding special items to net income (loss):

(Millions of dollars)	**1999**	1998	1997
Income (loss) from continuing operations excluding special items	**$ 296**	$ (24)	$343
Special items, net of taxes –			
Asset impairment	**—**	(299)	—
Merger costs	**(116)**	—	—
Equity affiliate's full-cost ceiling write-down	**—**	(27)	—
Net provision for environmental remediation and restoration of inactive sites	**—**	(26)	(13)
Restructuring	**(1)**	(25)	(1)
Pending/settled litigation	**(20)**	—	(1)
Transition costs	**(14)**	—	—
Settlement of prior years' income taxes	**1**	41	—
Settlements with insurance carriers	**—**	8	8
Effect of U.K. tax-rate change	**—**	8	—
Gains on the sales of equity securities	**—**	—	12
Other, net	**—**	(1)	3
Total	**(150)**	(321)	8
Discontinued operations, net of taxes	**—**	277	33
Extraordinary charge, net of taxes	**—**	—	(2)
Change in accounting principle, net of taxes	**(4)**	—	—
Net income (loss)	**$ 142**	$ (68)	$382

Effective January 1, 1999, the company adopted Statement of Position (SOP) No. 98-5, "Reporting on the Costs of Start-Up Activities." The SOP requires costs of start-up activities to be expensed as incurred. Unamortized start-up costs at the beginning of 1999 were required to be recognized as a cumulative effect of a change in accounting principle, which decreased 1999 after-tax income by $4 million.

The company sold its coal operations in 1998, resulting in an after-tax gain of $257 million. All amounts related to coal are shown in the Consolidated Statement of Income as discontinued operations.

In 1997, the company recognized an extraordinary loss of $2 million (net of $1 million of income taxes) from the write-off of unamortized debt issuance costs. These costs related to a $500 million credit facility that was replaced with a five-year, $500 million revolving credit agreement.

Income from continuing operations excluding special items for 1999 increased $320 million from 1998. This primarily resulted from a $343 million increase in exploration and production after-tax operating profit excluding special items, which was partially offset by a $29 million increase in net interest expense. In 1998, income (loss) from continuing operations excluding special items declined $367 million from the prior year, due primarily to a $368 million decline in exploration and production after-tax operating profit excluding special items, which was partially offset by a $20 million increase in chemical results.

Sales from continuing operations were $2.7 billion in 1999, $2.2 billion in 1998 and $2.6 billion in 1997. Sales for 1999 were higher than in 1998 due to higher average sales prices for oil and natural gas (37% and 11% increases, respectively), a 16% increase in oil volumes sold and an increase in titanium dioxide pigment sales volumes (mainly due to a full year of production from the company's European pigment operations, compared to nine months in 1998), partially offset by lower electrolytic and forest products sales volumes and lower European pigment sales prices. Sales in 1998 were lower than 1997 primarily due to declines in 1998 average sales prices for oil and natural gas, of 32% and 13%, respectively. In addition, natural gas volume decreases were partially offset by increased sales volumes and prices for pigment. The volume decreases in natural gas sales were primarily the result of damages to and repair times for pipeline systems, hurricane downtimes and normal production declines. Volume increases in pigment sales relate to the March 1998 purchase of the European pigment operations and the expansion of the pigment facility in Hamilton, Mississippi.

Costs and operating expenses totaled $1.1 billion in both 1999 and 1998 and $1 billion in 1997. The 1998 amount was higher than the prior year principally due to costs of the acquired European pigment operations and higher per-unit costs at the U.S. pigment and synthetic rutile facilities. This was partially offset by the absence of costs of natural gas purchased for resale.

Following are general and administrative expenses for 1999, 1998 and 1997:

(Millions of dollars)	**1999**	1998	1997
General and administrative expenses excluding special items	**$186**	$204	$194
Special items –			
Net provision for environmental remediation and restoration of inactive sites	**—**	41	20
Restructuring	**—**	36	2
Pending/settled litigation	**30**	—	2
Transition costs associated with the Oryx merger	**22**	—	—
Other, net	**—**	(3)	3
Total	**52**	74	27
General and administrative expenses	**$238**	$278	$221

The decrease in 1999 general and administrative expense compared with 1998 resulted from the synergies realized through the merger, principally by the exploration and production segment. The estimated general and administrative synergies of approximately $35 million were partially offset by slightly higher chemical costs and higher corporate charges primarily due to higher costs associated with improved employee benefit plans. The provision for pending or settled litigation is principally related to facilities or properties no longer operated or owned by the company. Transition costs were those associated with ongoing business during the time of the merger, which will not re-occur in 2000. The increase in 1998 over 1997 general and administrative expenses excluding special items primarily resulted from additional costs related to the company's European pigment operations. Net provisions for environmental remediation and restoration of inactive sites primarily represented additional provisions established for the removal of low-level radioactive materials from the company's inactive facility and offsite areas in West Chicago, Illinois. Restructuring charges were for a 1998 voluntary severance program for the former Oryx U.S. operations, a work process review and organizational restructuring of several groups, the 1996-1997 relocation of part of the exploration and production unit to Houston, Texas, and severance associated with the divestiture program and the merger of certain of the company's North American onshore properties into Devon Energy Corporation (Devon) effective December 31, 1996.

Asset impairments totaled $446 million in 1998 (see Note 17). Of this amount, $389 million were for write-downs associated with certain oil and gas fields located in the North Sea, China and United States. Asset impairment of $57 million was also recognized for certain chemical facilities in Idaho and Alabama. The impairments were recorded because these assets were no longer expected to recover their net book values through future cash flows.

Exploration costs for 1999, 1998 and 1997 were $140 million, $215 million and $139 million, respectively. The decrease for 1999 resulted from lower dry hole costs principally in the Gulf of Mexico, Kazakhstan and China, lower costs of geophysical projects primarily in the United States onshore area and lower exploration district expense in the United States, the North Sea and China. The primary reasons for the 1998 increase over the prior year were higher dry hole costs in the Gulf of Mexico, Kazakhstan, Thailand and onshore United States, higher undeveloped leasehold amortization in the Gulf of Mexico, higher geophysical expenses related to the Gulf of Mexico and higher district expense in China, the North Sea and Gulf of Mexico, partially offset by lower dry hole costs in China.

Taxes, other than income taxes, were $85 million in 1999, $53 million in 1998 and $103 million in 1997. The 1999 and 1998 variances from the prior year were both due principally to severance taxes, a direct result of changes in oil and gas prices.

Merger costs totaling $163 million were recognized in 1999 and represent costs incurred in connection with the Oryx merger, which have no future benefit to the combined operations. The major items included are severance and associated benefit plan adjustments; lease cancellation costs; transfer fees for seismic data; investment bankers, lawyers and accountants fees; and the write-off of duplicate computer systems and fixtures (see Note 22).

Interest and debt expense totaled $190 million in 1999, $157 million in 1998 and $141 million in 1997. The 1999 increase resulted from lower capitalized interest and higher borrowings related to the costs of the merger. Borrowings increased in 1998 due to the acquisitions of European chemical operations and North Sea oil and gas assets, partially offset by the proceeds from the sale of the coal assets.

Other income was as follows for each of the years in the three-year period ended December 31, 1999:

(Millions of dollars)	**1999**	1998	1997
Other income excluding special items	**$39**	$36	$43
Special items –			
Interest income from settlement of prior years' income taxes	**1**	19	—
Settlements with insurance carriers	**—**	12	12
Equity affiliate's full-cost ceiling write-down	**—**	(27)	—
Gains on the sale of nonstrategic oil and gas properties	**—**	2	2
Gains on sales of equity securities	**—**	—	18
Other, net	**—**	1	7
Total	**1**	7	39
Other income	**$40**	$43	$82

The increase in 1999 other income excluding special items compared with 1998 was due primarily to higher foreign currency gains, partially offset by lower interest income. Lower equity earnings from unconsolidated affiliates were the primary reason for the decline in 1998 other income excluding special items, compared with the prior year. Equity earnings from the company's investment in Devon were impacted by lower oil and gas prices and decreased $14 million for 1998 compared with 1997.

Segment Operations

Operating profit (loss) from each of the company's segments is summarized in the following table:

(Millions of dollars)	**1999**	1998	1997
Operating profit excluding special items –			
Exploration and production	**$ 562**	$ 62	$ 597
Chemicals –			
Pigment	**113**	89	49
Other	**15**	26	35
Total Chemicals	**128**	115	84
Total	**690**	177	681
Special items	**(21)**	(482)	(5)
Operating profit (loss)	**$ 669**	$ (305)	$ 676

Exploration and Production

Exploration and production sales, operating profit (loss) and production and sales statistics are shown in the following table:

(Millions of dollars, except per-unit amounts)	**1999**	1998	1997
Sales	**$1,770**	$1,267	$1,845
Operating profit excluding special items	**$ 562**	$ 62	$ 597
Special items	**(20)**	(423)	(2)
Operating profit (loss)	**$ 542**	$ (361)	$ 595
Net crude oil and condensate produced (thousands of barrels per day)	**197**	172	172
Average price of crude oil sold (per barrel)	**$17.15**	$12.52	$18.32
Natural gas sold (MMCF per day)	**580**	584	685
Average price of natural gas sold (per MCF)	**$ 2.35**	$ 2.12	$ 2.43

Special items in 1999 are transition costs associated with the work necessary to accomplish the Oryx merger. Asset impairment for certain oil and gas fields in the North Sea, China and the United States totaled $389 million in 1998 and is reflected in special items. Also in 1998, a $34 million restructuring reserve is shown as a special item. This amount was provided primarily for a voluntary severance program for employees of former Oryx U.S. operations. Special items in 1997 consisted primarily of additional costs for the segment's restructuring and relocation to Houston, Texas.

Chemicals

Chemical sales and operating profit are shown in the following table:

(Millions of dollars)	**1999**	1998	1997
Sales –			
Pigment	**$700**	$640	$470
Other	**226**	293	290
Total	**$926**	$933	$760
Operating profit excluding special items –			
Pigment	**$113**	$ 89	$ 49
Other	**15**	26	35
	128	115	84
Special items –			
Pigment	**—**	(33)	—
Other	**(1)**	(26)	(3)
Operating profit	**$127**	$ 56	$ 81

Severance charges of $1 million and $2 million were recorded as special items in 1999 and 1998, respectively. Also included in 1998 special charges are asset impairments totaling $57 million for noncore chemical assets in Alabama and Idaho. Special items in 1997 were primarily for the write-off of obsolete equipment.

Pigment – The increase in 1999 titanium dioxide pigment sales from the prior year was due principally to increased volumes in Europe as a result of a full year of sales after the company's acquisition at the end of March 1998 of the European pigment operations, partially offset by lower European pigment prices. Operating profit excluding special items increased in 1999 due to the higher sales in Europe and lower U.S. per-unit production costs. Pigment prices increased throughout 1998. This improvement in pricing, along with the company's acquisition of the European pigment operations and a full year's production from a 27,000 tonne-per-year expansion of the company's Hamilton, Mississippi, plant were the primary reasons for the $170 million increase in pigment sales in 1998. These sales increases were partially offset by higher per-unit production costs resulting in a $40 million increase in operating profit excluding special items.

Other – Other chemical sales were lower in 1999 as compared with 1998 principally due to lower forest products sales volumes, the company's withdrawal from the ammonium perchlorate business in 1998 and lower vanadium sales volumes. The decline in sales was the major reason for lower operating profit in 1999. The decline in 1998 operating profit from 1997 resulted primarily from the withdrawal from the ammonium perchlorate business in 1998 and higher sodium chlorate per-unit production costs.

Financial Condition

(Millions of dollars)	**1999**	1998	1997
Current ratio	**1.4**	0.8	1.0
Total debt	**$2,525**	$2,250	$1,766
Total debt less cash	**2,258**	2,129	1,574
Stockholders' equity	**$1,492**	$1,346	$1,558
Total debt less cash to total capitalization	**60%**	61%	50%
Floating-rate debt to total debt	**38**	33	15

Cash Flow

Net cash provided by operating activities was $713 million in 1999, compared with $385 million in 1998 and $1.1 billion in 1997. The rebound in crude oil prices and the resulting impact on net income were the primary reasons for the 1999 increase in net cash provided by operating activities and more than offset the reduction in cash from the costs of the merger.

Cash used in 1999 investing activities was primarily for $543 million of capital expenditures and $33 million of unsuccessful exploratory well costs. Additionally, the company invested in several oil and gas property acquisitions totaling $78 million, including the buy-out of the limited partners of Sun Energy Partners, L.P. Net other investing activities used $8 million of cash.

Dividends increased to $138 million in 1999 with the additional Kerr-McGee shares outstanding after the merger. These quarterly dividend payments of $.45 per share and the net cash used in investing activities were in excess of net cash provided by operating activities. To supplement this shortfall, borrowings and other financings increased by a net amount of $238 million, with a net result of $146 million increase in cash.

The decrease in 1998 net cash provided by operating activities resulted primarily from the low crude oil price environment, which contributed to the net loss and from increased working capital and other changes that used cash from operating activities. Net cash provided by operating activities was reduced by taxes paid related to the sale of the discontinued coal operations of $115 million.

In 1998, proceeds of approximately $600 million were received from the sale of the company's discontinued coal operations, $150 million from the sale of the marginal exploration and production properties and the ammonium perchlorate operations and $20 million from other investing activities. These sources of cash from investing activities and net proceeds from debt issuances of $481 million were used for capital expenditures of $981 million, acquisitions of the Gulf Canada North Sea assets and the European titanium dioxide pigment facilities totaling $518 million and dry hole costs of $92 million.

The company's Board of Directors authorized a stock purchase program in 1998. A total of 580,000 shares ($25 million) was purchased before the program was cancelled because of the company's merger.

Liquidity

At year-end 1999, total debt outstanding was $2.5 billion. The percentage of total debt less cash to total capitalization was 60% at December 31, 1999; 61% at December 31, 1998; and 50% at year-end 1997. The slight improvement at year-end 1999 resulted from the impact of the equity increase from 1999 net income on total capitalization. The impact of increased borrowings in 1998 accounted for the increase in the 1998 percentage. Borrowings increased because the level of the capital expenditure program and the two 1998 acquisitions were, in total, in excess of the proceeds from the sale of the coal operations.

Significant 1999 debt transactions included issuance of $327 million 5-1/2% debt exchangeable for common stock of Devon due August 2, 2004. (The company owns 9,954,000 shares of outstanding Devon common stock.) In addition, $150 million floating rate notes due 2001 were issued in November to institutional investors with interest payable quarterly at three-month LIBOR plus 0.5%.

The company believes it has the ability to provide for its operational needs and its long- and short-term capital programs through its operating cash flow, borrowing capacity and ability to raise capital. At December 31, 1999, the company had unused lines of credit and revolving credit agreements totaling $1.4 billion. Of this amount, $835 million and $400 million could be used to support the commercial paper borrowings of Kerr-McGee Credit LLC and Kerr-McGee Oil (U.K.) PLC, respectively, both wholly owned subsidiaries. Outstanding revolving credit borrowings at year-end 1999 totaled $85 million at varying rates of interest.

On February 26, 1999, the date of the merger, the company signed two revolving credit facilities replacing $75 million of a Kerr-McGee Oil (U.K.) PLC revolving credit facility and Oryx's $500 million, five-year revolving credit facility entered into October 20, 1997. The two agreements consist of a three-year, $500 million facility and a 364-day, $250 million facility. One-third of the borrowings under each of the agreements can be drawn by foreign subsidiaries. The borrowings can be made in British pound sterling, euros or other local European currencies. Interest for each of the revolving credit facilities is payable at varying rates. Effective February 25, 2000, the 364-day facility was renewed and increased to $350 million.

At December 31, 1999, the company classified $793 million of its short-term obligations as long-term debt. Final settlement of these obligations, consisting of revolving credit borrowings and commercial paper, is not expected to occur in 2000. The company has the intent and the ability, as evidenced by committed credit arrangements, to refinance this debt on a long-term basis. The company's practice has been to continually refinance its commercial paper, while maintaining levels believed to be appropriate.

The company increased its shelf registration with the Securities and Exchange Commission in January 2000 to offer up to $1.5 billion of debt securities, preferred stock, common stock or warrants. In February 2000, under this registration, the company issued 7.5 million shares of its common stock and $600 million of 5-1/4% convertible subordinated debentures due 2010, generating nearly $1 billion in net proceeds to the company. These proceeds will be used to redeem the short-term floating rate debt used for the $555 million acquisition of Repsol S.A.'s North Sea oil and gas operations in January 2000, the pending $403 million acquisition of Kemira Oyj's U.S. and Dutch titanium dioxide pigment operations and/or reduction of other debt.

In connection with these first quarter transactions and offerings, rating agencies confirmed the company's debt ratings. The ratings are "BBB+," "Baa1" and "BBB" for senior unsecured debt. See Note 8 for a discussion of the company's debt at year-end 1999.

The company finances capital expenditures through internally generated funds and various borrowings. Cash capital expenditures were $543 million in 1999, $981 million in 1998 and $836 million in 1997, a total of $2.4 billion. During this same three-year period, $2.9 billion of net cash was provided by operating activities (exclusive of working capital and other changes), which exceeded cash capital expenditures and dividends paid during the periods by approximately $200 million.

Management anticipates that 2000 cash capital requirements, currently estimated at $675 million, and the capital expenditures programs for the next several years can continue to be provided through internally generated funds and selective borrowings.

Market Risks

The company is exposed to a variety of market risks, including the effects of movements in foreign currency exchange rates, interest rates and certain commodity prices. The company addresses its risks through a controlled program of risk management that includes the use of derivative financial instruments. The company does not hold or issue derivative financial instruments for trading purposes. See Notes 1 and 18 for additional discussions of the company's financial instruments and hedging activities.

Foreign Currency Exchange

The U.S. dollar is the functional currency for the company's international operations, except for its European chemical operations. It is the company's intent to hedge a portion of its monetary assets and liabilities denominated in foreign currencies. Periodically, the company purchases foreign currency forward contracts to provide funds for operating and capital expenditure requirements that will be denominated in foreign currencies, primarily Australian dollars and British pound sterling. These contracts generally have durations of less than three years. The company also enters into forward contracts to hedge the sale of various foreign currencies, principally generated from accounts receivable for titanium dioxide pigment sales denominated in foreign currencies. These contracts are principally for European currencies and generally have durations of less than a year. Since these contracts qualify as hedges and correlate to currency movements, any gains or losses resulting from exchange rate changes are deferred and recognized as adjustments of the hedged transaction when it is settled in cash.

Following are the notional amounts at the contract exchange rates, weighted-average contractual exchange rates and estimated fair value by contract maturity for open contracts at year-end 1999 and 1998 to purchase (sell) foreign currencies. All amounts are U.S. dollar equivalents.

(Millions of dollars, except average contract rate)	Notional Amount	Weighted-Average Contract Rate	Estimated Fair Value
Open contracts at December 31, 1999 –			
Maturing in 2000 –			
Australian dollar	$48	.6306	$50
French franc	(1)	6.2908	(1)
British pound sterling	(1)	.6187	(1)
Italian lira	(1)	1839.8282	(1)
New Zealand dollar	(1)	1.9775	(1)
Japanese yen	(1)	102.4479	(1)
Maturing in 2001 –			
Australian dollar	32	.6499	32
Maturing in 2002 –			
Australian dollar	16	.6538	16
Open contracts at December 31, 1998 –			
Maturing in 1999 –			
Australian dollar	56	.7117	48
German mark	(1)	1.6745	(1)
British pound sterling	41	1.6355	42
Maturing in 2000 –			
Australian dollar	21	.6145	21

Interest Rates

The company's exposure to changes in interest rates relates primarily to long-term debt obligations. The company has participated in various interest rate hedging arrangements to help manage the floating-rate portion of its debt. There were no interest rate hedging contracts entered into during 1999 or 1998. At December 31, 1998, all interest rate hedging contracts had expired.

The table below presents principal amounts and related weighted-average interest rates by maturity date for the company's long-term debt obligations outstanding at year-end 1999. All borrowings are in U.S. dollars.

(Millions of dollars)	2000	2001	2002	2003	2004	There-after	Total	Fair Value 12/31/99
Fixed-rate debt –								
Principal amount	$20	$174	$ 33	$116	$491	$739	$1,573	$1,612
Weighted-average interest rate	8.54%	9.78%	8.85%	8.04%	6.45%	7.28%	7.40%	
Variable-rate debt –								
Principal amount	—	$416	$467	$ 60	—	—	$ 943	$ 943
Weighted-average interest rate	—	6.74%	6.70%	6.37%	—	—	6.69%	

At December 31, 1998, long-term debt included fixed-rate debt of $1,497 million (fair value – $1,648 million) with a weighted-average interest rate of 8.17% and $717 million of variable-rate debt, which approximated fair value, with a weighted-average interest rate of 5.92%.

Commodity Prices

The company periodically uses commodity futures and collar contracts to hedge a portion of its crude oil and natural gas sales and natural gas purchased for operations in order to minimize the price risks associated with the production and marketing of crude oil and natural gas. Since the contracts qualify as hedges and correlate to price movements of crude oil and natural gas, any gain or loss from these contracts is deferred and recognized as part of the hedged transaction.

The company did not enter into any hedging arrangements in 1999 and settled all open 1998 contracts during the year. At December 31, 1998, the company had open crude oil collar contracts that hedged 4% of its 1999 worldwide crude oil sales volumes at an average floor price of $15.85 per barrel and an average ceiling price of $17.35 per barrel. Also at December 31, 1998, the company had collar arrangements that hedged 21% of its 1999 worldwide natural gas sales volumes at an average floor price of $2.29 per MMBtu and an average ceiling price of $2.47 per MMBtu. The aggregate carrying value of these contracts at December 31, 1998, was $7 million, and the aggregate fair value, based on quotes from brokers, was approximately $22 million.

Environmental Matters

The company's operations are subject to various environmental laws and regulations. Under these laws, the company is or may be required to remove or mitigate the effects on the environment of the disposal or release of certain chemical, petroleum or low-level radioactive substances at various sites, including sites that have been designated Superfund sites by the U.S. Environmental Protection Agency (EPA) pursuant to the Comprehensive Environmental Response, Compensation, and Liability Act of 1980 (CERCLA), as amended. At December 31, 1999, the company had received notices that it has been named a potentially responsible party (PRP) with respect to 17 existing EPA Superfund sites that require remediation and may share liability at certain of these sites with approximately 300 other PRPs. In addition, the company and/or its subsidiaries have executed consent orders, operate under a license or have reached agreements to perform or have performed remediation or remedial investigations and feasibility studies on sites not included as EPA Superfund sites.

The company does not consider the number of sites for which it has been named a PRP to be a relevant measure of liability. The company is uncertain as to its involvement in many of the sites because of continually changing environmental laws and regulations; the nature of the company's businesses; the large number of other PRPs; the present state of the law, which imposes joint and several liability on all PRPs under CERCLA; and pending legal proceedings. Therefore, the company is unable to reliably estimate the potential liability and the timing of future expenditures that may arise from many of these environmental sites. Reserves have been established for the remediation and restoration of active and inactive sites where it is probable that future costs will be incurred and the liability is estimable. In 1999, $85 million was added to the reserve for active and inactive sites. At December 31, 1999, the company's reserve for these sites totaled $204 million. In addition, at year-end 1999, the company had a reserve of $257 million for the future costs of the abandonment and removal of offshore well and production facilities at the end of their productive lives. In the Consolidated Balance Sheet, $391 million of the total reserve is classified as a deferred credit, and the remaining $70 million is included in current liabilities.

20 Kerr-McGee Corporation 1999 Annual Report

Expenditures for the environmental protection and cleanup of existing sites for each of the last three years and for the three-year period ended December 31, 1999, are as follows:

(Millions of dollars)	**1999**	1998	1997	Total
Charges to environmental reserves	**$121**	$109	$ 96	$326
Recurring expenses	**17**	13	20	50
Capital expenditures	**5**	24	17	46
Total	**$143**	$146	$133	$422

The company has not recorded in the financial statements potential reimbursements from governmental agencies or other third parties, except for amounts due from the U.S. government under Title X of the Energy Policy Act of 1992 (see Notes 11 and 14). The following table reflects the company's portion of the known estimated costs of investigation and/or remediation that is probable and estimable. The table includes all EPA Superfund sites where the company has been notified it is a PRP under CERCLA and other sites for which the company believes it had some ongoing financial involvement in investigation and/or remediation at year-end 1999.

Location of Site	Stage of Investigation/Remediation	Total Known Estimated Cost	Total Expenditures Through 1999
		(Millions of dollars)	
EPA Superfund sites			
Milwaukee, Wis.	Executed consent decree to remediate the site of a former wood-treating facility. Awaiting approval of proposed remedy; installed and operating a free-product recovery system.	$ 15	$ 9
West Chicago, Ill., four sites outside the facility	Began cleanup of first site in 1995. Cleanup of second site began in 1997, and removal work neared completion at end of 1999. Two sites are under study (see Note 11).	85	82
12 sites individually immaterial	Various stages of investigation/remediation.	47	40
		147	131
Non-EPA Superfund sites under consent order, license or agreement			
West Chicago, Ill., facility	Decommissioning is in progress under State of Illinois supervision (see Note 11). Began shipments to a permanent disposal facility in 1994.	385	263
Cleveland/Cushing, Okla.	Began cleanup in 1996.	75	61
Henderson, Nev.	Entered consent agreement in 1999.	47	16
		507	340
Non-EPA Superfund sites individually immaterial	Various stages of investigation/remediation.	220	199
Total for all sites		$874	$670

Management believes adequate reserves have been provided for environmental and all other known contingencies. However, it is possible that additional reserves could be required in the future due to the previously noted uncertainties.

New Accounting Standards

In June 1998, the Financial Accounting Standards Board issued Statement No. 133, "Accounting for Derivative and Hedging Activities." The statement requires recording all derivative instruments as assets or liabilities, measured at fair value. The standard is effective for fiscal years beginning after June 15, 2000. The company is currently evaluating the impact the standard will have on income from continuing operations; however, management believes it will not be material due to the limited amount of derivative and hedging activities in which the company currently engages.

Year 2000 Readiness

In 1996, the company established a formal Year 2000 Program (Program) and expanded the Program to include Oryx systems at the time of the merger in February 1999. The Program was to assess and correct Year 2000 problems in both information technology and noninformation technology systems. The Program was organized into two major areas: Business Systems and Facilities Integrity. Business Systems included replacement applications such as purchasing, inventory, engineering, financial, human resources, etc. Facilities Integrity encompassed telecommunications, plant process controls, instrumentation and embedded chip systems as well as an assessment of third-party Year 2000 readiness. An integral part of the Program was communication with third parties to assess the extent and status of their Year 2000 efforts. The company contacted key suppliers, customers and partners requesting information regarding Year 2000 readiness.

No significant Year 2000 failures or events occurred within the company during the 1999 year-end rollover or during the first two months of 2000. It is believed the readiness of key third-party suppliers and partners has been validated and there will be no future material impact on the company due to Year 2000 problems. While there have been a few Year 2000 problems related to vendor-supplied items, none was considered significant to business operations. There are no significant, ongoing Year 2000 contingencies related to major customers or vendors. Company spending patterns have not been materially affected by Year 2000 remediation. Even though some information technology projects were postponed, none was considered significant.

As of December 31, 1999, total Program expenditures for the merged company's Year 2000 activities were approximately $53 million from inception to completion, which is not material to the company's consolidated results of operations, financial position or cash flows. The total cost of the Program was in line with the company's original estimate. Program expenditures were provided through internally generated funds. No further significant expenditures are expected for Year 2000 activities.

Cautionary Statement Concerning Forward-Looking Statements

Statements in this Financial Review regarding the company's or management's intentions, beliefs or expectations are forward-looking statements within the meaning of the Securities Litigation Reform Act. Future results and developments discussed in these statements may be affected by numerous factors and risks, such as the accuracy of the assumptions that underlie the statements, the success of the oil and gas exploration and production program, drilling risks, the market value of Kerr-McGee's products, uncertainties in interpreting engineering data, demand for consumer products for which Kerr-McGee's businesses supply raw materials, general economic conditions, and other factors and risks discussed in the company's SEC filings. Actual results and developments may differ materially from those expressed or implied in this Financial Review.

Responsibility for Financial Reporting

The company's management is responsible for the integrity and objectivity of the financial data contained in the financial statements. These financial statements have been prepared in conformity with generally accepted accounting principles appropriate under the circumstances and, where necessary, reflect informed judgments and estimates of the effects of certain events and transactions based on currently available information at the date the financial statements were prepared.

The company's management depends on the company's system of internal accounting controls to assure itself of the reliability of the financial statements. The internal control system is designed to provide reasonable assurance, at appropriate cost, that assets are safeguarded and transactions are executed in accordance with management's authorizations and are recorded properly to permit the preparation of financial statements in accordance with generally accepted accounting principles. Periodic reviews are made of internal controls by the company's staff of internal auditors, and corrective action is taken if needed.

The Board of Directors reviews and monitors financial statements through its audit committee, which is composed solely of directors who are not officers or employees of the company. The audit committee meets regularly with the independent public accountants, internal auditors and management to review internal accounting controls, auditing and financial reporting matters.

The independent public accountants are engaged to provide an objective and independent review of the company's financial statements and to express an opinion thereon. Their audits are conducted in accordance with generally accepted auditing standards, and their report is included on the following page.

Report of Independent Public Accountants

To the Stockholders and Board of Directors
of Kerr-McGee Corporation:

We have audited the accompanying consolidated balance sheet of Kerr-McGee Corporation (a Delaware corporation) and subsidiary companies as of December 31, 1999 and 1998, and the related consolidated statements of income, comprehensive income and stockholders' equity and cash flows for each of the three years in the period ended December 31, 1999. These financial statements are the responsibility of the company's management. Our responsibility is to express an opinion on these financial statements based on our audits. We did not audit the financial statements of Oryx Energy Company in 1998 or 1997, which was merged into the company during 1999 in a transaction accounted for as a pooling of interests, as discussed in Note 1. Such statements are included in the consolidated financial statements of Kerr-McGee Corporation and reflect total assets and total revenues of 36 percent and 37 percent in 1998, respectively, and total revenues of 47 percent in 1997, of the related consolidated totals, after restatement to reflect certain adjustments necessary to conform accounting policies and presentation as set forth in Note 23. The financial statements of Oryx Energy Company prior to those adjustments were audited by other auditors whose report has been furnished to us and our opinion, insofar as it relates to the amounts included for Oryx Energy Company, is based solely on the report of the other auditors.

We conducted our audits in accordance with auditing standards generally accepted in the United States. Those standards require that we plan and perform the audit to obtain reasonable assurance about whether the financial statements are free of material misstatement. An audit includes examining, on a test basis, evidence supporting the amounts and disclosures in the financial statements. An audit also includes assessing the accounting principles used and significant estimates made by management, as well as evaluating the overall financial statement presentation. We believe that our audits and the report of other auditors provide a reasonable basis for our opinion.

In our opinion, based on our audits and the report of other auditors, the financial statements referred to above present fairly, in all material respects, the financial position of Kerr-McGee Corporation and subsidiary companies as of December 31, 1999 and 1998, and the results of their operations and their cash flows for each of the three years in the period ended December 31, 1999, in conformity with accounting principles generally accepted in the United States.

Oklahoma City, Oklahoma,
February 25, 2000

ARTHUR ANDERSEN LLP

Report of Independent Accountants

To the Shareholders and Board of Directors,
Oryx Energy Company:

In our opinion, the accompanying consolidated balance sheets of Oryx Energy Company and its Subsidiaries and the related consolidated statements of income, cash flows and changes in shareholders' equity (not presented separately herein) present fairly, in all material respects, the consolidated financial position of Oryx Energy Company and its Subsidiaries as of December 31, 1998 and the consolidated results of their operations and their cash flows for each of the two years in the period ended December 31, 1998 in conformity with accounting principles generally accepted in the United States. These financial statements are the responsibility of the Company's management; our responsibility is to express an opinion on these financial statements in accordance with auditing standards generally accepted in the United States which require that we plan and perform the audit to obtain reasonable assurance about whether the financial statements are free of material misstatement. An audit includes examining, on a test basis, evidence supporting the amounts and disclosures in the financial statements, assessing the accounting principles used and significant estimates made by management and evaluating the overall financial statement presentation. We believe that our audits provide a reasonable basis for the opinion expressed above.

PricewaterhouseCoopers LLP
Dallas, Texas
February 26, 1999

Consolidated Statement of Income

(Millions of dollars, except per-share amounts)	1999	1998	1997
Sales	**$2,696**	$2,200	$2,605
Costs and Expenses			
Costs and operating expenses	**1,056**	1,053	1,003
General and administrative expenses	**238**	278	221
Depreciation and depletion	**607**	561	545
Asset impairment	**—**	446	—
Exploration, including dry holes and amortization of undeveloped leases	**140**	215	139
Taxes, other than income taxes	**85**	53	103
Merger costs	**163**	—	—
Interest and debt expense	**190**	157	141
Total Costs and Expenses	**2,479**	2,763	2,152
	217	(563)	453
Other Income	**40**	43	82
Income (Loss) from Continuing Operations before Income Taxes, Extraordinary Charge and Change in Accounting Principle	**257**	(520)	535
Taxes on Income	**(111)**	175	(184)
Income (Loss) from Continuing Operations before Extraordinary Charge and Change in Accounting Principle	**146**	(345)	351
Income from Discontinued Operations, net of taxes of $156 in 1998 and $12 in 1997	**—**	277	33
Income (Loss) before Extraordinary Charge and Change in Accounting Principle	**146**	(68)	384
Extraordinary Charge, net of taxes of $1	**—**	—	(2)
Cumulative Effect of Change in Accounting Principle, net of taxes of $2	**(4)**	—	—
Net Income (Loss)	**$ 142**	$ (68)	$ 382
Net Income (Loss) per Common Share			
Basic –			
Continuing operations	**$ 1.69**	$ (3.98)	$ 4.04
Discontinued operations	**—**	3.20	.38
Extraordinary charge	**—**	—	(.02)
Cumulative effect of accounting change	**(.05)**	—	—
Net income (loss)	**$ 1.64**	$ (.78)	$ 4.40
Diluted –			
Continuing operations	**$ 1.69**	$ (3.98)	$ 4.02
Discontinued operations	**—**	3.20	.38
Extraordinary charge	**—**	—	(.02)
Cumulative effect of accounting change	**(.05)**	—	—
Net income (loss)	**$ 1.64**	$ (.78)	$ 4.38

The accompanying notes are an integral part of this statement.

24 Kerr-McGee Corporation 1999 Annual Report

Consolidated Statement of Comprehensive Income and Stockholders' Equity

(Millions of dollars)	Comprehensive Income (Loss)	Common Stock	Capital in Excess of Par Value	Retained Earnings	Accumulated Other Comprehensive Income (Loss)	Treasury Stock	Deferred Compensation and Other	Total Stockholders' Equity
Balance December 31, 1996		$93	$1,241	$436	$(12)	$(306)	$(173)	$1,279
Net income	$382	—	—	382	—	—	—	382
Unrealized gains on securities, net of $1 income tax	2	—	—	—	2	—	—	2
Realized gains on securities, net of $6 income tax	(12)	—	—	—	(12)	—	—	(12)
Appreciation of securities donated, net of $1 income tax	(2)	—	—	—	(2)	—	—	(2)
Minimum pension liability adjustment	(5)	—	—	—	(5)	—	—	(5)
Shares issued	—	—	33	—	—	—	—	33
Shares acquired	—	—	—	—	—	(57)	—	(57)
Dividends declared ($1.80 per share)	—	—	—	(86)	—	—	—	(86)
Other	—	—	—	(1)	—	—	25	24
Total	$365							
Balance December 31, 1997		93	1,274	731	(29)	(363)	(148)	1,558
Net loss	$(68)	—	—	(68)	—	—	—	(68)
Foreign currency translation adjustment	(5)	—	—	—	(5)	—	—	(5)
Minimum pension liability adjustment	(2)	—	—	—	(2)	—	—	(2)
Shares issued	—	—	8	—	—	—	—	8
Shares acquired	—	—	—	—	—	(25)	—	(25)
Dividends declared ($1.80 per share)	—	—	—	(86)	—	—	—	(86)
Effect of equity affiliate's merger	—	—	—	(51)	—	—	—	(51)
Other	—	—	—	1	—	—	16	17
Total	$ (75)							
Balance December 31, 1998		93	1,282	527	(36)	(388)	(132)	1,346
Net income	$142	—	—	142	—	—	—	142
Unrealized gains on securities, net of $42 income tax	79	—	—	—	79	—	—	79
Foreign currency translation adjustment	(23)	—	—	—	(23)	—	—	(23)
Minimum pension liability adjustment	25	—	—	—	25	—	—	25
Shares issued	—	—	2	—	—	—	—	2
Dividends declared ($1.80 per share)	—	—	—	(156)	—	—	—	(156)
Effect of equity affiliate's merger	—	—	—	63	—	—	—	63
Other	—	—	—	—	—	—	14	14
Total	$223							
Balance December 31, 1999		$93	$1,284	$576	$ 45	$(388)	$(118)	$1,492

The accompanying notes are an integral part of this statement.

Consolidated Balance Sheet

(Millions of dollars)	**1999**	1998
ASSETS		
Current Assets		
Cash	**$ 267**	$ 121
Accounts receivable, net of allowance for doubtful accounts of $8 in 1999 and $5 in 1998	**501**	389
Inventories	**281**	247
Deposits, prepaid expenses and other	**112**	120
Total Current Assets	**1,161**	877
Investments		
Equity affiliates	**59**	170
Otherassets	**467**	87
Property, Plant and Equipment – Net	**4,085**	4,153
Deferred Charges	**127**	164
Total Assets	**$5,899**	$5,451
LIABILITIES AND STOCKHOLDERS' EQUITY		
Current Liabilities		
Accounts payable	**$ 404**	$ 385
Short-term borrowings	**9**	36
Long-term debt due within one year	**20**	236
Taxes on income	**70**	48
Taxes, other than income taxes	**40**	11
Accrued liabilities	**297**	334
Total Current Liabilities	**840**	1,050
Long-Term Debt	**2,496**	1,978
Deferred Credits and Reserves		
Income taxes	**401**	329
Other	**670**	748
Total Deferred Credits and Reserves	**1,071**	1,077
Stockholders' Equity		
Common stock, par value $1.00 – 300,000,000 shares authorized, 93,494,186 shares issued in 1999 and 93,378,069 shares issued in 1998	**93**	93
Capital in excess of par value	**1,284**	1,282
Preferred stock purchase rights	**1**	1
Retained earnings	**576**	527
Accumulated other comprehensive income (loss)	**45**	(36)
Common stock in treasury, at cost – 7,010,790 shares in both 1999 and 1998	**(388)**	(388)
Deferred compensation	**(119)**	(133)
Total Stockholders' Equity	**1,492**	1,346
Total Liabilities and Stockholders' Equity	**$5,899**	$5,451

The "successful efforts" method of accounting for oil and gas exploration and production activities has been followed in preparing this balance sheet.
The accompanying notes are an integral part of this balance sheet.

26 Kerr-McGee Corporation 1999 Annual Report

Consolidated Statement of Cash Flows

(Millions of dollars)	1999	1998	1997
Cash Flow from Operating Activities			
Net income (loss)	**$ 142**	$ (68)	$ 382
Adjustments to reconcile to net cash provided by operating activities –			
Depreciation, depletion and amortization	**648**	615	593
Deferred income taxes	**—**	(98)	31
Dry hole cost	**43**	100	53
Merger and transition costs	**131**	—	—
Asset impairment	**—**	446	—
Provision for environmental remediation and restoration of inactive sites	**—**	41	20
Gain on sale of coal operations, net of income taxes	**—**	(257)	—
Gain on sale of exploration and production properties	**(2)**	(20)	(3)
Realized gain on available-for-sale securities	**—**	—	(18)
Retirements and (gain) loss on sale of other assets	**(1)**	13	(4)
Noncash items affecting net income	**67**	13	23
Changes in current assets and liabilities and other, net of effects of operations acquired and sold –			
(Increase) decrease in accounts receivable	**(56)**	164	139
(Increase) decrease in inventories	**(34)**	(54)	40
(Increase) decrease in deposits and prepaids	**10**	(92)	17
Decrease in accounts payable and accrued liabilities	**(198)**	(103)	(53)
Increase (decrease) in taxes payable	**92**	(165)	66
Other	**(129)**	(150)	(189)
Net cash provided by operating activities	**713**	385	1,097
Cash Flow from Investing Activities			
Capital expenditures	**(543)**	(981)	(836)
Cash dry hole cost	**(33)**	(92)	(52)
Acquisitions	**(78)**	(518)	—
Purchase of long-term investments	**(39)**	(3)	(12)
Proceeds from sale of long-term investments	**27**	12	13
Proceeds from sale of discontinued operations	**—**	599	—
Proceeds from sale of chemical and exploration and production properties	**4**	150	17
Proceeds from sale of available-for-sale securities	**—**	—	21
Proceeds from sale of other assets	**—**	11	17
Net cash used in investing activities	**(662)**	(822)	(832)
Cash Flow from Financing Activities			
Issuance of long-term debt	**1,084**	563	390
Issuance of common stock	**4**	6	28
Increase (decrease) in short-term borrowings	**(27)**	11	(12)
Repayment of long-term debt	**(782)**	(93)	(464)
Dividends paid	**(138)**	(86)	(85)
Lease buyout	**(41)**	—	—
Treasury stock purchased	**—**	(25)	(60)
Net cash provided by (used in) financing activities	**100**	376	(203)
Effects of Exchange Rate Changes on Cash and Cash Equivalents	**(5)**	(10)	—
Net Increase (Decrease) in Cash and Cash Equivalents	**146**	(71)	62
Cash and Cash Equivalents at Beginning of Year	**121**	192	130
Cash and Cash Equivalents at End of Year	**$ 267**	$ 121	$ 192

The accompanying notes are an integral part of this statement.

Notes to Financial Statements

1 The Company and Significant Accounting Policies

Kerr-McGee is an energy and chemical company with worldwide operations. It explores for, develops, produces and markets crude oil and natural gas, and its chemical operations primarily produce and market titanium dioxide pigment. The exploration and production unit produces and explores for oil and gas in the United States, the United Kingdom sector of the North Sea, Indonesia, China, Kazakhstan and Ecuador. Exploration efforts are also extended to Australia, Algeria, Brazil, Gabon, Thailand, Yemen and the Danish sector of the North Sea. The chemical unit has operations in the United States, Australia, Germany and Belgium.

On February 26, 1999, the merger between Kerr-McGee and Oryx Energy Company (Oryx) was completed. Oryx was a worldwide independent oil and gas exploration and production company. Under the merger agreement, each outstanding share of Oryx common stock was effectively converted into the right to receive 0.369 shares of newly issued Kerr-McGee common stock. The merger qualified as a tax-free exchange to Oryx's shareholders and has been accounted for as a pooling of interests. In the aggregate, Kerr-McGee issued approximately 39 million shares of Kerr-McGee common stock, bringing the total shares outstanding to approximately 86 million. Kerr-McGee's consolidated financial statements have been restated for periods prior to the merger to include the operations of Oryx, adjusted to conform to Kerr-McGee's accounting policies and presentation.

Basis of Presentation

The consolidated financial statements include the accounts of all subsidiary companies that are more than 50% owned and the proportionate share of joint ventures in which the company has an undivided interest. Investments in affiliated companies that are 20% to 50% owned are carried as Investments - Equity affiliates in the Consolidated Balance Sheet at cost adjusted for equity in undistributed earnings. Except for dividends and changes in ownership interest, changes in equity in undistributed earnings are included in the Consolidated Statement of Income. All material intercompany transactions have been eliminated.

The preparation of financial statements in conformity with generally accepted accounting principles requires management to make estimates and assumptions that affect the reported amounts of assets and liabilities, the disclosure of contingent assets and liabilities at the date of the financial statements and the reported amounts of revenues and expenses during the reporting period. Actual results could differ from those estimates as additional information becomes known.

Foreign Currencies

The U.S. dollar is considered the functional currency for each of the company's international operations, except for its European chemical operations. Foreign currency transaction gains or losses are recognized in the period incurred. The company recorded net foreign currency transaction gains of $11 million in 1999. The net foreign currency transaction gains and losses in 1998 and 1997 were immaterial.

The euro is the functional currency for the European chemical operations. Translation adjustments resulting from translating the functional currency financial statements into U.S. dollar equivalents are reported separately in Accumulated Other Comprehensive Income (Loss) in the Consolidated Statement of Comprehensive Income and Stockholders' Equity.

Net Income (Loss) per Common Share

Basic net income (loss) per share includes no dilution and is computed by dividing income available to common stockholders by the weighted-average number of common shares outstanding for the period. Diluted net income per share is computed by dividing net income by the weighted-average number of common shares and common stock equivalents outstanding for the period.

The weighted-average number of shares used to compute basic net income (loss) per share was 86,414,373 in 1999, 86,688,026 in 1998 and 86,756,138 in 1997. After adding the dilutive effect of the conversion of options to the weighted-average number of shares outstanding, the shares used to compute diluted net income per share were 86,497,207 in 1999 and 87,113,906 in 1997. There was no dilution for 1998 since the company incurred a loss from continuing operations.

Not included in the calculation of the denominator for diluted net income per share were 2,063,079 and 494,063 employee stock options outstanding at year-end 1999 and 1997, respectively. The inclusion of these options would have been antidilutive since they were not "in the money" at the end of the respective years.

The company has reserved 1,791,646 shares of common stock for issuance to the owners of its 7-1/2% Convertible Subordinated Debentures due 2014 (Debentures). The Debentures are convertible into the company's common stock at any time prior to maturity at $106.03 per share of common stock. The conversion of the Debentures was not considered for purposes of calculating income (loss) per share, as the impact would have been antidilutive to net income (loss) per share for the periods presented.

Cash Equivalents

The company considers all investments with a maturity of three months or less to be cash equivalents. Cash equivalents totaling $156 million in 1999 and $58 million in 1998 were comprised of time deposits, certificates of deposit and U.S. government securities.

Inventories

The costs of the company's product inventories are determined by the first-in, first-out (FIFO) method. Inventory carrying values include material costs, labor and the associated indirect manufacturing expenses. Materials and supplies are valued at average cost.

28 Kerr-McGee Corporation 1999 Annual Report

Property, Plant and Equipment

Oil and Gas - Exploration expenses, including geological and geophysical costs, rentals and exploratory dry holes, are charged against income as incurred. Costs of successful wells and related production equipment and developmental dry holes are capitalized and amortized by field using the unit-of-production method as the oil and gas are produced.

Undeveloped acreage costs are capitalized and amortized at rates that provide full amortization on abandonment of unproductive leases. Costs of abandoned leases are charged to the accumulated amortization accounts, and costs of productive leases are transferred to the developed property accounts.

Other - Property, plant and equipment is stated at cost less reserves for depreciation, depletion and amortization. Maintenance and repairs are expensed as incurred, except that costs of replacements or renewals that improve or extend the lives of existing properties are capitalized. Costs of nonproducing mineral acreage surrendered or otherwise disposed of are charged to expense at the time of disposition.

Depreciation and Depletion - Property, plant and equipment is depreciated or depleted over its estimated life by the unit-of-production or the straight-line method. Capitalized exploratory drilling and development costs are amortized using the unit-of-production method based on total estimated proved developed oil and gas reserves. Amortization of producing leasehold, platform costs and acquisition costs of proved properties is based on the unit-of-production method using total estimated proved reserves. In arriving at rates under the unit-of-production method, the quantities of recoverable oil, gas and other minerals are established based on estimates made by the company's geologists and engineers.

Retirements and Sales - The costs and related depreciation, depletion and amortization reserves are removed from the respective accounts upon retirement or sale of property, plant and equipment. The resulting gain or loss is included in other income.

Interest Capitalized - The company capitalizes interest costs on major projects that require a considerable length of time to complete. Interest capitalized in 1999, 1998 and 1997 was $9 million, $28 million and $24 million, respectively.

Impairment of Long-Lived Assets

Proved oil and gas properties are reviewed for impairment on a field-by-field basis when facts and circumstances indicate that their carrying amounts may not be recoverable. In performing this review, future cash flows are estimated by applying estimated future oil and gas prices to estimated future production, less estimated future expenditures to develop and produce the reserves. If the sum of these estimated future cash flows (undiscounted and without interest charges) is less than the carrying amount of the property, an impairment loss is recognized for the excess of the carrying amount over the estimated fair value of the property.

Other assets are reviewed for impairment by asset group for which the lowest level of independent cash flows can be identified and impaired in the same manner as proved oil and gas properties.

Revenue Recognition

Except for natural gas sales and most crude oil, revenue is recognized when title passes to the customer. Natural gas revenues and gas-balancing arrangements with partners in natural gas wells are recognized when the gas is produced using the entitlements method of accounting and are based on the company's net working interests. At December 31, 1999 and 1998, both the quantity and dollar amount of gas balancing arrangements were immaterial. Crude oil sales are recognized when produced using the entitlements method if a contract exists for the sale of the production.

Lease Commitments

The company utilizes various leased properties in its operations, principally for office space and production facilities. Lease rental expense was $41 million in 1999, $37 million in 1998 and $39 million in 1997.

The aggregate minimum annual rentals under noncancelable leases in effect on December 31, 1999, totaled $92 million, of which $20 million is due in 2000, $21 million in 2001, $24 million in the period 2002 through 2004 and $27 million thereafter.

Income Taxes

Deferred income taxes are provided to reflect the future tax consequences of differences between the tax basis of assets and liabilities and their reported amounts in the financial statements.

Site Dismantlement, Remediation and Restoration Costs

The company provides for the estimated costs at current prices of the dismantlement and removal of oil and gas production and related facilities. Such costs are accumulated over the estimated lives of the facilities by the use of the unit-of-production method. As sites of environmental concern are identified, the company assesses the existing conditions, claims and assertions, generally related to former operations, and records an estimated undiscounted liability when environmental assessments and/or remedial efforts are probable and the associated costs can be reasonably estimated.

Employee Stock Option Plans

The company accounts for its employee stock option plans using the intrinsic value method in accordance with Accounting Principles Board Opinion (APB) No. 25, "Accounting for Stock Issued to Employees."

Futures, Forward and Option Contracts

The company hedges a portion of its monetary assets, liabilities and commitments denominated in foreign currencies. Periodically, the company purchases foreign currency forward contracts to provide funds for operating and capital expenditure requirements that will be denominated in foreign currencies and sells foreign currency forward contracts to convert receivables that will be paid in foreign currencies to U.S. dollars. Since these contracts qualify as hedges and correlate to currency movements, any gain or loss resulting from market changes is offset by gains or losses on the hedged receivable, capital item or operating cost.

From time to time the company enters into arrangements to hedge the impact of price fluctuations on anticipated crude oil and natural gas sales. Gains or losses on hedging activities are recognized in oil and gas revenues in the period in which the hedged production is sold.

The company periodically enters into interest rate hedging agreements to alter the floating rate portion of its underlying debt portfolio. Advance proceeds received under the agreements are included in deferred credits and are amortized as offsets to interest and debt expense over the relevant periods. The differentials paid or received during the terms of such agreements are accrued as interest rates change and are recorded as adjustments to interest and debt expense.

2 Cash Flow Information

Net cash provided by operating activities reflects cash payments for income taxes and interest as follows:

(Millions of dollars)	**1999**	1998	1997
Income tax payments	**$111**	$151	$ 89
Less refunds received	**(85)**	(40)	(37)
Net income tax payments	**$ 26**	$111	$ 52
Interest payments	**$191**	$180	$163

Noncash transactions not reflected in the Consolidated Statement of Cash Flows include capital expenditures for which payment will be made in the subsequent year totaling $28 million, $43 million and $19 million at year-end 1999, 1998 and 1997, respectively; the revaluation in 1999 to fair value of stock owned in a company previously accounted for using the equity method of accounting and the revaluation to fair value of the debt that is exchangeable into the stock of the investee; transactions during 1997 associated with the assignments of interest of certain North Sea oil and gas properties; the revaluation of certain other investments to fair value; and transactions affecting deferred compensation associated with the Employee Stock Ownership Plan in each of the three years (see Notes 18 and 20).

3 Inventories

Major categories of inventories at year-end 1999 and 1998 are:

(Millions of dollars)	**1999**	1998
Chemicals and other products	**$212**	$185
Materials and supplies	**67**	53
Crude oil	**2**	9
Total	**$281**	$247

4 Deferred Charges

Deferred charges are as follows at year-end 1999 and 1998:

(Millions of dollars)	**1999**	1998
Pension plan prepayments	**$ 79**	$101
Unamortized debt issue costs	**18**	8
Amounts pending recovery from third parties	**10**	8
Intangible assets	**6**	8
Nonqualified pension plan deposits	**—**	10
Preoperating and startup costs	**—**	6
Other	**14**	23
Total	**$127**	$164

Effective January 1, 1999, the company began expensing the cost of start-up activities in accordance with Statement of Position No. 98-5, "Reporting on the Costs of Start-Up Activities." The $6 million of unamortized costs at the end of 1998 was recognized in 1999 as the cumulative effect of a change in accounting principle ($4 million after taxes).

5 Investments – Equity Affiliates

At December 31, 1999 and 1998, investments in equity affiliates are as follows:

(Millions of dollars)	1999	1998
Devon Energy Corporation	**$—**	$108
Javelina Company	**27**	30
National Titanium Dioxide Company Limited	**18**	18
Other	**14**	14
Total	**$59**	$170

The company holds 9,954,000 shares of Devon common stock, a publicly traded oil and gas exploration and production company. The company accounted for this investment using the equity method until 1999 when its ownership interest of approximately 20% fell to approximately 12% as a result of Devon issuing additional common stock in its merger with a third party. The difference between the company's carrying amount of the investment before the merger and the underlying net book value of the investment after the merger was $63 million, net of deferred tax, and was reflected as a 1999 increase to retained earnings. The company's investment in Devon is now accounted for at market value (see Notes 6 and 18).

Javelina Company and National Titanium Dioxide Company Limited represent the company's investment of 40% and 25%, respectively, in nonexploration and production joint ventures or partnerships.

Following are financial summaries of the company's equity affiliates. Due to immateriality, investments shown as Other in the preceding table have been excluded from the information below.

(Millions of dollars)	1999	1998	1997
Results of operations –			
Net sales(1)	**$256**	$ 593	$570
Net income (loss)(2)	**33**	(41)	105
Financial position –			
Current assets	**125**	222	
Property, plant and equipment – net	**234**	1,334	
Total assets	**361**	1,582	
Current liabilities	**91**	170	
Total liabilities	**144**	844	
Stockholders' equity	**217**	738	

(1) Includes net sales to the company of $2 million and $26 million for 1998 and 1997, respectively. There were no sales to the company in 1999.

(2) The 1998 loss includes a full-cost write-down recorded by Devon. The company's proportionate share of the write-down was $27 million.

6 Investments – Other Assets

Investments in other assets consist of the following at December 31, 1999 and 1998:

(Millions of dollars)	1999	1998
Devon Energy Corporation common stock(1)	**$327**	$—
Long-term receivables, net of $9 allowance for doubtful notes in both 1999 and 1998	**105**	41
Net deferred tax asset	**12**	17
U.S. government obligations	**11**	17
Patents	**7**	6
Other	**5**	6
Total	**$467**	$87

(1) See Notes 5 and 18.

7 Property, Plant and Equipment

Fixed assets and related reserves at December 31, 1999 and 1998, are as follows:

	Gross Property		Reserves for Depreciation and Depletion		Net Property	
(Millions of dollars)	1999	1998	1999	1998	1999	1998
Exploration and production	**$ 9,689**	$ 9,359	**$6,245**	$5,837	**$3,444**	$3,522
Chemicals	**1,224**	1,162	**640**	588	**584**	574
Other	**136**	130	**79**	73	**57**	57
Total	**$11,049**	$10,651	**$6,964**	$6,498	**$4,085**	$4,153

8 Debt

Lines of Credit and Short-Term Borrowings

At year-end 1999, the company had available unused bank lines of credit and revolving credit facilities of $1,373 million. Of this amount, $835 million and $400 million can be used to support commercial paper borrowing arrangements of Kerr-McGee Credit LLC and Kerr-McGee Oil (U.K.) PLC, respectively.

The company has arrangements to maintain compensating balances with certain banks that provide credit. At year-end 1999, the aggregate amount of such compensating balances was immaterial, and the company was not legally restricted from withdrawing all or a portion of such balances at any time during the year.

Short-term borrowings at year-end 1999 consisted of notes payable totaling $9 million (4.88% average interest rate). The notes are denominated in foreign currency and represent approximately 361 million Belgian francs. Short-term borrowings outstanding at year-end 1998 were made up of commercial paper totaling $28 million (6.37% average effective interest rate) and notes payable totaling $8 million (3.63% average interest rate) or 281 million Belgian francs.

Long-Term Debt

The company's policy is to classify certain borrowings under revolving credit facilities and commercial paper as long-term debt since the company has the ability under certain revolving credit agreements and the intent to maintain these obligations for longer than one year. At year-end 1999 and 1998, debt totaling $793 million and $717 million, respectively, was classified as long-term consistent with this policy.

Long-term debt consisted of the following at year-end 1999 and 1998.

(Millions of dollars)	**1999**	1998
Debentures –		
7-1/2% Convertible subordinated debentures, $10 due annually May 15, 2000 through 2013 and $50 due May 15, 2014	**$ 190**	$ 200
7.125% Debentures due October 15, 2027 (7.01% effective rate)	**150**	150
7% Debentures due November 1, 2011, net of unamortized debt discount of $103 in 1999 and $105 in 1998 (14.25% effective rate)	**147**	145
5-1/2% Exchangeable notes due August 2, 2004	**327**	—
8-1/2% Sinking fund debentures due June 1, 2006	**—**	11
Notes payable –		
10% Notes due April 1, 2001	**150**	250
6.625% Notes due October 15, 2007 (6.54% effective rate)	**150**	150
8.375% Notes due July 15, 2004	**150**	150
8.125% Notes due October 15, 2005	**150**	150
8% Notes due October 15, 2003	**100**	100
9.5% Notes due November 1, 1999	**—**	100
Variable interest rate revolving credit agreements with banks (6.34% average rate at December 31, 1999), $25 due December 4, 2001 and $60 due March 6, 2003	**85**	598
Variable interest rate notes due November 1, 2001 (6.7% effective rate)	**150**	—
Medium-Term Notes (9.29% average effective interest rate at December 31, 1999), $11 due January 2, 2002 and $2 due February 1, 2002	**13**	28
Commercial paper (6.76% average effective interest rate at December 31, 1999)	**612**	119
Euro Commercial paper (6.54% average effective interest rate at December 31, 1999)	**96**	—
Guaranteed Debt of Employee Stock Ownership Plan 9.61% Notes due in installments through January 2, 2005	**43**	49
Other	**3**	14
	2,516	2,214
Long-term debt due within one year	**(20)**	(236)
Total	**$2,496**	$1,978

Maturities of long-term debt due after December 31, 1999, are $20 million in 2000, $590 million in 2001, $500 million in 2002, $176 million in 2003, $491 million in 2004 and $739 million thereafter.

Certain of the company's long-term debt agreements contain restrictive covenants, including a minimum tangible net worth requirement and a minimum cash flow to fixed charge ratio. At December 31, 1999, the company was in compliance with its debt covenants.

Additional information regarding the major changes in debt during the periods and unused commitments for financing is included in the Financial Condition section in Management's Discussion and Analysis.

9 Accrued Liabilities

Accrued liabilities at year-end 1999 and 1998 are as follows:

(Millions of dollars)	1999	1998
Interest payable	**$ 72**	$ 71
Current environmental reserves	**70**	83
Employee-related costs and benefits	**66**	59
Royalties payable	**22**	14
Merger reserve(1)	**20**	—
Litigation reserves	**18**	—
Drilling and operating costs	**15**	67
Restructuring reserve(1)	**—**	20
Other	**14**	20
Total	**$297**	$334

(1) See Note 22.

10 Common Stock

Changes in common stock issued and treasury stock held for 1999, 1998 and 1997 are as follows:

(Thousands of shares)	Common Stock	Treasury Stock
Balance December 31, 1996	92,601	5,569
Exercise of stock options and stock appreciation rights	627	—
Issuance of shares for achievement awards	—	(2)
Stock purchase program	—	867
Balance December 31, 1997	93,228	6,434
Exercise of stock options and stock appreciation rights	150	—
Issuance of shares for achievement awards	—	(3)
Stock purchase program	—	580
Balance December 31, 1998	93,378	7,011
Exercise of stock options and stock appreciation rights	112	—
Issuance of restricted stock	4	—
Balance December 31, 1999	93,494	7,011

The company has 40 million shares of preferred stock without par value authorized, and none is issued.

There are 1,107,692 shares of the company's common stock registered in the name of a wholly owned subsidiary of the company. These shares are not included in the number of shares shown in the preceding table or in the Consolidated Balance Sheet. These shares are not entitled to be voted.

In mid-1998, the Board of Directors authorized management to purchase up to $300 million of company common stock over the following three years. A total of 580,000 shares was acquired at a cost of $25 million before this stock purchase program was cancelled because of the merger with Oryx. The 1995 stock purchase program was completed in 1997 with a total of 4,829,000 shares of the company's stock acquired in open-market transactions at a cost of $300 million.

The company has granted restricted common shares to certain key employees under the 1998 Long-Term Incentive Plan. Shares are awarded in the name of the employee, who has all the rights of a shareholder, subject to certain restrictions on transferability and a risk of forfeiture. The forfeiture provisions on the 1999 awards expire on December 1, 2003.

The company has had a stockholders-rights plan since 1986. The current rights plan is dated July 6, 1996, and replaced the previous plan prior to its expiration. Rights were distributed under the original plan as a dividend at the rate of one right for each share of the company's common stock. Generally, the rights become exercisable the earlier of 10 days after a public announcement that a person or group has acquired, or a tender offer has been made for, 15% or more of the company's then-outstanding stock. If either of these events occurs, each right would entitle the holder (other than a holder owning more than 15% of the outstanding stock) to buy the number of shares of the company's common stock having a market value two times the exercise price. The exercise price is $215. Generally, the rights may be redeemed at $.01 per right until a person or group has acquired 15% or more of the company's stock. The rights expire in July 2006.

11 Contingencies

West Chicago

In 1973, a wholly owned subsidiary, Kerr-McGee Chemical Corporation, closed the facility at West Chicago, Illinois, that processed thorium ores. Kerr-McGee Chemical Corporation now operates as Kerr-McGee Chemical LLC (Chemical). Operations resulted in some low-level radioactive contamination at the site and, in 1979, Chemical filed a plan with the Nuclear Regulatory Commission (NRC) to decommission the facility. The NRC transferred jurisdiction of this site to the State of Illinois (the State) in 1990. Following is the current status of various matters associated with the West Chicago site.

Closed Facility - In 1994, Chemical, the City of West Chicago (the City) and the State reached agreement on the initial phase of the decommissioning plan for the closed West Chicago facility, and Chemical began shipping material from the site to a licensed permanent disposal facility.

In February 1997, Chemical executed an agreement with the City as to the terms and conditions for completing the final phase of decommissioning work. The State has indicated approval of this agreement and has issued license amendments authorizing much of the work. Chemical expects most of the work to be completed within four years.

In 1992, the State enacted legislation imposing an annual storage fee equal to $2 per cubic foot of byproduct material located at the closed facility. The storage fee cannot exceed $26 million per year, and any storage fee payments must be reimbursed to Chemical as decommissioning costs are incurred. Chemical has been fully reimbursed for all storage fees paid pursuant to this legislation. In June 1997, the legislation was amended to provide that future storage fee obligations are to be offset against decommissioning costs incurred but not yet reimbursed.

Offsite Areas - The U.S. Environmental Protection Agency (EPA) has listed four areas in the vicinity of the West Chicago facility on the National Priority List that the EPA promulgates under authority of the Comprehensive Environmental Response, Compensation, and Liability Act of 1980 (CERCLA) and has designated Chemical as a potentially responsible party in these four areas. Two of the four areas are presently being studied to determine the extent of contamination and the nature of any remedy. These two are known as the Sewage Treatment Plant and Kress Creek. The EPA previously issued unilateral administrative orders for the other two areas (known as the residential areas and Reed-Keppler Park), which require Chemical to conduct removal actions to excavate contaminated soils and ship the soils elsewhere for disposal. Without waiving any of its rights or defenses, Chemical is conducting the cleanup of the two areas for which unilateral administrative orders have been issued. Cleanup at one of the sites is nearly complete.

Judicial Proceedings - In December 1996, a lawsuit was filed against the company and Chemical in Illinois state court on behalf of a purported class of present and former West Chicago residents. The lawsuit seeks damages for alleged diminution in property values and the establishment of a medical monitoring fund to benefit those allegedly exposed to thorium wastes originating from the former facility. The case was removed to federal court and is being vigorously defended.

Government Reimbursement - Pursuant to Title X of the Energy Policy Act of 1992 (Title X), the U. S. Department of Energy is obligated to reimburse Chemical for certain decommissioning and cleanup costs in recognition of the fact that much of the facility's production was dedicated to United States government contracts. Title X was amended in 1998 to increase the amount authorized to $140 million plus inflation adjustments. Through January 31, 2000, Chemical has been reimbursed approximately $69 million under Title X. These reimbursements are provided by congressional appropriations.

Other Matters

The company's current and former operations involve management of regulated materials and are subject to various environmental laws and regulations. These laws and regulations will obligate the company to clean up various sites at which petroleum, chemicals, low-level radioactive substances or other regulated materials have been disposed of or released. Some of these sites have been designated Superfund sites by the EPA pursuant to CERCLA. The company is also a party to legal proceedings involving environmental matters pending in various courts and agencies. In addition, the company and/or its subsidiaries are also parties to a number of other legal proceedings pending in various courts or agencies in which the company and/or its subsidiaries appear as plaintiff or defendant.

The company provides for costs related to contingencies when a loss is probable and the amount is reasonably estimable.

It is not possible for the company to reliably estimate the amount and timing of all future expenditures related to environmental and legal matters because of:

- the difficulty of predicting cleanup requirements and estimating cleanup costs;
- the uncertainty in quantifying liability under environmental laws that impose joint and several liability on all potentially responsible parties;
- the continually changing nature of environmental laws and regulations and the uncertainty inherent in legal matters.

As of December 31, 1999, the company has recorded reserves totaling $204 million for cleaning up and remediating environmental sites, reflecting the reasonably estimable costs for addressing these sites. This includes $125 million for the West Chicago sites. Management believes, after consultation with general counsel, that currently the company has reserved adequately for contingencies. However, additions to the reserves may be required as additional information is obtained that enables the company to better estimate its liability, including any liability at sites now being studied, though management cannot now reliably estimate the amount of any future additions to the reserves. Historical expenditures at all sites from inception through December 31, 1999, total $670 million.

12 Income Taxes

The taxation of a company that has operations in several countries involves many complex variables, such as differing tax structures from country to country and the effect on U.S. taxation of international earnings. These complexities do not permit meaningful comparisons between the U.S. and international components of income before income taxes and the provision for income taxes, and disclosures of these components do not provide reliable indicators of relationships in future periods. Income (loss) from continuing operations before income taxes and extraordinary charge is composed of the following:

(Millions of dollars)	**1999**	1998	1997
United States	**$ (30)**	$(345)	$252
International	**287**	(175)	283
Total	**$257**	$(520)	$535

The corporate tax rate in the United Kingdom decreased to 30% from 31% effective April 1, 1999, and decreased to 31% from 33% effective April 1, 1997. The deferred income tax liability balance was adjusted to reflect the revised rates, which decreased the international deferred provision for income taxes by $10 million in 1998 and $13 million in 1997. The 1999, 1998 and 1997 taxes on income from continuing operations are summarized below:

(Millions of dollars)	**1999**	1998	1997
U.S. Federal –			
Current	**$ (38)**	$(159)	$ 2
Deferred	**38**	20	84
	—	(139)	86
International –			
Current	**147**	18	146
Deferred	**(37)**	(55)	(52)
	110	(37)	94
State	**1**	1	4
Total	**$111**	$(175)	$184

At December 31, 1999, the company had foreign operating loss carryforwards totaling $117 million – $9 million that expire in 2001, $8 million that expire in 2003, $15 million that expire in 2004 and $85 million that have no expiration date. Realization of these operating loss carryforwards is dependent on generating sufficient taxable income.

The net deferred tax asset, classified as Investments – Other assets in the Consolidated Balance Sheet, represents the net deferred taxes in certain foreign jurisdictions. Although realization is not assured, the company believes it is more likely than not that all of the net deferred tax asset will be realized. Deferred tax liabilities and assets at December 31, 1999 and 1998, are composed of the following:

(Millions of dollars)	**1999**	1998
Net deferred tax liabilities –		
Accelerated depreciation	**$564**	$593
Exploration and development	**34**	69
Undistributed earnings of foreign subsidiaries	**28**	28
Postretirement benefits	**(86)**	(86)
AMT credit carryforward	**(60)**	(60)
Foreign operating loss carryforward	**(35)**	(28)
Dismantlement, remediation, restoration and other reserves	**(30)**	(79)
Other	**(14)**	(108)
	401	329
Net deferred tax asset –		
Accelerated depreciation	**5**	5
Foreign operating loss carryforward	**(13)**	(14)
Other	**(4)**	(8)
	(12)	(17)
Total	**$389**	$312

In the following table, the U.S. Federal income tax rate is reconciled to the company's effective tax rates for income (loss) from continuing operations as reflected in the Consolidated Statement of Income.

	1999	1998	1997
U.S. statutory rate	**35.0%**	(35.0)%	35.0%
Increases (decreases) resulting from –			
Taxation of foreign operations	**4.8**	9.6	3.9
Adjustment of prior years' accruals	**—**	(.4)	(.8)
Refunds of prior years' income taxes	**—**	(5.6)	—
Adjustment of deferred tax balances due to tax rate changes	**—**	(2.0)	(2.4)
Other – net	**3.3**	(.2)	(1.3)
Total	**43.1%**	(33.6)%	34.4%

The Internal Revenue Service has examined the Kerr-McGee Corporation and subsidiaries' pre-merger Federal income tax returns for all years through 1994, and the years have been closed through 1994. The Oryx income tax returns have been examined through 1997, and the years have been closed through 1978. The company believes that it has made adequate provision for income taxes that may become payable with respect to open tax years.

13 Taxes, Other than Income Taxes

Taxes, other than income taxes, as shown in the Consolidated Statement of Income for the years ended December 31, 1999, 1998 and 1997, are composed of the following:

(Millions of dollars)	1999	1998	1997
Production/severance	**$52**	$26	$ 74
Payroll	**19**	12	11
Property	**11**	14	15
Other	**3**	1	3
Total	**$85**	$53	$103

14 Deferred Credits and Reserves – Other

Other deferred credits and reserves consist of the following at year-end 1999 and 1998:

(Millions of dollars)	1999	1998
Reserves for site dismantlement, remediation and restoration	**$391**	$376
Postretirement benefit obligations	**186**	269
Minority interest in subsidiary companies	**23**	39
Other	**70**	64
Total	**$670**	$748

The company provided for environmental remediation and restoration of former plant sites, net of authorized reimbursements, during each of the years 1999, 1998 and 1997 as follows:

(Millions of dollars)	1999	1998	1997
Provision, net of authorized reimbursements	**$ 3**	$47	$18
Reimbursements received	**15**	14	12
Reimbursements accrued	**67**	—	—

The reimbursements, which pertain to the former facility in West Chicago, Illinois, are authorized pursuant to Title X of the Energy Policy Act of 1992 (see Note 11).

15 Discontinued Operations

The company exited the coal business in 1998 with the sales of its mining operations at Galatia, Illinois, and Kerr-McGee Coal Corporation, which held Jacobs Ranch Mine in Wyoming. The cash sales resulted in proceeds of approximately $600 million. The 1998 gain on the sale was $257 million ($2.97 per share), net of $149 million for income taxes. The income from operations of the discontinued coal business totaled $20 million ($.23 per share), net of $7 million for income taxes, in 1998 and $33 million ($.38 per share), net of $12 million in income taxes, in 1997. Revenues applicable to the discontinued operations totaled $174 million in 1998 and $323 million in 1997.

16 Other Income

Other income was as follows during each of the years in the three-year period ended December 31, 1999:

(Millions of dollars)	**1999**	1998	1997
Income (loss) from unconsolidated affiliates	**$16**	$(12)	$32
Interest	**14**	38	14
Gain (loss) on foreign currency exchange	**11**	(2)	1
Gain on sale of assets	**3**	7	6
Settlements with insurance carriers	**—**	12	12
Gain on sale of available-for-sale securities	**—**	—	18
Other	**(4)**	—	(1)
Total	**$40**	$ 43	$82

17 Impairment of Long-Lived Assets and Long-Lived Assets to Be Disposed Of

Assets to Be Held and Used

At year-end 1998, certain oil and gas fields located in the North Sea, China and the United States and two U.S. chemical plants were deemed to be impaired because the assets were no longer expected to recover their net book values through future cash flows. Expectations of future cash flows were lower than those previously forecasted primarily as a result of weakness in crude oil, natural gas and certain chemical product prices at the end of 1998. Downward reserve revisions were also deemed necessary for certain fields. The impairment loss was determined based on the difference between the carrying value of the assets and the present value of future cash flows or market value, when appropriate. There was no impairment loss recognized in 1999 or 1997.

Following is the impairment loss for assets held and used by segment for the year ended December 31, 1998:

(Millions of dollars)	1998
Exploration and production	$389
Chemicals – pigment	32
Chemicals – other	25
Total	$446

Assets to Be Disposed Of

The company withdrew from the ammonium perchlorate business in 1998. The carrying value of these assets was approximately $9 million. The gain on the sale was immaterial.

During 1997, the company's exploration and production operating unit completed a program to divest a number of crude oil and natural gas producing properties considered to be non-strategic. Most of these properties were located onshore in the United States; however, some were located in the Gulf of Mexico, Canada and the North Sea. Net gains recognized on the sales of properties included in the divestiture program totaled $6 million in 1997. The divestiture program properties did not constitute a material portion of the company's oil and gas production or cash flows from operations for 1997.

Following are the sales and pretax income included in the Consolidated Statement of Income in 1998 and 1997 for assets sold during the two-year period ended December 31, 1998. Any impairment loss is included in the pretax income amounts. The company had no material assets held for disposal at year-end 1999 or 1998.

(Millions of dollars)	1998	1997
Sales –		
Chemicals – other	$ 11	$30
Income –		
Chemicals – other	—	3

18 Financial Instruments and Hedging Activities

Investments in Certain Debt and Equity Securities

The company has certain investments that are considered to be available for sale. The company also has debt that is exchangeable into equity securities of an investee that are considered available for sale. These financial instruments are carried in the Consolidated Balance Sheet at fair value, which is based on quoted market prices. The company had no securities classified as held to maturity or trading at December 31, 1999 and 1998. At December 31, 1999 and 1998, available-for-sale securities for which fair value can be determined are as follows:

	1999			1998		
	Fair Value	**Cost**	**Gross Unrealized Holding Gains**	Fair Value	Cost	Gross Unrealized Holding Gains
Equity securities	**$327**	**$209**	**$118**	$—	$—	$—
U.S. government obligations –						
Maturing within one year	**5**	**5**	**—**	13	13	—
Maturing between one year and four years	**11**	**11**	**—**	17	17	—
Exchangeable debt	**327**	**330**	**3**	—	—	—
Total			**$121**			$—

The equity securities represent the company's investment in Devon common stock that is no longer accounted for under the equity method of accounting (see Note 5). The securities are carried in the Consolidated Balance Sheet as Investments – Other assets. U.S. government obligations are carried as Current Assets or as Investments – Other assets, depending on their maturities.

The exchangeable debt represents 5-1/2% notes exchangeable into common stock (DECS) of Devon held by the company. The notes are due August 2, 2004, and holders of the notes will receive between .84746 and one common share of Devon per DECS, depending on the average trading price of Devon's common stock at that time, or the cash equivalent of such common stock. The DECS are carried at fair value in the Consolidated Balance Sheet as Long-Term Debt (see Note 8).

The change in unrealized holding gains, net of income taxes, as shown in accumulated other comprehensive income for the years ended December 31, 1999, 1998 and 1997, is as follows:

(Millions of dollars)	**1999**	1998	1997
Beginning balance –	**$—**	$—	$12
Net unrealized holding gains	**79**	—	2
Net realized gains	**—**	—	(12)
Net appreciation of donated securities	**—**	—	(2)
Ending balance	**$79**	$—	$—

During 1997, the company sold available-for-sale equity securities. Proceeds from the sales totaled $21 million in 1997. The average cost of the securities was used in the determination of the realized gains, which totaled $18 million before income taxes in 1997. Also during 1997, the company donated a portion of its available-for-sale equity securities to Kerr-McGee Foundation Corporation, a tax-exempt entity whose purpose is to contribute to not-for-profit organizations. The fair value of these donated shares totaled $3 million in 1997, which included appreciation of $3 million before income taxes.

Financial Instruments for Other than Trading Purposes

In addition to the financial instruments previously discussed, the company holds or issues financial instruments for other than trading purposes. At December 31, 1999 and 1998, the carrying amount and estimated fair value of these instruments for which fair value can be determined are as follows:

	1999		1998	
(Millions of dollars)	**Carrying Amount**	**Fair Value**	Carrying Amount	Fair Value
Cash and cash equivalents	**$ 267**	**$ 267**	$ 121	$ 121
Long-term notes receivable	**26**	**23**	9	9
Long-term receivables	**72**	**60**	—	—
Contracts to sell foreign currencies	**—**	**5**	—	2
Contracts to purchase foreign currencies	**—**	**98**	—	111
Oil and gas price hedging contracts	**—**	**—**	7	22
Short-term borrowings	**9**	**9**	36	36
Long-term debt, excluding DECS	**2,189**	**2,228**	2,214	2,365

The carrying amount of cash and cash equivalents approximates fair value of those instruments due to their short maturity. The fair value of notes receivable is based on discounted cash flows or the fair value of the note's collateral. The fair value of long-term receivables is based on discounted cash flows. The fair value of the company's short-term and long-term debt is based on the quoted market prices for the same or similar debt issues or on the current rates offered to the company for debt with the same remaining maturity. The fair value of foreign currency forward contracts represents the aggregate replacement cost based on financial institutions' quotes.

Hedging Activities

Most of the company's foreign currency contracts are hedges principally for chemical's accounts receivable generated from titanium dioxide pigment sales denominated in foreign currencies, the operating costs and capital expenditures of international pigment operations, and the operating costs and capital expenditures of U.K. oil and gas operations. The purpose of these foreign currency hedging activities is to protect the company from the risk that the functional currency amounts from sales to foreign customers and purchases from foreign suppliers could be adversely affected by changes in foreign currency exchange rates. The company recognized net foreign currency hedging losses of $5 million in 1999 and net foreign currency hedging gains of $4 million in 1997. The net foreign currency hedging loss recognized in 1998 was immaterial.

Net unrealized gains on foreign currency contracts totaled $2 million at year-end 1999. Net unrealized losses on foreign currency contracts totaled $7 million at year-end 1998 and $13 million at year-end 1997. The company's foreign currency contract positions at year-end 1999 and 1998 were as follows:

December 31, 1999 -

- Contracts maturing January 2000 through December 2002 to purchase $150 million Australian for $96 million
- Contracts maturing January through March 2000 to sell various foreign currencies (principally European) for $5 million

December 31, 1998 -

- Contracts maturing January 1999 through December 2000 to purchase $113 million Australian for $77 million and 25 million British pound sterling for $40 million
- Contracts maturing January through March 1999 to sell various foreign currencies (principally European) for $2 million

The company has periodically used oil or natural gas futures or collar contracts to reduce the effect of the price volatility of crude oil and natural gas. The futures contracts permitted settlement by delivery of commodities.

The company did not enter into any hedging arrangements in 1999 and settled all open 1998 contracts during the year. Net hedging gains recognized in 1999 totaled $28 million. The effect of the gains was to increase the company's 1999 average gross margin for crude oil and natural gas by $.11 per barrel and $.09 per MCF, respectively.

During 1998, the company entered into hedging arrangements for 7 million barrels of crude oil and 61 billion cubic feet of natural gas representing approximately 11% and 29% of its worldwide crude oil and natural gas sales volumes, respectively. Net hedging gains recognized in 1998 totaled $45 million. The effect of the gains was to increase the company's 1998 average gross margin for crude oil and natural gas by $.55 per barrel and $.05 per MCF, respectively. At year-end 1998, open crude oil and natural gas contracts had an aggregate value of $7 million, and the unrecognized gain on the contracts totaled $15 million.

During 1997, the company entered into hedging arrangements for 12 million barrels of crude oil and 75 billion cubic feet of natural gas representing approximately 18% and 30% of its worldwide crude oil and natural gas sales volumes, respectively. Net hedging losses recognized in 1997 totaled $27 million. The effect of the losses was to reduce the company's 1997 average gross margin for crude oil and natural gas by $.10 per barrel and $.08 per MCF, respectively. At year-end 1997, open crude oil and natural gas contracts had an aggregate value of $2 million, and the unrecognized loss on these contracts totaled $7 million.

Contract amounts do not quantify risk or represent assets or liabilities of the company but are used in the calculation of cash settlements under the contracts. These financial instruments limit the company's market risks, are with major financial institutions, expose the company to credit risks and at times may be concentrated with certain institutions or groups of institutions. However, the credit worthiness of these institutions is subject to continuing review, and full performance is anticipated. Additional information regarding market risk is included in Management's Discussion and Analysis.

Year-end hedge positions and activities during a particular year are not necessarily indicative of future activities and results.

19 Employee Benefit Plans

The company has both noncontributory and contributory defined-benefit retirement plans and company-sponsored contributory postretirement plans for health care and life insurance. Most employees are covered under the company's retirement plans, and substantially all U.S. employees may become eligible for the postretirement benefits if they reach retirement age while working for the company. Following are the changes in the benefit obligations during the past two years:

(Millions of dollars)	Retirement Plans 1999	Retirement Plans 1998	Postretirement Health and Life Plans 1999	Postretirement Health and Life Plans 1998
Benefit obligation, beginning of year	**$1,027**	$ 976	**$ 217**	$ 209
Service cost	**15**	16	**1**	3
Interest cost	**69**	66	**9**	13
Plan amendments	**70**	38	**4**	1
Net actuarial loss (gain)	**(15)**	15	**(2)**	8
Acquisitions	**—**	6	**—**	—
Assumption changes	**(50)**	—	**—**	—
Changes resulting from plan mergers	**14**	—	**(7)**	—
Dispositions, curtailments, settlements	**21**	5	**—**	(2)
Benefits paid	**(174)**	(95)	**(7)**	(15)
Benefit obligation, end of year	**$ 977**	$1,027	**$ 215**	$ 217

The benefit amount that can be covered by the retirement plans that qualify under the Employee Retirement Income Security Act of 1974 (ERISA) is limited by both ERISA and the Internal Revenue Code. Therefore, the company has unfunded supplemental plans designed to maintain benefits for all employees at the plan formula level and to provide senior executives with benefits equal to a specified percentage of their final average compensation. The benefit obligation for the unfunded retirement plans was $42 million and $109 million at December 31, 1999 and 1998, respectively. Although not considered plan assets, a grantor trust was established from which payments for certain of these supplemental plans are made. The trust had a balance of $5 million at year-end 1999 and $10 million at year-end 1998. The postretirement plans are also unfunded.

Following are the changes in the fair value of plan assets during the past two years and the reconciliation of the plans' funded status to the amounts recognized in the financial statements at December 31, 1999 and 1998:

(Millions of dollars)	Retirement Plans 1999	Retirement Plans 1998	Postretirement Health and Life Plans 1999	Postretirement Health and Life Plans 1998
Fair value of plan assets, beginning of year	**$1,404**	$1,138	**$ —**	$ —
Actual return on plan assets	**381**	351	**—**	—
Employer contribution	**35**	10	**—**	—
Changes resulting from plan mergers	**7**	—	**—**	—
Benefits paid	**(174)**	(95)	**—**	—
Fair value of plan assets, end of year	**1,653**	1,404	**—**	—
Benefit obligation	**(977)**	(1,027)	**(215)**	(217)
Funded status of plans – over (under)	**676**	377	**(215)**	(217)
Amounts not recognized in the Consolidated Balance Sheet–				
Transition asset	**(6)**	(13)	**—**	—
Prior service costs	**86**	33	**5**	—
Net actuarial loss (gain)	**(704)**	(366)	**11**	18
Prepaid expense (accrued liability)	**$ 52**	$ 31	**$(199)**	$(199)

40 Kerr-McGee Corporation 1999 Annual Report

Following is the classification of the amounts recognized in the Consolidated Balance Sheet at December 31, 1999 and 1998:

	Retirement Plans		Postretirement Health and Life Plans	
(Millions of dollars)	**1999**	1998	**1999**	1998
Prepaid benefits expense	**$ 79**	$ 102	**$ —**	$ —
Accrued benefit liability	**(39)**	(109)	**(199)**	(199)
Additional minimum liability –				
Intangible asset	**6**	7	**—**	—
Accumulated other comprehensive income	**6**	31	**—**	—
Total	**$ 52**	$ 31	**$(199)**	$(199)

Total costs recognized for employee retirement and postretirement benefit plans for each of the years ended December 31, 1999, 1998 and 1997 were as follows:

	Retirement Plans			Postretirement Health and Life Plans		
(Millions of dollars)	**1999**	1998	1997	**1999**	1998	1997
Net periodic cost –						
Service cost	**$ 15**	$ 16	$ 15	**$ 1**	$ 3	$ 2
Interest cost	**69**	66	64	**9**	13	15
Expected return on plan assets	**(98)**	(94)	(84)	**—**	—	—
Net amortization –						
Transition asset	**(6)**	(8)	(8)	**—**	—	—
Prior service cost	**12**	3	2	**—**	—	—
Net actuarial loss (gain)	**(3)**	1	1	**—**	(1)	—
	(11)	(16)	(10)	**10**	15	17
Dispositions, curtailments, settlements	**29**	26	6	**—**	(1)	—
Total	**$ 18**	$ 10	$ (4)	**$10**	$14	$17

The following assumptions were used in estimating the actuarial present value of the plans' benefit obligations and net periodic expense:

	1999	1998	1997
Discount rate	**5.50-7.75%**	6.75%	6.5-7.0%
Expected return on plan assets	**6.25-9.5**	9.0-9.5	9.0-9.5
Rate of compensation increases	**3.0-5.0**	4.0-5.0	4.0-5.0

The health care cost trend rates used to determine the year-end 1999 postretirement benefit obligation were 7.5% in 2000, gradually declining to 5% in the year 2010 and thereafter. A 1% increase in the assumed health care cost trend rate for each future year would increase the postretirement benefit obligation at December 31, 1999, by $14 million and increase the aggregate of the service and interest cost components of net periodic postretirement expense for 1999 by $1 million. A 1% decrease in the trend rate for each future year would reduce the benefit obligation at year-end 1999 by $14 million. It was not practical to calculate the effect of the percent decrease on net periodic expense in the health care cost trend rate.

20 Employee Stock Ownership Plan

In 1989, the company's Board of Directors approved a leveraged Employee Stock Ownership Plan (ESOP) into which is paid the company's matching contribution for the employees' contributions to the Kerr-McGee Corporation Savings Investment Plan (SIP). Most of the company's employees are eligible to participate in both the ESOP and the SIP. Although the ESOP and the SIP are separate plans, matching contributions to the ESOP are contingent upon participants' contributions to the SIP.

In 1989, the ESOP trust borrowed $125 million from a group of lending institutions and used the proceeds to purchase approximately 3 million shares of the company's treasury stock. The company used the $125 million in proceeds from the sale of the stock to acquire shares of its common stock in open-market and privately negotiated transactions. In 1996, a portion of the third-party borrowings was replaced with a note payable to the company (sponsor financing). The third-party borrowings are guaranteed by the company and are reflected in the Consolidated Balance Sheet as Long-Term Debt, while the sponsor financing does not appear in the company's balance sheet.

The Oryx Capital Accumulation Plan (CAP) was a combined stock bonus and leveraged employee stock ownership plan available to substantially all U.S. employees of the former Oryx operations. On August 1, 1989, Oryx privately placed $110 million of notes pursuant to the provisions of the CAP. Oryx loaned the proceeds to the CAP, which used the funds to purchase Oryx common stock that was placed in a trust. This loan was sponsor financing and does not appear in the accompanying balance sheet.

During 1999, the company merged the Oryx CAP into the ESOP and SIP. As a result, a total of 159,000 and 294,000 shares was transferred from the CAP into the ESOP and SIP, respectively.

The company stock owned by the ESOP trust is held in a loan suspense account. Deferred compensation, representing the unallocated ESOP shares, is reflected as a reduction of stockholders' equity. The company's matching contribution and dividends on the shares held by the ESOP trust are used to repay the loan, and stock is released from the loan suspense account as the principal and interest are paid. The expense is recognized and stock is then allocated to participants' accounts at market value as the participants' contributions are made to the SIP. Long-term debt is reduced as payments are made on the third-party financing. Dividends paid on the common stock held in participants' accounts are also used to repay the loans, and stock with a market value equal to the amount of dividends is allocated to participants' accounts.

Shares of stock allocated to the ESOP participants' accounts and in the loan suspense account are as follows:

(Thousands of shares)	**1999**	1998
Participants' accounts	**1,357**	1,398
Loan suspense account	**1,380**	1,610

The shares allocated to ESOP participants at December 31, 1999, included approximately 51,000 shares released in January 2000, and at December 31, 1998, included approximately 52,000 shares released in January 1999.

All ESOP shares are considered outstanding for net income per-share calculations. Dividends on ESOP shares are charged to retained earnings.

Compensation expense is recognized using the cost method and is reduced for dividends paid on the unallocated ESOP shares. The company recognized ESOP and CAP-related expense of $14 million, $17 million and $13 million in 1999, 1998 and 1997, respectively. These amounts include interest expense incurred on the third-party ESOP debt of $4 million in 1999 and $5 million in both 1998 and 1997. The company contributed $18 million, $16 million and $12 million to the ESOP and CAP in 1999, 1998 and 1997, respectively. The cash contributions are net of $4 million for the dividends paid on the company stock held by the ESOP trust in each of the years 1999, 1998 and 1997.

21 Employee Stock Option Plans

The 1998 Long Term Incentive Plan (1998 Plan) authorizes the issuance of shares of the company's common stock any time prior to December 31, 2007, in the form of stock options, restricted stock or long-term performance awards. The options may be accompanied by stock appreciation rights. A total of 2,300,000 shares of the company's common stock is authorized to be issued under the 1998 Plan.

In January 1998, the Board of Directors approved a broad-based stock option plan (BSOP) that provides for the granting of options to purchase the company's common stock to all full-time employees, except officers. A total of 1,500,000 shares of common stock is authorized to be issued under the BSOP.

The 1987 Long Term Incentive Program (1987 Program) authorized the issuance of shares of the company's stock over a 15-year period in the form of stock options, restricted stock or long-term performance awards. The 1987 Program was terminated when the stockholders approved the 1998 Plan. No options could be granted under the 1987 Program after that time, although options and any accompanying stock appreciation rights outstanding may be exercised prior to their respective expiration dates.

The company's employee stock options are fixed-price options granted at the fair market value of the underlying common stock on the date of the grant. Generally, one-third of each grant vests and becomes exercisable over a three-year period immediately following the grant date and expires 10 years after the grant date.

In connection with the merger with Oryx (see Note 1), outstanding stock options under the stock option plans maintained by Oryx were assumed by the company. Stock option transactions summarized below include amounts for the 1998 Plan, the BSOP, the 1987 Program and the Oryx plans using the merger exchange rate of 0.369 for each Oryx share under option.

	1999		1998		1997	
	Options	**Weighted-Average Exercise Price per Option**	Options	Weighted-Average Exercise Price per Option	Options	Weighted-Average Exercise Price per Option
Outstanding, beginning of year	**2,783,482**	**$58.77**	2,050,671	$56.84	2,241,136	$54.06
Options granted	**377,000**	**46.53**	1,105,043	61.97	481,213	68.04
Options exercised	**(110,521)**	**42.20**	(127,576)	44.34	(580,605)	50.49
Options surrendered upon exercise of stock appreciation rights	**(14,000)**	**45.25**	(4,000)	38.06	(5,000)	32.38
Options forfeited	**(45,929)**	**60.73**	(24,928)	60.26	(6,703)	57.46
Options expired	**(166,698)**	**72.95**	(215,728)	65.65	(79,370)	93.43
Outstanding, end of year	**2,823,334**	**56.78**	2,783,482	58.77	2,050,671	56.84
Exercisable, end of year	**2,003,138**	**57.63**	1,497,753	55.38	1,249,055	53.96

The following table summarizes information about stock options issued under the plans described above that are outstanding and exercisable at December 31, 1999:

Options Outstanding				Options Exercisable	
Options	Range of Exercise Prices per Option	Weighted-Average Remaining Contractual Life (years)	Weighted-Average Exercise Price per Option	Options	Weighted-Average Exercise Price per Option
142,883	$ 30.00-$ 39.99	2.8	$ 35.01	142,883	$ 35.01
625,172	40.00- 49.99	5.0	43.98	430,172	45.69
1,035,573	50.00- 59.99	7.1	57.00	499,851	56.33
803,023	60.00- 69.99	4.7	65.87	752,886	65.93
212,639	70.00- 79.99	5.0	72.69	173,302	72.51
2,679	90.00- 99.99	.9	97.56	2,679	97.56
1,365	110.00- 120.00	.1	119.58	1,365	119.58
2,823,334	30.00- 120.00	5.5	56.78	2,003,138	57.63

Statement of Financial Accounting Standards (SFAS) No. 123, "Accounting for Stock-Based Compensation," prescribes a fair-value method of accounting for employee stock options under which compensation expense is measured based on the estimated fair value of stock options at the grant date and recognized over the period that the options vest. The company, however, chooses to account for its stock option plans under the optional intrinsic value method of APB No. 25, "Accounting for Stock Issued to Employees," whereby no compensation expense is recognized for fixed-price stock options. Compensation cost for stock appreciation rights, which is recognized under both accounting methods, was immaterial for 1999, 1998 and 1997.

Had compensation expense been determined in accordance with SFAS No. 123, the resulting compensation expense would have affected net income and per-share amounts as shown in the following table. These amounts may not be representative of future compensation expense using the fair-value method of accounting for employee stock options as the number of options granted in a particular year may not be indicative of the number of options granted in future years, and the fair-value method of accounting has not been applied to options granted prior to January 1, 1995.

(Millions of dollars, except per-share amounts)	**1999**	1998	1997
Net income (loss) –			
As reported	**$ 142**	$ (68)	$ 382
Pro forma	**136**	(76)	376
Net income (loss) per share –			
Basic –			
As reported	**1.64**	(.78)	4.40
Pro forma	**1.57**	(.88)	4.34
Diluted –			
As reported	**1.64**	(.78)	4.38
Pro forma	**1.57**	(.88)	4.32

The fair value of each option granted in 1999, 1998 and 1997 was estimated as of the date of grant using the Black-Scholes option pricing model with the following weighted-average assumptions. The use of ranges in prior years was necessitated by the Oryx merger.

	Assumptions				
	Risk-Free Interest Rate	Expected Dividend Yield	Expected Life (years)	Expected Volatility	Weighted Average Fair Value of Options Granted
1999	5.4%	3.1%	5.8	25.2%	$11.33
1998	5.0 - 5.4	0 - 3.0	5.8 - 10	17.3 - 30.3	$ 9.78 - 11.20
1997	6.3 - 7.0	0 - 3.1	5.8 - 10	17.5 - 30.8	11.65 - 14.37

22 Merger and Restructuring Charges

In 1999, the company recorded an accrual of $163 million for items associated with the Oryx merger. Included in this charge were transaction costs, severance and other employee-related costs, contract termination costs, lease cancellations, write-off of redundant systems and equipment and other merger-related costs. The merger resulted in approximately 550 employees being terminated during 1999 under an involuntary termination program.

Oryx initiated a voluntary severance program in 1998, prior to the agreement to merge with Kerr-McGee, for its U.S. operations. The company also completed a work process review during 1998 which resulted in the elimination of nonessential work processes, organizational restructuring and employee reductions in both the operating and staff units. The programs resulted in the notification of approximately 260 employees that their positions would be eliminated.

During the three-year period ended December 31, 1999, the company accrued a total of $206 million for the cost of special termination benefits for retiring employees to be paid from retirement plan assets, future compensation, relocation, transaction costs related to the merger, lease cancellation and outplacement.

The merger reserve at December 31, 1999, includes $16 million for costs associated with an office lease obligation that has no economic benefit to the company but will be paid through the cancellation date in 2001 and the remaining severance costs, which are expected to be paid and charged to the reserve during 2000. The accruals, expenditures and reserve balances are set forth below:

	Merger Reserve	Restructuring Reserve	
(Millions of dollars)	**1999**	**1999**	1998
Beginning balance	**$ —**	**$ 20**	$ 12
Accruals	**163**	**1**	40
Benefits to be paid from employee benefit plans	**(31)**	**—**	(23)
Payments	**(126)**	**(15)**	(9)
Transfer to merger reserve from restructuring reserve and other accrued liabilities[(1)]	**14**	**(6)**	—
Ending balance	**$ 20**	**$ —**	$ 20

(1) In a prior Oryx reduction in force, a $6 million reserve was established for lease cancellation costs on a portion of the Dallas, Texas, office space. Additionally, Oryx had planned to cancel the remainder of the lease and had established an accrued liability of $8 million. These liabilities were combined with the 1999 merger reserve since Kerr-McGee also plans to cancel the lease at the date of the first option to cancel.

44 Kerr-McGee Corporation 1999 Annual Report

23 Merger with Oryx Energy Company

On February 26, 1999, the merger between Kerr-McGee and Oryx was completed. The following table provides a reconciliation of sales reported by Kerr-McGee to the combined amounts presented in the Consolidated Statement of Income:

(Millions of dollars)	**1999**	1998	1997
Sales –			
Pre-Merger –			
Kerr-McGee	**$ 199**	$1,396	$1,388
Oryx	**103**	820	1,197
Merger reclassifications	**—**	(16)	20
Post-Merger	**2,394**	—	—
Total	**$2,696**	$2,200	$2,605

Merger reclassifications primarily represent the reclassification of Oryx's other income to Kerr-McGee's presentation.

The following table provides a reconciliation of net income reported by Kerr-McGee to the combined amounts presented for the three years ended December 31, 1999, 1998 and 1997:

(Millions of dollars)	Income (Loss) From Continuing Operations (net of taxes)	Discontinued Operations (net of taxes)	Extraordinary Charge (net of taxes)	Accounting Change (net of taxes)	Net Income (Loss)
1999 –					
Pre-Merger –					
Kerr-McGee	$ 6	$ —	$—	$ (4)	$ 2
Oryx	(14)	—	—	—	(14)
Post-Merger	154	—	—	—	154
Total	$ 146	$ —	$—	$ (4)	$142
1998 –					
Pre-Merger –					
Kerr-McGee	$(227)	$277	$—	$—	$ 50
Oryx	(95)	—	—	—	(95)
Merger adjustments	(23)	—	—	—	(23)
Total	$(345)	$277	$—	$—	$ (68)
1997 –					
Pre-Merger –					
Kerr-McGee	$ 161	$ 33	$—	$—	$194
Oryx	172	—	(2)	—	170
Merger adjustments	18	—	—	—	18
Total	$ 351	$ 33	$ (2)	$—	$382

The merger adjustments were to conform accounting policy changes primarily related to the following: (1) accounting for postretirement benefits other than pensions; (2) different methods for recognizing Petroleum Revenue Tax for U.K. operations; and (3) different methods of providing for nonproducing leasehold cost impairments.

24 Reporting by Business Segments and Geographic Locations

The company has three reportable segments: oil and gas exploration and production and manufacturing and marketing of titanium dioxide pigment and other chemicals. The exploration and production unit produces and explores for oil and gas in the United States, United Kingdom sector of the North Sea, Indonesia, China, Kazakhstan and Ecuador. Exploration efforts are also extended to Australia, Algeria, Brazil, Gabon, Thailand, Yemen and the Danish sector of the North Sea. The chemical unit primarily produces and markets titanium dioxide pigment and has operations in the United States, Australia, Germany and Belgium. Other chemicals include the company's electrolytic manufacturing and marketing operations and forest products treatment business. All of these operations are in the United States.

Crude oil sales to an individually significant customer totaled $420 million in 1999. There were no individually significant customers in 1998 or 1997. Sales to subsidiary companies are eliminated as described in Note 1.

(Millions of dollars)	**1999**	1998	1997
Sales –			
Exploration and production	**$1,770**	$1,267	$1,845
Chemicals –			
Pigment	**700**	640	470
Other	**226**	293	290
Total Chemicals	**926**	933	760
Total	**$2,696**	$2,200	$2,605
Operating profit (loss)(1) –			
Exploration and production	**$ 542**	$ (361)	$ 595
Chemicals –			
Pigment	**113**	56	49
Other	**14**	—	32
Total Chemicals	**127**	56	81
Total	**$ 669**	$ (305)	$ 676
Net operating profit (loss)(1) –			
Exploration and production	**$ 338**	$ (266)	$ 375
Chemicals –			
Pigment	**73**	35	31
Other	**9**	—	21
Total Chemicals	**82**	35	52
Total	**420**	(231)	427
Net interest expense(1)	**(117)**	(77)	(84)
Net nonoperating income (expense)(1)	**(157)**	(37)	8
Income from discontinued operations, net of taxes	**—**	277	33
Extraordinary charge, net of taxes	**—**	—	(2)
Cumulative effect of change in accounting principle, net of taxes	**(4)**	—	—
Net income (loss)	**$ 142**	$ (68)	$ 382
Depreciation, depletion and amortization –			
Exploration and production	**$ 578**	$ 527	$ 508
Chemicals –			
Pigment	**45**	49	34
Other	**18**	19	21
Total Chemicals	**63**	68	55
Other	**7**	6	5
Discontinued operations	**—**	14	25
Total	**$ 648**	$ 615	$ 593

(1) Includes special items. Refer to Management's Discussion and Analysis.

46 Kerr-McGee Corporation 1999 Annual Report

(Millions of dollars)	**1999**	1998	1997
Cash capital expenditures –			
Exploration and production	**$ 447**	$ 871	$ 708
Chemicals –			
Pigment	**76**	69	64
Other	**14**	23	27
Total Chemicals	**90**	92	91
Other	**6**	8	10
Discontinued operations	**—**	10	27
Total	**543**	981	836
Cash exploration expenses –			
Exploration and production –			
Dry hole costs	**43**	100	53
Amortization of undeveloped leases	**41**	40	23
Other	**56**	75	63
Total exploration expenses	**140**	215	139
Less – Amortization of leases and other noncash expenses	**(51)**	(42)	(23)
Total cash exploration expenses	**89**	173	116
Total cash capital expenditures and cash exploration expenses	**$ 632**	$1,154	$ 952
Identifiable assets –			
Exploration and production	**$4,013**	$4,083	$3,924
Chemicals –			
Pigment (1)	**921**	863	601
Other	**229**	235	274
Total Chemicals	**1,150**	1,098	875
Total	**5,163**	5,181	4,799
Corporate and other assets	**736**	270	270
Discontinued operations	**—**	—	270
Total	**$5,899**	$5,451	$5,339
Sales –			
U.S. operations	**$1,471**	$1,311	$1,635
International operations –			
North Sea – exploration and production	**752**	472	644
Other – exploration and production	**78**	67	105
Europe – pigment	**210**	163	—
Australia – pigment	**185**	178	185
Other – pigment	**—**	9	36
	1,225	889	970
Total	**$2,696**	$2,200	$2,605
Operating profit (loss) (2) –			
U.S. operations	**$ 364**	$ (116)	$ 400
International operations –			
North Sea – exploration and production	**270**	(146)	85
Other – exploration and production	**—**	(85)	178
Australia – pigment	**21**	19	13
Europe – pigment	**14**	23	—
	305	(189)	276
Total	**$ 669**	$ (305)	$ 676

(1) Includes net deferred tax asset of $12 million, $17 million and $22 million at December 31, 1999, 1998 and 1997, respectively (see Note 12).
(2) Includes special items. Refer to Management's Discussion and Analysis.

(Millions of dollars)	**1999**	1998	1997
Net property, plant and equipment –			
U.S. operations	**$2,106**	$2,095	$2,382
International operations –			
North Sea – exploration and production	**1,547**	1,617	1,101
Other – exploration and production	**219**	213	303
Australia – pigment	**132**	129	133
Europe – pigment	**81**	99	—
	1,979	2,058	1,537
Total	**$4,085**	$4,153	$3,919

25 Subsequent Events

On January 18, 2000, the company completed the purchase of Repsol S.A.'s upstream oil and gas operations in the United Kingdom sector of the North Sea for $555 million. The cash transaction was financed initially through transition financing, which will be replaced with the permanent financing discussed below. Additionally, on February 14, 2000, the company reached definitive agreements with Kemira Oyj of Finland to purchase its titanium dioxide pigment operations in Savannah, Georgia, and Boltek, the Netherlands, for $403 million.

To provide financing for these two acquisitions, the company completed in February a public offering of 7.5 million shares of its common stock at $50.0625 per share and a separate offering of $600 million of 5-1/4%, 10-year convertible subordinated debentures. The conversion price of the debentures is $61.0763.

26 Costs Incurred in Crude Oil and Natural Gas Activities

Total expenditures, both capitalized and expensed, for crude oil and natural gas property acquisition, exploration and development activities for the three years ended December 31, 1999, are reflected in the following table:

(Millions of dollars)	Property Acquisition Costs(1)	Exploration Costs(2)	Development Costs(3)
1999 –			
United States	$ 81	$ 92	$224
North Sea	30	18	106
Other international	8	32	9
Total	$119	$142	$339
1998 –			
United States	$117	$136	$347
North Sea	423	38	311
Other international	5	75	29
Total	$545	$249	$687
1997 –			
United States	$ 70	$110	$360
North Sea	2	18	146
Other international	2	61	50
Total	$ 74	$189	$556

(1) Includes $49 million, $280 million and $11 million applicable to purchases of reserves in place in 1999, 1998 and 1997, respectively.
(2) Exploration costs include delay rentals, exploratory dry holes, dry hole and bottom hole contributions, geological and geophysical costs, costs of carrying and retaining properties and capital expenditures, such as costs of drilling and equipping successful exploratory wells.
(3) Development costs include costs incurred to obtain access to proved reserves (surveying, clearing ground, building roads), to drill and equip development wells, and to acquire, construct and install production facilities and improved recovery systems. Development costs also include costs of developmental dry holes.

48 Kerr-McGee Corporation 1999 Annual Report

27 Results of Operations from Crude Oil and Natural Gas Activities

The results of operations from crude oil and natural gas activities for the three years ended December 31, 1999, consist of the following:

(Millions of dollars)	Gross Revenues	Production (Lifting) Costs	Other Related Costs[1]	Exploration Expenses	Depreciation and Depletion Expenses	Asset Impairment	Income Tax Expenses (Benefits)	Results of Operations, Producing Activities
1999 –								
United States	$ 938	$178	$ 73	$ 97	$316	$ —	$ 96	$ 178
North Sea	731	231	22	22	205	—	99	152
Other international	77	23	18	21	15	—	2	(2)
Total crude oil and natural gas activities	1,746	432	113	140	536	—	197	328
Other[2]	24	6	—	—	1	—	7	10
Total	$1,770	$438	$113	$140	$537	$ —	$204	$ 338
1998 –								
United States	$ 721	$184	$126	$141	$285	$114	$ (36)	$ (93)
North Sea	450	195	7	21	170	160	(20)	(83)
Other international	67	12	9	52	31	115	(45)	(107)
Total crude oil and natural gas activities	1,238	391	142	214	486	389	(101)	(283)
Other[2]	29	5	1	—	—	—	6	17
Total	$1,267	$396	$143	$214	$486	$389	$ (95)	$(266)
1997 –								
United States	$1,045	$211	$101	$ 82	$316	$ —	$120	$ 215
North Sea	615	207	11	19	140	—	94	144
Other international	101	29	12	36	29	—	(6)	1
Total crude oil and natural gas activities	1,761	447	124	137	485	—	208	360
Other[2]	84	55	2	—	—	—	12	15
Total	$1,845	$502	$126	$137	$485	$ —	$220	$ 375

(1) Includes transition and restructuring charges of $20 million, $34 million and $2 million in 1999, 1998 and 1997, respectively (see Note 22).
(2) Includes gas marketing, gas processing plants, pipelines and other items that do not fit the definition of crude oil and natural gas activities but have been included above to reconcile to the segment presentations.

The table below presents the company's average per-unit sales price of crude oil and natural gas and production costs per barrel of oil equivalent for each of the past three years. Natural gas production has been converted to a barrel of oil equivalent based on approximate relative heating value (6 MCF equals 1 barrel).

	1999	1998	1997
Average sales price –			
Crude oil (per barrel) –			
United States	**$16.70**	$12.73	$18.34
North Sea	**17.88**	12.93	18.93
Other international	**14.34**	9.90	15.36
Average	**17.15**	12.52	18.32
Natural gas (per MCF) –			
United States	**2.38**	2.09	2.43
North Sea	**2.12**	2.46	2.44
Average	**2.35**	2.12	2.43
Production costs –			
(Per barrel of oil equivalent)			
United States	**2.92**	3.23	3.25
North Sea	**5.57**	5.62	6.25
Other international	**4.32**	1.78	4.33
Average	**4.01**	3.97	4.27

28 Capitalized Costs of Crude Oil and Natural Gas Activities

Capitalized costs of crude oil and natural gas activities and the related reserves for depreciation, depletion and amortization at the end of 1999 and 1998 are set forth in the table below.

(Millions of dollars)	**1999**	1998
Capitalized costs –		
Proved properties	**$9,153**	$8,701
Unproved properties	**438**	583
Other	**98**	75
Total	**9,689**	9,359
Reserves for depreciation, depletion and amortization –		
Proved properties	**6,100**	5,734
Unproved properties	**102**	69
Other	**43**	34
Total	**6,245**	5,837
Net capitalized costs	**$3,444**	$3,522

50 Kerr-McGee Corporation 1999 Annual Report

29 Crude Oil, Condensate, Natural Gas Liquids and Natural Gas Net Reserves (Unaudited)

The estimates of proved reserves have been prepared by the company's geologists and engineers in accordance with the Securities and Exchange Commission definitions. Such estimates include reserves on certain properties that are partially undeveloped and reserves that may be obtained in the future by improved recovery operations now in operation or for which successful testing has been demonstrated. The company has no proved reserves attributable to long-term supply agreements with governments or consolidated subsidiaries in which there are significant minority interests.

The following table summarizes the changes in the estimated quantities of the company's crude oil, condensate, natural gas liquids and natural gas reserves for the three years ended December 31, 1999.

	Crude Oil, Condensate and Natural Gas Liquids (Millions of barrels)				Natural Gas (Billions of cubic feet)			
	United States	North Sea	Other Interna-tional	Total	United States(1)	North Sea	Other Interna-tional	Total
Proved developed and undeveloped reserves –								
Balance December 31, 1996(2)	251	211	101	563	1,481	197	39	1,717
Revisions of previous estimates	12	11	1	24	1	22	3	26
Purchases of reserves in place	5	—	—	5	19	—	—	19
Sales of reserves in place	—	(1)	—	(1)	(30)	—	—	(30)
Extensions, discoveries and other additions	28	1	9	38	227	—	214	441
Production	(26)	(30)	(7)	(63)	(235)	(16)	—	(251)
Balance December 31, 1997	270	192	104	566	1,463	203	256	1,922
Revisions of previous estimates	6	6	(15)	(3)	(4)	7	13	16
Purchases of reserves in place	—	45	—	45	4	46	—	50
Sales of reserves in place	(13)	—	—	(13)	(88)	—	—	(88)
Extensions, discoveries and other additions	14	9	21	44	129	3	103	235
Production	(24)	(32)	(7)	(63)	(197)	(17)	—	(214)
Balance December 31, 1998	253	220	103	576	1,307	242	372	1,921
Revisions of previous estimates	5	14	1	20	14	9	5	28
Purchases of reserves in place	4	7	—	11	18	36	—	54
Sales of reserves in place	(1)	(5)	—	(6)	(1)	—	—	(1)
Extensions, discoveries and other additions	1	34	13	48	101	2	138	241
Production	(29)	(38)	(5)	(72)	(191)	(23)	—	(214)
Balance December 31, 1999	233	232	112	577	1,248	266	515	2,029
Proved developed reserves –								
December 31, 1997	166	115	55	336	919	161	—	1,080
December 31, 1998	148	141	38	327	812	163	—	975
December 31, 1999	166	167	32	365	837	157	—	994

(1) 1998 and 1997 U.S. gas volumes have been restated to be consistent with the current year's presentation.
(2) Includes 1 million barrels of oil and 3 billion cubic feet of natural gas held for sale at December 31, 1996 (see Note 17).

The following presents the company's barrel of oil equivalent proved developed and undeveloped reserves based on approximate relative heating value (6 MCF equals 1 barrel).

(Millions of equivalent barrels)	United States(1)	North Sea	Other International	Total
December 31, 1997	514	226	147	887
December 31, 1998	471	260	165	896
December 31, 1999	441	276	198	915

(1) 1998 and 1997 U.S. gas volumes have been restated to be consistent with the current year's presentation.

30 StandardizedMeasure of and Reconciliation of Changesin Discounted Future Net CashFlows (Unaudited)

The standardized measure of future net cash flows presented in the following table was computed using year-end prices and costs and a 10% discount factor. The future income tax expense was computed by applying the appropriate year-end statutory rates, with consideration of future tax rates already legislated, to the future pre-tax net cash flows less the tax basis of the properties involved. However, the company cautions that actual future net cash flows may vary considerably from these estimates. Although the company's estimates of total reserves, development costs and production rates were based on the best information available, the development and production of the oil and gas reserves may not occur in the periods assumed. Actual prices realized and costs incurred may vary significantly from those used. Therefore, such estimated future net cash flow computations should not be considered to represent the company's estimate of the expected revenues or the current value of existing proved reserves.

(Millions of dollars)	Future Cash Inflows	Future Development and Production Costs	Future Income Taxes	Future Net Cash Flows	10% Annual Discount	Standardized Measure of Discounted Future Net Cash Flows
1999 –						
United States	$ 7,928	$3,332	$1,398	$3,198	$1,343	$1,855
North Sea	6,146	2,608	1,245	2,293	665	1,628
Other international	3,693	1,665	783	1,245	717	528
Total	$17,767	$7,605	$3,426	$6,736	$2,725	$4,011
1998 –						
United States	$ 4,780	$2,108	$ 718	$1,954	$ 713	$1,241
North Sea	3,121	2,474	82	565	160	405
Other international	1,499	977	181	341	264	77
Total	$ 9,400	$5,559	$ 981	$2,860	$1,137	$1,723
1997 –						
United States	$ 8,006	$2,936	$1,584	$3,486	$1,310	$2,176
North Sea	4,026	2,678	282	1,066	356	710
Other international	2,291	1,471	236	584	283	301
Total	$14,323	$7,085	$2,102	$5,136	$1,949	$3,187

The changes in the standardized measure of future net cash flows are presented below for each of the past three years:

(Millions of dollars)	**1999**	1998	1997
Net change in sales, transfer prices and production costs	**$ 4,310**	$(2,156)	$(3,704)
Changes in estimated future development costs	**(318)**	(377)	(283)
Sales and transfers less production costs	**(1,314)**	(847)	(1,314)
Purchases of reserves in place	**117**	159	26
Changes due to extensions, discoveries, etc.	**592**	173	478
Changes due to revisions in quantity estimates	**272**	43	81
Changes due to sales of reserves in place	**(104)**	(107)	(9)
Current period development costs	**339**	687	556
Accretion of discount	**231**	437	759
Changes in income taxes	**(1,414)**	693	1,242
Timing and other	**(423)**	(169)	37
Net change	**2,288**	(1,464)	(2,131)
Total at beginning of year	**1,723**	3,187	5,318
Total at end of year	**$ 4,011**	$ 1,723	$ 3,187

52 Kerr-McGee Corporation 1999 Annual Report

31 Quarterly Financial Information (Unaudited)

A summary of quarterly consolidated results for 1999 and 1998 is presented below. In periods in which there was a loss from continuing operations, the conversion of stock options was not assumed since the loss per-share amount would have been lower. Therefore, the quarterly per-share amounts may not add to the annual amounts. Refer to Management's Discussion and Analysis for information about special items.

(Millions of dollars, except per-share amounts)	Sales	Operating Profit (Loss)	Income (Loss) from Continuing Operations	Net Income (Loss)	Diluted Income (Loss) per Common Share: Continuing Operations	Diluted Income (Loss) per Common Share: Net Income
1999 Quarter Ended –						
March 31	$ 486	$ 49	$(107)	$(111)	$(1.23)	$(1.28)
June 30	657	135	45	45	.52	.52
September 30	753	239	98	98	1.13	1.13
December 31	800	246	110	110	1.27	1.27
Total	$2,696	$ 669	$ 146	$ 142	$ 1.69	$ 1.64
1998 Quarter Ended –						
March 31	$ 507	$ 55	$ 16	$ 24	$ .18	$ .27
June 30	601	73	32	83	.36	.95
September 30	556	11	(68)	150	(.77)	1.73
December 31	536	(444)	(325)	(325)	(3.75)	(3.74)
Total	$2,200	$(305)	$(345)	$ (68)	$(3.98)	$ (.78)

The company's common stock is listed for trading on the New York Stock Exchange and at year-end 1999 was held by approximately 32,000 Kerr-McGee stockholders of record and Oryx owners who have not yet exchanged their stock. The ranges of market prices and dividends declared during the last two years for Kerr-McGee Corporation are as follows:

	Market Prices 1999 High	Market Prices 1999 Low	Market Prices 1998 High	Market Prices 1998 Low	Dividends per Share 1999	Dividends per Share 1998
Quarter Ended –						
March 31	**41 7/16**	**28 1/2**	73 3/16	55 7/8	**$.45**	$.45
June 30	**52 3/8**	**32 1/2**	70 1/4	56 5/8	**.45**	.45
September 30	**60 1/16**	**49 5/16**	60 1/2	38	**.45**	.45
December 31	**62**	**52**	47 9/16	36 3/16	**.45**	.45

Six-Year Financial Summary

(Millions of dollars, except per-share amounts)	**1999**	1998	1997	1996	1995	1994
Summary of Net Income (Loss)						
Sales	**$2,696**	$2,200	$2,605	$2,740	$2,419	$ 2,359
Costs and operating expenses	**2,289**	2,606	2,011	2,122	2,305	2,185
Interest and debt expense	**190**	157	141	145	193	211
Total costs and expenses	**2,479**	2,763	2,152	2,267	2,498	2,396
	217	(563)	453	473	(79)	(37)
Other income	**40**	43	82	110	147	15
Taxes on income	**(111)**	175	(184)	(225)	42	(9)
Income (loss) from continuing operations	**146**	(345)	351	358	110	(31)
Income from discontinued operations	**—**	277	33	56	27	55
Extraordinary charge	**—**	—	(2)	—	(23)	(12)
Cumulative effect of change in accounting principle	**(4)**	—	—	—	—	(948)
Net income (loss)	**$ 142**	$ (68)	$ 382	$ 414	$ 114	$ (936)
Common Stock Information, per Share						
Diluted net income (loss) –						
Continuing operations	**$ 1.69**	$ (3.98)	$ 4.02	$ 4.05	$ 1.23	$ (.36)
Discontinued operations	**—**	3.20	.38	.63	.30	.63
Extraordinary charge	**—**	—	(.02)	—	(.26)	(.14)
Cumulative effect of accounting change	**(.05)**	—	—	—	—	(10.82)
Net income (loss)	**$ 1.64**	$ (.78)	$ 4.38	$ 4.68	$ 1.27	$(10.69)
Dividends declared	**$ 1.80**	$ 1.80	$ 1.80	$ 1.64	$ 1.55	$ 1.52
Stockholders' equity	**17.19**	15.58	17.88	14.59	12.47	12.33
Market high for the year	**62.00**	73.19	75.00	74.13	64.00	51.00
Market low for the year	**28.50**	36.19	55.50	55.75	44.00	40.00
Market price at year-end	**$62.00**	$38.25	$63.31	$72.00	$63.50	$ 46.25
Shares outstanding at year-end (thousands)	**86,483**	86,367	86,794	87,032	89,613	90,143
Balance Sheet Information						
Working capital	**$ 321**	$ (173)	$ —	$ 161	$ (106)	$ (254)
Property, plant and equipment – net	**4,085**	4,153	3,919	3,693	3,807	4,497
Total assets	**5,899**	5,451	5,339	5,194	5,006	5,918
Long-term debt	**2,496**	1,978	1,736	1,809	1,683	2,219
Total debt	**2,525**	2,250	1,766	1,849	1,938	2,704
Total debt less cash	**2,258**	2,129	1,574	1,719	1,831	2,612
Stockholders' equity	**1,492**	1,346	1,558	1,279	1,124	1,112
Cash Flow Information						
Net cash provided by operating activities	**713**	385	1,097	1,169	728	678
Cash capital expenditures	**543**	981	836	875	745	611
Dividends paid	**138**	86	85	83	79	79
Treasury stock purchased	**$ —**	$ 25	$ 60	$ 195	$ 45	$ —
Ratios and Percentage						
Current ratio	**1.4**	.8	1.0	1.2	.9	.8
Average price/earnings ratio	**27.6**	NM	14.9	13.9	42.5	NM
Total debt less cash to total capitalization	**60%**	61%	50%	57%	62%	70%
Employees						
Total wages and benefits	**$ 327**	$ 359	$ 367	$ 367	$ 402	$ 422
Number of employees at year-end	**3,653**	4,400	4,792	4,827	5,176	6,724

54 Kerr-McGee Corporation 1999 Annual Report

Six-Year Operating Summary

	1999	1998	1997	1996	1995	1994
Exploration and Production						
Net production of crude oil and condensate – (thousands of barrels per day)						
United States	**79.3**	66.2	70.6	73.8	74.8	73.4
North Sea	**102.9**	87.4	83.3	86.5	91.9	88.7
Other international	**14.7**	18.4	18.1	16.8	17.4	26.4
Total	**196.9**	172.0	172.0	177.1	184.1	188.5
Average price of crude oil sold (per barrel) –						
United States	**$16.70**	$12.73	$18.34	$19.45	$15.73	$14.25
North Sea	**17.88**	12.93	18.93	19.60	16.56	15.33
Other international	**14.34**	9.90	15.36	15.85	14.70	14.58
Average	**$17.15**	$12.52	$18.32	$19.18	$16.05	$14.80
Natural gas sales (MMCF per day)	**580**	584	685	781	809	872
Average price of natural gas sold (per MCF)	**$ 2.35**	$ 2.12	$ 2.43	$ 2.10	$ 1.63	$ 1.82
Net exploratory wells drilled –						
Productive	**1.70**	4.40	7.65	6.91	4.71	11.61
Dry	**3.75**	14.42	7.42	5.52	11.16	13.47
Total	**5.45**	18.82	15.07	12.43	15.87	25.08
Net development wells drilled –						
Productive	**46.23**	62.30	95.78	143.33	135.86	69.27
Dry	**5.89**	9.00	7.00	13.04	11.95	9.63
Total	**52.12**	71.30	102.78	156.37	147.81	78.90
Undeveloped net acreage (thousands) –						
United States	**1,560**	1,487	1,353	1,099	1,280	1,415
North Sea	**861**	908	523	560	570	629
Other international	**19,039**	14,716	14,630	4,556	4,031	7,494
Total	**21,460**	17,111	16,506	6,215	5,881	9,538
Developed net acreage (thousands) –						
United States	**796**	810	830	871	1,190	1,270
North Sea	**105**	115	70	79	58	68
Other international	**785**	612	201	198	207	1,015
Total	**1,686**	1,537	1,101	1,148	1,455	2,353
Estimated proved reserves (millions of equivalent barrels)	**915**	896	886	849	864	1,059
Chemicals						
Industrial and specialty chemical sales (thousands of tonnes)	**518**	481	443	405	404	346

STOCKHOLDER AND INVESTOR INFORMATION

Stock Exchange Listing
Kerr-McGee (KMG) common stock is listed on the New York Stock Exchange and also is traded on the Boston, Chicago, Pacific and Philadelphia stock exchanges.

Stockholder Assistance
Contact UMB Bank, N.A., of Kansas City, Missouri, toll-free at (877) 860-5820 or (800) 884-4225 for assistance with:

- Direct deposit of cash dividends
- Direct stock purchase and dividend reinvestment plan
- Transfer of stock certificates
- Replacement of lost or destroyed stock certificates and dividend checks

Stockholder Information and Publications
Contact the Office of the Corporate Secretary at (800) STOCK KM (800-786-2556) for general information and assistance or to request the company's annual report on Form 10-K and quarterly reports on Form 10-Q, as filed with the Securities and Exchange Commission, and the company's annual report to stockholders.

Direct Purchase and Dividend Reinvestment Plan
This plan allows stockholders to buy Kerr-McGee common stock directly from the company and to reinvest quarterly dividends in additional shares. The company pays all fees and commissions for these services. For a prospectus, please call (800) 786-2556.

Investor Information
Richard C. Buterbaugh, Vice President, Investor Relations, is available at (405) 270-3561 to answer questions from stockholders, security analysts and other interested parties.

Transfer Agent and Registrar
UMB Bank, N.A.
Securities Transfer Division
P.O. Box 410064
Kansas City, MO 64141-0064
Toll-free telephone: (877) 860-5820
or (800) 884-4225

Corporate Headquarters
Kerr-McGee Corporation
Kerr-McGee Center
123 Robert S. Kerr Avenue
Oklahoma City, OK 73102

Mailing address:
P.O. Box 25861
Oklahoma City, OK 73125

Telephone: (405) 270-1313

Forward-Looking Information

Statements in this annual report regarding the company's or management's intentions, beliefs or expectations are forward-looking statements within the meaning of the Securities Litigation Reform Act. Future results and developments discussed in these statements may be affected by numerous factors and risks, such as the accuracy of the assumptions that underlie the statements, the success of the oil and gas exploration and production program, drilling risks, the market value of Kerr-McGee's products, uncertainties in interpreting engineering data, demand for consumer products for which Kerr-McGee's businesses supply raw materials, general economic conditions, and other factors and risks discussed in the company's SEC filings. Actual results and developments may differ materially from those expressed or implied in this annual report.

56 Kerr-McGee Corporation 1999 Annual Report

abbreviations

AICPA	American Institute of Certified Public Accountants
AFE	Authority for Expenditure
AIPN	Association of International Petroleum Negotiators
API	American Petroleum Institute
ARB	Accounting Research Bulletin
ASR	Accounting Series Release (SEC)
bbl	Barrels
BOE	Barrels of Oil Equivalent
BS&W	Basic Sediment and Water
BTU	British Thermal Unit
C	Capitalize
CNG	Compressed Natural Gas
COPAS	Council of Petroleum Accountants Societies
CT	Corporation Tax

CV	Carrying Value
DD&A	Depreciation, Depletion, and Amortization
E	Expense
EDQ	Equal Daily Quantities
Eq	Equivalent
E&F	Equipment and Facilities
FASB	Financial Accounting Standards Board
FC	Full Cost
FERC	Federal Energy Regulatory Commission
FMV	Fair Market Value
GAAP	Generally Accepted Accounting Principles
G&G	Geological and Geophysical
IASC	International Accounting Standards Committee
IDC	Intangible Drilling and Development Costs
ITC	Investment Tax Credit
JI	Joint Interest
JIS	Joint Interest Suspense
JOA	Joint Operating Agreement
LACT	Lease Automatic Custody Transfer Unit
LCM	Lower of Cost or Market
LDC	Local Distribution Company
LOE	Lease Operating Expense
L&WE	Lease and Well Equipment
Mcf	Thousand Cubic Feet of Natural Gas
MMcf	Million Cubic Feet of Natural Gas
MEOR	Microbial Enhanced Oil Recovery
MER	Maximum Efficiency Rate
MI	Mineral Interest
MMBTU	Million British Thermal Units

NA	Not Applicable
NI	Net Income
NOL	Net Operating Loss
NPI	Net Profits Interest
NWI	Nonworking Interest
OCS	Outer Continental Shelf
OIAC	Oil Industry Accounting Committee
ORI	Overriding Royalty Interest
PDR	Proved Developed Reserves
P/P	Proved Property
PPI	Production Payment Interest
PR	Proved Reserves
PRT	Petroleum Revenue Tax
PSC	Production Sharing Contract
Psia	Pressure per Square Inch Absolute
PV	Present Value
Reg	Regulation
RI	Royalty Interest
RRA	Reserve Recognition Accounting
SE	Successful Efforts
SEC	Securities and Exchange Commission
SFAS	Statement of Financial Accounting Standards
SL	Straight Line
SM	Standardized Measure
SP	Selling Price
SPS	State Profit Share
SV	Salvage Value
U/P	Unproved Property
WI	Working Interest
W/P	Wells in Progress

index

A

B

C

D

E

F

G

H

I

J

K

L

M

N

O

P

R

S

T

U

V

W

Y